ATLAS OF SKELETAL DYSPLASIAS

ATLAS OF SKELETAL DYSPLASIAS

Ruth Wynne-Davies PhD FRCS
Formerly Reader in Orthopaedic Genetics Research, University of
Edinburgh, Edinburgh, UK

Christine M. Hall FRCR
Consultant Radiologist, Hospital for Sick Children, London, UK

A. Graham Apley FRCS
Honorary Consulting Orthopaedic Surgeon, St Thomas' Hospital,
London, UK

Churchill Livingstone ▦

EDINBURGH LONDON MELBOURNE AND NEW YORK 1985

CHURCHILL LIVINGSTONE
Medical Division of Longman Group UK Limited

Distributed in the United States of America by Churchill Livingstone
Inc., 1560 Broadway, New York, N.Y. 10036, and by associated
companies, branches and representatives throughout the world.

First published 1985
  Reprinted 1987

ISBN 0 443 03047 2

British Library Cataloguing in Publication Data
Wynne-Davies, Ruth.
    Atlas of skeletal dysplasias.
    Includes bibliographical references and index.
    1. HUMAN SKELETON—ABNORMALITIES—ATLASES.
2. Abnormalities, Human—Atlases.   I. Hall, Christine M.   II.
Apley, A. Graham (Alan Graham).   III. Title. [DNLM: 1. Bone
Diseases, Developmental—atlases. WE 17 W988a]
RD761.W95   1984   616.7′1043   84–9442

Library of Congress Cataloging in Publication Data
Wynne-Davies, Ruth
    Atlas of skeletal dysplasias.
    1. Bones—Abnormalities   2. Bones—Radiography
    I. Title   II. Hall, Christine M.
    III. Apley, A. Graham
    616.7′10757   RC930.5

Typeset, printed and bound in Great Britain by
William Clowes Limited, Beccles and London

PREFACE

Who needs an atlas of skeletal dysplasias? A few years ago the answer might well have been 'hardly anyone'—but not now. Today, little more than three decades after Sir Thomas Fairbank's pioneer effort, the whole subject has 'taken off' with dramatic speed. Meetings, international conferences, publications, societies, groups, all testify to a lively interest on the part of clinicians from many disciplines—orthopaedic surgeons, radiologists, paediatricians and clinical geneticists, as well as biochemists, molecular biologists and no doubt many others.

Why this sudden interest in so abstruse a field? Perhaps the most obvious factor for clinicians was the virtual disappearance from the developed countries of poliomyelitis, tuberculosis and other serious infective disorders; this liberated time, energy and funds which it seemed logical to apply to other crippling disorders. Then came the thalidomide disaster with its painfully compelling reminder that we ourselves, and not some spiteful Providence, could be responsible for multiple deformities; it seemed that the least that could be done was to study the development of apparently 'natural' deformities. At the same time, developing knowledge of the structure of DNA and of cytogenetics generated a whole world of new ideas, important among them being the possibility that inherited disorders, such as the skeletal dysplasias, might be brought under control by means other than merely the avoidance of having children.

This present atlas is the culmination of about 10 years study of patients: their clinical history, radiographs, natural history, progress with surgical and other treatment, their families, and of course the very considerable and widely scattered literature. Skeletal Dysplasia Clinics were established in various parts of the country for the sole purpose of investigating and caring for these patients, both as children and adults, and concerned not only with genetic advice but also with management of the progressive deformities and joint degeneration which are so often an accompaniment.

We have tried to take a fresh look at the entire subject, surveying the scene widely and hoping to paint a fairly complete picture. Early studies concentrated mainly on affected children, but we have extended the range to the entire span of life: it had soon became apparent in the Skeletal Dysplasia Clinics that many grown children felt abandoned once they had passed the age when paediatricians and children's hospitals looked after them. Also, of course, advances in orthopaedic surgery were opening up possibilities of joint replacement, leg lengthening and other operative procedures. Nowadays these operations are so successful, and the correction of deformity so precise, that many patients can be helped to an extent that, 20 years ago, would have seemed like science fiction.

One of our hoped for objectives has been the more accurate clinical delineation of these disorders, although a similar clinical and radiographic picture readily obscures fundamental pathological differences—this inevitably hampering progress. Ultimate understanding and logical classification will have to await continuing biochemical advances, but we hope that the assembling of material from the large number of patients in this study, each having had a careful genetic, clinical and radiographic investigation, will contribute further to their delineation.

The greater part of the book describes the true skeletal dysplasias—a term taken to mean generalised developmental disorders of skeletal growth, which may affect both bone and cartilage (osteochondrodysplasias). Most, but not all, of these appear to be of simple Mendelian inheritance.

Also included are certain malformation syndromes (dysostoses) with predominantly appendicular skeletal manifestations. Here, by definition, the disorders are not of generalised bone and cartilage growth but are structural defects of one or more bones and involve more than one system (for example acrocephalosyndactyly, or the nail-patella syndrome). Again, many are of simple Mendelian inheritance. (Syndromes associated with chromosome anomalies are not discussed.)

These two major groups, of course, tend to overlap, in that minor bony structural defects may accompany the skeletal dysplasia (for example, the polydactyly of chondroectodermal dysplasia, or the congenital vertebral anomalies of Conradi's disease); and the major limb reduction defects in the severe mesomelic disorders could perhaps be more properly classified as malformation syndromes. However, in general the distinction between the skeletal dysplasias and malformation syndromes is clear, valid, and a helpful guide to classification and diagnosis.

Although these disorders are individually rare, the collective total is substantial: greater, for example than the number of haemophiliac patients in Britain. Prevalence figures quoted in this atlas were derived primarily from the south-east of Scotland, based on Edinburgh. These were compared with figures obtained from other parts of Scotland and from the Skeletal Dysplasia Clinics in different parts of England and Wales, and since no significant differences were observed, it was felt that a simple process of extrapolation was justified—in the absence, that is, of any more accurate data. It is estimated (Wynne-Davis & Gormley 1984) that there are upwards of 10 000 people in England, Scotland and Wales with one of the skeletal dysplasias described in this atlas. Many, of course, are only mildly affected and require little or no clinical care other than appropriate genetic advice (if such data is available, which it may not be); probably some 6000 of them are more seriously affected with some physical handicap throughout life, and/or requiring repeated periods of hospital or institutional care.

We hope this atlas will provide a starting point for future research; only a few recent key references have been included with each subject, from which others can, of course, be traced. The pedigrees and original radiographs are in most instances available for study; and a continuing index of all these skeletal dysplasia patients and their affected relatives is maintained in order that advantage can be taken of future advances in biochemistry and surgery.

We acknowledge with gratitude the advice, help and funds we have received. Much of the work was completed from the Department of Orthopaedic Surgery, University of Edinburgh (RWD) and the Department of Paediatric Radiology, Hospital for Sick Children, Great Ormond Street, London (CMH). Grants have been provided at various times since 1975 from Action for the Crippled Child, St Thomas' Hospital Endowment Fund, London, and the Laming Evans Fund of the Royal College of Surgeons of England. The hospitals at which Skeletal Dysplasia Clinics were held are listed below; the facilities and assistance they provided constituted the indispensable background to the work and we hope that some small part of our debt to them has been discharged by the educational opportunities which the Clinics offer to their staff. Thanks are due also to Dr Roger Smith (Nuffield Orthopaedic Centre, Oxford) for the help and expertise he provided in the chapter on metabolic bone disorders.

But most of all we are grateful to the patients who form the subject of this book. They and their relatives have been unfailingly co-operative and their fortitude in the face not only of severe disability, but (until recently) very little medical interest or community care, has been quite remarkable. If this atlas can contribute to an improved outlook for themselves or their successors, we shall feel amply rewarded.

Oxford and London, 1984

RWD
CMH
AGA

Hospitals
Princess Margaret Rose Orthopaedic Hospital, Edinburgh
Hospital for Sick Children, Great Ormond Street, London
St Thomas' Hospital, London
Western Infirmary, Glasgow
Victoria Infirmary, Glasgow
Harlow Wood Orthopaedic Hospital, Mansfield, Nottingham
Robert Jones and Agnes Hunt Orthopaedic Hospital, Oswestry, Shropshire

ACKNOWLEDGEMENTS

Some of the illustrations have been previously published and the authors gratefully acknowledge permission from principal co-authors, editors and publishers for the following:

Wynne-Davies R, Fairbank T J 1976 Fairbank's atlas of general affections of the skeleton, 2nd edn. Churchill Livingstone, Edinburgh
  Figs 3.13, 12.2, 12.6, 27.4, 27.41, 39.7, 40.2, 40.3, 41.2, 50.2, 50.3, 51.4, 73.1

American Journal of Medical Genetics
  Figs 3.10, 3.11, 3.21, 3.29, 3.37, 3.38

Annales de Radiologie
  Figs 17.1, 17.3, 73.6, 73.7, 73.8, 73.10

British Journal of Radiology
  Figs 24.2, 24.3, 24.4, 24.5, 24.9

Journal of Bone and Joint Surgery
  Figs 5.3, 5.54, 5.55, 5.85, 6.1, 6.27, 6.31, 7.1, 9.2, 9.3, 9.5, 9.6, 9.20, 9.39, 9.42, 9.45, 19.2, 20.1, 20.2, 20.3, 20.4, 20.10, 20.11, 20.19, 22.1, 22.2, 22.4, 22.7, 22.10, 22.11, 39.6, 39.15, 39.17, 39.18, 40.1, 40.2, 40.3, 40.6, 40.8, 40.13, 40.14, 40.15, 40.16

Journal of Medical Genetics
  Fig 15.2, 15.3, 15.5, 15.7, 15.8, 15.9, 15.10, 25.4, 25.11, 25.12, 83.10, 83.14

Pediatric Radiology
  Figs 22.6, 22.8, 22.9

Radiology
  Figs 35.2, 35.4, 35.5, 35.6, 35.7, 35.8, 78.1, 78.5, 78.7, 78.9, 78.10, 78.13

CONTENTS

How to use the atlas   xi
Prenatal diagnosis in the skeletal dysplasias   xiii

SECTION ONE **Normal appearances**
  1. Normal appearances   3

SECTION TWO **Epiphyseal disorders**
  2. Multiple epiphyseal dysplasia   19
  3. Hereditary progressive arthro-
     ophthalmopathy (Stickler syndrome)   36
  4. Chondrodysplasia calcificans punctata
     (Conradi's disease)   49

SECTION THREE **Spondylo-epiphyseal disorders**
  5. Spondylo-epiphyseal dysplasia congenita
     with severe coxa vara   61
  6. Spondylo-epiphyseal dysplasia congenita
     with mild coxa vara   86
  7. Spondylo-epiphyseal dysplasia tarda—X-
     linked type   99
  8. Spondylo-epiphyseal dysplasia tarda—
     autosomal dominant and recessive types   104
  9. Spondylo-epiphyseal dysplasia tarda with
     progressive arthropathy   116

SECTION FOUR **Metaphyseal disorders**
 10. Metaphyseal chondrodysplasia—type
     Schmid   131
 11. Metaphyseal chondrodysplasia—type
     McKusick   145
 12. Metaphyseal chondrodysplasia—type
     Jansen   151
 13. Metaphyseal chondrodysplasia with
     malabsorption and neutropenia   156
 14. Metaphyseal chondrodysplasia—type Peña   161
 15. Metaphyseal chondrodysplasia with
     retinitis pigmentosa and brachydactyly   164

SECTION FIVE **Spondylometaphyseal dysplasia**
 16. and 17. The spondylometaphyseal
     dysplasias (including types Kozlowski and
     Sutcliffe)   171

SECTION SIX **Short limbs—normal trunk**
 18. Achondroplasia   181
 19. Hypochondroplasia   200
 20. Dyschondrosteosis   213
 21. Fetal face syndrome (Robinow)   220
 22. Mesomelic dysplasia (Werner type:
     absent/dysplastic tibiae with five-fingered
     hand)   223
 23. Mesomelic dysplasia (severe dominant
     form)   227
 24. Acromesomelic dwarfism   229
 25. Acrodysostosis (peripheral dysostosis)   233

SECTION SEVEN **Short limbs and trunk**
 26. Pseudoachondroplasia   239
 27. Diastrophic dysplasia (dwarfism)   258
 28. Metatropic dwarfism   274
 29. Kniest disease   289
 30. Dyggve-Melchior-Clausen disease   295
 31. Chondro-ectodermal dysplasia (Ellis-van
     Creveld syndrome)   301
 32. Asphyxiating thoracic dysplasia (Jeune's
     disease)   304

SECTION EIGHT **Lethal forms of short-limbed dwarfism**
 33.–38. Lethal forms of short-limbed dwarfism   311
     33. Thanatophoric dwarfism; 34. Short
     rib/polydactyly syndrome Type I
     (Saldino–Noonan); 35. Type II
     (Majewski); 36. Type III (Naumoff);
     37. A short rib/polydactyly syndrome
     (unnamed); 38. Achondrogenesis

SECTION NINE **Increased limb length**
 39. Marfan syndrome   325
 40. Homocystinuria   335

SECTION TEN **Storage diseases**
 41. Hurler syndrome   343
 42. Hunter syndrome   350
 43. Scheie syndrome   356

44. Sanfilippo syndrome 360
45. Mannosidosis 363
46. Maroteaux-Lamy syndrome 366
47. Morquio syndrome 372
48. and 49. The mucolipidoses 385

SECTION ELEVEN **Metabolic bone disease**
50. Hypophosphatasia 393
51. Idiopathic hyperphosphatasia 400
52. Hypophosphataemic rickets (vitamin D resistant) 404
53. Pseudohypoparathyroidism 407

SECTION TWELVE **Decreased bone density**
54. Osteogenesis imperfecta 411
55. Idiopathic juvenile osteoporosis 431
56. Idiopathic osteolysis (Hajdu–Cheney type) 434

SECTION THIRTEEN **Sclerosing bone dysplasias**
57. Osteopetrosis 439
58. Pycnodysostosis 454
59. Metaphyseal dysplasia (Pyle's disease) 465
60. Craniometaphyseal dysplasia 469
61. Frontometaphyseal dysplasia 475
62. Osteodysplasty (Melnick-Needles syndrome) 484
63. Engelmann's disease (progressive diaphyseal dysplasia) 488
64. Craniodiaphyseal dysplasia 495
65. Osteopathia striata (with cranial sclerosis) 501
66. Osteopoikilosis 506
67. Melorheostosis 508
68. Familial hypertrophic osteoarthropathy (pachydermoperiostitis) 512

SECTION FOURTEEN **Tumour-like bone dysplasias**
69. Diaphyseal aclasis (multiple hereditary exostoses) 519
70. Ollier's disease (enchondromatosis) 533
71. Maffucci's disease 537
72. Dysplasia epiphysealis hemimelica 539

SECTION FIFTEEN **Fibrous disorders**
73. Fibrodysplasia ossificans progressiva (myositis ossificans progressiva) 547
74. Polyostotic fibrous dysplasia (including Albright's syndrome) 557
75. Neurofibromatosis (von Recklinghausen's disease) 566

SECTION SIXTEEN **Malformation syndromes**
76. Acrocephalo-syndactyly (and similar) 579
77. Cleidocranial dysplasia (craniocleido dysostosis) 584
78. Cranio-carpo-tarsal dysplasia (Freeman-Sheldon syndrome—'whistling face' syndrome) 596
79. Cornelia de Lange syndrome 600
80. Larsen's syndrome 604
81. Menkes' syndrome (kinky hair disease) 609
82. Nail-patella syndrome (hereditary osteo-onychodysplasia) 614
83. Oral-facial-digital syndrome (including Mohr syndrome) 617
84. Oto-palato-digital syndrome 623
85. Rubinstein-Taybi syndrome 625
86. Tricho-rhino-phalangeal syndrome 629

Index 639

HOW TO USE THE ATLAS

The opening chapter includes normal growth charts of height and span, followed by representative radiographs of normal children from birth to just under 15 years of age.

Subsequent chapters describe single disorders (or occasionally groups of disorders). The material is arranged systematically: first a brief genetic, clinical and radiological description using headings whose order does not vary; then clinical photographs and, where relevant (or available), growth charts from patients known to us, where the diagnosis is as certain as it can be. Finally come the radiographs arranged anatomically and, within each anatomical region, chronologically. It was judged preferable to present the material in this way, rather than patient-by-patient, because the sometimes wide range in clinical expression is thus better illustrated. In general, the Paris nomenclature is used (McKusick & Scott 1971), but not their detailed classification. Instead we have used the simpler one related to stature, body proportions and radiographic appearances, which is of greater clinical assistance. The material and index have been arranged to facilitate reference: the following scheme is suggested to aid diagnosis:

First obtain

1. *Clinical data*, in particular the patient's measurements, the three most important being height, span and head to pubis/pubis to heel.
2. *Radiographs*. A total skeletal survey is unnecessary at this stage. For diagnostic purposes eight views only are required:
   —Lateral skull
   —AP and lateral spine (T1 to S1 on one film)
   —Chest (which will also show the shoulders)
   —Pelvis/hips
   —AP one knee
   —AP one forearm
   —PA one hand/wrist

Second, use the contents list to establish the major group to which the patient belongs:
—Epiphyseal or spondylo-epiphyseal
—Metaphyseal or spondylometaphyseal (check also, metabolic bone disease here)
—Mesomelic and other forms of short stature (check also storage diseases here)
—Lethal forms of dwarfism
—Increased limb length
—Storage diseases
—Metabolic bone disease
—Decreased bone density
—Sclerosing bone dysplasias
—Tumour-like bone dysplasias
—Fibrous disorders of bone growth

The combination of clinical measurements and the predominant radiographic features should suffice to identify the correct section in the atlas.

Third, use the index to look up additional physical signs (e.g. cleft palate, pathological fracture, platybasia). However, some signs are too common and too unreliable as diagnostic features for complete lists to be included. Thus 'scoliosis' may be a feature of almost any skeletal dysplasia in which the spine is involved; and 'genu valgum' may occur in almost any skeletal dysplasia involving the lower limbs.

Finally, carry out further investigations—radiographic, biochemical or other. These should be apparent from the 'differential diagnosis' section in the relevant chapter.

REFERENCES

McKusick V A, Scott C I 1971 A nomenclature for constitutional disorder of bone. Journal of Bone and Joint Surgery 53A: 978.
Wynne-Davies R, Gormley J 1985 The prevalence of skeletal dysplasias. Journal of Bone and Joint Surgery 67B: 133–137

PRENATAL DIAGNOSIS IN THE SKELETAL DYSPLASIAS

The possibility of early prenatal diagnosis for some of the conditions described in this Atlas is now firmly established. This is a rapidly changing field of practice and research and so details are not given, but the reader is referred to Ferguson-Smith (1983) for a very full review.

In summary, there are basically two methods of prenatal diagnosis:
(i) biochemical tests
(ii) ultrasonic examination of the fetus.

Biochemical diagnosis

Material for biochemical tests, usually enzyme assay, may be obtained from amniocentesis and cell culture; from fetal blood sampling (or from other fetal tissues such as skin) and from chorionic biopsy. Amniocentesis for cell culture is usually carried out around the fifteenth to sixteenth week of pregnancy, chorionic biopsy rather earlier, around the eighth to tenth week, and fetal tissue sampling a little later, at the seventeenth to eighteenth week. Of the conditions described in this book, the enzyme disorders of the mucopolysaccaride and muco-lipidosis groups and of homocystinuria can be identified at this early stage of pregnancy; the disordered ^{64}Cu metabolism of Menkes' syndrome is the only non-enzymatic assay at present available—all these disorders are of recessive inheritance (autosomal or X-linked). The collagen diseases, particularly some forms of osteogenesis imperfecta, are likely to become detectable in the near future using DNA techniques (Table 1).

Ultrasonic examination of the fetus

The detection of fetal anomalies by ultrasound techniques is becoming increasingly sophisticated and reliable; thus diagnosis by direct inspection (fetoscopy), with its attendant risks of haemorrhage and abortion, is

Table 1 Inherited metabolic disorders

Disorder	Enzyme or other assay
Mucopolysaccaride and related disorders	
Hurler's syndrome	α-iduronidase
Hunter's syndrome	Iduronate sulphatase
Scheie's syndrome	α-iduronidase
Sanfilippo's syndrome	Heparin sulphamidase
Maroteaux–Lamy syndrome	Arylsulphatase B
Morquio's syndrome	N-acetylgalactosamine 6-sulphatase
Mannosidosis	α-mannosidase
GM₁ gangliosidosis type I	β-galactosidase
Mucolipidosis II (I-cell disease)	Multiple lysosomal enzymes
Amino acid disorder	
Homocystinuria	Cystathione β-synthase
Other disorder	
Menkes' syndrome	^{64}Cu uptake

Table 2 Diagnosis by ultrasound

(Lower) limb shortening
Chondrodysplasia punctata (*severe*)
Metaphyseal chondrodysplasia (*severe*)
Achondroplasia (*including homozygous*)
Hypochondroplasia (*severe*)
Mesomelic dysplasias (*including severe dyschondrosteosis*)
Lethal forms of short-limbed dwarfism
Pseudo-achondroplasia
Diastrophic dysplasia
Chondro-ectodermal dysplasia (*short tibiae and fibulae*)
Asphyxiating thoracic dysplasia (*severe*)
Osteogenesis imperfecta (*severe, with multiple fractures*)

Congenital malformations
Cranial defects, including hydrocephaly, microcephaly
Cleft palate, cleft lip
Congenital vertebral anomalies
Congenital limb anomalies—'absence' defects, severe hypoplasia, polydactyly of hands or feet
Some congenital heart, gastrointestinal and urinary defects

being superseded. Two groups of disorders may be identified:

(i) congenital shortening of the (lower) limbs
(ii) congenital malformations, such as polydactyly, cleft palate, absent radius or ulna.

Detection of these abnormalities becomes increasingly reliable as the fetus grows; thus, while a very severe limb shortening such as in achondrogenesis may be identified before the twentieth week, the relatively milder (and commoner) condition of achondroplasia is unlikely to be certain before the twenty-second week or even later. Measurement of the lower limb bones is compared with a recently established range of normal values rather than by the more difficult technique of noting disproportion in length between the limbs and trunk in the same fetus. Table 2 shows the disorders and malformations which *may* be diagnosed, while recognising that absolute certainty is only possible in the more severe cases and that the older the fetus, the more reliable the finding.

Any of the congenital malformations may occur as isolated anomalies, thus their presence does not necessarily mean a generalised skeletal dysplasia is present. Conversely, several skeletal dysplasias may be accompanied by congenital malformations (e.g. the cleft palate of spondylo-epiphyseal dysplasia congenita or diastrophic dysplasia, the congenital vertebral anomalies of Conradi's disease or the polydactyly of the Ellis–van Creveld syndrome) but these signs are not invariably part of the disorder; thus their absence is not necessarily indicative of a normal fetus.

REFERENCE

Ferguson-Smith M A (ed) 1983 Early prenatal diagnosis. British Medical Bulletin 39: 301–408

Section One
NORMAL APPEARANCES

Chapter One
NORMAL APPEARANCES

This first chapter comprises normal growth charts and a representative collection of radiographs of normal children, covering 10 anatomical sites: skull, cervical spine, thoracic and lumbar spine, chest, pelvis and hips, knee, foot, elbow, forearm, wrist and hand. Only four age groups are illustrated for each site, and neither the sex nor precise age is given, since individual variation can be so wide. The age groups are:

Birth to 8 weeks
$2\frac{1}{2}$–$3\frac{1}{2}$ years
$7\frac{1}{2}$–$8\frac{1}{2}$ years
$12\frac{1}{2}$–$13\frac{1}{2}$ years

Clearly, this small collection cannot be considered as a replacement for major works on normal radiological development, and the reader is referred to the appropriate textbook. However, the radiographs included here should prove to be of some assistance as a comparison with the frequently grossly abnormal skeletal structure described in subsequent chapters.

REFERENCES

Greulich WW, Pyle SI 1950 Radiographic atlas of skeletal development of the hand and wrist. Stanford University Press, Stanford, Ca
Keats TE 1973 An atlas of normal Roentgen variants that may simulate disease. Year Book Medical Publishers, Chicago
Keats TE 1978 An atlas of normal developmental Roentgen anatomy. Year Book Medical Publishers, Chicago, London
Pyle SI, Hoerr NL 1969 A radiographic standard of reference for the growing knee. Charles C. Thomas, Springfield, Ill

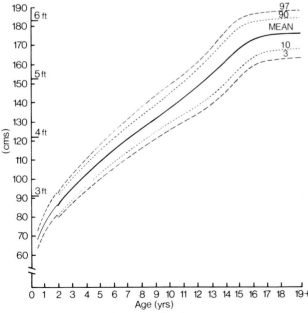

Fig. 1.1   Percentile height—normal males.

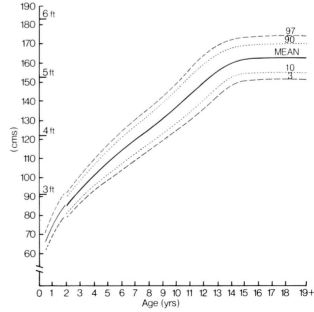

Fig. 1.2   Percentile height—normal females.

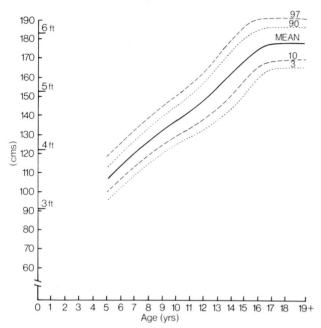

Fig. 1.3   Percentile span—normal males.

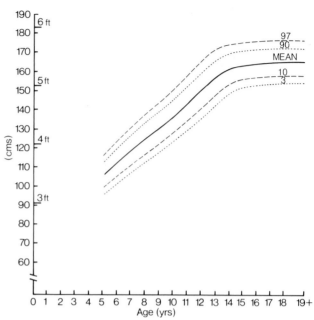

Fig. 1.4   Percentile span—normal females.

Skull: AP

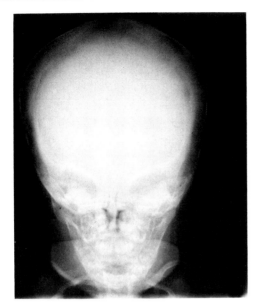

Fig. 1.5 Birth–8 weeks.

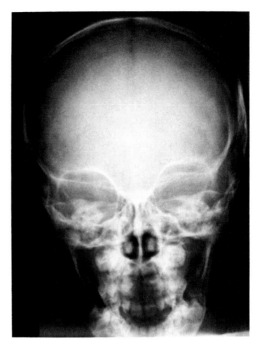

Fig. 1.6 2½–3½ years.

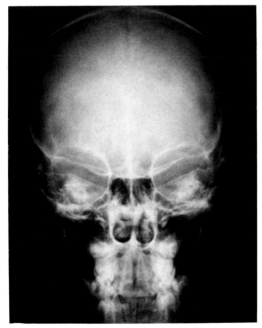

Fig. 1.7 7½–8½ years.

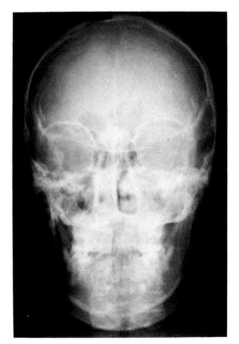

Fig. 1.8 12½–13½ years.

Skull : Lateral

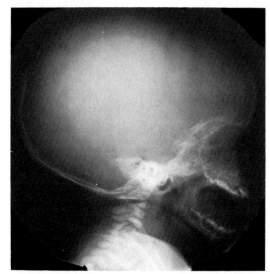

Fig. 1.9   Birth—8 weeks.

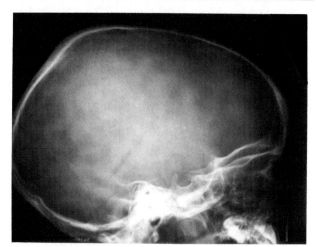

Fig. 1.10 $2\frac{1}{2}$–$3\frac{1}{2}$ years.

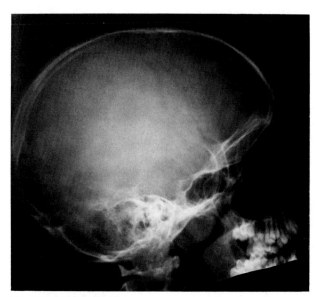

Fig. 1.11 $7\frac{1}{2}$–$8\frac{1}{2}$ years.

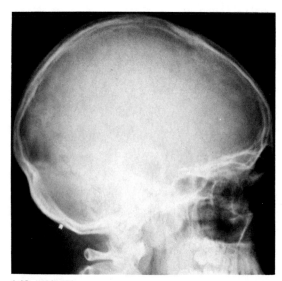

Fig. 1.12 $12\frac{1}{2}$–$13\frac{1}{2}$ years.

Cervical spine : Lateral

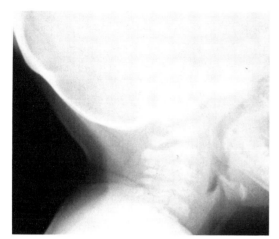

Fig. 1.13 Birth—8 weeks.

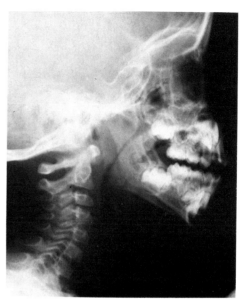

Fig. 1.14 $2\frac{1}{2}$–$3\frac{1}{2}$ years.

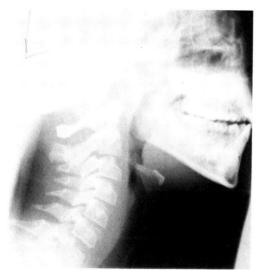

Fig. 1.15 $7\frac{1}{2}$–$8\frac{1}{2}$ years.

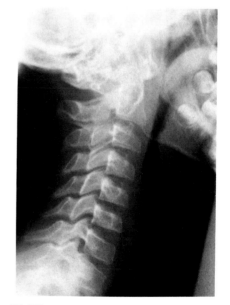

Fig. 1.16 $12\frac{1}{2}$–$13\frac{1}{2}$ years.

Thoracic and lumbar spine: AP

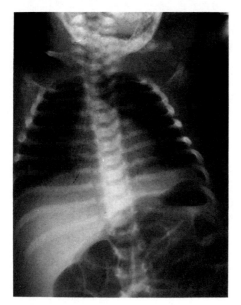

Fig. 1.17 Birth—8 weeks.

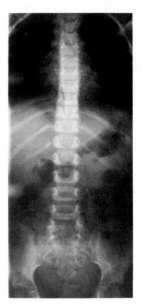

Fig. 1.18 2½–3½ years.

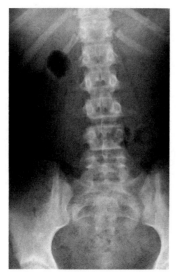

Fig. 1.19 7½–8½ years.

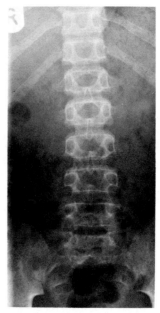

Fig. 1.20 12½–13½ years.

Thoracic and lumbar spine : Lateral

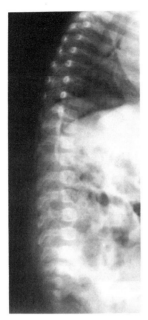

Fig. 1.21   Birth–8 weeks.

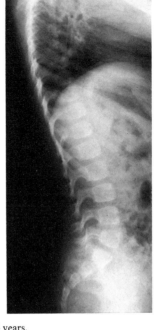

Fig. 1.22 $2\frac{1}{2}$–$3\frac{1}{2}$ years.

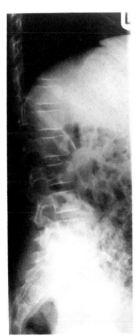

Fig. 1.23 $7\frac{1}{2}$–$8\frac{1}{2}$ years.

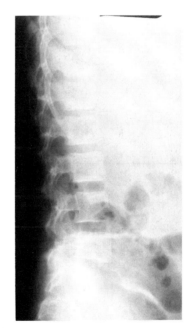

Fig. 1.24 $12\frac{1}{2}$–$13\frac{1}{2}$ years.

Chest : AP

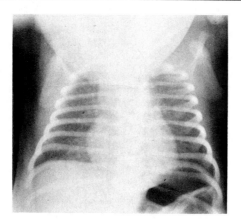

Fig. 1.25   Birth–8 weeks.

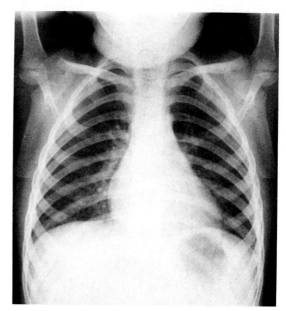

Fig. 1.26   2½–3½ years.

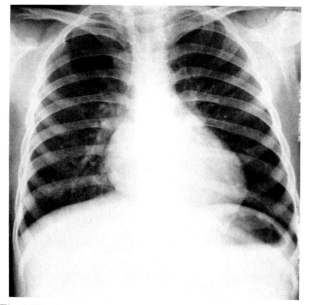

Fig. 1.27   7½–8½ years.

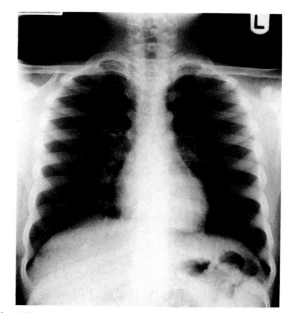

Fig. 1.28   12½–13½ years.

Pelvis

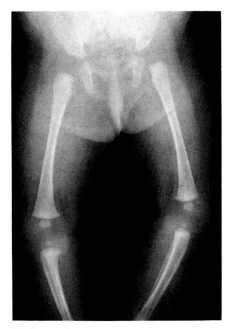

Fig. 1.29   Birth–8 weeks.

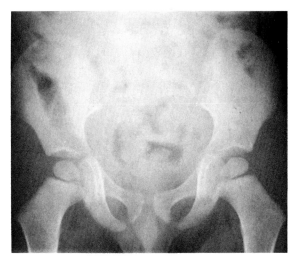

Fig. 1.30   2½–3½ years.

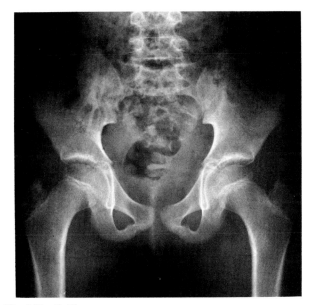

Fig. 1.31   7½–8½ years.

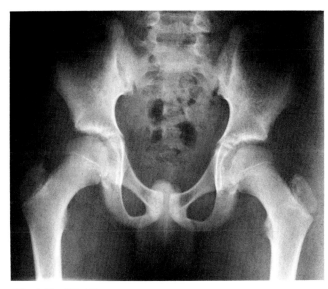

Fig. 1.32   12½–13½ years.

Knee

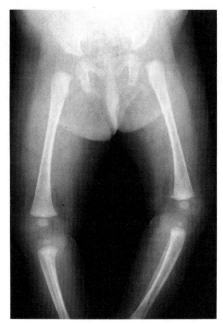

Fig. 1.33   Birth–8 weeks.

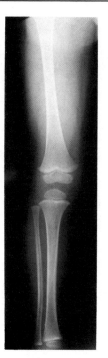

Fig. 1.34   2½–3½ years.

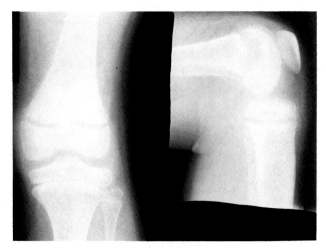

Fig. 1.35   7½–8½ years.

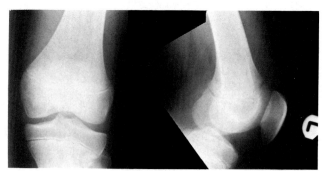

Fig. 1.36   12½–13½ years.

Foot: AP

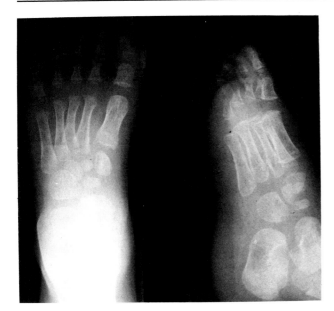

Fig. 1.37 2½–3½ years.

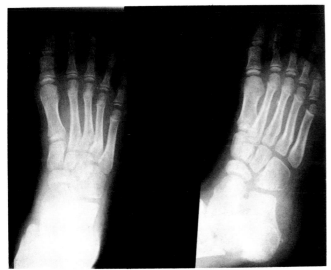

Fig. 1.38 7½–8½ years.

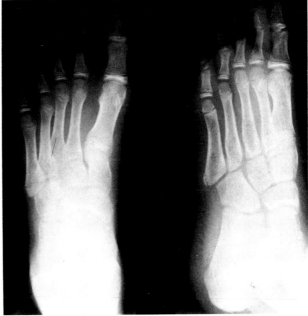

Fig. 1.39 12½–13½ years.

Elbow

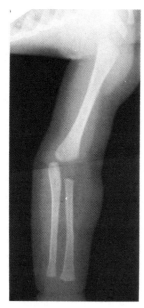

Fig. 1.40   Birth–8 weeks.

Fig. 1.41   2½–3½ years.

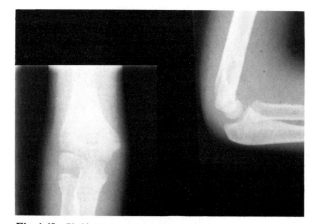

Fig. 1.42   7½–8½ years.

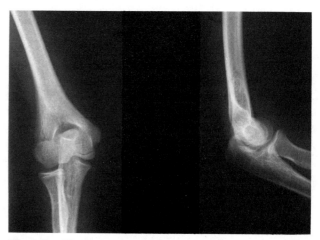

Fig. 1.43   12½–13½ years.

Hand

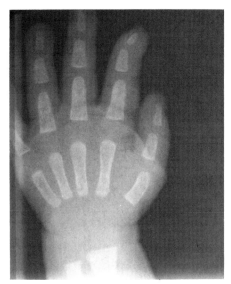

Fig. 1.44 Birth–8 weeks.

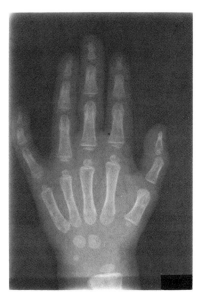

Fig. 1.45 2½–3½ years.

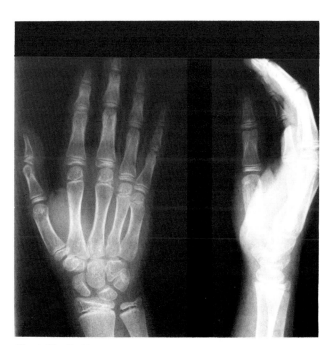

Fig. 1.46 7½–8½ years.

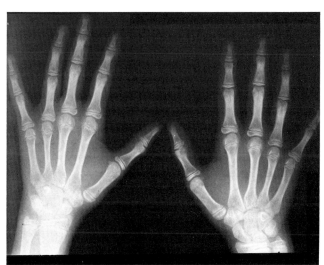

Fig. 1.47 12½–13½ years.

Section Two
EPIPHYSEAL DISORDERS

Chapter Two
MULTIPLE EPIPHYSEAL DYSPLASIA

Some shortness of stature. Irregular epiphyseal growth, usually involving many joints symmetrically, but occasionally only one pair, leading to premature degenerative changes. Minor vertebral involvement at lower dorsal spine, mainly in childhood. Wide range of severity.

Inheritance

Nearly always autosomal dominant, but autosomal recessive cases have been described: clinical delineation between them is impossible.

Frequency

One of the commonest skeletal dysplasias, possible prevalence 11 per million index patients; 16 per million including affected relatives.

Clinical features

Facial appearance—normal.

Intelligence—normal.

Stature—see graphs. Some shortness of stature, principally of limbs.

Presenting feature/age—pain and stiffness of affected joints. Rarely in early infancy, usually in childhood, occasionally only found incidentally in adult life.

Deformities—joint contractures. Occasionally scoliosis. Valgus ankles.

Range of severity—from one pair of joints, discovered only incidentally, to widespread involvement with crippling osteoarthritis and short stature. Wrists and hands often normal in the mild (Ribbing) type and short and stubby in the severe (Fairbank) type.

Associated anomalies—none.

Radiographic features

Skull—normal.

Spine—irregularity of end-plates mainly in lower dorsal spine.

Limb epiphyses—a pair of joints is nearly always symmetrically involved. Centres of ossification are late in appearing, becoming fragmented and irregular. In severe cases the adjacent metaphyses are also splayed and irregular.

Hip joints—capital femoral epiphyses are nearly always involved, and the femoral head subsequently flattened—sometimes with poor coverage and subluxation, sometimes deeply set with protrusio acetabuli in later life. The acetabulum is occasionally normal, but more usually there is some loss of definition with a scalloped outline.

Hands/wrists—there may be cone-shaped epiphyses in one or more areas, associated with premature fusion of the epiphyseal plates and resulting in shortening of these metacarpals and phalanges.

Bone maturation

Generalised delay for up to 4 or 5 years.

Biochemistry

Not known.

Differential diagnosis

The spondylo-epiphyseal dysplasias have severe vertebral changes and platyspondyly, the limbs usually being less involved.

Pseudoachondroplasia—there is severe dwarfing with very short limbs and marked metaphyseal involvement. Some patients have striking joint laxity, but differential diagnosis from the severe form of multiple epiphyseal dysplasia is made most readily from a pelvic radiograph. In pseudoachondroplasia development of the pelvis is disorganised and delayed, with poor formation of the acetabulum and widening of the tri-radiate cartilage.

Untreated hypothyroidism is differentiated by general clinical and biochemical features. Bone maturation is very retarded: there is one hook-shaped vertebra or more in the lumbar region. Limb epiphyses may be identical.

Bilateral Perthes' disease—affects only the capital femoral epiphyses, although growth and bone maturation can be more generally delayed. It differs from multiple epiphyseal dysplasia in being always asymmetrical, exhibiting patches of increased density in the epiphyses and often showing marked metaphyseal involvement with cyst-like changes. The acetabulum is normal, or only slightly affected (secondarily) by changes in the shape of the developing femoral head. Perthes' is also characterised by an active phase of worsening, followed by improvement.

Progress/complications

Related to the premature degeneration of affected joints.

REFERENCES

Crossan J F, Wynne-Davis R, Fulford G E 1983 Bilateral failure of the capital femoral epiphysis. Journal of Pediatric Orthopedics 3:297–301

Maroteaux P, Stanescu R, Cohen-Solal D 1975 Dysplasie poly-epiphysaire probablement recessive autosomique. Apport de l'étude ultra-structurale dans l'isolement de cette forme autonome. Nouveau presse médicale 4:2169–2172

Spranger J 1976 The epiphyseal dysplasias. Clinical Orthopaedics and Related Research 114:46–60

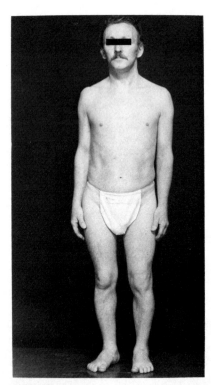 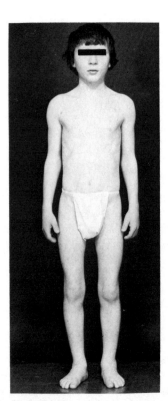

Figs 2.1 and 2.2   Father and son with some disproportionate limb shortening compared with the trunk.

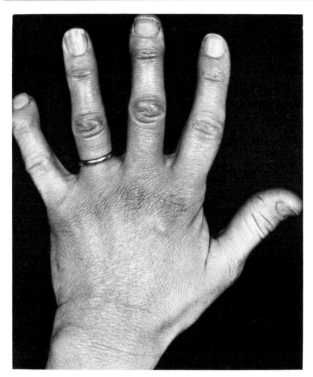 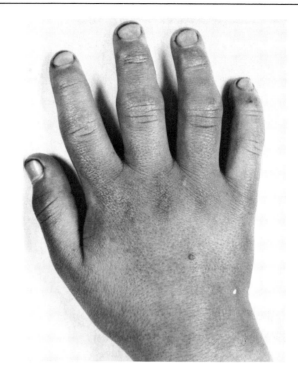

Figs 2.3 and 2.4 Their hands; stubby, with particularly short metacarpals and terminal phalanges.

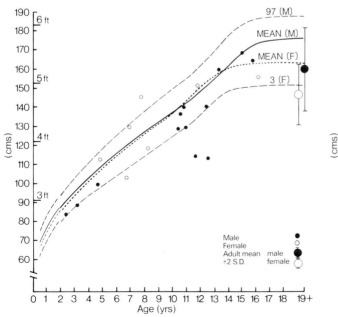 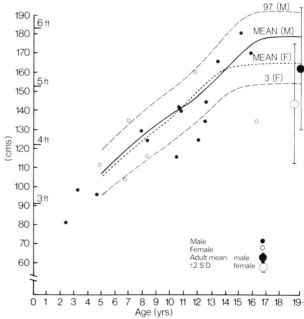

Fig. 2.5 Multiple epiphyseal dysplasia—percentile height.

Fig. 2.6 Multiple epiphyseal dysplasia—percentile span.

Spine: 4 years–adult

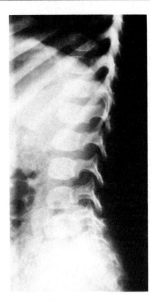

Fig. 2.7   D.W.—4 years—Lateral spine
Normal appearance.

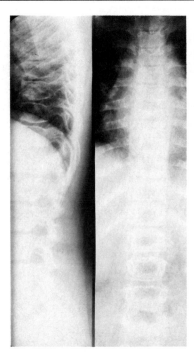

Fig. 2.8   K.L.—5 years—AP and lateral dorso-lumbar spine
There are vertebral end-plate irregularities in the mid-dorsal region.
Minimal scoliosis.

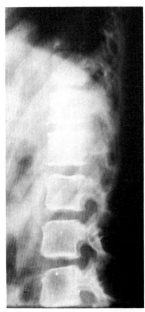

Fig. 2.9   D.H.—11 years 10 months—Lateral lumbar spine
The pedicles are rather short. The vertebral bodies appear normal.
Same patient: Figs 2.15, 2.21, 2.34, 2.51

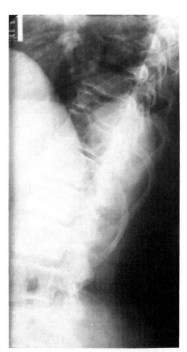

Fig. 2.10   P.L.—Adult—Lateral dorso-lumbar spine (mother
of child in Fig. 2.8)
Irregularity of surfaces of two adjacent vertebrae.

Pelvis and hip: 7 years

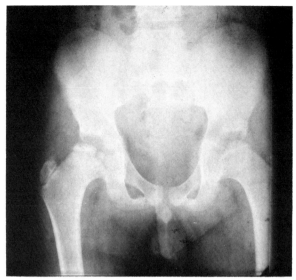

Fig. 2.11   M.A.—7 years—AP pelvis
The capital femoral epiphyses are symmetrically flattened, fragmented and sclerotic. The femoral necks are rather broad and the metaphyses a little irregular.

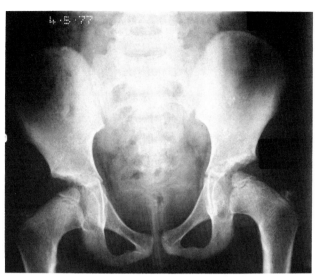

Fig. 2.12   J.P.—7 years—AP pelvis
The femoral heads are small and triangular. There is a medial slope to the femoral metaphyses. The joint spaces here are wide, suggesting unossified epiphyses in this region.

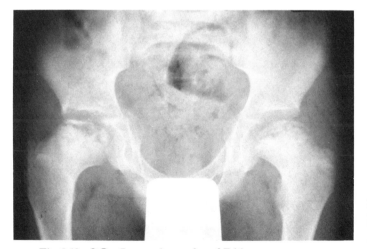

Fig. 2.13   S.C.—7 years 4 months—AP hips
The femoral heads are symmetrically flattened and fragmented. The femoral necks are short and broad.
Same patient: Figs 2.14, 2.45

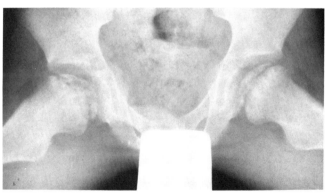

Fig. 2.14   S.C.—7 years 4 months—Hips in abduction (same patient as Fig. 2.13)
The femoral heads are fragmented, flattened and slightly sclerotic. Metaphyseal irregularity is pronounced for this condition.

Pelvis and hip: 8–17 years

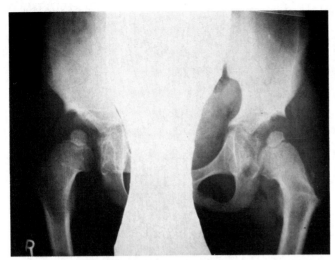

Fig. 2.15   D.H.—8 years 9 months—AP hips
The acetabular roofs are irregular (scalloped) and femoral heads small for this age. The femoral necks are elongated and there is coxa valga.
Same patient: Figs 2.9, 2.21, 2.34, 2.51

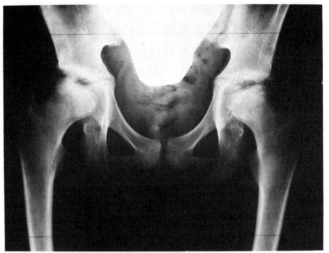

Fig. 2.16   D.K.—11 years 11 months—AP hips
Both capital femoral epiphyses are flattened, fragmented and sclerotic. There is mild coxa valga. The acetabular roofs are horizontal. This child had no symptoms. She was X-rayed because her mother was affected, and she herself had slight limitation of abduction only.

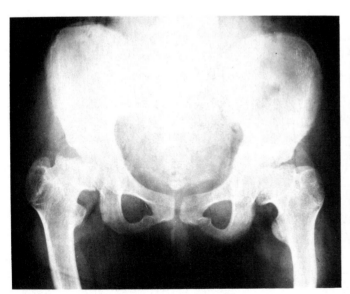

Fig. 2.17   A.W.—12 years 6 months—AP pelvis
The capital femoral epiphyses are flattened and the articular margin is irregular. There is coxa vara and the femoral necks are short. This child's mother was also affected.

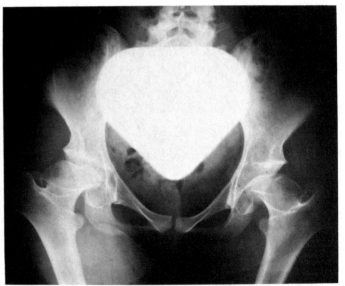

Fig. 2.18   L.W.—17 years 5 months—AP pelvis
The femoral heads are flat and mushroomed, and poorly covered by the acetabular roofs. There is coxa vara and the femoral necks appear short.

Pelvis and hip: young adult

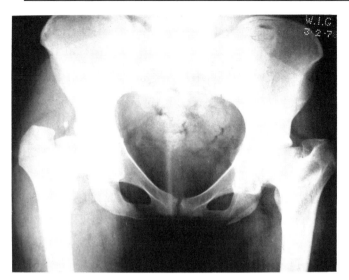

Fig. 2.19 S.McG.—20 years 2 months—AP pelvis
There is marked coxa vara and the greater trochanters are displaced upwards. The femoral heads are flattened and poorly covered.
Same patient: Fig. 2.20

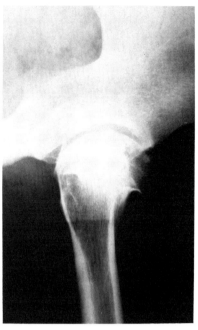

Fig. 2.20 S.McG.—20 years 3 months—Lateral right hip
(same patient as Fig. 2.19)
The joint space is narrowed and the femoral head flattened.

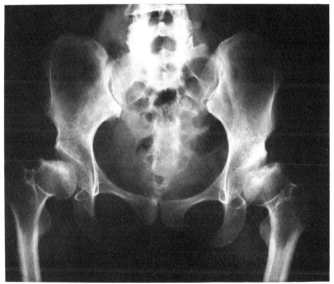

Fig. 2.21 D.H.—24 years—AP pelvis
There is bilateral coxa vara. The femoral heads are flattened with irregular articular margins. Subchondral cysts are present. The acetabula are shallow and slope steeply. There is poor coverage of the femoral heads (cf. Fig. 2.22) and evidence of osteoarthritis in the supra-acetabular sclerosis on the left.
Same patient: Figs 2.9, 2.15, 2.34, 2.51

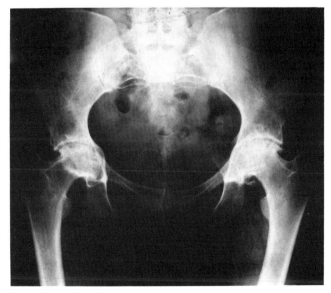

Fig. 2.22 A.W.—28 years—AP pelvis (aunt of the patient in Fig. 2.16)
Both hip joint spaces are reduced. There is subchondral cyst formation above the acetabula and in the femoral heads. There is mild coxa vara. In contrast to Figure 2.21 the heads are well covered, but there is already marked osteoarthritis.

Pelvis and hip : older adult

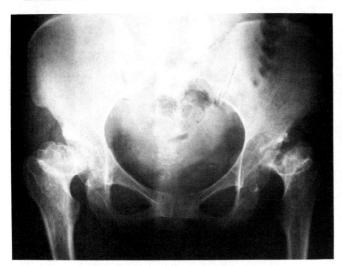

Fig. 2.23   J.B.W.—36 years 3 months—AP pelvis
The femoral heads are flattened with irregular articular margins and subchondral cysts. The joint spaces are narrowed and irregular. There is coxa valga.

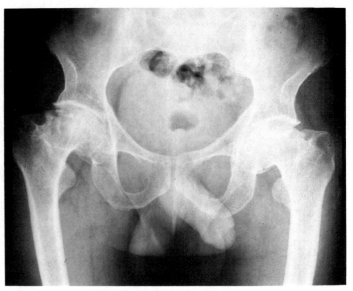

Fig. 2.24   J.M.—46 years—AP hips
There is coxa vara. The femoral necks are short. The femoral heads are flattened and mushroomed with subchondral cyst formation. The joint spaces are narrow.

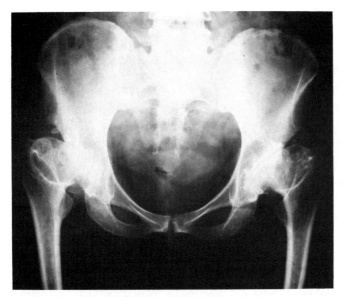

Fig. 2.25   A.T.—49 years—AP pelvis
There is coxa vara. The joint spaces are greatly narrowed and there is periarticular sclerosis. The femoral heads are flattened.

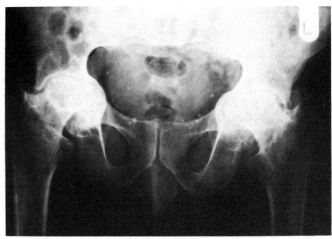

Fig. 2.26   H.W.—65+ years—AP hips
The femoral necks are short. The femoral heads are virtually destroyed. The joint spaces are narrow with much periarticular sclerosis. There is protrusio acetabuli, probably related to osteoporosis developing with age.

Knee: 7–10 years

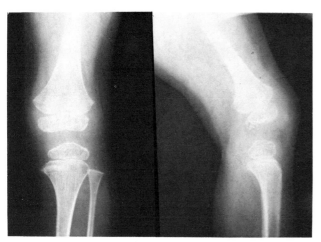

Fig. 2.27   J.M.—About 7 years—AP and lateral knees
There is flaring of the metaphyses. The epiphyses are small.
(Radiological bone age = 3–4 years.) This child's father and brother
(Fig. 2.28) were affected.

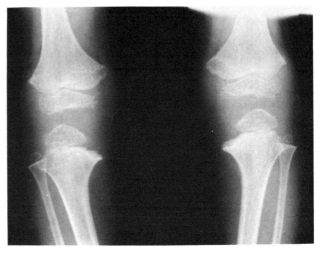

Fig. 2.28   A.M.—8 years—AP knees (brother of patient in Fig.
2.27)
The epiphyses are small and irregular. There is flaring of the
metaphyses and some irregularity. There is mild tibia vara.
(Radiological bone age = 4 years.)
Same patient: Figs 2.38, 2.39, 2.54, 2.58

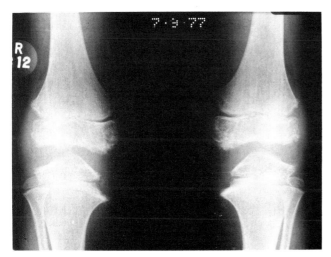

Fig. 2.29   N.S.—10 years—AP knees
The epiphyses are small and irregular. The metaphyses are flared and
irregular. There is mild medial beaking of the proximal tibial
metaphyses. (Radiological bone age = 6–7 years.)
Same patient: Figs 2.30, 2.57

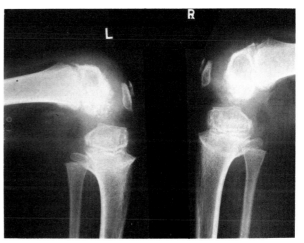

Fig. 2.30   N.S.—10 years—Lateral knees (same patient as in Fig.
2.29)
The epiphyses are fragmented. The metaphyses are flared and
irregular. There is more than one ossification centre for the patellae.

Knee: 10–11 years

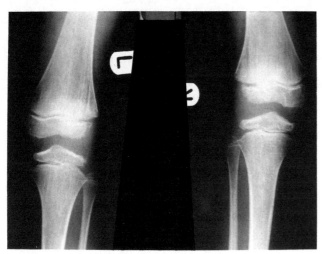

Fig. 2.31   O.G.—10 years—AP knees
The epiphyses are small and irregular. The metaphyses are normally
modelled. There is some striation in the distal femur. (Radiological
bone age = 5 years.)

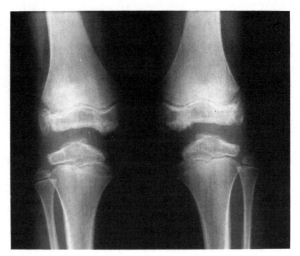

Fig. 2.32   K.G.—11 years 5 months—AP knees (same family as
patient in Fig. 2.29)
The epiphyses are small and fragmented. The articular margins are
irregular. (Radiological bone age = 5–6 years.)
Same patient: Figs 2.40, 2.55

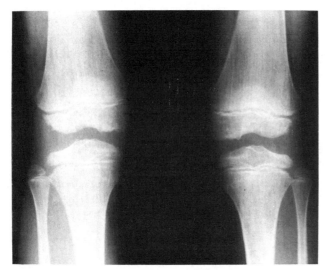

Fig. 2.33   R.A.—11 years 7 months—AP knees
The epiphyses are small and irregular. (Radiological bone age = 8
years.)
Same patient: Fig. 2.41

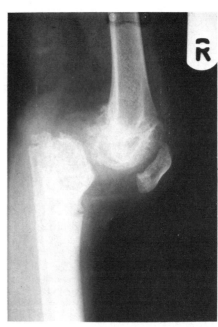

Fig. 2.34   D.H.—11 years 10 months—Lateral right knee
The knee joint is dislocated. The epiphyses are irregular and
fragmented.
Same patient: Figs 2.9, 2.15, 2.21, 2.51

Knee: adult

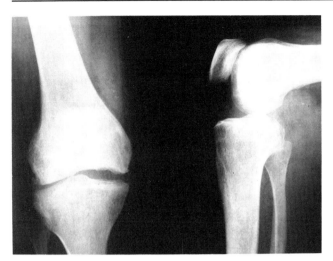

Fig. 2.35   T.H.—32 years—AP and lateral knee
There is now evidence of osteoarthritis with narrowing of the joint space, irregularity of the articular surfaces, adjacent sclerosis and osteophyte formation.

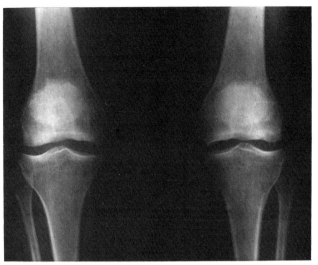

Fig. 2.36   J.G.—32 years—AP knees (same family as patient in Figs 2.29, 2.32)
The articular margins are irregular and flattened. Both patellae have irregular outlines.

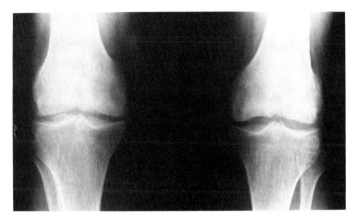

Fig. 2.37   A.E.—56½ years—AP knees (same family as Fig. 2.36)
The joint spaces are reduced. The articular margins are irregular. There is spiking of the tibial spines and lateral osteophyte formation.
Same patient: Fig. 2.44

Ankle and foot: 8–11 years

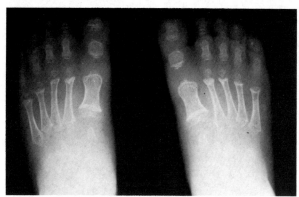

Fig. 2.38   A.M.—8 years—PA feet
The long bones of both feet are short, with irregular, frayed metaphyses. The epiphyses are small and irregular. There is widening of the epiphyseal plates.
Same patient: Figs 2.28, 2.39, 2.54, 2.58

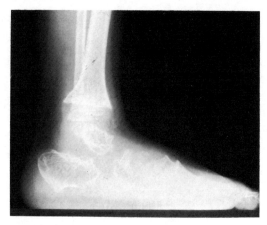

Fig. 2.39   A.M.—8 years—Lateral ankle (same patient as in Fig. 2.38)
The distal tibial metaphysis is flared and flat and the epiphysis is small. The tarsal centres are all small and irregular.

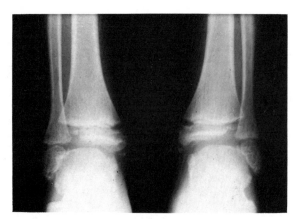

Fig. 2.40   K.G.—11 years 5 months—AP ankles
The epiphyses are irregular and there is apparent widening of the epiphyseal plates.
Same patient: Figs 2.32, 2.55

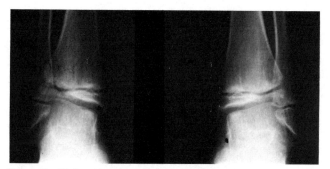

Fig. 2.41   R.A.—11 years 7 months—AP ankles
There is a valgus tilt at the ankle joint with a long distal fibula. The distal tibial epiphyses are flattened particularly on the lateral side.
Same patient: Fig. 2.33

Ankle and foot : adult

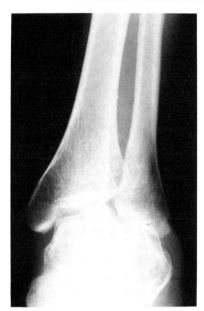

Fig. 2.42   JG—28 years—AP ankle
There is a valgus deformity at the ankle joint.
Same patient: Fig. 2.60

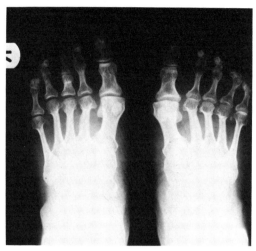

Fig. 2.43   M.W.—40 years 10 months—PA both feet
There is shortening of the metatarsals and some of the phalanges, probably with premature epiphyseal fusion. The MP joint spaces of the big toes are narrowed with marginal osteophyte formation.
Same patient: Fig. 2.47

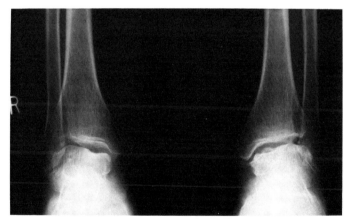

Fig. 2.44   A.E.—45 years—AP ankles
There is valgus deformity at the ankles and articular irregularity, especially of the tali.
Same patient: Fig. 2.37

Shoulder : 7 years–adult

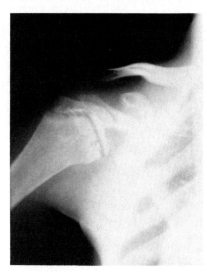

Fig. 2.45   S.C.—7 years 4 months—AP right shoulder
There is flattening and irregularity of the humeral epiphysis and mild
irregularity of the adjacent metaphysis.
Same patient : Figs 2.13, 2.14

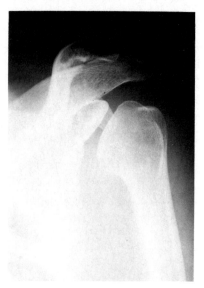

Fig. 2.46   M.P.—29 years—AP left shoulder
There is flattening of the humeral head and subluxation.

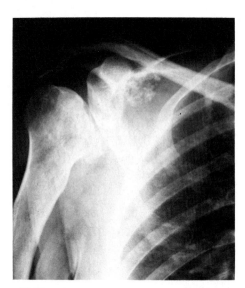

Fig. 2.47   M.W.—38 years 2 months—AP right shoulder
There is a 'hatchet-shaped' deformity of the humeral head, with
narrowing of the joint space. The glenoid fossa is flattened and the
articular margins are not apparent. Some soft tissue calcification is
present in the region of the coracoid process.
Same patient : Fig. 2.43

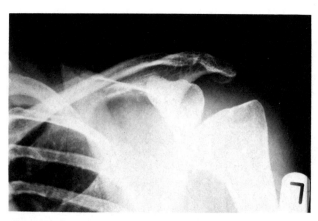

Fig. 2.48   E.W.—45 years—AP left shoulder (mother of patient
in Fig. 2.17)
There is flattening of the humeral head and the joint space appears
narrowed inferiorly due to subluxation.

Elbow : 8 years–adult

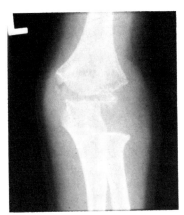

Fig. 2.49 A.M.—8 years—AP left elbow
There is some metaphyseal irregularity. The epiphyses are delayed and those which are present are small with irregular ossification.
Same patient : Figs 2.28, 2.38, 2.39, 2.54, 2.58

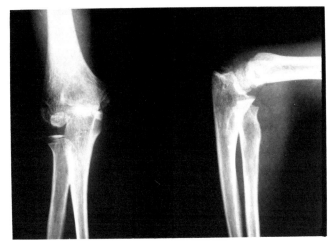

Fig. 2.50 S. van D.—10 years—AP and lateral elbow
The epiphyses all show flattening and irregular ossification with some sclerosis. There is minor irregularity of the adjacent metaphyses.
Same patient : Fig. 2.56

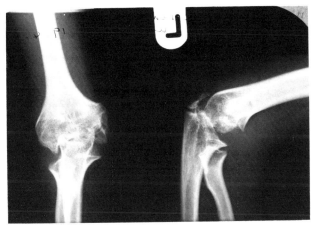

Fig. 2.51 D.H.—11 years 10 months—AP and lateral left elbow
The epiphyses are all very small and fragmented. There is marked irregularity of the articular surfaces and the joint space is narrowed.
Same patient : Figs 2.9, 2.15, 2.21, 2.34

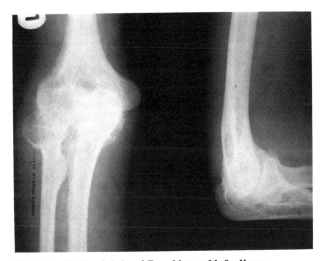

Fig. 2.52 R.W.—Adult—AP and lateral left elbow
There is metaphyseal flaring. The epiphyses have fused. There is mushrooming and enlargement of the radial head. The joint space is reduced and there is a flexion contracture.

Hand : 5–10 years

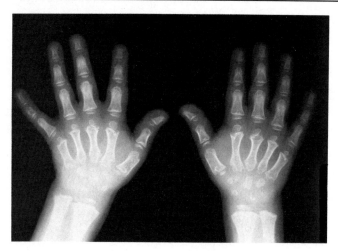

Fig. 2.53   C.G.—5 years 4 months—PA hands (daughter of patient in Fig. 2.42)
The carpal centres are small and irregular, but without significant delay in bone maturation. The distal ulna is rather long. The metacarpals are all short and have irregular metaphyses. The adjacent epiphyses are deformed. The phalanges are normal. (Radiological bone age = 4 years for carpals but 2 years 6 months for phalanges.)

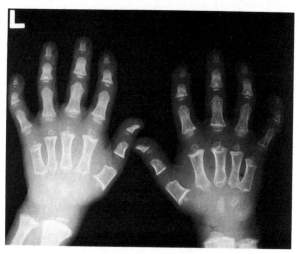

Fig. 2.54   A.M.—8 years 2 months—PA hands
The phalanges and metacarpals are short and there is marked delay in bone maturation. There is generalised metaphyseal irregularity and the bases of the middle phalanges have an inverted cone-shaped deformity. The second to fifth metacarpals appear to have a proximal as well as distal epiphysis. The carpals are small, irregular and delayed (severe Fairbank type). (Radiological bone age = 3 years 6 months.)
Same patient : Figs 2.28, 2.38, 2.39, 2.49, 2.58

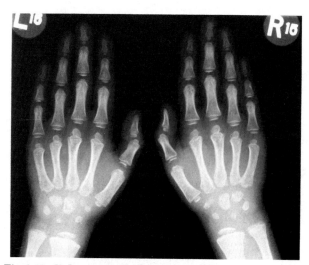

Fig. 2.55   K.G.—9½ years—PA hands
The carpal centres are small and retarded. There is some flattening of the distal radial epiphyses but little other abnormality (probably the mild Ribbing type). (Radiological bone age = 3 years 6 months for carpal centres but 2 years 6 months for phalanges.)
Same patient : Figs 2.32, 2.40

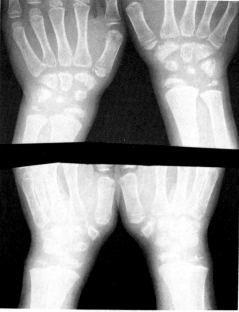

Fig. 2.56   S. van D.—10 years—PA and oblique wrists
The carpal centres are small with marked irregularity. The apparent trapezium/trapezoid fusion on the PA view is not confirmed on the oblique. The distal radial and ulnar epiphyses are small and fragmented with streaks of ossification surrounding them. There is apparent widening of the epiphyseal plates. (Radiological bone age = 6 years.)
Same patient : Fig. 2.50

Hand: 11 years—adult

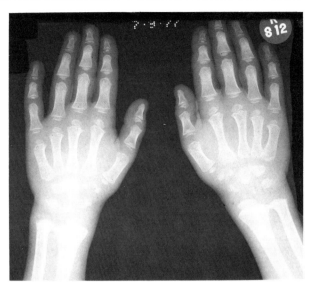

Fig. 2.57   N.S.—11 years—PA hands
The carpal centres are small, retarded and irregular. There is
fragmented and sclerotic ossification of the epiphyses of the terminal
phalanges of the index fingers. There are cone epiphyses of the
proximal phalanges of the thumbs. (Radiological bone age = about 5
years.)
Same patient: Figs 2.29, 2.30

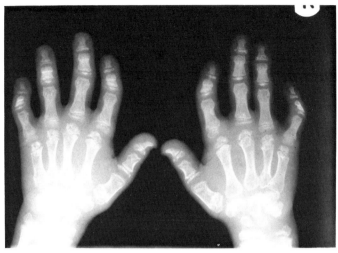

Fig. 2.58   A.M.—13 years 4 months—PA hands
There is clinodactyly and generalised shortening of the phalanges and
metacarpals, but the middle and terminal phalanges are maximally
involved. The epiphyses are all small, fragmented and irregular. The
carpal centres are irregular. (Radiological bone age = 11 years.)
Same patient: Figs 2.28, 2.38, 2.39, 2.49, 2.54

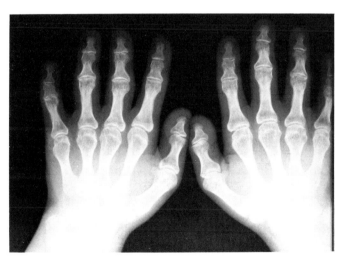

Fig. 2.59   M.MacD.—23 years 5 months—PA hands
There is shortening of all the long bones, but especially the middle
and terminal phalanges. The distal ulna is relatively long, with
subluxation at the wrist.

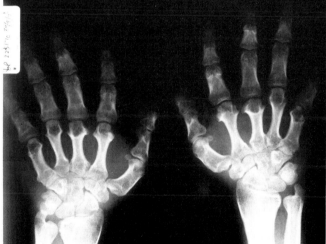

Fig. 2.60   J.G.—28 years—PA hands
There is irregular shortening of all the long bones, particularly the
terminal phalanges and metacarpals of the middle and little fingers;
this appears to have been due to premature fusion at these sites. Both
ulnar styloid processes are long.
Same patient: Fig. 2.42

Chapter Three
HEREDITARY PROGRESSIVE ARTHRO-OPHTHALMOPATHY
(Stickler syndrome)

There are variable manifestations including enlargement of joints, epiphyseal changes, mild platyspondyly and severe myopia associated with retinal detachment. Sometimes cleft palate and conductive hearing loss.

Inheritance

Autosomal dominant inheritance.

Frequency

Extremely rare, perhaps 0.1–0.3 per million prevalence.

Clinical features

Facial appearance—high degree of myopia, flat nasal bridge, sometimes micrognathos in infancy.

Intelligence—normal, allowing for blindness and deafness in some patients.

Stature—normal.

Presenting feature/age—'Pierre Robin' syndrome at birth: later the eye signs are noted. Epiphyseal changes may not cause symptoms and are probably discovered incidentally.

Deformities—probably none.

Associated anomalies—none apart from cleft palate, myopia and sometimes hearing loss.

Radiographic features

Skull—normal vault, but the facial bones are small.

Spine—mild irregular platyspondyly. Anterior wedging sometimes, similar to Scheuermann's disease.

Long bones—in the neonatal period there is a 'dumb-bell' appearance, with enlarged epiphyses and metaphyses. With growth, the shafts become normal, but epiphyseal dysplasia develops, with small irregular epiphyses and later degenerative changes.

Bone maturation

Not known.

Biochemistry

Not known.

Differential diagnosis

In infancy—flared metaphyses are similar to those in *metatropic dwarfism* and *Kniest disease*, but these signs diminish with age in the Stickler syndrome.

Later, *multiple epiphyseal dysplasia* has similar joint changes, but is without the myopia and other associated features.

Spondylo-epiphyseal dysplasia congenita is differentiated by its more severe spinal and hip changes, delayed ossification and dwarfism.

Progress/complications

The eye symptoms are the most serious complication requiring treatment. The epiphyseal changes are not severe, but premature osteoarthritis is likely to develop.

REFERENCES

Baraitser M 1982 Marshall/Stickler syndrome. Journal of Medical Genetics 19:139–140

Kelly T E, Wells H H, Tuck K B 1982 The Weissenbacher–Zweymüller syndrome. Possible neonatal expression of the Stickler syndrome. American Journal of Medical Genetics 11:113–119

Haller J O, Berdon W E, Robinow M, Slovis T L, Baker D H, Johnson G F 1975 The Weissenbacher–Zweymüller syndrome of micrognathia and rhizomelic chondrodysplasia at birth with subsequent normal growth. American Journal of Roentgenology, Radium Therapy and Nuclear Medicine 125:936–943

Winter R M, Baraitser M, Laurence K M, Donnai D, Hall C M 1983 The Weissenbacher-Zweymuller, Stickler and Masrshall syndromes. Further evidence for their identity. American Journal of Medical Genetics 16:189–199

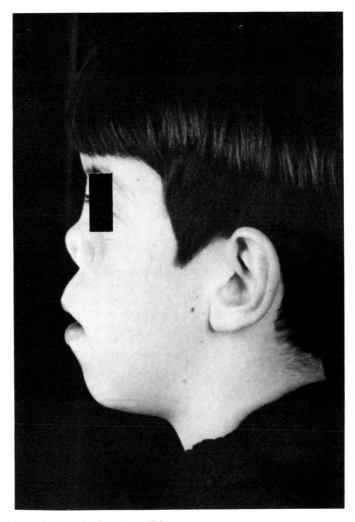

Fig. 3.1 Markedly flat nasal bridge and under-developed mandible.

Skull and cervical spine : 2–10 years

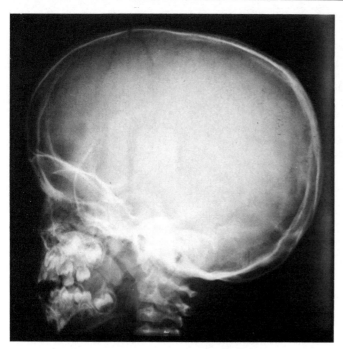

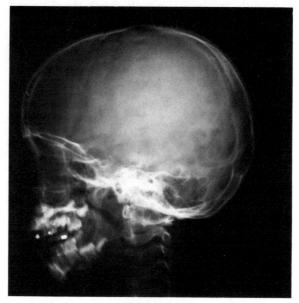

Fig. 3.3   K.M.—6 years—Lateral skull
Similar to Fig. 3.2, but the disproportion is less marked.

Fig. 3.2   B.D.—2 years 6 months—Lateral skull
The facial bones and mandible are small in relation to the vault.
Same patient : Figs 3.10, 3.11, 3.21, 3.22, 3.29, 3.30, 3.37, 3.38

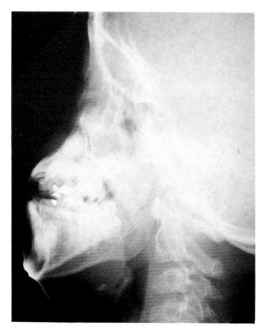

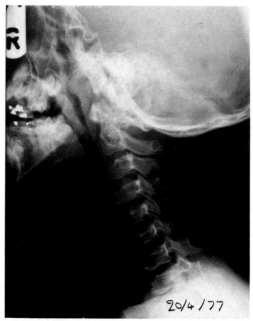

Fig. 3.4   S.S.—9 years—Lateral face
The mandible is now of normal size but the facial bones are
hypoplastic (dish face).
Same patient : Figs 3.5, 3.6, 3.7, 3.15, 3.16, 3.27, 3.40, 3.42

Fig. 3.5   S.S.—10 years—Lateral cervical spine (same patient
as in Fig. 3.4)
There is platyspondyly throughout.

Skull: 9–13 years

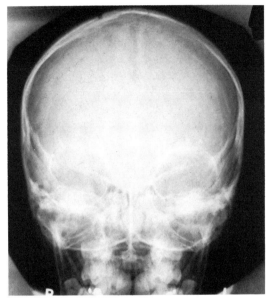

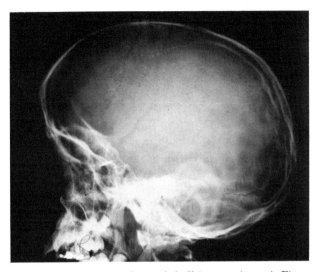

Fig. 3.7   S.S.—9 years—Lateral skull (same patient as in Fig. 3.4)
The hypoplastic facial bones (dish face) are now obvious.

Fig. 3.6   S.S.—9 years—AP skull (same patient as in Fig. 3.4)
Appears normal.

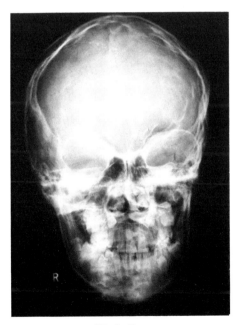

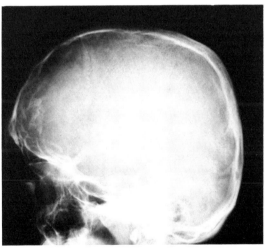

Fig. 3.9   S.D.—13 years—Lateral skull (same patient as in Fig. 3.8)
There is premature fusion of the coronal suture with a small frontal bone and pronounced convolutional markings. The hypoplastic facial bones are less obvious than in Figure 3.2.

Fig. 3.8   S.D.—13 years—AP skull
The skull and mandible appear normal, but there is a right lateral cleft palate with deviation of the nasal septum to the left.
Same patient: Figs 3.9, 3.35

Thoracic and lumbar spine : Birth–3 years

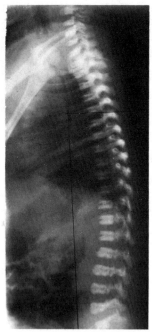

Fig. 3.10   B.D.—Neonate—Lateral thoracic and lumbar spine
Coronal cleft vertebrae are present in the lumbar region.
Same patient : Figs 3.2, 3.11, 3.21, 3.22, 3.29, 3.30, 3.37, 3.38

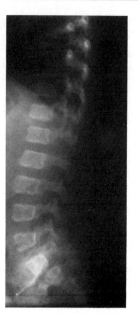

Fig. 3.11   B.D.—2 months—Lateral thoracic and lumbar spine (same patient as in Fig. 3.10)
Traces only of the coronal clefts can be seen.

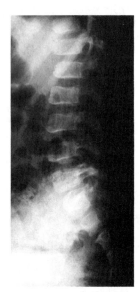

Fig. 3.12   L.K.—3 years—Lateral lumbar spine
Platyspondyly is now developing.
Same patient : Figs 3.23, 3.32, 3.39, 3.41

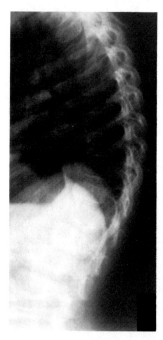

Fig. 3.13   K.M.—3 years—Lateral thoracic and lumbar spine
In addition to the platyspondyly there is thoraco-lumbar kyphosis with backward displacement of T12.
Same patient : Figs 3.3, 3.17–3.20, 3.25, 3.26, 3.33, 3.34

Thoracic and lumbar spine : 7–14 years

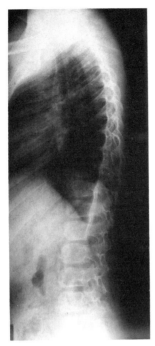

Fig. 3.14   K.McD—7 years—Lateral thoracic and lumbar spine
These vertebrae have an immature oval configuration rather than platyspondyly and there is loss of the normal lumbar lordosis.
Same patient: Figs 3.24, 3.28, 3.31, 3.36

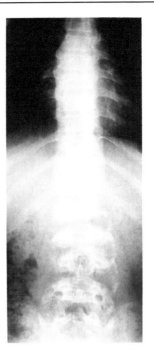

Fig. 3.15   S.S.—10 years—AP lower thoracic and lumbar spine
In addition to the universal platyspondyly there is extensive soft-tissue ossification adjacent to the vertebrae (not a recognised feature of this disease).
Same patient: Figs 3.4–3.7, 3.16, 3.27, 3.40, 3.42

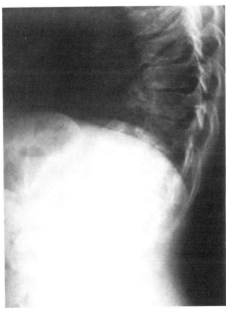

Fig. 3.16   S.S.—14 years—Lateral thoraco-lumbar spine
(same patient as in Fig. 3.15)
The soft-tissue ossification can be seen anteriorly, as well as anterior fusion of the vertebrae.

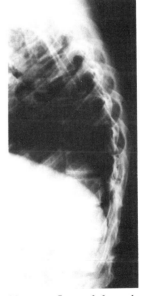

Fig. 3.17   K.M.—14 years—Lateral thoracic spine
Less severely affected than the patient in Figures 3.15 and 3.16, but there is mild, irregular platyspondyly.
Same patient: Figs 3.3, 3.13, 3.18–3.20, 3.25, 3.26, 3.33, 3.34

Chest: 2–11 years

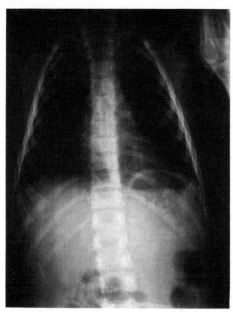

Fig. 3.18   K.M.—2 years—AP trunk (same patient as in Fig. 3.17) There is little abnormality apart from slight irregularity of vertebral end plates.

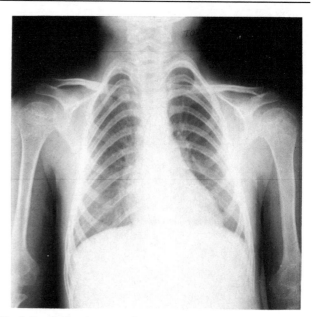

Fig. 3.19   K.M.—6 years—AP chest (same patient as in Fig. 3.17) The thoracic cage is narrow in its superior part. The humeral epiphyses show only mild dysplasia.

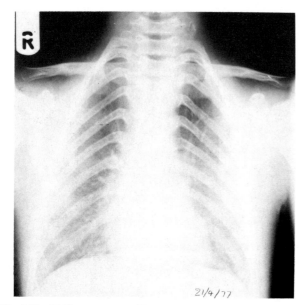

Fig. 3.20   K.M.—11 years—AP chest (same patient as in Fig. 3.17) The thoracic cage is narrow with steeply sloping ribs.

Pelvis: Birth–3 years

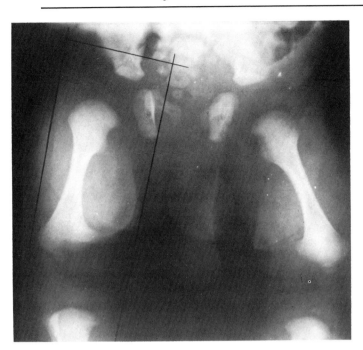

Fig. 3.21 B.D.—Neonate—AP pelvis and femora
There is a striking dumb-bell shape of both femora, with greatly expanded upper and lower ends, but the mid-shaft is normal.
Same patient: Figs 3.2, 3.10, 3.11, 3.22, 3.29, 3.30, 3.37, 3.38

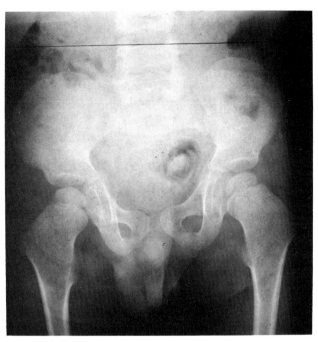

Fig. 3.22 B.D.—2 years 6 months—Pelvis (same patient as in Fig. 3.21)
There is no longer a dumb-bell appearance of the femora, but the capital femoral epiphyses are small and irregular. The femoral necks are very broad. The pelvis is normal.

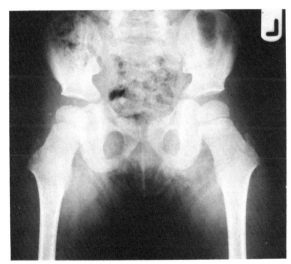

Fig. 3.23 L.K.—3 years—Pelvis
The ilia are rather square with horizontal acetabular roofs. The capital femoral epiphyses are large and only slightly irregular. The femoral necks are broadened in comparison with the shafts.
Same patient: Figs 3.12, 3.32, 3.39, 3.41

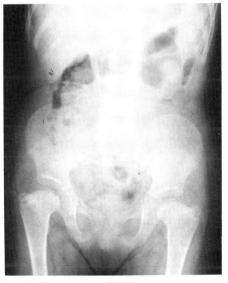

Fig. 3.24 K.McD.—3 years—Pelvis
The ilia are rather square, but the main feature is dysplasia of the upper femoral epiphyses, which are very small for this age; coxa vara is present.
Same patient: Figs 3.14, 3.28, 3.31, 3.36

Pelvis : 5–10 years

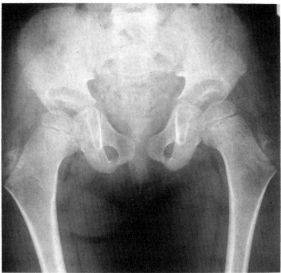

Fig. 3.25   K.M.—5 years—Pelvis
The femoral epiphyses and necks are disproportionately large in comparison with both the pelvis and femoral shafts. The acetabulum is probably tilted abnormally and the capital femoral epiphyses somewhat flat.
Same patient : Figs 3.3, 3.13, 3.17–3.20, 3.26, 3.33, 3.34

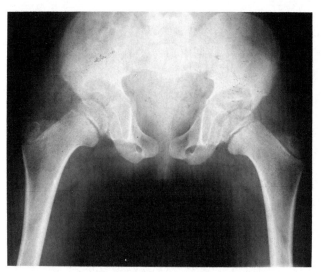

Fig. 3.26   K.M.—6 years—Pelvis (same patient as in Fig. 3.25)
Little change from Figure 3.25.

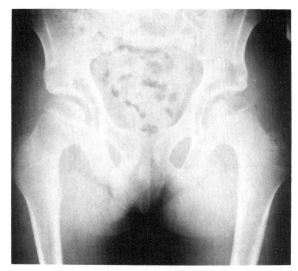

Fig. 3.27   S.S.—10 years—Pelvis
Similar disproportion to Figure 3.25, but no real abnormality of the capital femoral epiphyses.
Same patient : Figs 3.4–3.7, 3.15, 3.16, 3.40, 3.42

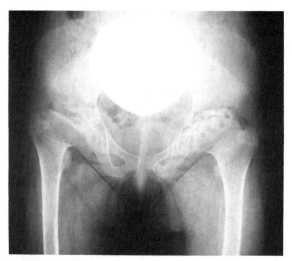

Fig. 3.28   K.McD.—10 years—Pelvis
Severe epiphyseal dysplasia, with broadened metaphyses and coxa vara. Premature osteoarthritis is developing.
Same patient : Figs 3.14, 3.24, 3.31, 3.36

Lower limb : 2–3 years

Fig. 3.29   B.D.—2 years—AP left lower limb
Slight 'dumb-bell' appearance of long bones—but not so marked as in
metatropic dwarfism.
Same patient : Figs 3.2, 3.10, 3.11, 3.21, 3.22, 3.30, 3.37, 3.38

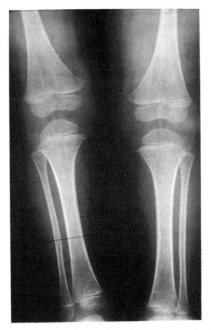

Fig. 3.30   B.D.—2 years 6 months—AP knees and ankles
(same patient as in Fig. 3.29)
By this age there is only mild flaring of the metaphyses. All epiphyses
are flattened, but not irregular.

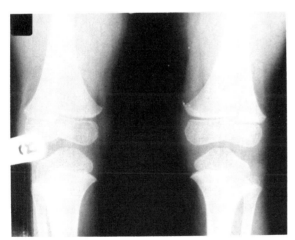

Fig. 3.31   K. McD.—3 years—AP knees and ankles
Similar to Figure 3.30.
Same patient : Figs 3.14, 3.24, 3.28, 3.36

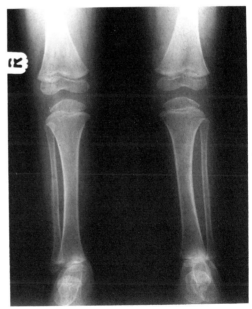

Fig. 3.32   L.K.—3 years—AP knees
Very mild metaphyseal flaring, with some epiphyseal flattening, is
present.
Same patient : Figs 3.12, 3.23, 3.39, 3.41

Lower limb: 7–13 years

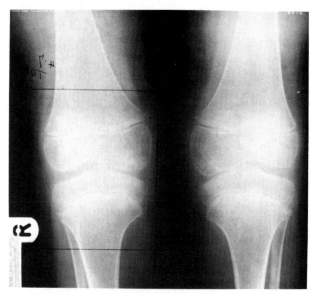

Fig. 3.33   K.M.—8 years—AP knees
There is little abnormality apart from possibly some narrowing of the joint space medially.
Same patient: Figs 3.3, 3.13, 3.17–3.20, 3.25, 3.26, 3.34

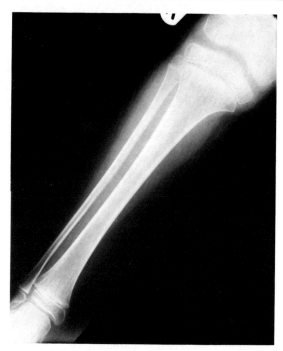

Fig. 3.34   K.M.—8 years—AP right tibia and fibula (same patient as in Fig. 3.33)
There is mild metaphyseal flaring.

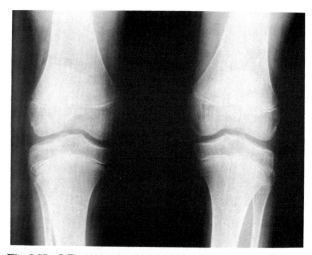

Fig. 3.35   S.D.—13 years—AP knees
There is only slight flattening of the epiphyses, and bony modelling is normal.
Same patient: Figs 3.8, 3.9

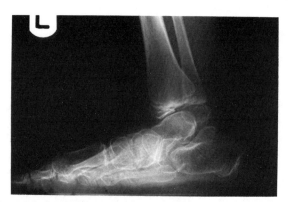

Fig. 3.36   K.McD—7 years—Standing lateral ankle and foot
There is marked flattening and irregularity of the lower tibial epiphyses, with narrowing of the joint space.
Same patient: Figs 3.14, 3.24, 3.28, 3.31

Upper limb: 2–10 years

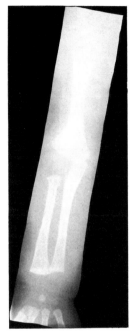

Fig. 3.37   B.D.—2 years—AP upper limb
The humerus is 'dumb-bell' shaped but the radius is expanded in the whole of its distal half.
Same patient: Figs 3.2, 3.10, 3.11, 3.21, 3.22, 3.29, 3.30, 3.38

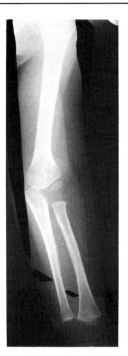

Fig. 3.38   B.D.—2 years 6 months—AP upper limb (same patient as in Fig. 3.37)
There is no longer any abnormality of bone modelling but the lower radial epiphysis is rather flattened.

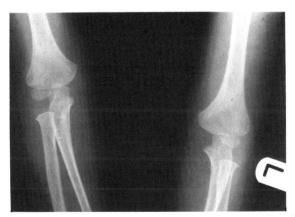

Fig. 3.39   L.K.—3 years—AP elbows
There is slight flaring of metaphyses but no other abnormality.
Same patient: Figs 3.12, 3.23, 3.32, 3.41

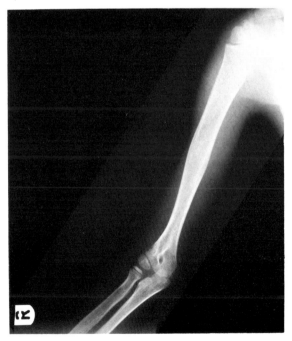

Fig. 3.40   S.S.—10 years—AP right humerus and elbow
The mid-shaft of the humerus is expanded and there is unusual density of the upper radial epiphysis.
Same patient: Figs 3.4–3.7, 3.15, 3.16, 3.27, 3.42

Hand : 3–10 years

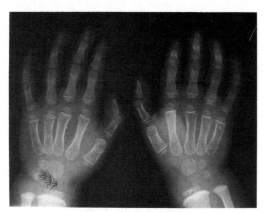

Fig. 3.41   L.K.—3 years—PA hands
The hands are normal, but the lower radial epiphyses are flattened and irregular.
Same patient: Figs 3.12, 3.23, 3.32, 3.39

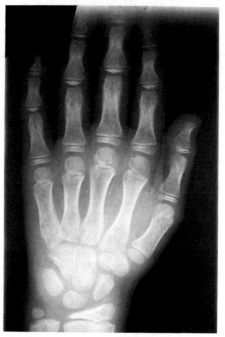

Fig. 3.42   S.S.—10 years—PA left hand
The hands are normal but the lower radial epiphysis is flattened and wedge-shaped.
Same patient: Figs 3.4–3.7, 3.15, 3.16, 3.27, 3.40

Chapter Four
CHONDRODYSPLASIA CALCIFICANS PUNCTATA
(Conradi's disease)

Probably a group of disorders, characterised by stippling of epiphyses and extra-epiphyseal calcification, in infancy only. Ranges from a severe rhizomelic form, often stillborn or dying in first year of life, to a milder disorder sometimes characterized by asymmetrical limb shortening. Wide range of associated congenital defects.

Inheritance

Autosomal recessive and dominant forms are described. Difficult to exclude dominance since mild cases may be unidentified until adult life.

Frequency

Possible prevalence 1.7 per million, but likely to be more as minor cases may be unrecognised.

Clinical features

Facial appearance—flat face, depressed bridge of nose.
Intelligence—mental retardation usual in the severe forms.
Stature—varies from severe short-limbed short stature to normal, or with shortening of one limb or one side of the body.
Presenting feature/age—severe forms apparent at birth, with short limbs, congenital cataracts and skin lesions (usually dry, scaling). Later onset may also present with cataract, limb inequality or congenital or 'idiopathic' scoliosis.
Deformities—often vertebral anomalies and congenital scoliosis. Leg inequality. Contractures.
Associated anomalies—congenital cataracts, skin lesions, congenital heart and other internal defects.

Radiographic features

Diagnostic feature is punctate stippling of epiphyses or extra-cartilaginous calcification—frequently around the vertebral column and pelvis. This disappears by the age of 3–4 years, but there may be residual epiphyseal deformity.
Skull—sometimes craniostenosis. May be normal.
Spine—congenital vertebral anomalies and scoliosis or kyphosis. Lateral views may show separate centres of ossification, anterior and posterior, seperated by a wide translucent band (coronal cleft vertebrae).
Long bones—severe cases have chiefly proximal limb shortening with metaphyseal cupping and splaying. Mild cases normal, or shortening of one limb only.

Bone maturation

No studies available.

Biochemistry

Unknown.

Differential diagnosis

From other short limbed dwarfs—but epiphyseal stippling differentiates Conradi's. In other disorders with stippled epiphyses the vertebral column is unaffected. In later years the epiphyseal deformity is indistinguishable from that of the epiphyseal dysplasias.

Progress/complications

Poor prognosis in severe forms, but those surviving the first year of life tend to improve. Scoliosis and leg inequality may require treatment.

REFERENCES

Heselson N G, Cremin B J, Beighton P 1978 Lethal chondrodysplasia punctata. Clinical Radiology 29/6:679–684
Sheffield L J, Danks D M, Mayne V, Hutchinson L A 1976 Chondrodysplasia punctata—23 cases of a mild and relatively common variety. Journal of Pediatrics 89: 916–923
Tasker W G, Mastri A R, Gold A P 1970 Chondrodystrophia calcificans congenita (dysplasia epiphysealis punctata). American Journal of Diseases of Children 119: 122

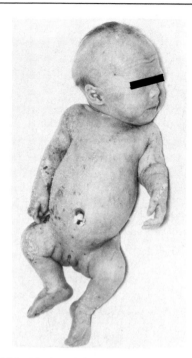 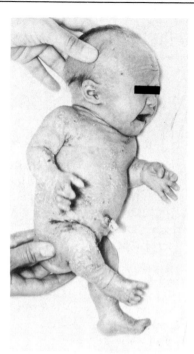

Figs 4.1 and 4.2   Newborn baby with a dry, scaling skin lesion, and probably some asymmetrical limb shortening.

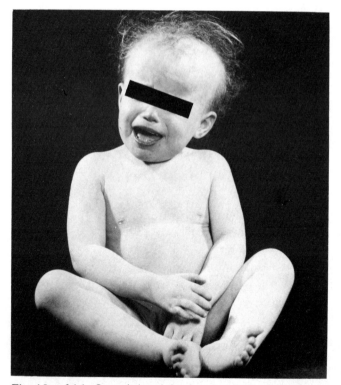 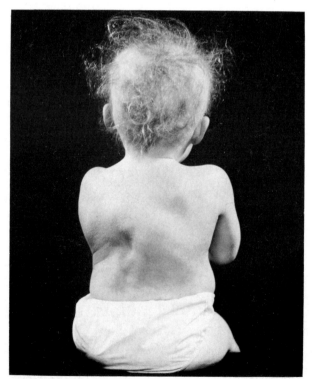

Figs 4.3 and 4.4   Sparse hair and already quite severe scoliosis (due to vertebral anomalies in this instance).

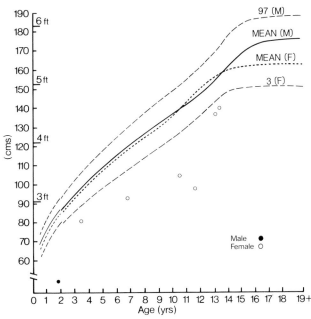

Fig. 4.5 Chondrodysplasia calcificans punctata—percentile height.

Whole body : Birth–2 years

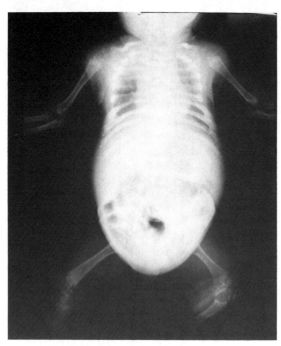

Fig. 4.6   R.W.—Neonate—Whole body
There is stippled ossification around the shoulders, elbows and knees and at the anterior ends of the ribs. The ribs are short and the thoracic cage rather small (cf. Figs 4.7 and 4.8). There is asymmetrical shortening of the left femur. (The tibiae and fibulae are not clearly shown.)
Same patient : Figs 4.21, 4.29

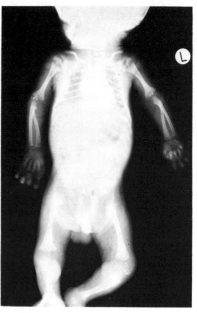

Fig. 4.7   M.E.—6 months—Whole body
There is stippled ossification at the shoulders, elbows, wrists, hips and knees, and shortening of both humeri.

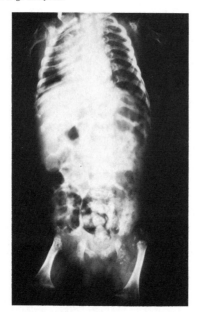

Fig. 4.8   J.W.—1 year—Whole body
There is abnormal stippled calcification around the femoral necks, hip joints and ischia. Both humeri are short with abnormal modelling of the metaphyses. The femora are short and the femoral necks are virtually absent. There is no ossification of the capital femoral epiphyses.

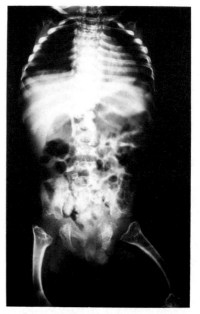

Fig. 4.9   M.E.—2 years—Whole body (same patient as in Fig. 4.7)
The only regions still with stippled ossification are the pubic rami and ischia. Gross modelling deformities are present in the pelvis and upper femur. The femoral heads are not ossified and the femoral necks are virtually absent. A radiolucent defect extends from the distal left femoral metaphysis towards the diaphysis. Congenital vertebral anomalies are present at D10 and D11.

Skull: Birth–10 years

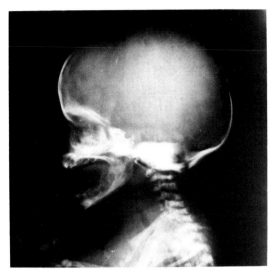

Fig. 4.10   Baby L.—Neonate—Skull
The vault bones appear normal. There is extensive stippled calcification in the region of the vertebral bodies of the cervical and upper dorsal spine.

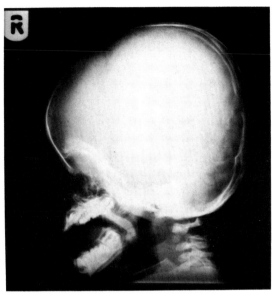

Fig. 4.11   L.T.—1 year—Skull
There is brachycephaly associated with a steep floor to the anterior and middle fossae and mild elongation of the sella.
Same patient: Figs 4.18, 4.19, 4.22, 4.23, 4.26

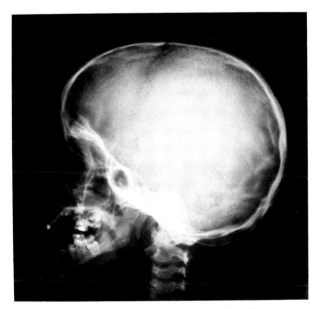

Fig. 4.12   J.M.—10 years 11 months—Lateral skull
Normal.
Same patient: Figs 4.13, 4.28

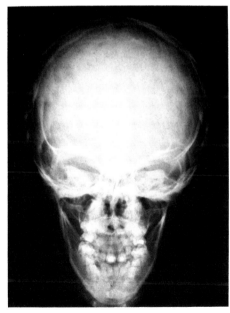

Fig. 4.13   J.M.—10 years 11 months—AP skull (same patient as in Fig. 4.12)
Normal.

Spine : Birth–3 years

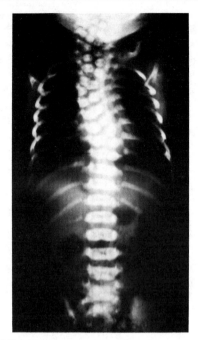

Fig. 4.14   S.B.—Neonate—AP spine
There is scoliosis in the upper dorsal region with multiple congenital anomalies. There are anterior cleft vertebral bodies (butterfly vertebrae) in the cervical and dorsal spine and one right and one left hemivertebra. Eleven ribs are present on the left and twelve on the right.
Same patient : Fig. 4.20

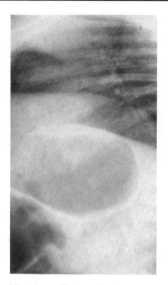

Fig. 4.15   J.H.—Neonate—Lateral spine
There are coronal clefts in the upper lumbar vertebral bodies associated with some posterior wedging.
Same patient : Figs 4.17, 4.25

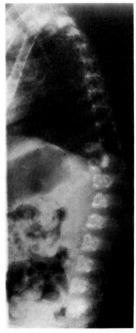

Fig. 4.16   N.I.—1 year—Lateral spine
Coronal clefts are still visible in the lower dorsal and upper lumbar regions. They are seen as marked indentations on the upper and lower vertebral end-plates.
Same patient : Fig. 4.27.

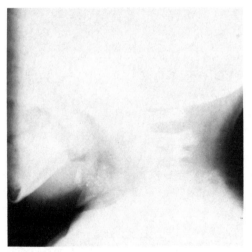

Fig. 4.17   J.H.—3 years—Lateral cervical spine (same patient as in Fig. 4.15)
Punctate calcification is present in the thyroid cartilage.

Spine: 3–10 years

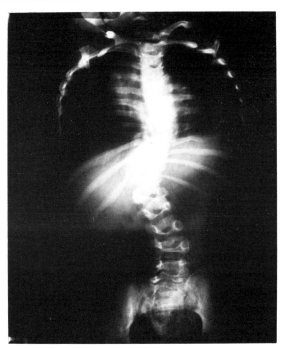

Fig. 4.18 L.T.—3 years—AP spine
There is congenital kypho-scoliosis with an anterior cleft vertebral body at L1 and rib deformities.
Same patient: Figs 4.11, 4.19, 4.22, 4.23, 4.26

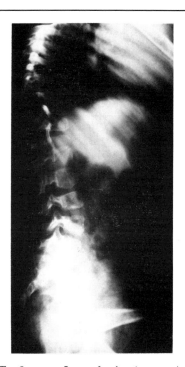

Fig. 4.19 L.T.—3 years—Lateral spine (same patient as in Fig. 4.18)
There is a mid dorsal kyphosis. The vertebral bodies of L1–L3 have irregular contours.

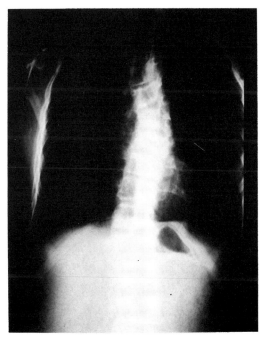

Fig. 4.20 S.B.—10 years 5 months—AP dorsal spine
The scoliosis has changed little since this child was a neonate (Fig. 4.14). Two hemivertebrae are present. The vertebral bodies otherwise appear normal.
Same patient: Fig. 4.14

Pelvis : Birth–3 years

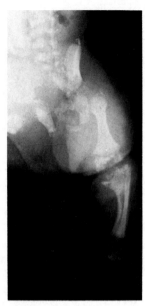

Fig. 4.21   R.W.—Neonate—Pelvis
There is extensive stippled calcification in the spine, pelvis and whole
lower limb. The femur is short with metaphyseal splaying. The tibia
is short with lateral bowing and the fibula is long in relation to it.
Same patient : Figs 4.6, 4.29

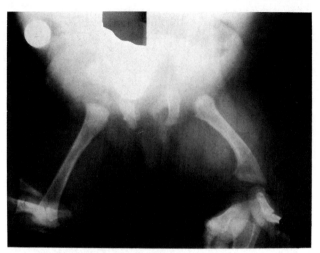

Fig. 4.22   L.T.—3 months—Pelvis and femora
There is marked shortening and bowing of the left femur with
abnormal modelling of the distal femoral metaphysis. There are a few
scattered areas of stippled ossification at the right hip and left knee
beneath the adult finger.
Same patient : Figs 4.11, 4.18, 4.19, 4.23, 4.26

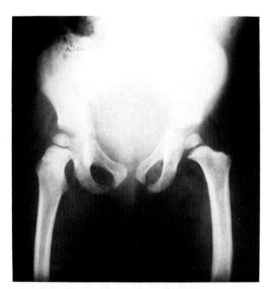

Fig. 4.23   L.T.—3 years—Pelvis (same patient as in Fig. 4.22)
All stippling has now disappeared. There is slight rotation of the
pelvis to the right. Mild lateral bowing of the left femoral shaft is
present. Bone modelling is otherwise normal. The whole of the
femora are not included but the left femur appears short.

Lower limb: Birth–3 years

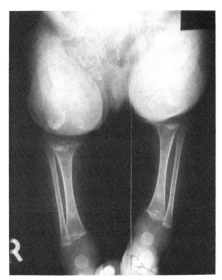

Fig. 4.24 Baby S.—Neonate—AP lower limbs
There is symmetrical deformity and marked shortening of the
femoral shafts. Punctate ossification is present around the knee joints.
Both tibiae are short and the distal metaphyses irregular. The fibulae
are relatively long.

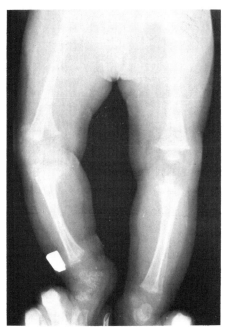

Fig. 4.25 J.H.—1 year—AP lower limbs
Left side appears normal. On the right there is punctate ossification
around the knee joint clearly outside the joint capsule. There is
irregularity of the distal femoral metaphysis and a radiolucent defect
extends towards the diaphysis. The tibia is short and bowed.
Same patient: Figs 4.15, 4.17

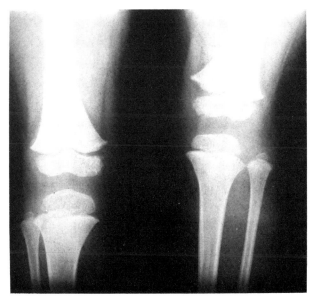

Fig. 4.26 L.T.—3 years 1 month—AP knees
There is now no evidence of bony abnormality and the epiphyses are
normal for her age.
Same patient: Figs 4.11, 4.18, 4.19, 4.22. 4.23

Upper limb: 1–6 years

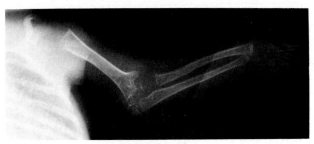

Fig. 4.27   N.I.—1 year—Left upper limb
There is humeral (rhizomelic) shortening with splaying of the distal
metaphysis. Punctate calcification is present in the regions of the
shoulder and elbow joints.
Same patient: Fig. 4.16

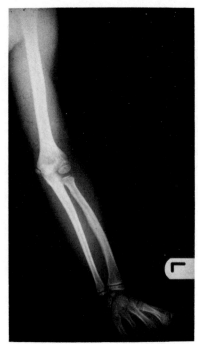

Fig. 4.28   J.M.—6 years—Left upper limb
Normal modelling of the long bones. There are areas of increased
bone density within the distal ulnar epiphysis and the head of the
radius is rather flat and sclerotic.
Same patient: Figs 4.12, 4.13

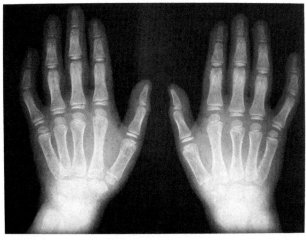

Fig. 4.29   R.W.—6½ years—Hands
Normal.
Same patient: Figs 4.6, 4.21

Section Three
SPONDYLO-EPIPHYSEAL DISORDERS

Chapter Five
SPONDYLO-EPIPHYSEAL DYSPLASIA CONGENITA
with severe coxa vara

In this more severe form of spondylo-epiphyseal dysplasia congenita the patients have very short stature, a disproportionately short trunk, gross disorganisation of the femoral heads and necks, with mild to severe epi- and metaphyseal changes in other joints, in general becoming normal towards the periphery of the limb.

Cleft palate, deafness and myopia are the commonest associated features.

Inheritance

In view of the severity, most cases are sporadic, but autosomal dominance has been reported.

Frequency

Possibly 1–2 per million prevalence in the population.

Clinical features

Facial appearance—not strikingly abnormal. Some reports state 'flat'.

Intelligence—normal.

Stature—markedly reduced, although this is not an obvious feature before the age of 2–3 years. Adult height 104–127 cm (see graphs).

Presenting feature/age—at birth—sometimes short trunk is noted, stiffness of hips or, less often, shoulder joints. Pectus carinatum. Cleft palate is not uncommon, sometimes clubfoot.

Deformities

Pectus carinatum.

Coxa vara of gross degree, sometimes with discontinuity of the femoral neck, developing during infancy and fully established by 4–7 years. Fixed flexion at the hips and lumbar lordosis.

Scoliosis or kyphoscoliosis, mild to severe, in most patients and developing before the age of 10 years.

Genu valgum, varum, recurvatum, talipes equinovarus or calcaneo-valgus are all reported, although less commonly.

Associated anomalies—many reported: cleft palate, severe myopia and later retinal detachment, deafness, cataract, herniae and clubfoot.

Radiographic features

A wide range of abnormalities, including complete normality at the periphery of the limbs, but always including epi- and metaphyseal disorder at the hip joint, and platyspondyly.

Skull—normal (excepting cleft palate, if present).

Spine—odontoid dysplasia and delay in ossification. Rarely, cervical vertebral body hypoplasia. Vertebral bodies reduced in height, with an increased anteroposterior diameter. In infancy they are oval ('pear-shaped') with anterior beaking; later, irregular platyspondyly develops.

Limb epiphyses—vary from normality to severe fragmentation, with delay in appearance—most marked at the capital femoral epiphyses.

Metaphyses—in spite of its name, metaphyses are also disordered and irregular. This feature disappears with epiphyseal fusion.

Shoulder joints—vary from normality to varus deformity with irregular epiphyses and metaphyses.

Pelvis—ilium small. Irregular acetabula with horizontal roofs. Delay in ossification of public rami.

Hip joints—capital femoral epiphyses do not appear until 4–5 years, if at all, and remain small and deformed. Coxa vara develops early, and with discontinuity of the femoral neck the greater trochanters displace upwards. At first it appears these children have bilateral dislocation of the hips, but *arthrography* indicates that the largely unossified femoral head is in fact contained within the acetabulum.

Hands/wrists—the hands are likely to be normal, but the carpal bones are small and irregular, with delay in ossification on the radial side.

Bone maturation

Generalised delay, but particularly at the pelvis and hip joint.

Biochemistry

Not known.

Differential diagnosis

1. *SED congenita with mild coxa vara* (see Chapter 6). It differs from the severe form described here in its lesser degree both of short stature and coxa vara.

2. *Morquio's disease*—superficial similarity only, all radiological signs differ.

3. *Pseudoachondroplasia*—easily differentiated by reason of its short-limbed stature, and marked epi- and metaphyseal involvement throughout the skeleton.

4. *Rarer disorders such as metatropic dwarfism and Kniest disease* have greatly expanded proximal and distal ends of the long bones.

Progress/complications

Premature osteoarthritis in affected joints is inevitable, particularly hips and shoulders. Scoliosis is probable, and paraplegia has occurred. Retinal detachment, associated with myopia. Atlanto-axial instability associated with odontoid dysplasia is a possibility, but these patients do not usually have joint laxity, and in general this is not a complication.

REFERENCES

Kozlowski K, Masel J, Nolte K 1977 Dysplasia spondylo-epiphysealis congenita Spranger–Weidemann: a critical analysis. Australasian Radiology 21: 260–280

Spranger J W, Langer L O 1970 Spondylo-epiphyseal dysplasia congenita. Radiology 94: 313–322

Wynne-Davies R, Hall C 1982 Two clinical variants of spondylo-epiphyseal dysplasia congenita. Journal of Bone and Joint Surgery 64B: 435–441

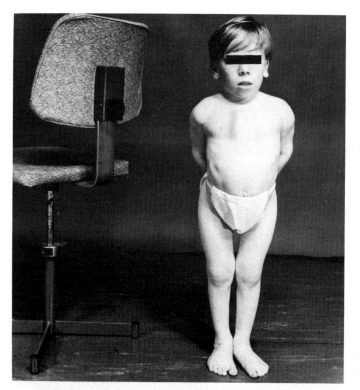
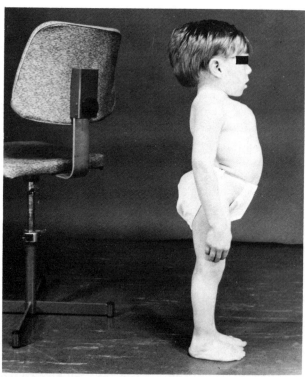

Figs 5.1 and 5.2   12-year-old boy with a normal face and head; protrusion of the manubrio-sternal angle, severe lumbar lordosis and obvious trunk shortening by this age—note that his finger tips reach nearly to the knee joint.

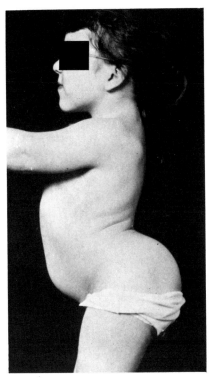

Fig. 5.3 Marked fixed flexion at the hip causes this degree of lumbar lordosis on standing.

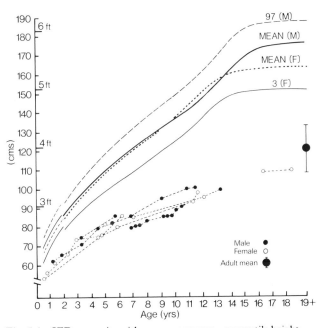

Fig. 5.4 SED congenita with severe coxa vara—percentile height.

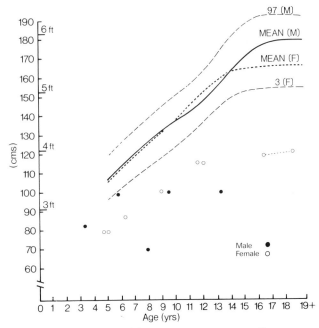

Fig. 5.5 SED congenita with severe coxa vara—percentile span.

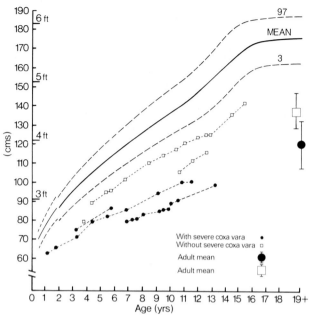

Fig. 5.6 Spondylo-epiphyseal dysplasia congenita—percentile height (males).

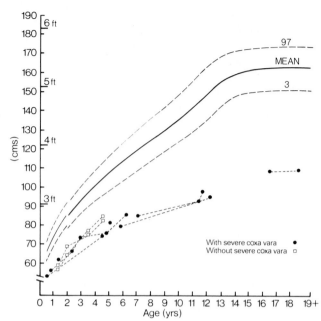

Fig. 5.7 Spondylo-epiphyseal dysplasia congenita—percentile height (females).

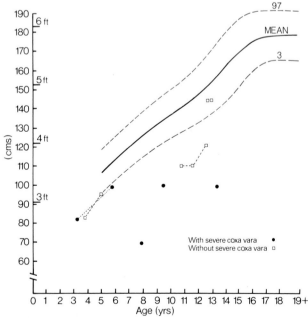

Fig. 5.8 Spondylo-epiphyseal dysplasia congenita—percentile span (males).

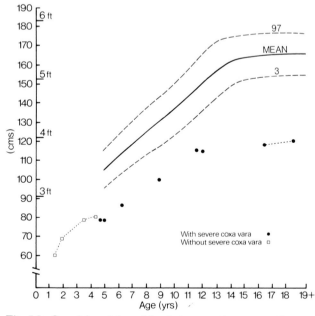

Fig. 5.9 Spondylo-epiphyseal dysplasia congenita—percentile span (females).

Whole body : Neonate

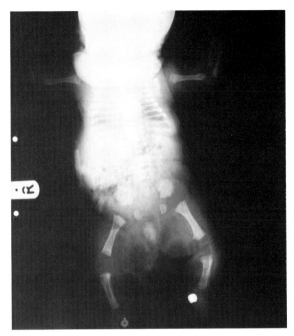

Fig. 5.10   Neonate—Whole body
All long bones and ribs are short. The iliac wings are small and square, with horizontal acetabular roofs. The ischia are small. Bone age is delayed, with no ossification seen in the pubic rami or knee epiphyses.

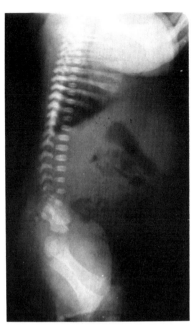

Fig. 5.11   Neonate—Lateral spine
The vertebral bodies have an oval configuration. They vary in size, being largest in the upper lumbar region and smallest at L5.

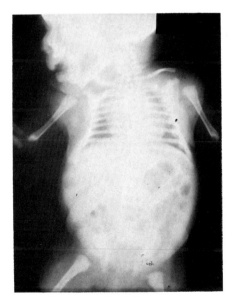

Fig. 5.12   A.D.—Neonate—Whole body
The symphysis pubis is wide. Signs are otherwise similar to Figure 5.10.
Same patient: Figs 5.14, 5.20, 5.33, 5.34, 5.37, 5.38, 5.45, 5.51, 5.89

Skull: 3–11 years

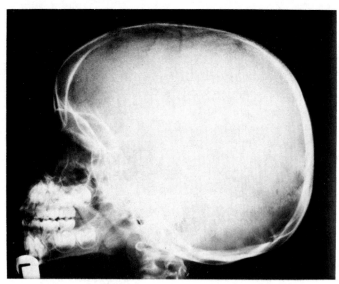

Fig. 5.13   J.D.—3 years 4 months—Lateral skull
No abnormality.
Same patient: Figs 5.25, 5.26, 5.64, 5.75, 5.82

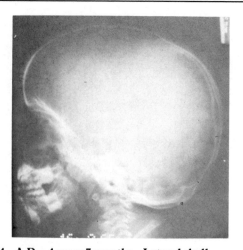

Fig. 5.14   A.D.—4 years 7 months—Lateral skull
No abnormality.
Same patient: Figs 5.12, 5.20, 5.33, 5.34, 5.37, 5.38, 5.45, 5.51, 5.89.

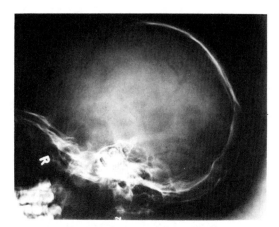

Fig. 5.15   D.H.—9 years 3 months—Lateral skull
No abnormality.
Same patient: Figs 5.17, 5.18, 5.21, 5.43, 5.44, 5.46, 5.49, 5.53, 5.57, 5.58, 5.68, 5.72, 5.79

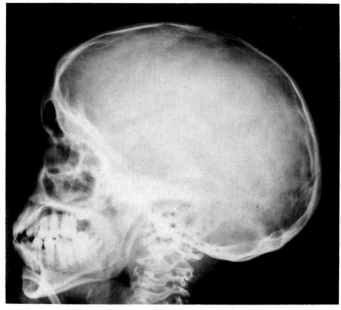

Fig. 5.16   R.V.—11 years 10 months—Lateral skull
The odontoid peg is absent.
Same patient: Figs 5.22, 5.23, 5.31, 5.32, 5.59, 5.60, 5.69, 5.73, 5.74, 5.76, 5.78, 5.82, 5.87, 5.88

Cervical spine—3–12 years

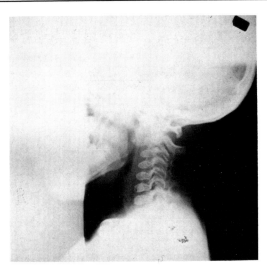

Fig. 5.17   D.H.—3 years—Lateral cervical spine
Oval vertebral bodies with an increase in their AP diameters. (C1 and C2 are not clearly shown.)
Same patient : Figs 5.15, 5.18, 5.21, 5.43, 5.44, 5.46, 5.49, 5.53, 5.57, 5.58, 5.68, 5.72, 5.79.

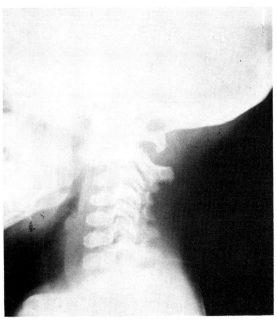

Fig. 5.18   D.H.—6 years—Lateral cervical spine (same patient as in Fig. 5.17)
There is platyspondyly and the vertebral bodies are increased in their AP diameters. The vertebral end-plates appear smooth. (C1 and C2 are not clearly shown.)

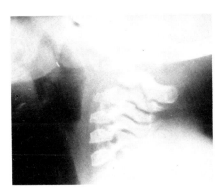

Fig. 5.19   J.S.—6 years 10 months—Lateral cervical spine
There is marked posterior subluxation of C2 on C3 with failure of formation of the vertebral body. The odontoid process is not seen. There is an increase in the vertebral bodies in their AP diameter.
Same patient : Figs 5.27, 5.28, 5.67, 5.71, 5.84.

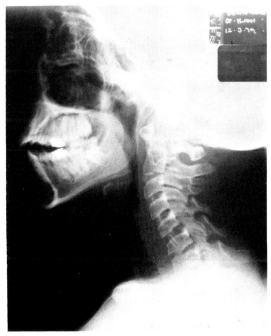

Fig. 5.20   A.D.—12 years 8 months—Lateral cervical spine
There is platyspondyly and the odontoid peg is hypoplastic.
Same patient : Figs 5.12, 5.14, 5.33, 5.34, 5.37, 5.38, 5.45, 5.51, 5.89.

Spine : Birth–2 years

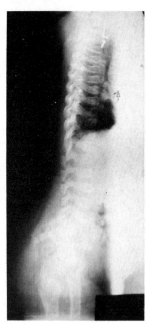

Fig. 5.21   D.H.—11 months—Lateral spine
The vertebral bodies are smooth and oval, with mild posterior wedging in the dorsal region, giving rise to the 'pear-shaped' appearance. The sacrum is approaching the horizontal.
Same patient: Figs 5.15, 5.17, 5.18, 5.43, 5.44, 5.46, 5.49, 5.53, 5.57, 5.58, 5.68, 5.72, 5.79

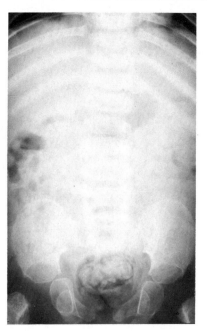

Fig. 5.22   R.V.—1 year 7 months—AP spine
Smooth vertebral end-plates. Wide interpedicular distances in the lumbar region.
Same patient: Figs 5.16, 5.23, 5.31, 5.32, 5.59, 5.60, 5.69, 5.73, 5.74, 5.76, 5.78, 5.82, 5.87, 5.88

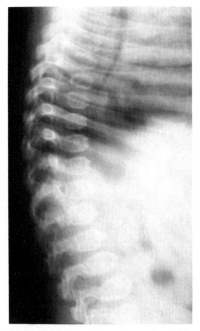

Fig. 5.23   R.V.—1 year 7 months—Lateral spine (same patient as in Fig. 5.22)
Oval vertebral bodies with anterior tongues in the lumbar region. There is anterior and posterior wedging in the dorsal region.

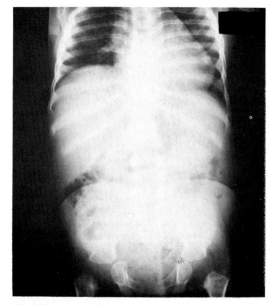

Fig. 5.24   C.J.—1 year 10 months—AP spine
The beginning of dorsal scoliosis.
Same patient: Figs 5.50, 5.66, 5.80, 5.83

Spine: 3–6 years

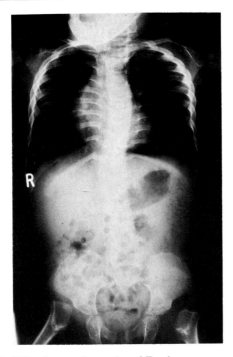

Fig. 5.25   J.D.—3 years 4 months—AP spine
There is a triple curve.
Same patient: Figs 5.13, 5.26, 5.64, 5.75, 5.81

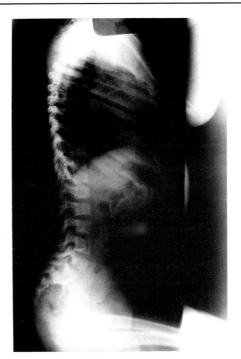

Fig. 5.26   J.D.—3 years 4 months—Lateral spine (same patient as in Fig. 5.25)
There is abnormal modelling of the vertebral bodies in the lumbar region; several show constrictions of the mid-end-plates suggesting previous coronal clefts. The sacrum is abnormally tilted backwards.

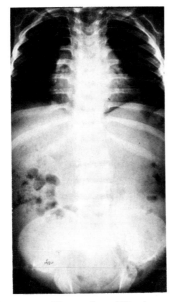

Fig. 5.27   J.S.—6 years 10 months—AP spine
There is slight scoliosis, and in the upper lumbar region there is widening of the interpedicular distances.
Same patient: Figs 5.19, 5.28, 5.67, 5.71, 5.84

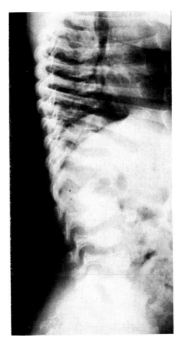

Fig. 5.28   J.S.—6 years 10 months—Lateral dorsal and lumbar spine (same patient as in Fig. 5.27)
The vertebral bodies have an oval configuration and there is mild platyspondyly, with hypoplasia and a kyphos at L2.

Spine : 7–11 years

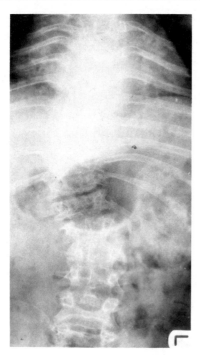

Fig. 5.29   J.D.—7 years 8 months—AP spine
There is a scoliosis convex to the right in the lower dorsal region,
platyspondyly and some irregularity of the vertebral end-plates.
Same patient : Figs 5.35, 5.36, 5.40–5.42, 5.61, 5.70, 5.90

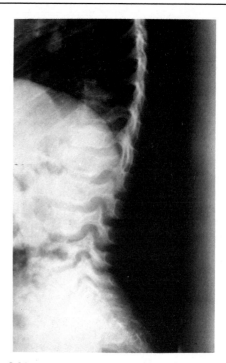

**Fig. 5.30   S.M.—9 years 6 months—Lateral dorsal and
lumbar spine**
There is vertebral end-plate irregularity, with some platyspondyly
and disc space narrowing.
Same patient : Figs 5.55, 5.86

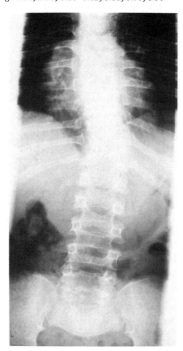

**Fig. 5.31   R.V.—11 years 10 months—AP dorsal and lumbar
spine**
There is a scoliosis convex to the left in the lower dorsal region, and
universal platyspondyly, but the vertebral end-plates appear smooth
and the disc spaces are normal.
*Same patient : Figs 5.16, 5.22, 5.23, 5.32, 5.59, 5.60, 5.69, 5.73, 5.74, 5.76,
5.78, 5.82, 5.87, 5.88*

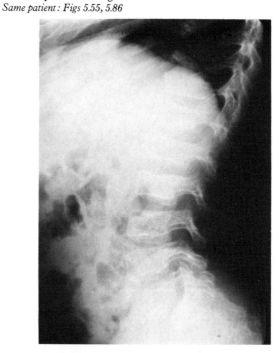

Fig. 5.32   R.V.—11 years 10 months—Lateral lumbar spine
(same patient as in Fig. 5.31)
There is mild platyspondyly with an increase in the AP diameters of
the vertebral bodies.

Spine: 12–14 years

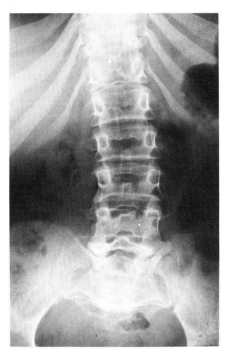

Fig. 5.33   A.D.—12 years 8 months—AP lumbar spine
There is some platyspondyly with disc space narrowing, and scoliosis.
Same patient: Figs 5.12, 5.14, 5.20, 5.34, 5.37, 5.38, 5.45, 5.51, 5.89

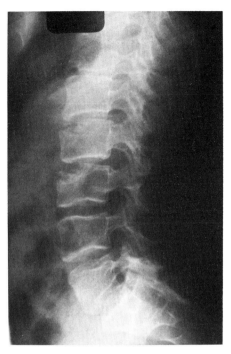

Fig. 5.34   A.D.—12 years 8 months—Lateral lumbar spine
(same patient as in Fig. 5.33)
There is vertebral end-plate irregularity of L2 and L3 with a defect of
the antero-superior margins. These are associated with disc space
narrowing and mild anterior elongation of the vertebral bodies. The
appearances suggest anterior disc herniation.

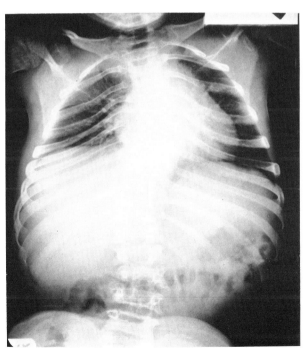

Fig. 5.35   J.D.—14 years 3 months—AP spine
There is a triple curve and universal platyspondyly, giving a short
trunk.
Same patient: Figs 5.29, 5.36, 5.40–5.42, 5.61, 5.70, 5.90

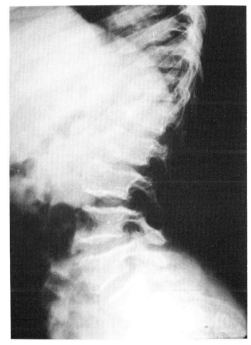

**Fig. 5.36   J.D.—14 years 2 months—Lateral dorsolumbar
spine** (same patient as in Fig. 5.35)
There is a pronounced lumbar lordosis, and platyspondyly with
vertebral end-plate irregularity. Anterior wedging is present in the
lower dorsal region where there is a mild kyphosis.

Spine: 16 years–Adult

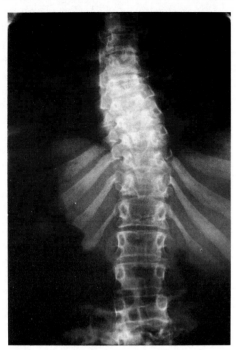

Fig. 5.37   A.D.—16 years 6 months—AP spine
There is scoliosis convex to the right in the lower dorsal region.
Same patient: Figs 5.12, 5.14, 5.20, 5.33, 5.34, 5.38, 5.45, 5.51, 5.89

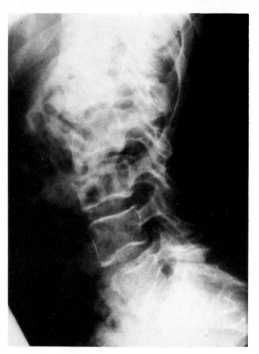

Fig. 5.38   A.D.—16 years 6 months—Lateral lumbar spine
(same patient as in Fig. 5.37)
There is mild platyspondyly with antero-superior defects of the upper lumbar vertebral bodies.

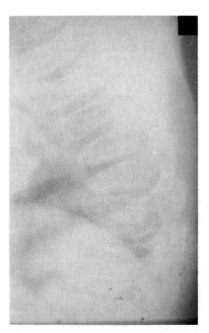

Fig. 5.39   M.G.—Adult—Lateral dorsal spine
There is marked platyspondyly with variable anterior and posterior wedging and vertebral end-plate irregularity.
Same patient: Figs 5.47, 5.62

Spine : Adult

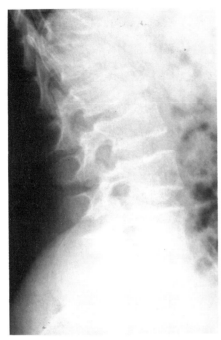

Fig. 5.40   J.D.—21 years 4 months—Lateral lumbar spine
The vertebral bodies are flattened with anterior wedging in the upper lumbar region. The pedicles appear rather short with a reduction in the size of the spinal canal. He presented at this time with paraplegia.
Same patient; Figs 5.29, 5.35, 5.36, 5.41, 5.42, 5.61, 5.70, 5.90

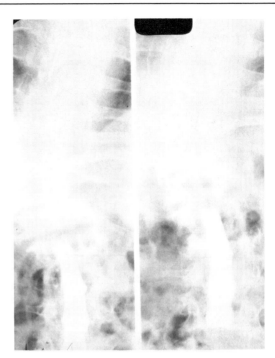

Fig. 5.41   J.D.—21 years 6 months—AP spine—Myelogram
(same patient as in Fig. 5.40)
Scoliosis convex to the right in the lower dorsal region. There is a hold-up of the contrast medium at D12.

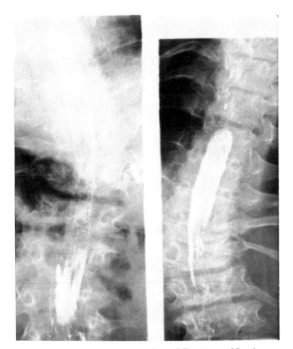

Fig. 5.42   J.D.—21 years 6 months—AP spine—Myelogram
(same patient as in Fig. 5.40)
The upper limit of the extrinsic compression is at D10 which is the apex of the curve. Neurological recovery followed decompression.

Pelvis : Birth–3 years

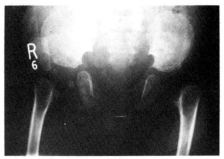

Fig. 5.43   D.H.—11 months—Pelvis
The iliac wings are small and square and acetabular roofs horizontal.
The symphysis pubis is wide. There is bilateral coxa vara and the
femoral necks are short and irregular with an inferior bony fragment.
The femoral heads are just ossified, and are in the normal position.
*Same patient : Figs 5.15, 5.17, 5.18, 5.21, 5.44, 5.46, 5.49, 5.53, 5.57, 5.58,
5.68, 5.72, 5.79*

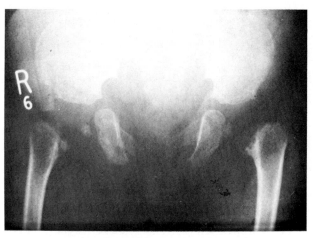

Fig. 5.44   D.H.—2 years 3 months—Pelvis (same patient as in
Fig. 5.43)
Small iliac wings. Wide symphysis pubis and horizontal acetabular
roofs. Small ossification centres of normally located capital femoral
epiphyses are present and there is marked coxa vara with virtually no
ossification of the femoral necks.

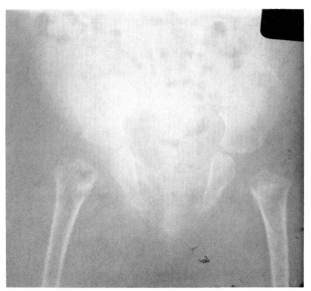

Fig. 5.45   A.D.—2 years 3 months—Pelvis
Coxa vara is present. There is no ossification of the capital femoral
epiphyses yet.
Same patient : Figs 5.12, 5.14, 5.20, 5.33, 5.34, 5.37, 5.38, 5.51, 5.89

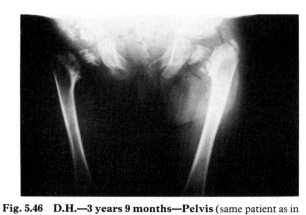

Fig. 5.46   D.H.—3 years 9 months—Pelvis (same patient as in
Fig. 5.43)
There is marked coxa vara and metaphyseal irregularity. The capital
femoral epiphyses are ossified and well located, but the greater
trochanters are displaced upwards.

Pelvis: 4–7 years

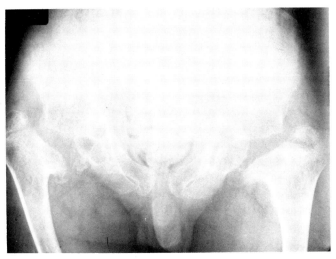

Fig. 5.47 M.G.—4 years—Pelvis
Severe coxa vara with upwardly displaced greater trochanters and
very short femoral necks (unossified). The capital femoral epiphyses
show patchy ossification and there is early posterior dislocation of the
femoral heads.
Same patient: Figs 5.39, 5.62

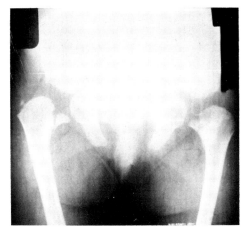

Fig. 5.48 M.F.—4 years 10 months—Pelvis
There is coxa vara with triangular bony fragments inferiorly. The
capital femoral epiphyses have not yet ossified.
Same patient: Fig. 5.65

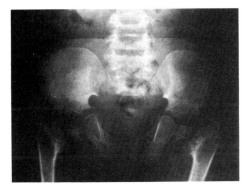

Fig. 5.49 D.H.—5 years 7 months—Pelvis
There is patchy ossification of the capital femoral epiphyses and
severe coxa vara. The epiphyseal plates are wide and femoral necks
short with irregular metaphyses. The symphysis pubis is wide with
short pubic and ischial rami.
*Same patient: Figs 5.15, 5.17, 5.18, 5.21, 5.43, 5.44, 5.46, 5.53, 5.57, 5.58,
5.68, 5.72, 5.79*

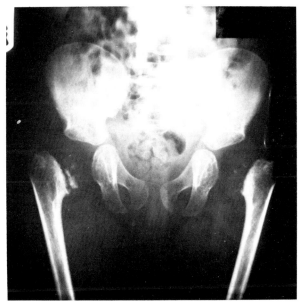

Fig. 5.50 C.J.—7 years—Pelvis
There is only a little patchy ossification of the capital femoral
epiphyses. The femoral necks are virtually absent and the greater
trochanters displaced upwards.
Same patient: Figs 5.24, 5.66, 5.80, 5.83

Pelvis: 8 years

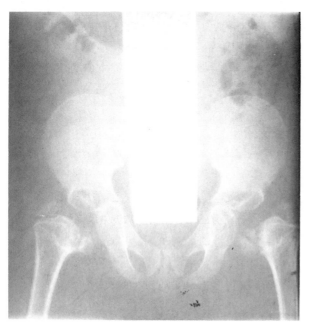

Fig. 5.51   A.D.—8 years—Pelvis
There is only a little irregular ossification of the capital femoral epiphyses. The greater trochanters are high and femoral necks short and irregular.
Same patient: Figs 5.12, 5.14, 5.20, 5.33, 5.34, 5.37, 5.38, 5.45, 5.89

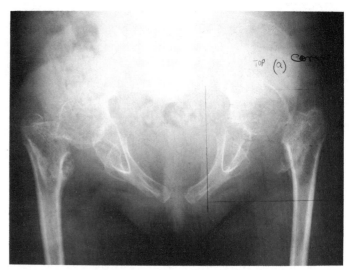

Fig. 5.52   T.M.—8 years 10 months—Pelvis
There is no ossification at all of the capital femoral epiphyses. The greater trochanters are high. There is also a pronounced pelvic tilt giving rise to a 'brim' view.
Same patient: Figs 5.54, 5.56, 5.77, 5.85

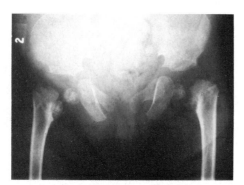

Fig. 5.53   D.H.—8 years 6 months—Pelvis
The capital femoral epiphyses show patchy ossification and femoral necks are short and irregular with coxa vara. The greater trochanters are high.
Same patient: Figs 5.15, 5.17, 5.18, 5.21, 5.43, 5.44, 5.46, 5.49, 5.57, 5.58, 5.68, 5.72, 5.79

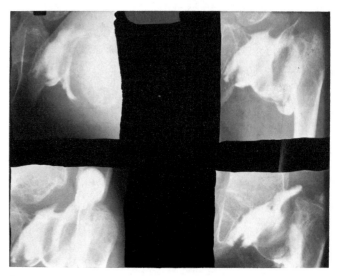

Fig. 5.54   T.M.—8 years 11 months—Arthrogram left hip
(same patient as in Fig. 5.52)
This illustrates the large unossified capital femoral epiphysis, well placed within the acetabulum, and its relationship to the upwardly displaced greater trochanter, showing severe coxa vara.

Pelvis: 9–11 years

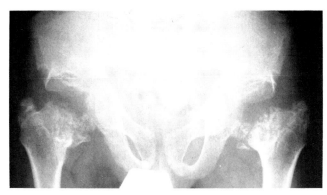

Fig. 5.55   S.M.—9 years 6 months—Pelvis
There is patchy ossification of the small capital femoral epiphyses, and coxa vara with short and irregular femoral necks. The greater trochanters are large and high. There are supero-lateral acetabular notches, where the greater trochanters impinge on abduction of the hip.
Same patient: Figs 5.30, 5.86

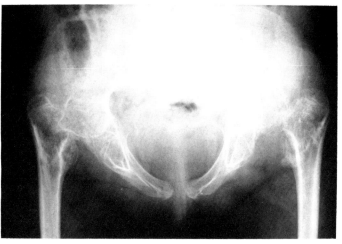

Fig. 5.56   T.M.—11 years—Pelvis (compare Fig. 5.52)
There has been a progressive tilting of the pelvis. There is still virtually no ossification of the capital femoral epiphyses.
Same patient: Figs 5.52, 5.54, 5.77, 5.85

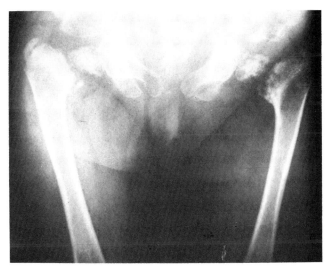

Fig. 5.57   D.H.—11 years 9 months—Pelvis
There is coxa vara with patchy ossification of the capital femoral epiphyses and high greater trochanters. The femoral necks are short and fragmented. The symphysis pubis is wide and the ischia are short.
Same patient: Figs 5.15, 5.17, 5.18, 5.21, 5.43, 5.44, 5.46, 5.49, 5.53, 5.58, 5.68, 5.72, 5.79

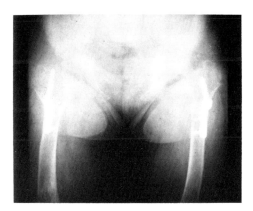

Fig. 5.58   D.H.—11 years 9 months—Pelvis (same patient as in Fig. 5.57)
This shows improved alignment following upper femoral rotation osteotomies with plate and screw fixation.

Pelvis : 11 years–Young adult

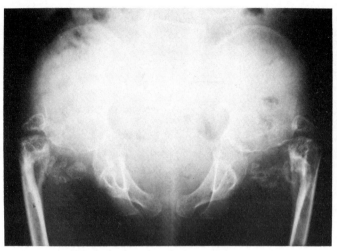

Fig. 5.59   R.V.—11 years 10 months—Pelvis
There is severe coxa vara and only a little patchy ossification of the
capital femoral epiphyses and femoral necks. The pelvis is tilted and
the symphysis pubis wide, with shortening of both pubic and ischial
rami.
*Same patient : Figs 5.16, 5.22, 5.23, 5.31, 5.32, 5.60, 5.69, 5.73, 5.74, 5.76,
5.78, 5.82, 5.87, 5.88*

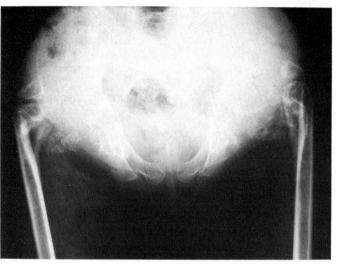

Fig. 5.60   R.V.—13 years 3 months—Pelvis (same patient as in
Fig. 5.59)
The capital femoral epiphyses are now better ossified, although still
small. There appears to be discontinuity of the femoral necks.

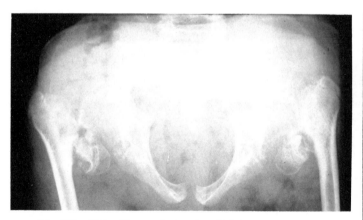

Fig. 5.61   J.D.—14 years 2 months—Pelvis
The femoral heads are small, but ossified. There is discontinuity of
the femoral necks and marked upward displacement of the femoral
shaft.
Same patient : Figs 5.29, 5.35, 5.36, 5.40–5.42, 5.70, 5.90

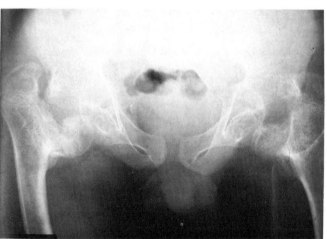

Fig. 5.62   M.G.—Young adult—Pelvis
The capital femoral epiphyses are small, but ossified. There is
discontinuity of the femoral necks with severe coxa vara.
Same patient : Figs 5.39, 5.47

Lower limb: 1–5 years

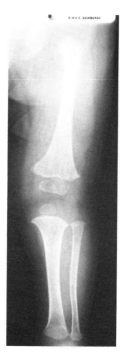

Fig. 5.63   J.V.—1 year 6 months—Left lower limb
There is mild metaphyseal flaring and irregularity. The capital
femoral epiphysis is not ossified and coxa vara is present. The knee
epiphysis is only mildly retarded.

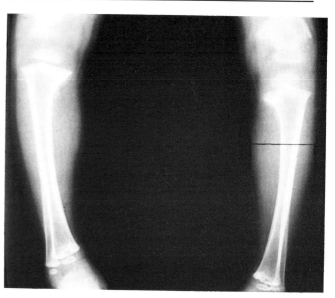

Fig. 5.64   J.D.—3 years 4 months—AP both legs
There is metaphyseal irregularity and splaying and tibia vara. All
epiphyses are small. The fibulae appear long both proximally and
distally.
Same patient: Figs 5.13, 5.25, 5.26, 5.64, 5.75, 5.81

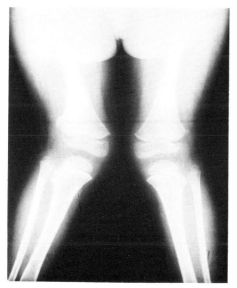

Fig. 5.65   M.F.—4 years 10 months—AP lower limbs
There is a severe genu valgum. The epiphyses are rather small and
medially placed, and the lateral aspect of both femora and tibiae is
deficient. The fibulae are abnormally long at their proximal end.
Same patient: Fig. 5.48

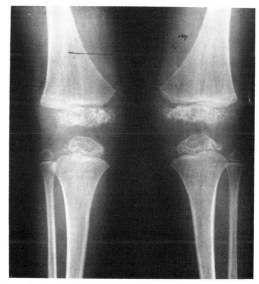

Fig. 5.66   C.J.—5 years—AP knees
There is irregular and patchy ossification of the epiphyses. Modelling
of the metaphyses is relatively normal in this patient.
Same patient: Figs 5.24, 5.50, 5.80, 5.83

Lower limb: 6 years–Adult

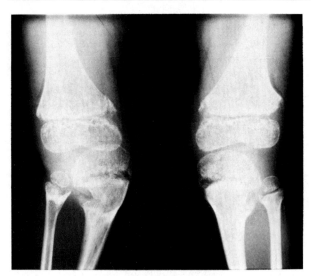

Fig. 5.67   J.S.—6 years 10 months—AP knees
The metaphyses are flared and there are irregular notches and
fragments at their periphery. The epiphyses are large and rounded
with irregular patchy ossification. Genu valgum is present.
Same patient: Figs 5.19, 5.27, 5.28, 5.67, 5.71, 5.84

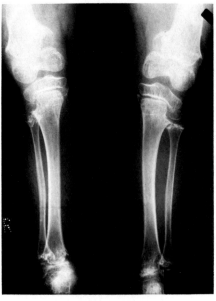

Fig. 5.68   D.H.—8 years 3 months—AP legs
There is metaphyseal splaying with marked irregularity and
peripheral defects associated with mild genu valgum. The epiphyses
are large and the proximal fibulae short.
*Same patient: Figs 5.15, 5.17, 5.18, 5.21, 5.43, 5.44, 5.46, 5.49, 5.53, 5.57,
5.58, 5.72, 5.79*

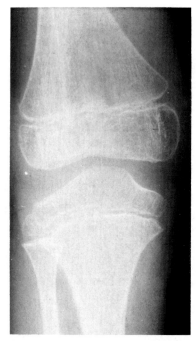

Fig. 5.69   R.V.—13 years 3 months—AP knee
Only mild abnormal modelling defects are present.
*Same patient: Figs 5.16, 5.22, 5.23, 5.31, 5.32, 5.59, 5.60, 5.73, 5.74, 5.76,
5.78, 5.82, 5.87, 5.88*

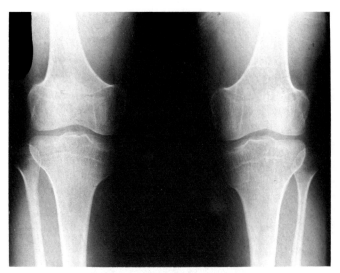

Fig. 5.70   J.D.—24 years—AP knees
Metaphyses are splayed, but there is little other abnormality.
Same patient: Figs 5.29, 5.35, 5.36, 5.40, 5.41, 5.42, 5.61, 5.90

Foot: 6–13 years

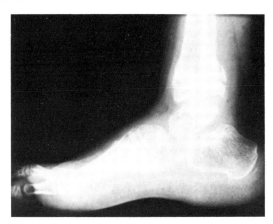

Fig. 5.71   J.S.—6 years 10 months—Lateral foot
The distal tibia and fibula show splaying of the metaphyses with fragmentation peripherally. There is a tendency towards a vertical talus.
Same patient: Figs 5.19, 5.27, 5.28, 5.67, 5.84

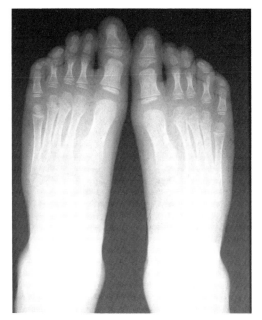

Fig. 5.72   D.H.—8 years 3 months—PA feet
Normal.
Same patient: Figs 5.15, 5.17, 5.18, 5.21, 5.43, 5.44, 5.46, 5.49, 5.53, 5.57, 5.58, 5.68, 5.79

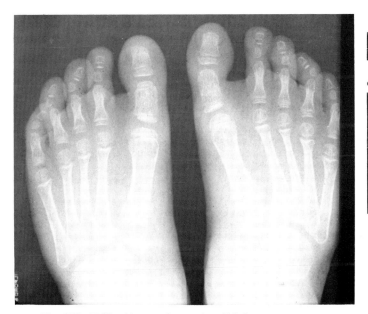

Fig. 5.73   R.V.—13 years 3 months—PA feet
The tarsal bones are irregular and there are pseudo-epiphyses of the distal first metatarsal and first proximal phalanx. Several 'ivory' epiphyses are present; (this can be a normal finding, but is unusual at this age).
Same patient: Figs 5.16, 5.22, 5.23, 5.31, 5.32, 5.59, 5.60, 5.69, 5.74, 5.76, 5.78, 5.82, 5.87, 5.88

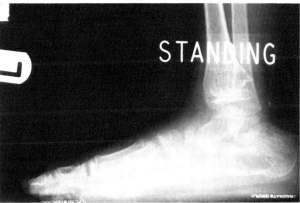

Fig. 5.74   R.V.—13 years 3 months—Lateral left foot (standing) (same patient as in Fig. 5.73)
There is metaphyseal splaying and flattening of the epiphyses, with a somewhat displaced talus.

Upper limb: 3–13 years

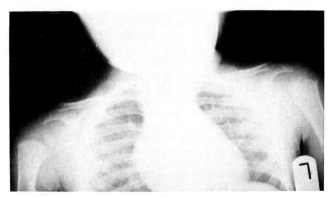

Fig. 5.75   J.D.—3 years 4 months—AP shoulders
There is abnormal modelling of the proximal humeral metaphyses
with some irregularity, and small fragmented epiphyses.
Same patient: Figs 5.13, 5.25, 5.26, 5.64, 5.81

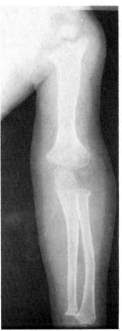

Fig. 5.76   R.V.—4 years 6 months—AP left upper limb
There is a short thick humerus with splayed metaphyses and patchy
ossification of the proximal humeral epiphyses.
*Same patient: Figs 5.16, 5.22, 5.23, 5.31, 5.32, 5.59, 5.60, 5.69, 5.73, 5.74,
5.78, 5.82, 5.87, 5.88*

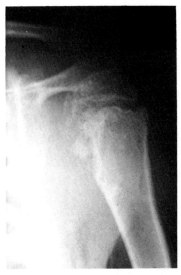

Fig. 5.77   T.M.—8 years 10 months—AP left shoulder
There is abnormal modelling of the proximal humeral metaphysis
with fragmentation and the diaphysis is also affected. The epiphysis
shows irregular ossification with fragmentation and there is some
discontinuity of the medial diaphyseal cortex suggesting a pseudo-
fracture.
Same patient: Figs 5.52, 5.54, 5.56, 5.85

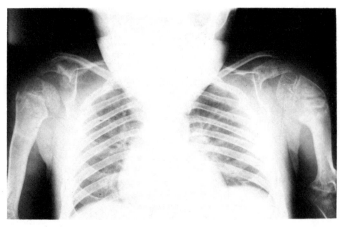

Fig. 5.78   R.V.—13 years 3 months—AP shoulders (same
patient as in Fig 5.76)
There is shortening with modelling deformities of both proximal
humeri, more marked on the left, with splaying of the metaphyses
and mild irregularity here. On the left the diaphysis appears to have
developed from several ossification centres and the appearance
suggests a longstanding, un-united fracture or pseudo-fracture.

There is a varus deformity of the upper humerus and on the right a
sclerotic diaphyseal band suggestive of a healed fracture or possibly a
separate ossification centre.

Hand: Birth–4 years

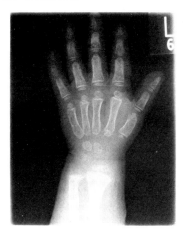

Fig. 5.79   D.H.—11 months—PA left hand
The distal radial and ulnar metaphyses are irregular and there is
flattening of the distal radial epiphysis. There is a proximal pseudo-
epiphysis of the second metacarpal and a distal pseudo-epiphysis of
the first. The bone age corresponds to 2 years.
*Same patient: Figs 5.15, 5.17, 5.18, 5.21, 5.43, 5.44, 5.46, 5.49, 5.53, 5.57,
5.58, 5.68, 5.72*

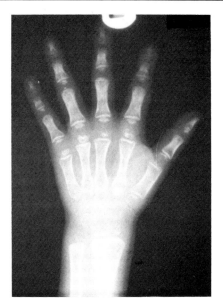

Fig. 5.80   C.J.—1 year 10 months—PA left hand
All the epiphyses are irregular and fragmented with patchy sclerosis.
Pseudo-epiphyses are present. Bone maturation (digits) is normal.
Carpal centres are delayed, with an approximate age of only 3 months.
Same patient: Figs 5.24, 5.50, 5.66, 5.83

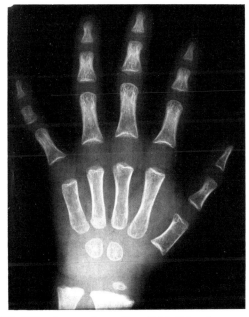

Fig. 5.81   J.D.—3 years 4 months—PA left hand
Phalangeal bone age is retarded at 1 year 3 months. The metaphyses
are irregular, with cup-shaped deformities.
Same patient: Figs 5.13, 5.25, 5.26, 5.64, 5.75

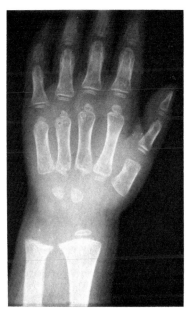

Fig. 5.82   R.V.—4 years 6 months—PA left hand
The first metacarpal is short and there are pseudo-epiphyses. The
distal metacarpal epiphyses are irregular in outline. Bone age (digits)
corresponds to 2 years 8 months. The carpal centres are more
retarded.
*Same patient: Figs 5.16, 5.22, 5.23, 5.31, 5.32, 5.59, 5.60, 5.69, 5.73, 5.74,
5.76, 5.78, 5.87, 5.88*

Hand : 6–9 years

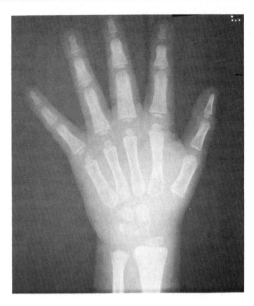

Fig. 5.83   C.J.—6 years 7 months–PA left hand
Bone modelling is unremarkable. The phalangeal bone age is only
about 2 years.
Same patient : Figs 5.24, 5.50, 5.66, 5.80

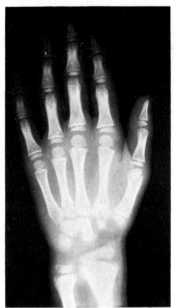

Fig. 5.84   J.S.—6 years 10 months—PA left hand
There is splaying and irregularity of the distal radial and ulnar
metaphyses. The adjacent epiphyses are remarkably well developed
(about an 8-year standard). Bone age at the phalanges is 4 years 2
months. The carpus is also retarded, particularly on the radial side
(the lunate, scaphoid and trapezium centres have not yet appeared).
The remaining carpal bones are rather angular.
Same patient : Figs 5.19, 5.27, 5.28, 5.67, 5.71

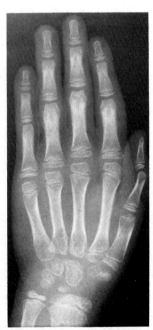

Fig. 5.85   T.M.—8 years 10 months—PA left hand
Phalangeal bone age corresponds to the chronological age. However,
the carpal centres are small, angular and retarded, and the scaphoid
has not ossified. The distal metacarpal epiphyses are angular.
Same patient : Figs 5.52, 5.54, 5.56, 5.77

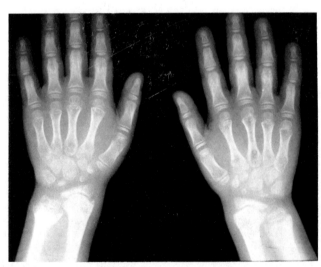

Fig. 5.86   S.M.—9 years 6 months—PA both hands
The distal radial and ulnar metaphyses are splayed and irregular with
a central hump and peripheral fragmentation. Phalangeal bone age
corresponds to 6 years. The carpal centres are small and irregular,
especially on the radial side. There is some sclerosis of the trapezium.
Metacarpals and phalanges are remarkably well modelled.
Same patient : Figs 5.30, 5.55

Hand: 11 years–Adult

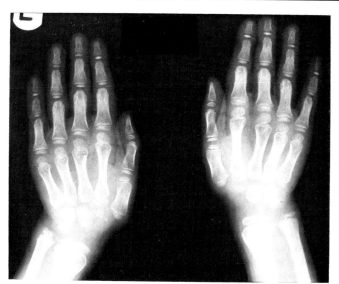

Fig. 5.87 R.V.—11 years 10 months—PA both hands
The first metacarpal is short and proximally placed. Phalangeal bone age is about 6 years and there is even greater retardation of the carpal centres, with no ossification of the trapezium and scaphoid.
Same patient: Figs 5.16, 5.22, 5.23, 5.31, 5.32, 5.59, 5.60, 5.69, 5.73, 5.74, 5.76, 5.78, 5.82, 5.88

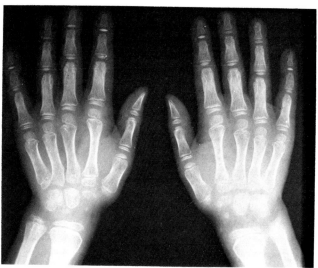

Fig. 5.88 R.V.—13 years 3 months—PA both hands (same patient as in Fig. 5.87)
The first metacarpals are proximally placed. Bone age corresponds to 8 years. Multiple sclerotic phalangeal epiphyses are present; these represent a normal stage of bone development, but are not usually seen at this age. Carpal centres on the radial side are very small and pseudo-epiphyses are present.

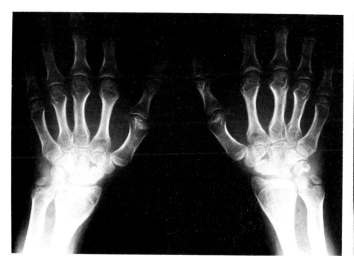

Fig. 5.89 A.D.—16 years 6 months—PA both hands
The first metacarpal is proximally placed; as a result the articular surface of the trapezium slopes almost vertically. Accessory carpal ossicles are present overlying both lunates. Bone age (digital) corresponds to the chronological age and epiphyses have all fused. The long bones in the hand are short with flared metaphyseal areas. The carpus is small and the centres angular and deformed. There is separation of the ulnar styloid processess bilaterally—? separate ossicles.
Same patient: Figs 5.12, 5.14, 5.20, 5.33, 5.34, 5.37, 5.38, 5.45, 5.51

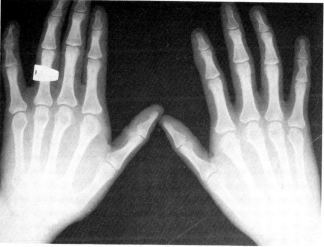

Fig. 5.90 J.D.—24 years—PA both hands
The epiphyses have all fused and both metacarpal and phalangeal bone modelling is remarkably normal.
Same patient: Figs 5.29, 5.35, 5.36, 5.40–5.42, 5.61, 5.70

Chapter Six
SPONDYLO-EPIPHYSEAL DYSPLASIA CONGENITA
with mild coxa vara

There is much overlap between this condition and SED congenita with severe coxa vara, but it is never so severe as the type with gross hip disorganisation and markedly short stature.

It is not possible to differentiate the two in early infancy, but thereafter the disparity in growth rate and lack of severe coxa vara differentiate them. This and the previous chapter should be considered together.

Inheritance

Possibly autosomal dominant.

Frequency

Possibly 1–2 per million prevalence.

Clinical features

Facial appearance—normal.

Intelligence—normal.

Stature—after the age of about 4 years, the height of these patients averages 20 cm more than those with SED congenita and severe coxa vara—far more than can be accounted for by the hip deformity itself (see graphs).

Presenting feature/age—at birth, short trunk, or more likely stiffness of hips or shoulders. Pectus carinatum, cleft palate and myopia may occur.

Deformities—pectus carinatum. Coxa vara developing during childhood, with associated lumbar lordosis. Scoliosis, developing during childhood.

Associated anomalies—as in SED congenita with severe coxa vara: myopia, cleft palate and deafness.

Radiographic features

A wide range of abnormalities from complete normality at the periphery of the limbs, to extensive epi- and metaphyseal dysplasia particularly of the large proximal joints; all patients have platyspondyly.

Skull—normal.

Spine—may be odontoid dysplasia. Platyspondyly throughout; 'pear-shaped' vertebrae in early infancy.

Limb epiphyses—vary from normality to severe fragmentation and delay in appearance, most marked at the capital femoral epiphyses.

Metaphyses—again, vary from normality to disordered, irregular growth, but with epiphyseal fusion the signs disappear.

Pelvis—small ilia, horizontal acetabular roofs, delayed ossification of pubic rami.

Hip joints—delay in appearance of the capital femoral epiphyses, but only by one or two years. They become irregular and fragmented, and coxa vara develops with the typical triangular bony fragment inferiorly—however, this is never so severe in degree as to proceed to discontinuity of the femoral neck with upward displacement of the trochanters.

Bone maturation

Generalised delay, perhaps of about two years.

Biochemistry

Not known.

Differential diagnosis

1. SED congenita with severe coxa vara, but diagnosis is not possible in the first 2–3 years of life. After this, the lesser degree of hip disorganisation and of short stature differentiates the milder form.

2. SED tarda (X-linked or other inheritance patterns). These have a later onset, perhaps at 10 years of age or after. Patients are not markedly short, do not have coxa vara and rarely any radiological signs other than in vertebrae, shoulders and hips. The platyspondyly of the X-linked type is characteristic.

3. Multiple epiphyseal dysplasia. This lacks any vertebral signs other than minimal involvement at the thoraco-lumbar junction during childhood.

In both SED tarda and multiple epiphyseal dysplasia, the hip joints may be similar to SED congenita as described here, at a similar age.

Progress/complications

Premature osteoarthritis in affected joints. Scoliosis may occur.

REFERENCE

Wynne-Davies R, Hall C 1982 Two clinical variants of spondylo-epiphyseal dysplasia congenita. Journal of Bone and Joint Surgery 64B: 435–441

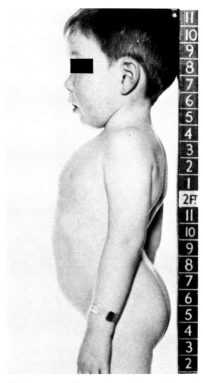

Fig. 6.1 In this less severe form of spondylo-epiphyseal dysplasia congenita the hip flexion deformity and hence lumbar lordosis are less severe. Protrusion of the manubrio-sternal angle is present, but again, less marked than in Figure 5.2.

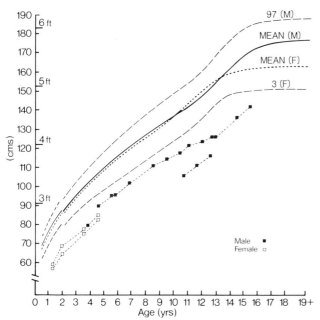

Fig. 6.2 SED congenita without severe coxa vara—percentile height.

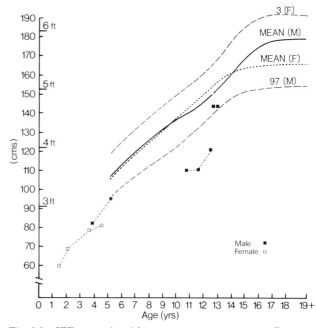

Fig. 6.3 SED congenita without severe coxa vara—percentile span.

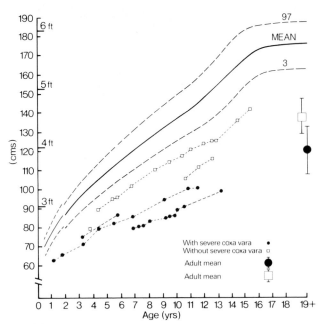

Fig. 6.4 Spondylo-epiphyseal dysplasia congenita—percentile height (males).

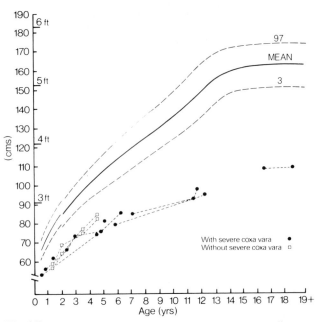

Fig. 6.5 Spondylo-epiphyseal dysplasia congenita—percentile height (females).

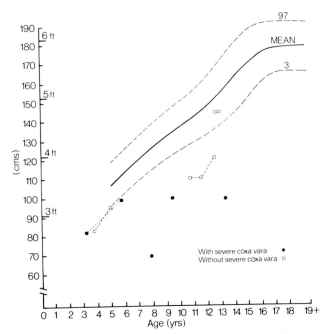

Fig. 6.6 Spondylo-epiphyseal dysplasia congenita—percentile span (males).

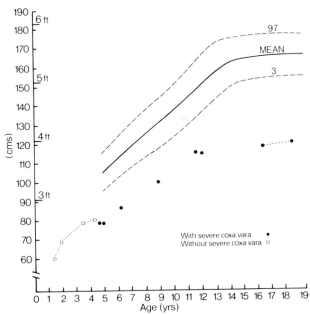

Fig. 6.7 Spondylo-epiphyseal dysplasia congenita—percentile span (females).

Spine: 1 year

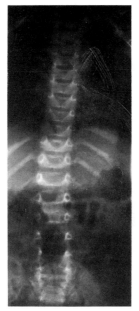

Fig. 6.8 A.B.—1 year 4 months—AP spine
There is mild widening of the interpedicular distances in the lumbar region.
Same patient: Figs 6.9, 6.20, 6.22, 6.25, 6.34, 6.37, 6.39, 6.41

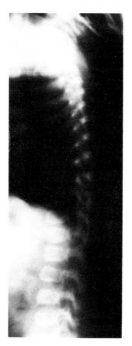

Fig. 6.9 A.B.—1 year 4 months—Lateral lumbar spine (same patient as in Fig. 6.8)
The vertebral bodies have an oval configuration with mild posterior wedging and an increase in the AP diameters ('pear-shaped').

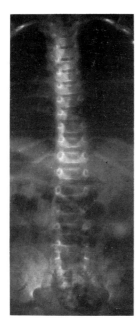

Fig. 6.10 F.B.—1 year 4 months—AP spine (identical twin of the child in Fig. 6.8)
See Fig. 6.8.
Same patient: Figs 6.11, 6.13, 6.21, 6.23, 6.26, 6.27, 6.35, 6.38, 6.40, 6.42

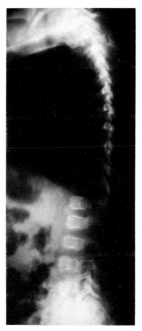

Fig. 6.11 F.B.—1 year 4 months—Lateral lumbar spine (same patient as in Fig. 6.10)
See Fig. 6.9.

Spine : 1–5 years

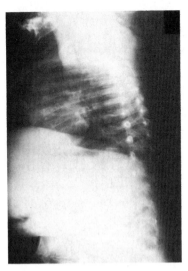

Fig. 6.12   L.W.—1 year 4 months—Lateral spine
Oval vertebral bodies. Posterior wedging in dorsal region.
Same patient : Figs 6.15, 6.16, 6.29

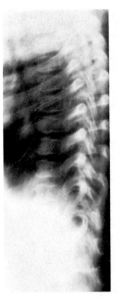

Fig. 6.13   F.B.—2 years—Lateral dorsal spine
Smooth vertebral end-plates. Posterior wedging—'pear-shaped'.
Same patient : Figs 6.10, 6.11, 6.21, 6.23, 6.26, 6.27, 6.35, 6.38, 6.40, 6.42

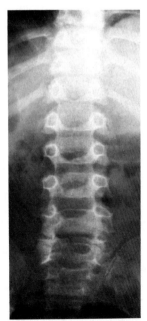

Fig. 6.14   M.B.—4 years 3 months—AP spine
Smooth vertebral end-plates. Wide interpedicular distances.
Same patient : Figs 6.17–6.19, 6.28, 6.33

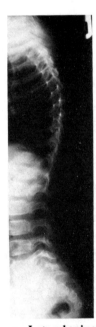

Fig. 6.15   L.W.—5 years—Lateral spine (same patient as in Fig. 6.12)
The vertebral bodies are elongated in their AP diameters. Vertebral end-plates remain smooth.

Spine: 5–10 years

Fig. 6.16 L.W.—5 years—AP spine
There is mild scoliosis, convex to the right in the dorsal region, and a mid-dorsal anterior cleft vertebral body (butterfly vertebra).
Same patient: Figs 6.12, 6.15, 6.29

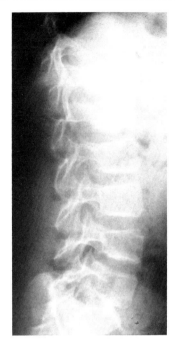

Fig. 6.17 M.B.—9 years 4 months—Lateral lumbar spine
There is irregularity of the vertebral end-plates of D12 and L1 with some disc space narrowing, and mild posterior wedging.
Same patient: Figs 6.14, 6.18, 6.19, 6.28, 6.33

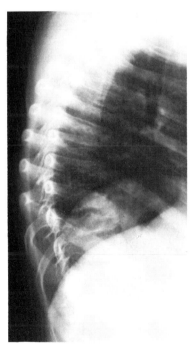

Fig. 6.18 M.B.—9 years 4 months—Lateral dorsal spine
(same patient as in Fig. 6.17)
There is marked platyspondyly with elongation of the vertebral bodies in their AP diameters and vertebral end-plate irregularity.

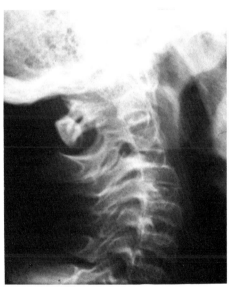

Fig. 6.19 M.B.—10 years 4 months—Lateral cervical spine
There is hypoplasia of the odontoid peg. The vertebral bodies are flattened with some increase in their AP diameters.

Pelvis : 1–2 years

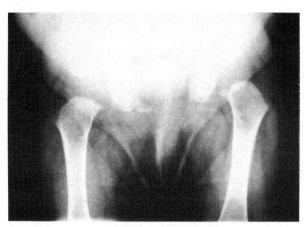

Fig. 6.20   A.B.—1 year 4 months—Pelvis
The proximal femoral metaphyses are irregular and medial triangular bony fragments are present. The capital femoral epiphyses are not present. The symphysis pubis is wide and superior pubic rami virtually absent.
Same patient : Figs 6.8, 6.9, 6.22, 6.25, 6.34, 6.37, 6.39, 6.41

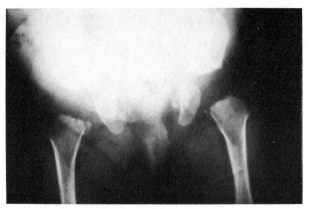

Fig. 6.21   F.B.—1 year 4 months—Pelvis (identical twin of the child in Fig. 6.20)
The acetabular roofs are horizontal. Capital femoral epiphyses are not yet visible. As in her twin, an infero-medial bony fragment is present in the neck. The symphysis pubis is wide, and the rami developing from several ossification centres.
Same patient : Figs 6.10, 6.11, 6.13, 6.23, 6.26, 6.27, 6.35, 6.38, 6.40, 6.42

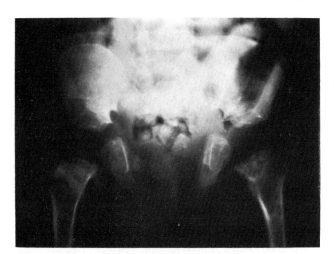

Fig. 6.22   A.B.—2 years—Pelvis (same patient as in Fig. 6.20)
The capital femoral epiphyses have not yet ossified. The acetabular roofs are horizontal. There is virtually no ossification of the left pubic ramus. On the right the ramus is developing from at least two ossification centres.

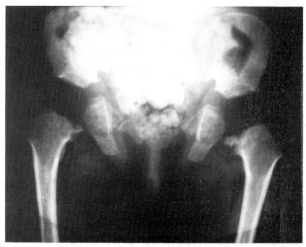

Fig. 6.23   F.B.—2 years—Pelvis (same patient as in Fig. 6.21)
The proximal femoral metaphyses are irregular and the capital femoral epiphyses have not yet ossified. The acetabular roofs are horizontal, and iliac wings small and square. The symphysis is wide and there are two ossification centres to each pubic ramus.

Pelvis: 2–4 years

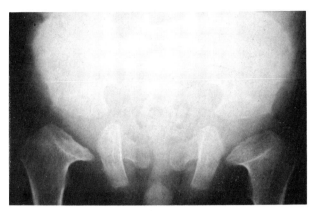

Fig. 6.24 K.M.—2–3 years—Pelvis
Symmetrical coxa vara with no ossification of capital femoral
epiphyses. Reasonably well formed femoral necks, but with splayed
and irregular metaphyses. These milder changes of the femoral necks
indicate that the overall height of the child will be only slightly
reduced.
Same patient: Fig. 6.30

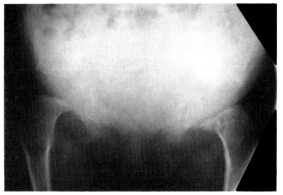

Fig. 6.25 A.B.—3 years 6 months—Pelvis
Coxa vara is now obvious and there is still no ossification of capital
femoral epiphyses. Horizontal acetabular roofs. The femoral necks are
reasonably well developed, indicating this patient has the milder type
of SED congenita (*without* severe coxa vara).
Same patient: Figs. 6.8, 6.9, 6.20, 6.22, 6.34, 6.37, 6.39, 6.41

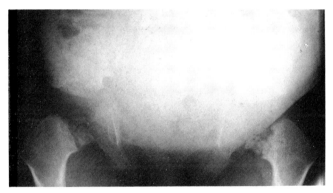

Fig. 6.26 F.B.—3 years 6 months—Pelvis (identical twin of the
child in Fig. 6.25)
There is now a very small capital femoral epiphysis present on each
side. Coxa vara is seen and the femoral necks are reasonably well
formed.
Same patient: Figs 6.10, 6.11, 6.13, 6.21, 6.23, 6.27, 6.35, 6.38, 6.40, 6.42

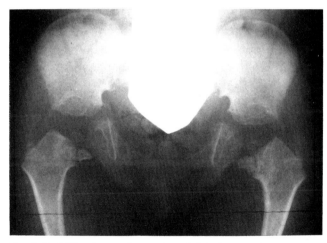

Fig. 6.27 F.B.—4 years 6 months—Pelvis (same patient as in
Fig. 6.26)
Horizontal acetabular roofs with small iliac wings. There is irregular
ossification of very small capital femoral epiphyses. The triangular
bony fragments of the medial femoral necks indicate coxa vara. The
symphysis remains rather wide and the ischia are short.

Pelvis: 4–9 years

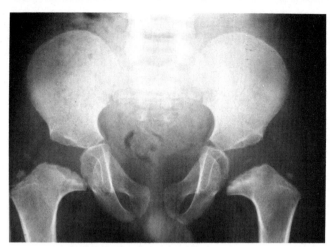

Fig. 6.28   M.B.—4 years 5 months—Pelvis
The acetabular roofs are horizontal. The proximal femoral metaphyses are irregular and there is a little ossification of the capital femoral epiphyses. The femoral necks are reasonably well developed indicating the mild type of SED congenita (*without* severe coxa vara).
Same patient: Figs 6.14, 6.17–6.19

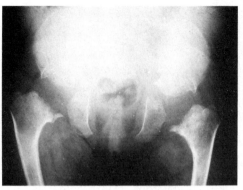

Fig. 6.29   L.W.—5 years—Pelvis
There are changes of 'congenital' coxa vara. The capital femoral epiphyses have not ossified. The acetabular roofs are horizontal and the iliac wings small. The pubic rami have not ossified, and the symphysis is very wide. The greater trochanters are displaced upwards and the femoral necks short.
Same patient: Figs 6.12, 6.15, 6.16

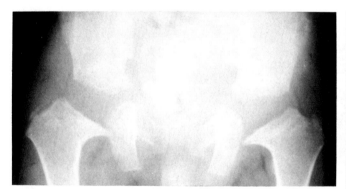

Fig. 6.30   K.M.—5 years 1 month—Pelvis
The acetabular roofs are horizontal. The capital femoral epiphyses have not ossified. There is mild coxa vara. The pubic rami and ischia are short and the symphysis wide.
Same patient: Fig. 6.24

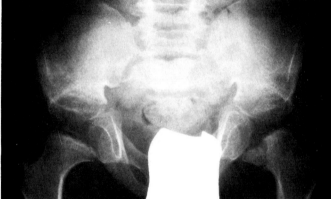

Fig. 6.31   J.D.—9 years 6 months—Pelvis
There is virtually no ossification of the capital femoral epiphyses. Coxa vara is present but the femoral necks are reasonably well developed (milder variety). The iliac wings are small with horizontal acetabular roofs.
Same patient: Figs 6.32, 6.36, 6.43

Pelvis : 11–12 years

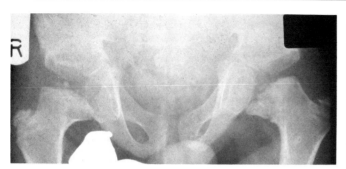

Fig. 6.32   J.D.—11 years 7 months—Pelvis
There is coxa vara with irregular ossification of the capital femoral epiphyses. The femoral necks are reasonably well developed. Acetabular roofs are horizontal.
Same patient : Figs 6.31, 6.36, 6.43

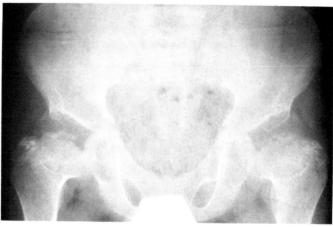

Fig. 6.33   M.B.—12 years 8 months—Pelvis
The capital femoral epiphyses are poorly ossified and flattened. Femoral necks are short and there is coxa vara.
Same patient : Figs 6.14, 6.17–6.19, 6.28

Knee: 1–9 years

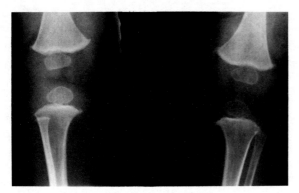

Fig. 6.34   A.B.—1 year 4 months—AP knees
Bony fragments are present at the angles of the metaphyses.
Modelling appears normal, but bone age is retarded.
Same patient: Figs 6.8, 6.9, 6.20, 6.22, 6.25, 6.37, 6.39, 6.41

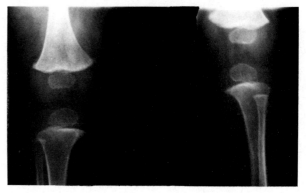

Fig. 6.35   F.B.—1 year 4 months—AP knees (Identical twin of
child in Fig. 6.34)
As Figure 6.34.
Same patient: Figs 6.10, 6.11, 6.13, 6.21, 6.23, 6.26, 6.27, 6.38, 6.40, 6.42

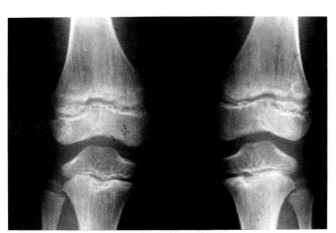

Fig. 6.36   J.D.—9 years 6 months—AP knees
There is metaphyseal irregularity and unusual modelling of the
epiphyses, but changes are mild.
Same patient: Figs 6.31, 6.32, 6.43

Hand: 1–2 years

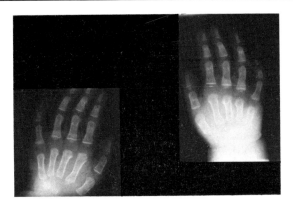

Fig. 6.37 A.B.—1 year 4 months—Hands
The first metacarpal is short. There are epiphyses at the bases of the second metacarpals but no other abnormality is seen at this age.
Same patient: Figs 6.8, 6.9, 6.20, 6.22, 6.25, 6.34, 6.39, 6.41

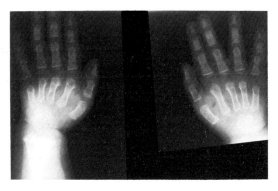

Fig. 6.38 F.B.—1 year 4 months—Hands (identical twin of child in Fig. 6.37)
Similar to Figure 6.37.
Same patient: Figs 6.10, 6.11, 6.13, 6.21, 6.23, 6.26, 6.27, 6.35, 6.40, 6.42

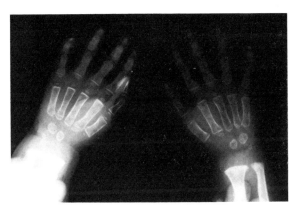

Fig. 6.39 A.B.—2 years—Hands (same patient as in Fig. 6.37)
In addition there are now epiphyses at the distal ends of the first metacarpals.

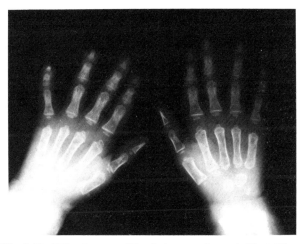

Fig. 6.40 F.B.—2 years—Hands (same patient as in Fig. 6.38)
Similar to Figure 6.39.

Hand: 3–9 years

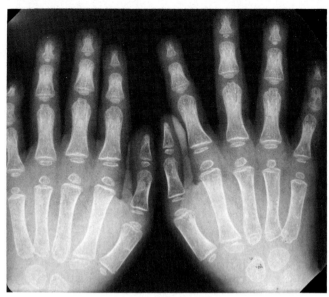

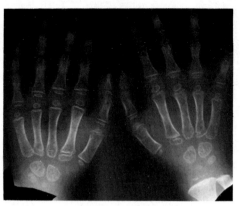

Fig. 6.42   F.B.—3 years 6 months—Hands (identical twin of child in Fig. 6.41)
Similar to Figure 6.41.
Same patient: Figs 6.10, 6.11, 6.13, 6.21, 6.23, 6.26, 6.27, 6.35, 6.38, 6.40

Fig. 6.41   A.B.—3 years 6 months—PA both hands
Bone age corresponds to only 2 years 6 months. The carpal centres and epiphyses are angular. There is some distal pointing of the proximal phalanges, and additional metacarpal epiphyses.
Same patient: Figs 6.8, 6.9, 6.20, 6.22, 6.25, 6.34, 6.37, 6.39

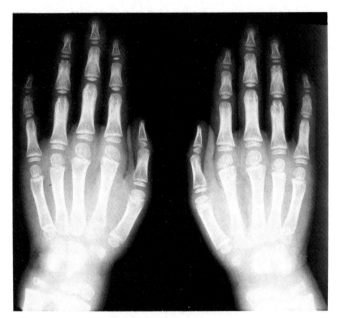

Fig. 6.43   J.D.—9 years 6 months—PA both hands
Phalangeal bone age corresponds to only 5 years. The scaphoid has not developed at all. Additional metacarpal epiphyses are present.
Same patient: Figs 6.31, 6.32, 6.36

Chapter Seven
SPONDYLO-EPIPHYSEAL DYSPLASIA TARDA—X-linked type

A disorder of vertebral and epiphyseal growth, principally affecting the large proximal joints, characterized by onset late in childhood, and not of great severity. Males only.

Inheritance

X-linked recessive.

Frequency

Possible prevalence of 3–4 per million population, but males only affected.

Clinical features

Facial appearance—normal.
Intelligence—normal.
Stature—some shortness, principally of the trunk, associated with platyspondyly.
Presenting feature/age—usually between 5 and 10 years of age, perhaps later, with pain and stiffness in the back or hips.
Deformities—those associated with secondary osteoarthritis. Sometimes scoliosis.
Range of severity—not very wide—there appear to be no very severe cases.
Associated anomalies—none.

Radiographic features

Skull—normal.
Spine—the vertebrae are flattened and disc spaces narrowed. On the lateral view the shape is quite distinctive, with a posterior hump and absence of ossification at the upper and lower anterior margins of the bodies.
Pelvis—sometimes reported as small.

Hips/shoulders—mild dysplastic changes, indistinguishable from some cases of multiple epiphyseal dysplasia at a similar age. Early onset of osteoarthritis.
Other epiphyses—probably normal, or minimal changes in knees.

Bone maturation

There may be some delay in maturation of the capital femoral epiphyses, but there is none at the carpus/hand.

Biochemistry

Not known.

Differential diagnosis

Principally from spondylo-epiphyseal tarda *not* of X-linked inheritance. The vertebrae (lateral view) distinguish this X-linked variety.

In *multiple epiphyseal dysplasia* the hip and shoulder joints are indistinguishable, but spinal changes in MED are minimal, and only affect the thoraco-lumbar junction.

Progress/complications

Limitation of movement in the back and hip, perhaps also the shoulder, is slowly progressive, and secondary osteoarthritis is established by the second or third decade.

REFERENCES

Maroteaux P, Lamy M, Bernard J 1957 La dysplasie spondylo-épiphysaire tardive; description clinique et radiologique. Presse médicale 65:1205
Bannerman R M, Ingall G B, Mohn J F 1971 X-linked spondyloepiphyseal dysplasia tarda: clinical and linkage data. Journal of Medical Genetics 8:291

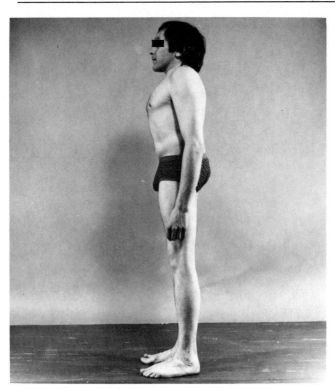

Fig. 7.1   This adult shows no abnormality apart from shortening of the trunk in comparison with the lower segment (pubis to heel).

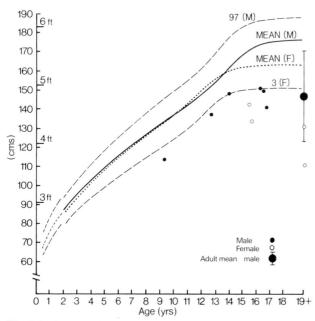

Fig. 7.2   Spondylo-epiphyseal tarda (all types)—percentile height.

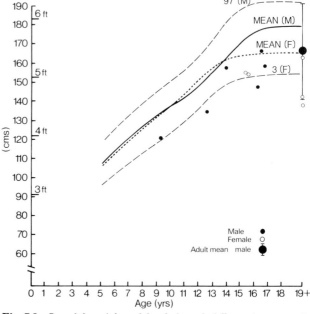

Fig. 7.3   Spondylo-epiphyseal dysplasia tarda (all types)—percentile span.

Lumbar spine 16 years–Adult

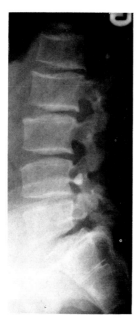

Fig. 7.4 Unknown—16 years—Lateral lumbar spine
There are bony mounds on the posterior two-thirds of the vertebral
end-plates with some sclerosis. This is characteristic of the X-linked
variety. There is also a Grade I spondylolisthesis of L5 on S.1 with
bilateral defects in the pars interarticulares. (Myodil remnants are
present.)
Same patient: Fig. 7.7

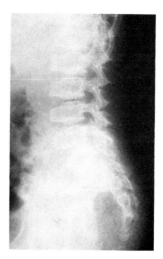

Fig. 7.5 N.E.—17 years—Lateral lumbar spine
There is platyspondyly, with mounds of bone over the posterior two-
thirds of the vertebral end-plates with some sclerosis.
Same patient: Fig. 7.9

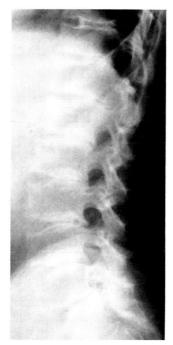

Fig. 7.6 G.R.—30 years 5 months—Lateral lumbar spine
There are dense mounds of bone over the posterior two-thirds of the
vertebral end-plates, with disc space narrowing.

Pelvis : 16 years–Adult

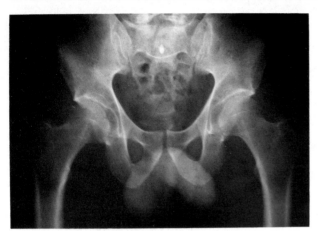

Fig. 7.7   Unknown—16 years—AP pelvis
There is slight joint space narrowing of the superior parts of both hip joints, but no other obvious defect.
Same patient : Fig. 7.4

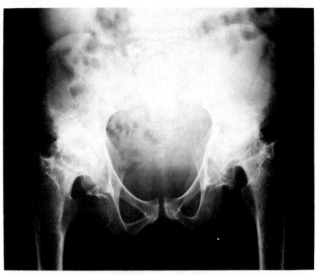

Fig. 7.8   R.C.—36 years—AP pelvis/hips
The joint spaces are narrowed and femoral heads flattened. There is marked sclerosis, subarticular cyst formation and marginal osteophytes; severe osteoarthritis for this age.
Same patient : Fig. 7.11

Limbs: 12 years–Adult

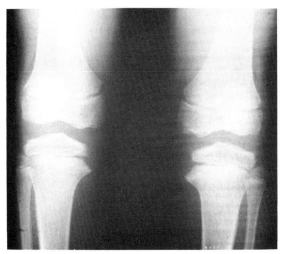

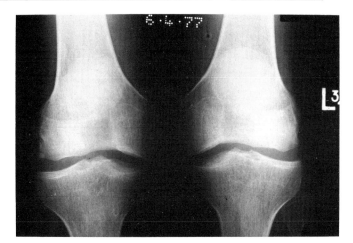

Fig. 7.9 N.E.—12 years—AP knees
There is some metaphyseal splaying and irregularity and the
epiphyses are small and irregular.
Same patient: Fig. 7.5

Fig. 7.10 D.E.—Adult—AP knees (brother of patient in Fig. 7.9)
There is joint space narrowing with abnormal modelling and
irregularity of the distal femoral articular surfaces. Mild metaphyseal
flaring is also present.

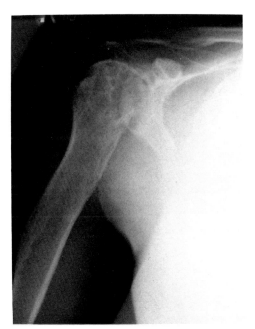

Fig. 7.11 R.C.—38 years—AP right shoulder
The joint space is virtually obliterated and subarticular cyst formation
is present in the humeral head—signs of osteoarthritis which are
extremely rare in the normal individual in this joint.
Same patient: Fig. 7.8

Chapter Eight
SPONDYLO-EPIPHYSEAL DYSPLASIA TARDA—autosomal dominant and recessive types

Apart from the pattern of inheritance and radiological appearance of the vertebrae, the features are similar to those described for X-linked SED tarda (Chapter 7).

Inheritance

Autosomal dominant and recessive types occur, perhaps several varieties, but they are not clinically delineated.

Frequency

Possible prevalence of 3–4 per million population, both sexes affected.

Clinical features

As for the X-linked type, with some shortness of stature, principally of the trunk, and onset in late childhood of back and hip pain and stiffness. Brachydactyly may occur.

Deformities—Scoliosis frequent. Other deformities related to secondary osteoarthritis.

Radiological features

Skull—normal.

Spine—irregular platyspondyly, sometimes affecting the whole spine, sometimes the thoracic region only. No posterior hump, as seen in the X-linked variety.

Peripheral joints—the dysplastic changes of the hip and shoulder are indistinguishable from those in the X-linked variety, or from multiple epiphyseal dysplasia at a similar age, and may include a degree of acetabular dysplasia with the early onset of osteoarthritis.

Bone maturation

May be some delay at the capital femoral epiphyses, but hand and wrist are normal.

Biochemistry

Not known.

Differential diagnosis

From the *X-linked* variety, but here the vertebrae have a characteristic shape.

From *multiple epiphyseal dysplasia*, but here the spine is only minimally affected, if at all.

From *SED tarda with progressive arthropathy* (autosomal recessive), but here it is particularly the wrist/finger joints which are involved with a progressive rheumatoid-like disease.

Progress/complications

Early onset of osteoarthritis, particularly of the spine and hip, perhaps also the shoulder.

REFERENCES

Carter C, Sutcliffe J 1970 Genetic varieties of spondyloepiphyseal dysplasia. Symposium Ossium. E & S Livingstone, Edinburgh

Crossan J F, Wynne-Davies R, Fulford G E 1983 Bilateral failure of the capital femoral epiphysis. Journal of Pediatric Orthopedics 3 : 297–301

Harper P S, Jenkins P, Laurence K M 1973 Spondyloepiphyseal dysplasia tarda : a report of four cases in two families. British Journal of Radiology 46 : 676

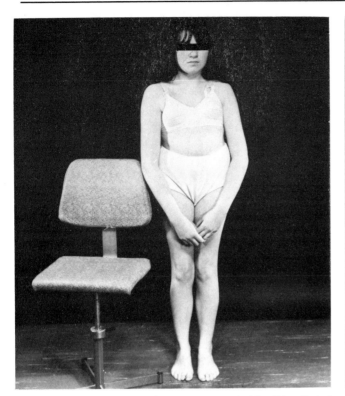 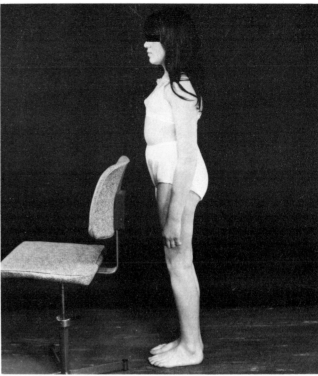

Figs 8.1 and 8.2   This girl of 14 years presented with mild scoliosis, but the shortening of the trunk compared with the lower segment (pubis to heel) is very obvious, and more than could be accounted for by the scoliosis.

Spine: 1–12 years

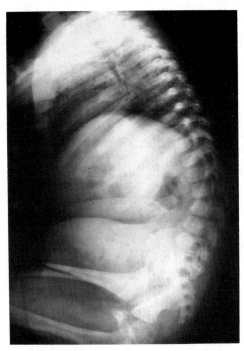

Fig. 8.3   R.W.—1 year—Lateral spine
There is a dorso-lumbar kyphosis and a small inferior hook on the body of L1 and L2, but no other abnormality.
Same patient: Figs 8.4, 8.15, 8.16, 8.20

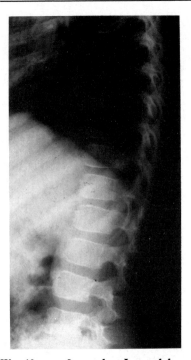

Fig. 8.4   R.W.—10 years 9 months—Lateral dorso-lumbar spine (same patient as in Fig. 8.3)
The vertebral bodies are increased in their AP diameters, but there is no irregularity of the vertebral plates.

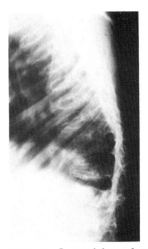

Fig. 8.5   M.v.K.—12 years—Lateral dorso-lumbar spine
There is marked platyspondyly with anterior wedging and irregularity of the vertebral end-plates.

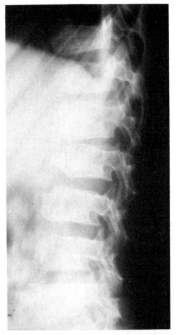

Fig. 8.6   B.S.—12 years 4 months—Lateral lumbar spine
There is an increase in the AP diameters of the vertebral bodies and an anterior tongue at L2 with disc space narrowing at L1 and L2. The pedicles are short with posterior scalloping.
Same patient: Figs 8.17, 8.18, 8.33–8.35, 8.37

Spine: 13–21 years

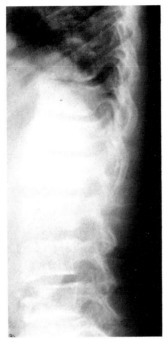

Fig. 8.7   P.St A.—13 years—Lateral lumbar spine
There is platyspondyly with an increase in the AP diameter of the vertebral bodies and end-plate irregularity at the dorso-lumbar region. There is a defect of the antero-superior margin of the body of L4.
Same patient: Fig. 8.21

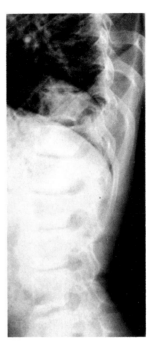

Fig. 8.8   R.D.—17 years 9 months—Lateral spine
There is only mild vertebral end-plate irregularity with disc space narrowing.
Same patient: Figs 8.19, 8.31, 8.32

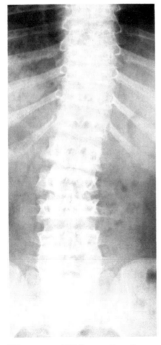

Fig. 8.9   A.H.—21 years—AP dorso-lumbar spine
There is mild scoliosis with platyspondyly and vertebral end-plate irregularity.
Same patient: Figs 8.10, 8.25, 8.30, 8.40

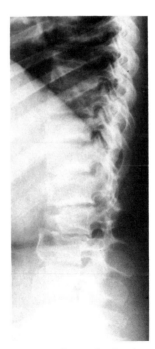

Fig. 8.10   A.H.—21 years—Lateral spine (same patient as in Fig. 8.9)
There is platyspondyly with disc space narrowing and vertebral end-plate irregularity, also anterior wedging of L2. The pedicles are short and there is posterior scalloping.

Spine: Adult

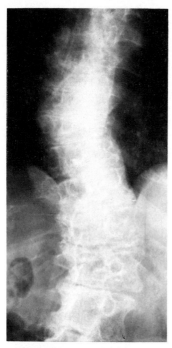

Fig. 8.11   A.P.—54 years—AP dorsal spine
There is a double scoliosis and platyspondyly is present, with disc space narrowing and marked spondylosis.
Same patient: Figs 8.12, 8.24

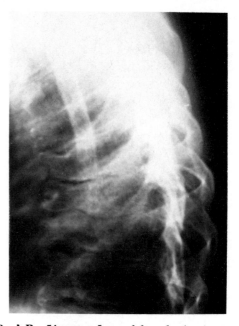

Fig. 8.12   A.P.—54 years—Lateral dorsal spine (same patient as in Fig. 8.11)
There is platyspondyly with anterior elongation of the vertebral bodies. The disc spaces are narrow, vertebral end-plates irregular and there is anterior osteophyte formation.

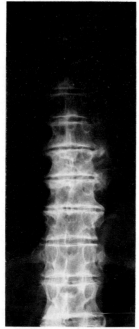

Fig. 8.13   P.H.—61 years—AP dorsal spine
There is disc space narrowing with pronounced marginal osteophyte formation. (It is possible this is the autosomal dominant type, as this patient's son is affected.)
Same patient: Figs 8.14, 8.22, 8.26, 8.36

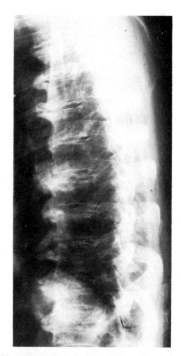

Fig. 8.14   P.H.—61 years—Lateral dorsal spine (same patient as in Fig. 8.13)
Marked anterior osteophytes are present with some bridging and the disc spaces are narrowed.

Pelvis and hip: 3–14 years

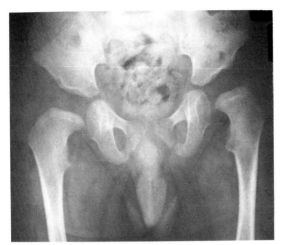

Fig. 8.15   R.W.—3–4 years—Pelvis/hips
The capital femoral epiphyses have not yet ossified. The acetebula are shallow and irregular and there is mild coxa valga. (No symptoms at this age—incidental finding.)
Same patient: Figs 8.3, 8.4, 8.16, 8.20

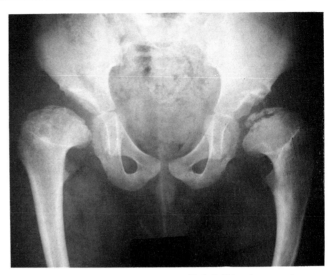

Fig. 8.16   R.W.—7 years 6 months—Pelvis/hips (same patient as in Fig. 8.15)
The acetabular roofs are shallow, irregular and slope steeply. There is flattening and fragmentation of the capital femoral epiphyses and the femoral necks are broad. Bone maturation appears to be delayed.

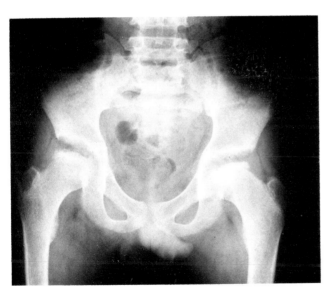

Fig. 8.17   B.S.—12 years 4 months—Pelvis/hips
Both capital femoral epiphyses are symmetrically flattened, fragmented and sclerotic, but in this patient the acetabular roofs appear almost normal and the capital femoral epiphyses are well covered.
Same patient: Figs 8.6, 8.18, 8.33–8.35, 8.37

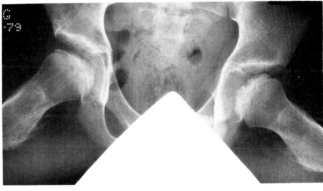

Fig. 8.18   B.S.—13 years 10 months—Hips in abduction (same patient as in Fig. 8.17)
There is symmetrical flattening, fragmentation and sclerosis of the capital femoral epiphyses.

Pelvis and hip : 17 years–Adult

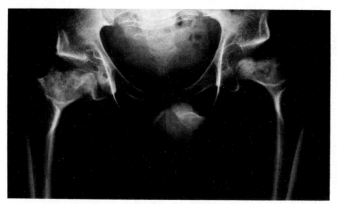

Fig. 8.19   R.D.—17 years—AP pelvis/hips
There is very poor ossification of the capital femoral epiphyses and the femoral necks are broad with patchy ossification and coxa vara. The acetabular roofs are horizontal, perhaps related to the poorly developing femoral heads.
Same patient : Figs 8.8, 8.31, 8.32

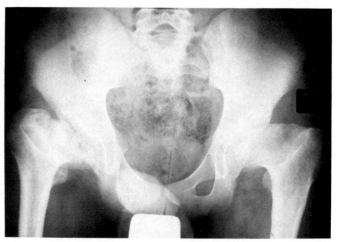

Fig. 8.20   R.W.—18 years—AP pelvis/hips
The acetabular roofs are irregular and slope steeply. The femoral heads are flattened and there is coxa vara with poor coverage of the femoral head on the left.
Same patient : Figs 8.3, 8.4, 8.15, 8.16

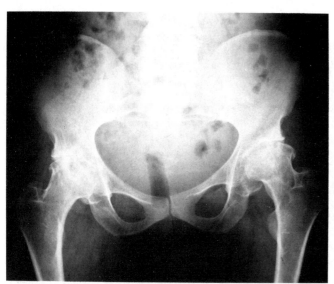

Fig. 8.21   P.St.A.—29 years—AP pelvis/hips
The joint spaces are narrow; the articular margins irregular and sclerotic with subarticular cyst formation and marginal osteophytes—severe osteoarthritis for this age.
Same patient : Fig. 8.7

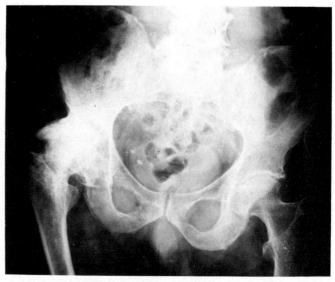

Fig. 8.22   P.H.—61 years—AP pelvis/hips
There is marked pelvic tilt. The right hip shows severe joint space narrowing with sclerosis and lateral osteophytes. The osteoarthritic changes in the left hip are less marked.
Same patient : Figs 8.13, 8.14, 8.26, 8.36

Knee: 16 years–Adult

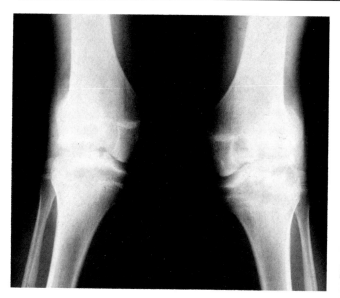

Fig. 8.23   J.H.—16 years—AP knees
There is genu valgum and a fibrous cortical defect in the proximal right tibial shaft, but no obvious epiphyseal disorder.

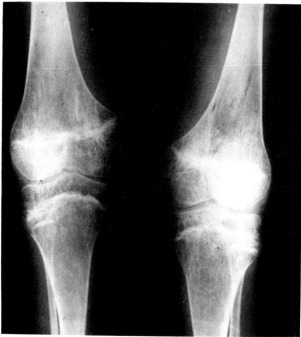

Fig. 8.24   A.P.—17 years 6 months—AP knees
There is coarse trabeculation of the epiphyses and metaphyses with mild metaphyseal flaring. The joint spaces appear narrow and premature osteoarthritis is developing.
Same patient: Figs 8.11, 8.12

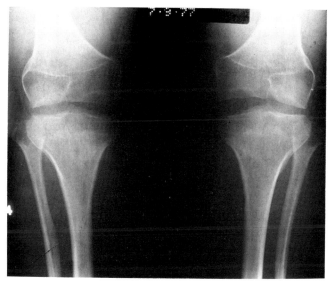

Fig. 8.25   A.H.—21 years—AP knees
There is abnormal modelling and irregularity of the articular surfaces. The fibulae are long.
Same patient: Figs 8.9, 8.10, 8.30, 8.40

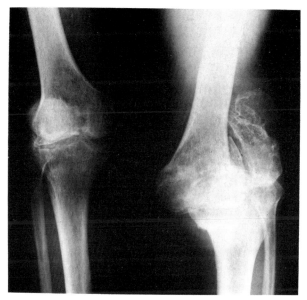

Fig. 8.26   P.H.—63 years—AP knees (father of the patient in Fig. 8.23)
There is subluxation of the left knee with disorganisation of the joint. Osteoarthritic changes are present on the right.
Same patient: Figs 8.13, 8.14, 8.22, 8.36

Lower limb: Adult

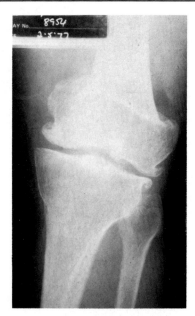

Fig. 8.27   R.C.—Adult—AP right knee
There is joint space narrowing with sclerosis of the articular surfaces
and marginal osteophyte formation. Metaphyseal flaring is present.

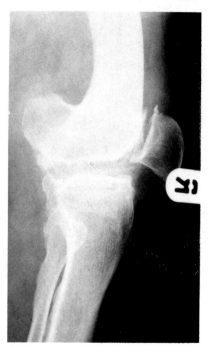

Fig. 8.28   R.C.—Adult—Lateral right knee (same patient as in
Fig. 8.27)

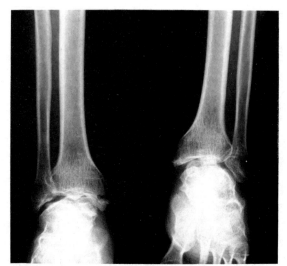

Fig. 8.29   N.H.—Adult—AP ankles
There is splaying of the distal tibial metaphyses. The joint spaces are
narrow and the articular surfaces irregular indicating osteoarthritis.
Loose bodies are also present.

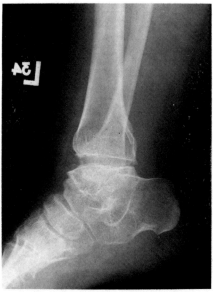

Fig. 8.30   A.H.—21 years—Lateral left ankle
There is joint space narrowing with flattening and articular
irregularity of the distal tibial articular surface and of the talus.
Same patient: Figs 8.9, 8.10, 8.25, 8.40

Shoulder : 17 years

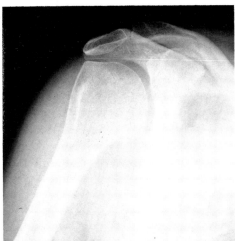

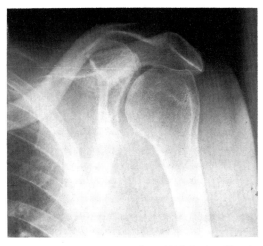

Fig. 8.31   R.D.—17 years 9 months—AP right shoulder
There is irregularity of the glenoid fossa and some flattening of the humeral head.
Same patient : Figs 8.8, 8.19, 8.32

Fig. 8.32   R.D.—17 years 9 months—AP left shoulder (same patient as in Fig. 8.31)
The articular surface of the glenoid fossa is irregular.

Elbow: 10 years–Adult

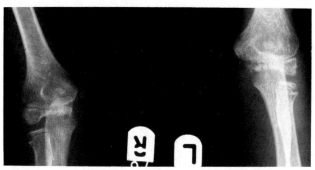

Fig. 8.33   B.S.—10 years 6 months—AP elbow
There is sclerosis and fragmentation of the trochlear epiphyses and
the proximal radial metaphyses are irregular.
Same patient: Figs 8.6, 8.17, 8.18, 8.34, 8.35, 8.37

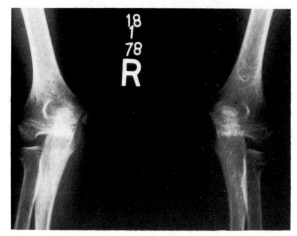

Fig. 8.34   B.S.—12 years 4 months—AP elbows (same patient as
in Fig. 8.33)
The articular surfaces are irregular with some trochlear sclerosis.

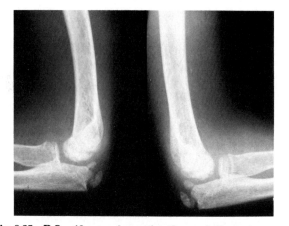

Fig. 8.35   B.S.—12 years 4 months—Lateral elbows (same
patient as in Fig. 8.33)
There is some joint space narrowing with fragmentation and sclerosis
of the distal humeral epiphyses.

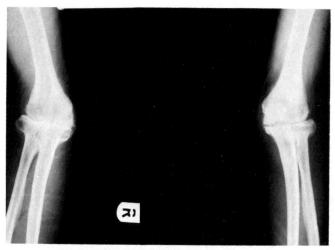

Fig. 8.36   P.H.—63 years—AP both elbows
There is marked joint space narrowing and subluxation of the radial
heads, with flaring of the proximal radial metaphyses.
Same patient: Figs 8.13, 8.14, 8.22, 8.26

Hand: 10 years–Adult

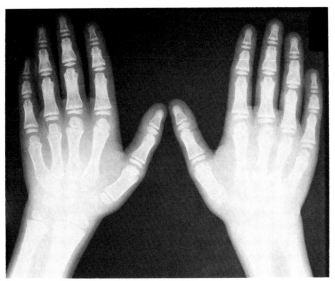

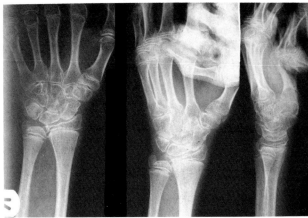

Fig. 8.38   E.C.—14 years—Right wrist AP, lateral and oblique
There appears to be some osteoporosis and joint space narrowing around the carpus and wrist joint. (It is possible that this patient has the autosomal recessive form as her brother is affected and the parents normal.)
Same patient: Fig. 8.39

Fig. 8.37   B.S.—10 years 6 months—PA hands
The carpal centres are small and irregular. There is little sign of epiphyseal dysplasia here apart from delay in bone maturation which corresponds to 6 to 7 years.
Same patient: Figs 8.6, 8.17, 8.18, 8.33–8.35

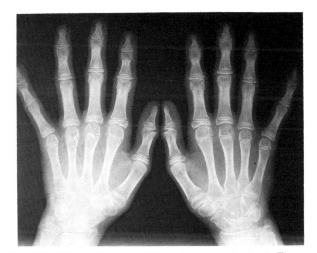

Fig. 8.39   E.C.—14 years—PA hands (same patient as in Fig. 8.38)
Generalised joint space narrowing but otherwise normal.

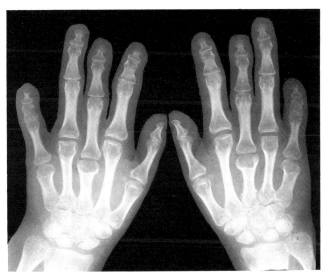

Fig. 8.40   A.H.—21 years—PA hands
There is shortening of the third and fifth metacarpals and all the terminal phalanges are short. The whole carpus is small.
Same patient: Figs 8.9, 8.10, 8.25, 8.30

Chapter Nine
SPONDYLO-EPIPHYSEAL DYSPLASIA TARDA WITH PROGRESSIVE ARTHROPATHY

This is an inherited bone dysplasia with platyspondyly and a clinical, although not radiological resemblance to rheumatoid arthritis.

Inheritance

Autosomal recessive.

Frequency

Not known, since it has only recently been delineated—perhaps about 1 per million prevalence.

Clinical features

Facial appearance—normal.

Intelligence—normal.

Stature—moderately reduced, part of this being accounted for by contractures of hips and knees.

Presenting feature/age—pain and joint swelling at any time during childhood after about 3 years. Usually affects the hands, but frequently multiple joints. There may be soft tissue swellings, as in rheumatoid arthritis.

Deformities—Progressive contractures develop in all affected joints. Mild scoliosis sometimes.

Associated anomalies—none.

Radiographic features

Skull—none.

Spine—universal platyspondyly. Thoraco-lumbar kyphosis in younger patients. The end-plates are irregular during growth but symmetrical when mature.

Hip joints—(if affected) the capital femoral epiphyses are large during childhood, developing cyst-like irregularities in the articular surfaces. Subsequently the head is flattened with progressive loss of joint space and superimposed osteoarthritis.

Hands—enlargement of epiphyses and metaphyses is striking, with later peri-articular osteoporosis. It should be noted that signs of rheumatoid arthritis (erosions, periostitis and soft tissue swelling) are absent.

Other affected joints—show similar loss of joint space, adjacent osteoporosis and irregular articular surfaces.

Bone maturation

Probably normal.

Biochemistry

Not known.

Differential diagnosis

From other forms of *spondylo-epiphyseal tarda* but the rheumatoid-like appearance of SED with progressive arthropathy is unmistakable. Also, most forms of SED tarda do not become apparent until 12–14 years of age.

Juvenile rheumatism is eliminated on both radiological evidence and serological tests.

Progress/complications

The disease appears to be progressive, with contractures developing in all affected joints and secondary osteoarthritis supervening. Severely affected individuals may become (virtually) immobile necessitating joint replacement.

REFERENCE

Wynne-Davies R, Hall C, Ansell B M 1982 Spondylo-epiphyseal dysplasia tarda with progressive arthropathy. Journal of Bone and Joint Surgery 64B : 442–445

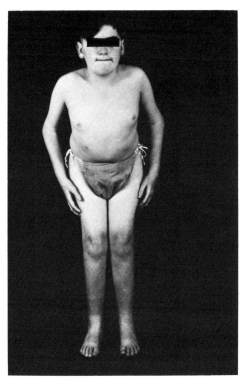 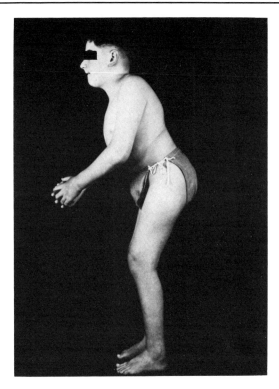

Figs 9.1 and 9.2   12-year-old boy of normal stature, but marked flexion contractures of most joints.

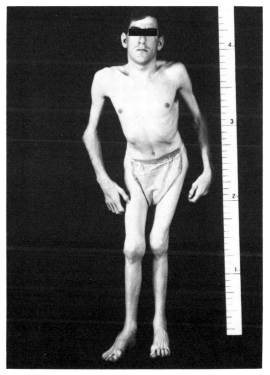 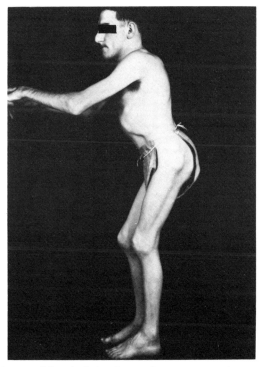

Figs 9.3 and 9.4   His elder brother (adult); the deformities have worsened with age and there is obvious bony enlargement around most joints.

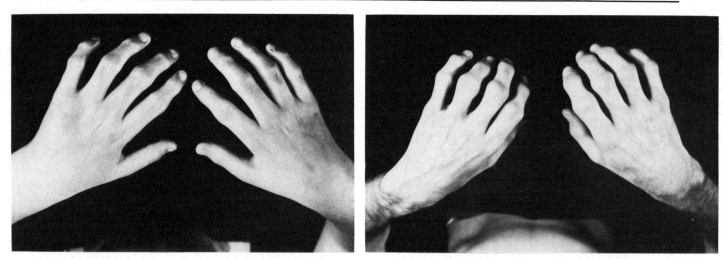

Figs 9.5 and 9.6   The hands of the two brothers show bony swelling around the small joints of the hands, more severe at the later age.

Cervical spine: 14 years–Adult

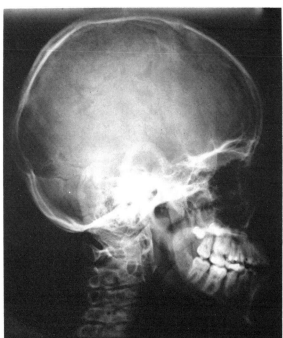

Fig. 9.7 N.R.—14 years 6 months—Lateral skull and cervical spine
The skull is normal. The cervical vertebrae are flattened and lengthened in their antero-posterior diameter.
Same patient: Figs 9.11, 9.15, 9.20, 9.23, 9.24, 9.30, 9.35, 9.39, 9.41, 9.43, 9.45

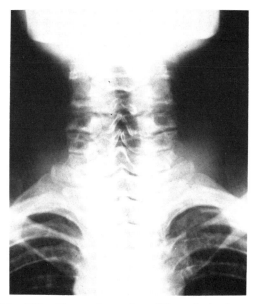

Fig. 9.8 C.R.—16 years 3 months—AP cervical spine
There is platyspondyly and also spina bifida occulta of C3, C4 and C5.
Same patient: Figs 9.13, 9.19, 9.28

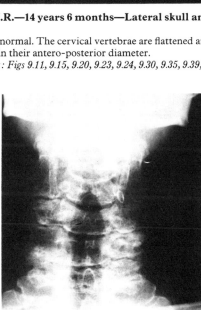

Fig. 9.9 S.M.—39 years—AP cervical spine
There is platyspondyly, narrowing of the intervertebral disc spaces and marked cervical spondylosis by this age.
Same patient: Figs 9.10, 9.21, 9.22, 9.33, 9.46

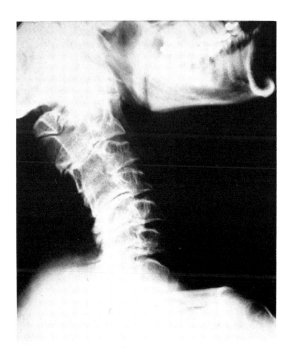

Fig. 9.10 S.M.—39 years—Lateral cervical spine (same patient as in Fig. 9.9.)
Similar to Figure 9.9.

Spine : 6 years–Adult

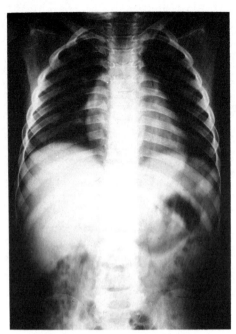

Fig. 9.11   N.R.—6 years 2 months—AP spine
The thorax, ribs and shoulder girdle are normal. Mild platyspondyly only.
Same patient : Figs 9.7, 9.15, 9.20, 9.23, 9.24, 9.30, 9.35, 9.39, 9.41, 9.43, 9.45

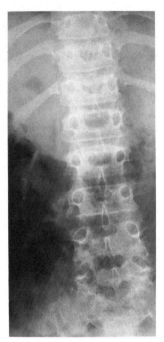

Fig. 9.12   T.Z.—11 years—AP lumbar spine (same patient as in Fig. 9.17)
Marked platyspondyly by this age.

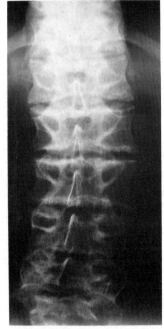

Fig. 9.13   C.R.—16 years 3 months—AP lumbar spine
Platyspondyly, lumbar scoliosis and narrowed disc spaces are obvious.
Same patient : Figs 9.8, 9.19, 9.28

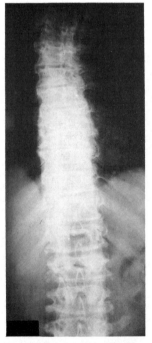

Fig. 9.14   E.D.—32 years 6 months—AP thoracic and lumbar spine
The platyspondyly and narrowed disc spaces are now accompanied by obvious degenerative changes with osteophytic lipping.
Same patient : Figs 9.34, 9.38, 9.44

Spine : 6–13 years

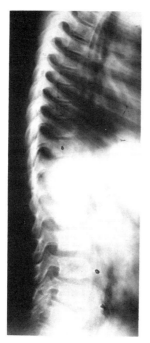

Fig. 9.15   N.R.—6 years 2 months—Lateral thoracic and lumbar spine
At this age there is only minimal platyspondyly and some irregularity of the end-plates, particularly anteriorly.
Same patient : Figs 9.7, 9.11, 9.20, 9.23, 9.24, 9.30, 9.35, 9.39, 9.41, 9.43, 9.45

Fig. 9.16   D.C.R.—10 years 11 months—Lateral thoracic and lumbar spine (brother of child in Fig. 9.15)
The irregular platyspondyly is more obvious at this stage.
Same patient : Figs 9.18, 9.26

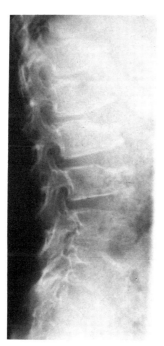

Fig. 9.17   T.Z.—11 years—Lateral lumbar spine
The vertebrae are flattened and there is some posterior scalloping of the bodies. The failure of ossification anteriorly is reminiscent of S.E.D. tarda, X-linked.
Same patient : Fig. 9.12

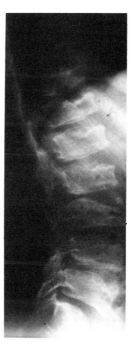

Fig. 9.18   D.C.R.—13 years 11 months—Lateral lumbar spine
(same patient as in Fig. 9.16)
There is irregular platyspondyly.

Spine : 16 years–Adult

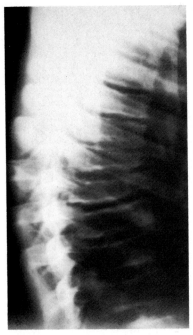

Fig. 9.19   C.R.—16 years—Lateral thoracic spine
There is gross platyspondyly, but the end-plates are relatively
smooth.
Same patient : Figs 9.8, 9.13, 9.28

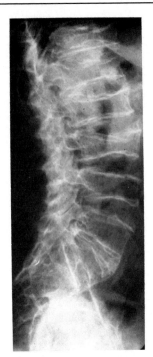

Fig. 9.20   N.R.—35 years—Lateral lumbar spine
Again there is gross platyspondyly, but with much irregularity of
surfaces.
*Same patient : Figs 9.7, 9.11, 9.15, 9.23, 9.24, 9.30, 9.35, 9.39, 9.41, 9.43,
9.45*

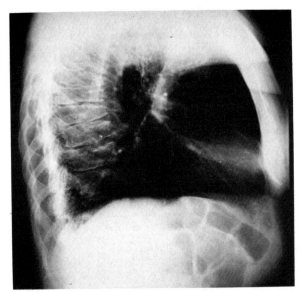

Fig. 9.21   S.M.—39 years—Lateral thoracic spine
There is irregular platyspondyly with secondary wedging of
vertebrae.
Same patient : Figs 9.9, 9.10, 9.22, 9.33, 9.46

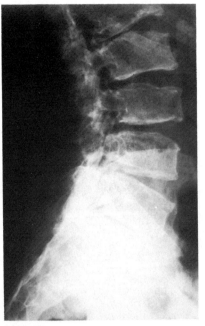

Fig. 9.22   S.M.—39 years—Lateral lumbar spine
The platyspondyly is less severe in the lumbar region, but anterior
defects are obvious, particularly at L2.
Same patient : Fig. 9.21

Hip: 5–11 years

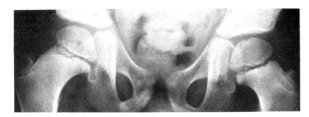

Fig. 9.23 N.R.—5 years 3 months—AP hips
The capital femoral epiphyses are large for his age, but only slightly irregular.
Same patient: Figs 9.7, 9.11, 9.15, 9.20, 9.23, 9.24, 9.30, 9.35, 9.39, 9.41, 9.43, 9.45

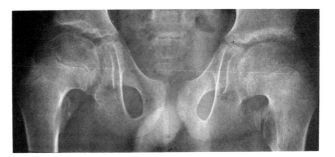

Fig. 9.24 N.R.—9 years—AP hips (same patient as in Fig. 9.23)
The capital femoral epiphyses are large, and both they and the acetabula are markedly irregular with osteoporosis. The adjacent metaphyses are also broadened.

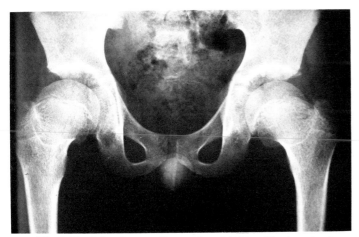

Fig. 9.25 S.A.—9 years 10 months—AP hips (Brother of patient in Fig. 9.29 and child in Fig. 9.31)
A less severe example than Figure 9.24, but the capital femoral epiphyses are large.
Same patient: Fig. 9.37

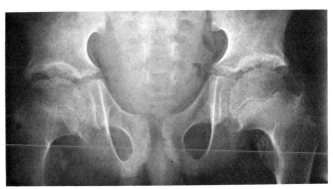

Fig. 9.26 D.C.R.—11 years—AP hips
Large capital femoral epiphyses with horizontal acetabular roofs, both being irregular and osteoporotic.
Same patient: Figs 9.16, 9.18

Hip: 12 years–Adult

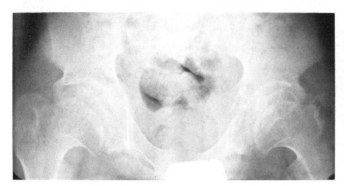

Fig. 9.27   M.C.—12 years 3 months—AP hips
There are enormous capital femoral epiphyses with osteoporosis.
There is some coxa vara also.
Same patient: Figs 9.32, 9.42

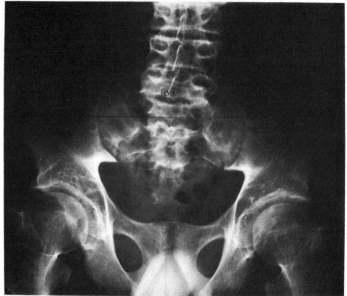

Fig. 9.28   C.R.—16 years 3 months—AP hips
There is only some smooth enlargement of the femoral heads.
Same patient: Figs 9.8, 9.13, 9.19

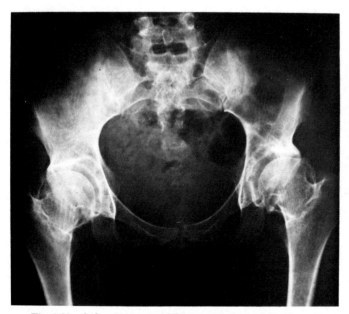

Fig. 9.29   A.A.—20 years—AP hips (Sister of child in Fig. 9.25
and child in Fig. 9.31)
There are obvious signs of degenerative arthritis by this age, with
osteophytic lipping, joint space narrowing and sclerosis at the lateral
acetabular margin.

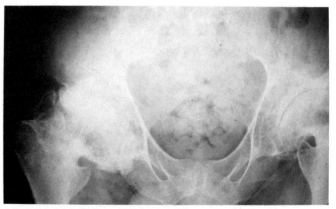

Fig. 9.30   N.R.—35 years—AP hips
There is almost total absence of the joint spaces and a gross degree of
degenerative change.
*Same patient: Figs 9.7, 9.11, 9.15, 9.20, 9.23, 9.24, 9.35, 9.39, 9.41, 9.43,
9.45*

Lower limb: 7 years–Adult

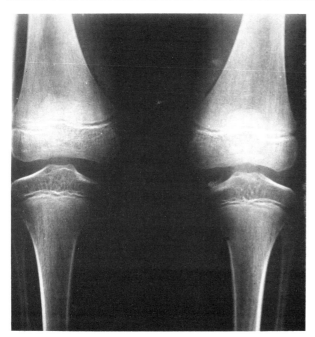

Fig. 9.31 G.A.—7 years 3 months—AP knees (brother of child in Fig. 9.25 and patient in Fig. 9.29)
The condylar surfaces are somewhat flattened and irregular.
Same patient: Figs 9.36, 9.40

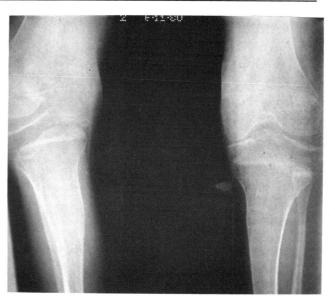

Fig. 9.32 M.C.—12 years 3 months—AP knees
There is epiphyseal irregularity and deepening of the intercondylar notches, but the main signs are of generalised osteoporosis, with thin cortices and bending of bone.
Same patient: Figs 9.27, 9.42

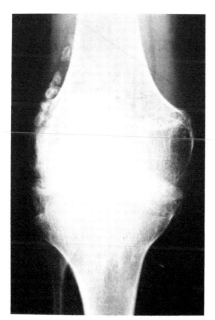

Fig. 9.33 S.M.—39 years—AP right knee
There are gross degenerative changes with loose bodies in the joint.
Same patient: Figs 9.9, 9.10, 9.21, 9.22, 9.46

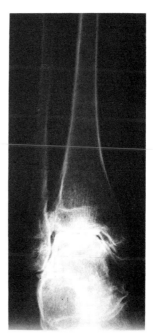

Fig. 9.34 E.D.—32 years—AP right ankle
There are some degenerative changes with osteoporosis and probable fusion of the inferior tibiofibular joint.
Same patient: Figs 9.14, 9.38, 9.44

Upper limb: 14 years–Adult

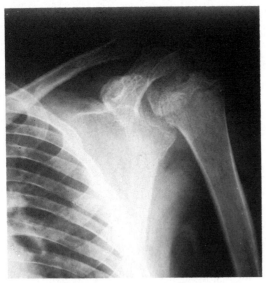

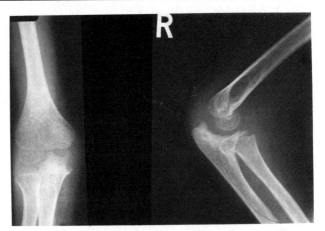

Fig. 9.36   G.A.—7 years 3 months—AP and lateral right elbow
There is similar enlargement of epiphyses with osteoporosis, and also some irregularity of joint surfaces.
Same patient: Figs 9.31, 9.40

Fig. 9.35   N.R.—14 years 6 months—AP left shoulder
The upper humeral epiphysis is much enlarged and there is adjacent osteoporosis.
Same patient: Figs 9.7, 9.11, 9.15, 9.20, 9.23, 9.24, 9.30, 9.39, 9.41, 9.43, 9.45

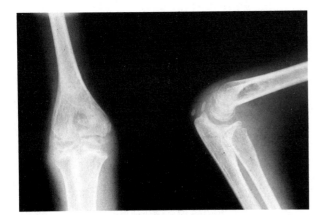

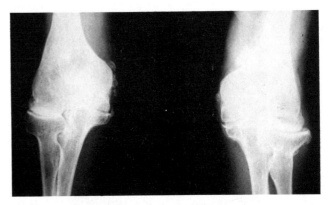

Fig. 9.38   E.D.—32 years 6 months—AP both elbows
There is bony enlargement, loss of joint space, osteophytic lipping and loose bodies in the joints.
Same patient: Figs 9.14, 9.34, 9.44

Fig. 9.37   S.A.—9 years 10 months
Similar to Figure 9.36.
Same patient: Fig. 9.25

Hand: 5–12 years

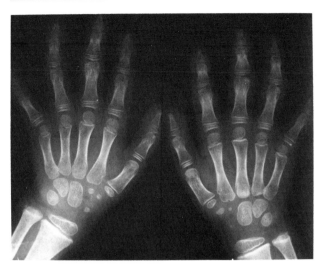

Fig. 9.39   N.R.—5 years 9 months—PA hands
There is some enlargement of the metacarpal and phalangeal
epiphyses and metaphyses, but this is not yet very marked. There is
no associated soft tissue swelling and bone maturation is not delayed.
*Same patient: Figs 9.7, 9.11, 9.15, 9.20, 9.23, 9.24, 9.30, 9.35, 9.39, 9.41,
9.43, 9.45*

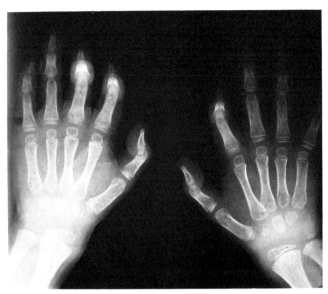

Fig. 9.40   G.A.—7 years 3 months—PA hands
Finger contractures are developing. The metacarpal epiphyses are
curiously square, and there is osteoporosis around the joints. Bone
maturation is delayed at the wrists.
Same patient: Figs 9.31, 9.36

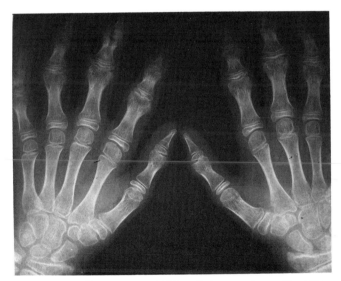

Fig. 9.41   N.R.—10 years—PA hands (same patient as Fig. 9.39)
The bony enlargement of epiphyses and metaphyses (particularly of
the middle phalanges) is obvious. Osteoporosis is present but there are
no soft tissue swellings nor erosions, as would be seen in juvenile
rheumatism.

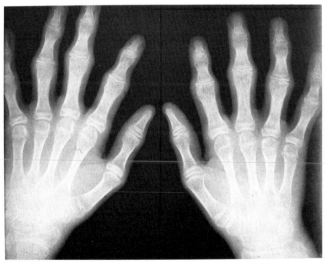

Fig. 9.42   M.C.—12 years 3 months—PA hands
A similar picture to Figure 9.41 at a later stage.
Same patient: Figs 9.27, 9.32

Hand: 14 years–Adult

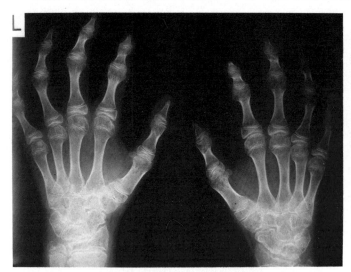

Fig. 9.43   N.R.—14 years 3 months—PA hands
There is enlargement and irregularity of all epiphyses, and the
adjacent metaphyses are also enlarged and osteoporotic. Degenerative
changes in the carpus and wrist joint are apparent.
*Same patient: Figs 9.7, 9.11, 9.15, 9.20, 9.23, 9.24, 9.30, 9.35, 9.39, 9.41,
9.43, 9.45*

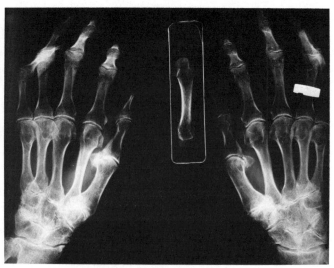

Fig. 9.44   E.D.—32 years—PA hands
There is narrowing or fusion of all joint spaces with osteoporosis:
(normal bone inset). The bony enlargement around the
metacarpophalangeal joints is particularly obvious.
Same patient: Figs 9.14, 9.34, 9.38

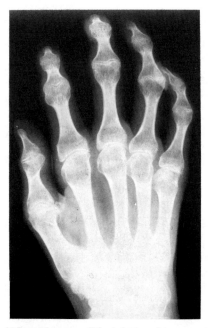

Fig. 9.45   N.R.—35 years—PA right hand (same patient as in
Fig. 9.43)
There is enormous bony enlargement around all hand joints, but no
soft tissue swelling. The osteoporosis is confined to the areas around
the joints. Some finger contracture is present.

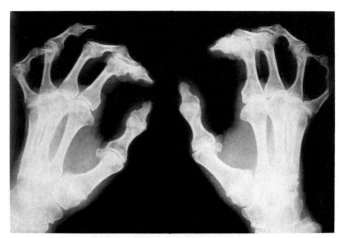

Fig. 9.46   S.M.—39 years—PA hands
There are severe contractures of both hands with marked bony
enlargement around all joints.
Same patient: Figs 9.9, 9.10, 9.21, 9.22, 9.33.

Section Four
METAPHYSEAL DISORDERS

Chapter Ten
METAPHYSEAL CHONDRODYSPLASIA—
Type Schmid

Probably the commonest of the generalised metaphyseal disorders, principally affecting the upper end of the femora, associated with coxa vara and some shortness of stature.

Inheritance

Autosomal dominant.

Frequency

Possible prevalence of 3–6 per million population.

Clinical features

Facial appearance—normal.
Intelligence—normal.
Stature—may be normal, but most are somewhat short, due principally to shortening of the lower limbs.
Presenting feature/age—usually a waddling gait in infancy (coxa vara), subsequently lumbar lordosis and shortness of stature.
Deformities—coxa vara, often with lateral bowing of the femora and genu/tibia varum; sometimes genu valgum.
Range of severity—wide: some patients have involvement of all metaphyses with severe progressive coxa vara, others only minimal findings.
Associated anomalies—none.

Radiographic features

Skull—probably normal.
Spine—may be normal, but in some cases the vertebral bodies are oval in the lateral view, this persisting long after infancy.
Thorax—normal.
Pelvis—normal.
Limb epiphyses—normal.

Limb metaphyses—expanded, irregular, perhaps cupped (simulating rickets).
Upper femoral metaphysis—medial beaking, irregularity and widening. In more severe cases the developing coxa vara is associated with a separated triangular fragment on the inferior aspect of the neck.

Bone maturation

Normal.

Biochemistry

No Ca/P imbalance.

Differential diagnosis

Other metaphyseal chondrodysplasias, in particular Type McKusick, but there the upper femur is less involved, and most other metaphyses more so, including the phalanges.
'*Rickets*'—always needs differentiating by chemical means. Radiographic signs of rickets include apparent widening of the epiphyseal plates and there is often associated osteoporosis, as well as the cupped, irregular metaphyses.

Progress/complications

Apart from some shortness of stature and residual coxa vara with lower-limb deformity, there are no complications. Since the epiphyses are not involved, premature degenerative arthritis is not a feature. Radiological signs in the metaphyses disappear after epiphyseal fusion.

REFERENCES

Kaufmann H J 1973 Intrinsic Diseases of Bones. Progress in Pediatric Radiology 4. W B Saunders, Philadelphia
Kozlowski K 1976 Metaphyseal and spondylometaphyseal chondrodysplasias. Clinical Orthopaedics and Related Research 114;83–93

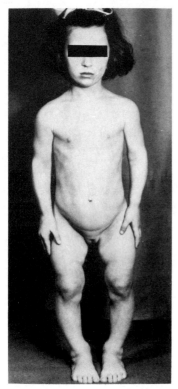

Figs 10.1 and 10.2   Little abnormality of the head, trunk or upper limbs, but the lower segment is short compared with the trunk and the thighs are far apart (due to the coxa vara), associated with a waddling gait. There is also lateral bowing of the tibiae (tibia varum).

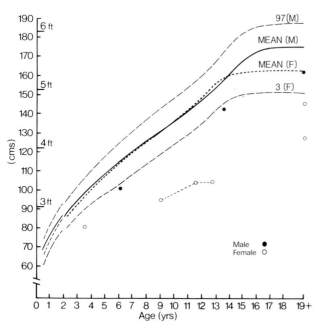

Fig. 10.3   Metaphyseal chondrodysplasia (Schmid)—percentile height.

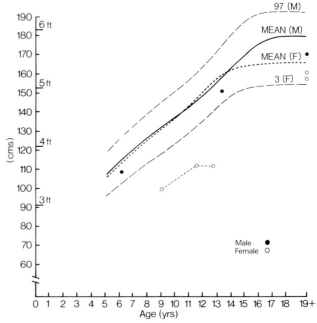

Fig. 10.4   Metaphyseal chondrodysplasia (Schmid)—percentile span.

Skull: 2 years–Adult

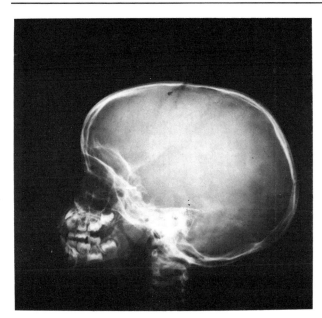

Fig. 10.5 J.B.—2 years 2 months—Lateral skull
Normal.
Same patient: Figs 10.9, 10.10, 10.14, 10.24, 10.36

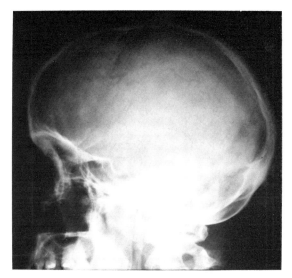

Fig. 10.6 G.B.—28 years—Lateral skull (father of child in Fig. 10.5)
Normal.
Same patient: Figs 10.38, 10.39, 10.43, 10.47

Spine: 1–2 years

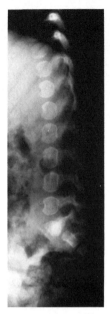

Fig. 10.7   P.R.—18 months—Lateral spine
The vertebral bodies are tall and have an oval configuration.
Same patient: Figs 10.13, 10.19, 10.29, 10.40

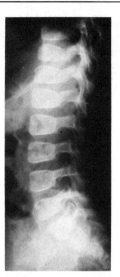

Fig. 10.8   H.H.—2 years—Lateral spine
The vertebral bodies have an immature, oval configuration.
Same patient: Figs 10.16, 10.25

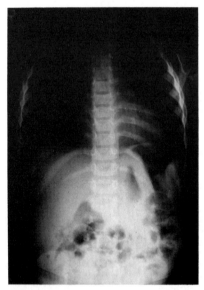

Fig. 10.9   J.B.—2 years 2 months—AP spine (including ribs)
Entirely normal appearances.
Same patient: Figs 10.5, 10.10, 10.14, 10.24, 10.36

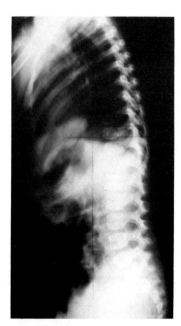

Fig. 10.10   J.B.—2 years 2 months—Lateral spine (same patient as in Fig. 10.9)
Slightly oval, immature configuration of the vertebral bodies.

Spine: 3–4 years

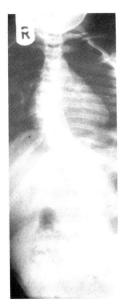

Fig. 10.11   M.S.—3 years—AP spine
There is a scoliosis convex to the right in the dorsal region and to the left in the lumbar region. The vertebrae are otherwise normal.
Same patient: Fig. 10.15

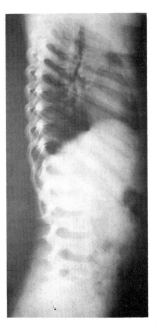

Fig. 10.12   E.S.—4 years 6 months—Lateral spine (first cousin of patient in Fig. 10.5)
Normal appearance.
Same patient: Figs 10.26, 10.27, 10.30

Pelvis: 1–3 years

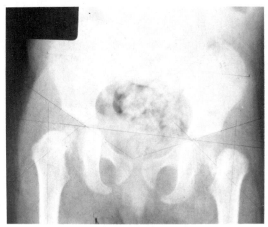

Fig. 10.13   P.R.—1 year 3 months—AP pelvis
Normal shape of ilia, ischia and pubic bones. The femoral heads are
small for this age and rounded. There is widening of the epiphyseal
plate, and the metaphyses are irregular. Coxa vara is most marked on
the left but on both sides a separated inferior fragment of the femoral
neck can be seen.
Same patient: Figs 10.7, 10.19, 10.29, 10.40

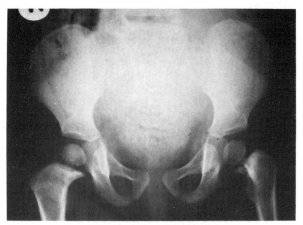

Fig. 10.14   J.B.—2 years 1 months—AP pelvis
The pelvis is normal. Both upper femoral metaphyses are abnormal,
the right showing a medial 'beak' and the left a separated fragment.
The capital femoral epiphyses are unusually large and round for this
age.
Same patient: Figs 10.5, 10.9, 10.10, 10.24, 10.36

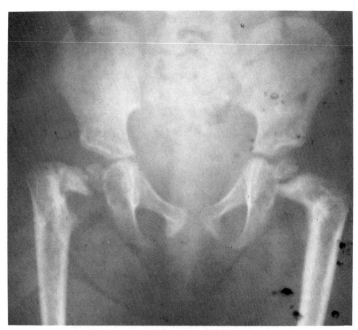

Fig. 10.15   M.S.—3 years—AP pelvis
The pelvis is normal. There are marked bilateral coxa vara
deformities with angulation. On the right there is deficient
ossification suggesting an un-united fracture. The metaphyses are
irregular and there is widening of the epiphyseal plates, similar to that
seen in rickets.
Same patient: Fig. 10.11

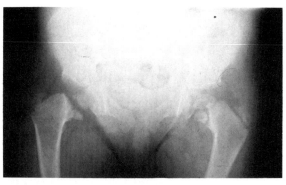

Fig. 10.16   H.H.—3 years 6 months—AP pelvis
The pelvis is normal. There is bilateral coxa vara with the
characteristic triangular fragments of bone infero-medially. The
metaphyses are irregular.
Same patient: Figs 10.8, 10.25

Pelvis: 5–9 years

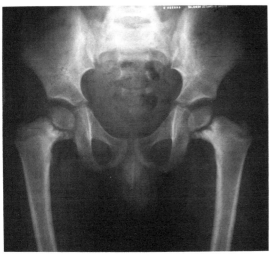

Fig. 10.17 S.B.—5 years 2 months—AP pelvis
The pelvis is normal. There are symmetrical metaphyseal irregularities with coxa vara. The femoral heads are relatively large.

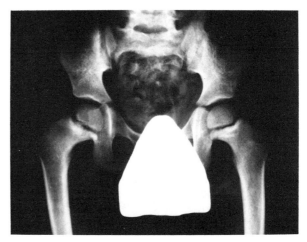

Fig. 10.18 S.G.—6 years 6 months—AP pelvis
The pelvis is normal. There is mild coxa vara and minimal metaphyseal irregularity.
Same patient: Figs 10.21, 10.31, 10.37, 10.41, 10.44, 10.46

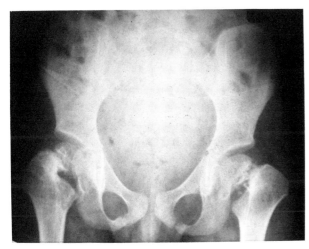

Fig. 10.19 P.R.—6 years 9 months—AP pelvis
The pelvis is normal. The metaphyseal irrregularity has increased (see Fig. 10.13). The femoral necks are virtually absent. The capital femoral epiphyses are large and well formed.
Same patient: Figs 10.7, 10.13, 10.29, 10.40

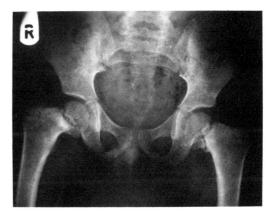

Fig. 10.20 N.R.—9 years—AP pelvis
The pelvis is normal. There is bilateral coxa vara and metaphyseal irregularity. The femoral necks are short with infero-medial triangular fragments and there is also lateral bowing of the shafts.
Same patient: Fig 10.42

Pelvis : 11 years–Adult

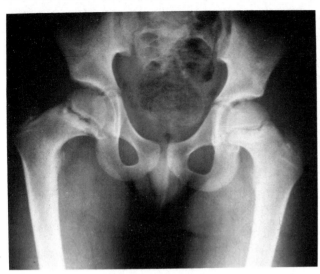

Fig. 10.21   S.G.—11 years 4 months—AP pelvis
The pelvis and capital femoral epiphyses are normal. There is
metaphyseal flaring with mild irregularity and some widening of the
epiphyseal plate, more marked on the right.
Same patient: Figs 10.18, 10.31, 10.37, 10.41, 10.44, 10.46

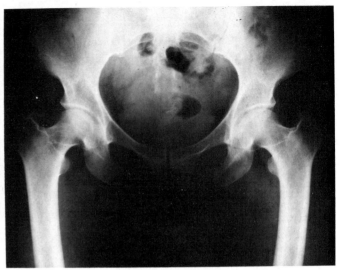

Fig. 10.22   C.S.—41 years—AP pelvis (mother of child in Fig.
10.12)
The pelvis is normal. There is bilateral coxa vara, and mild lateral
bowing of the femoral shafts.
Same patient: Fig. 10.35

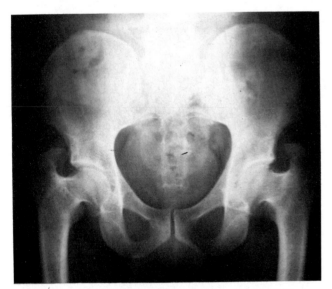

Fig. 10.23   D.B.—37 years—AP pelvis (father of child in Fig.
10.17)
There is marked coxa vara with large greater trochanters, impinging
on the pelvis and greatly limiting abduction.

Lower limb: 2–4 years

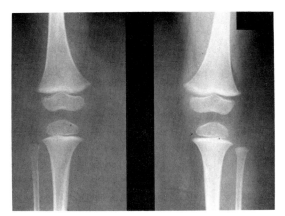

Fig. 10.24 J.B.—2 years 2 months—AP both knees
No abnormality.
Same patient: Figs 10.5, 10.9, 10.10, 10.14, 10.36

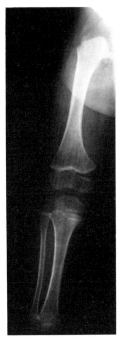

Fig. 10.25 H.H.—2 years 6 months—AP lower limb
There is mild metaphyseal flaring and slight irregularity, most pronounced in the proximal femur.
Same patient: Figs 10.8, 10.16

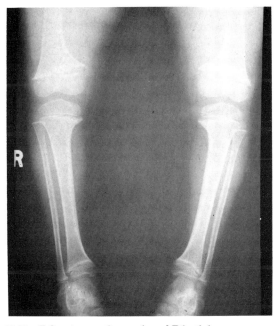

Fig. 10.26 E.S.—4 years 6 months—AP both legs
There is lateral femoral and tibial bowing with flared metaphyses and widening of the epiphyseal plates. This is most pronounced medially at the knees.
Same patient: Figs 10.12, 10.27, 10.30

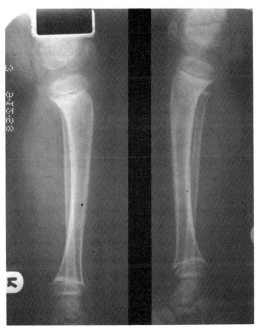

Fig. 10.27 E.S.—4 years 6 months—Lateral both tibiae and fibulae (same patient as in Fig. 10.26)
There is metaphyseal splaying of the proximal and distal tibial metaphyses with minimal irregularity. The epiphyses are normal.

Lower limb: 4–6 years

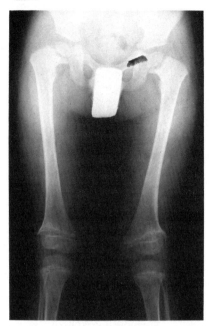

Fig. 10.28   R.A.—4 years 6 months—AP hips and knees
There is genu valgum and no bowing of the shafts of the long bones
here. The metaphyses are splayed and irregular with widening of the
epiphyseal plates. This is especially pronounced laterally at the distal
femoral metaphyses.

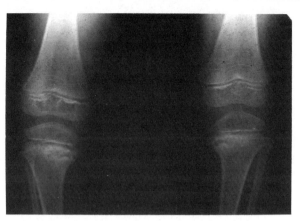

Fig. 10.29   P.R.—5 years 9 months—AP knees
The proximal tibial metaphyses are principally affected on the medial
sides, and appear compressed.
Same patient: Figs 10.7, 10.13, 10.19, 10.40

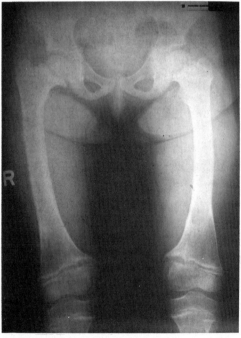

Fig. 10.30   E.S.—6 years 4 months—AP femora and knees
There is lateral bowing of the femoral shafts and bilateral coxa vara.
Metaphyseal splaying and irregularity is present and there is
widening of the epiphyseal plates especially medially at the knee
joints.
Same patient: Figs 10.12, 10.26, 10.27

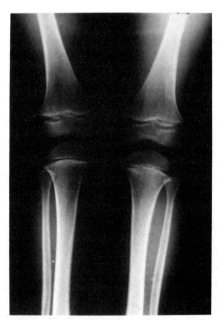

Fig. 10.31   S.G.—6 years 6 months—AP knees
There is mild genu valgum, but the metaphyses appear normal in this
case.
Same patient: Figs 10.18, 10.21, 10.37, 10.41, 10.44, 10.46

Lower limb: 7 years–Adult

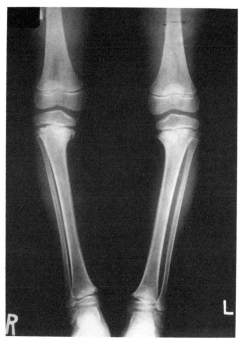

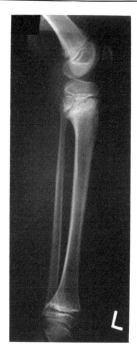

Fig. 10.32   M.R.—7 years 5 months—AP both lower limbs
(childhood radiographs of mother of child in Fig. 10.7)
There is genu varum with medial beaking of the upper tibial shafts.
Metaphyseal irregularity of the proximal and distal tibiae is present.
Same patient: Fig. 10.33

**Fig. 10.33   M.R.—7 years 5 months—Left tibia and fibula
(lateral view)** (same patient as in Fig. 10.32)

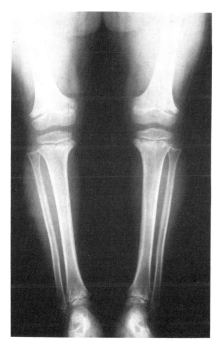

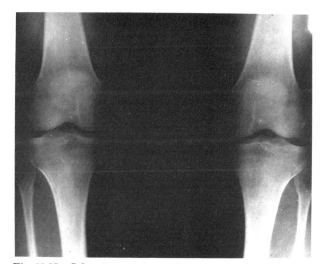

Fig. 10.35   C.S.—41 years 6 months—AP knees
Normal.
Same patient: Fig. 10.22

Fig. 10.34   S.T.—7 years 4 months—AP tibiae and fibulae
There is mild lateral bowing of the femora. Metaphyseal splaying and
iregularity is present and this is associated with abnormal modelling
and flattening of the adjacent epiphyses.
Same patient: Fig. 10.45

Ankle and foot: 2 years–Adult

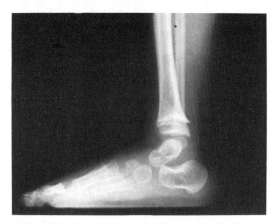

Fig. 10.36   J.B.—2 years 3 months—Lateral right foot and ankle (standing)
The only abnormality is mild metaphyseal splaying of the distal tibia.
Same patient: Figs 10.5, 10.9, 10.10, 10.14, 10.24

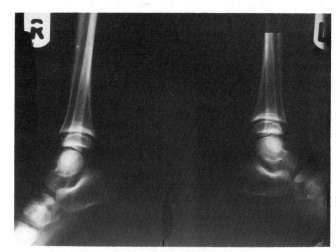

Fig. 10.37   S.G.—6 years 6 months—Both ankles
Minimal metaphyseal irregularity of the distal tibiae with slight widening of the epiphyseal plates.
Same patient: Figs 10.18, 10.21, 10.31, 10.41, 10.44, 10.46

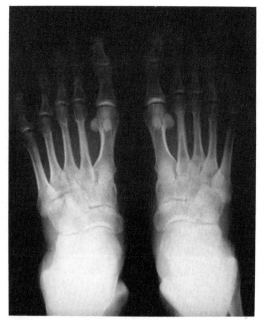

Fig. 10.38   G.B.—28 years—PA feet
Normal.
Same patient: Figs 10.6, 10.39, 10.43, 10.47

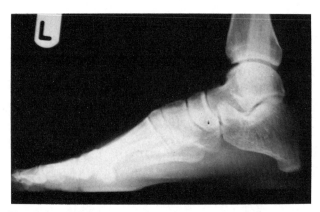

Fig. 10.39   G.B.—28 years—Lateral foot (standing) (same patient as in Fig. 10.38)
Normal.

Upper limb : 5 years–Adult

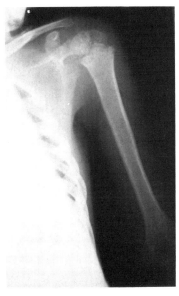

Fig. 10.40 P.R.—5 years 9 months—AP shoulder and humerus
There is metaphyseal irregularity of the proximal humerus but the epiphysis and shaft modelling are normal.
Same patient: Figs 10.7, 10.13, 10.19, 10.29

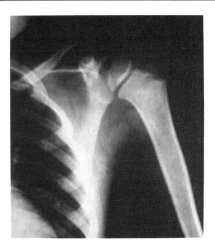

Fig. 10.41 S.G.—6 years 6 months—AP shoulder
Normal.
Same patient: Figs 10.18, 10.21, 10.31, 10.37, 10.44, 10.46

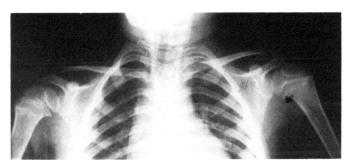

Fig. 10.42 N.R.—7 years—AP both shoulders
There is mild metaphyseal flaring of the proximal humeri with some widening of the epiphyseal plates.
Same patient: Fig. 10.20

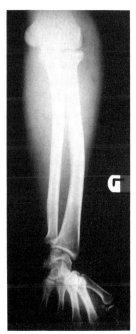

Fig. 10.43 G.B.—28 years—AP left forearm
Normal.
Same patient: Figs 10.6, 10.38, 10.39, 10.47

Hand: 6 years–Adult

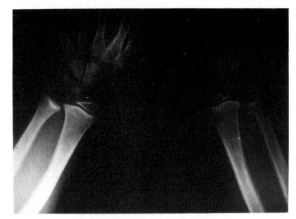

Fig. 10.44   S.G.—6 years 6 months—PA wrists
There is mild cupping of the distal ulnar metaphyses and the distal
radial epiphyses are slightly flattened and deficient medially.
Same patient: Figs 10.18, 10.21, 10.31, 10.37, 10.41, 10.46

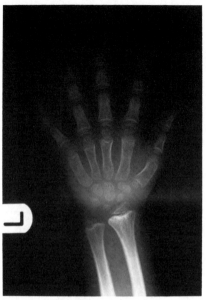

Fig. 10.45   S.T.—7 years 4 months—PA left hand
There is mild metaphyseal irregularity of both distal radius and ulna
and cupping of the metaphyses of the phalanges, but this is mild.
Same patient: Fig. 10.34

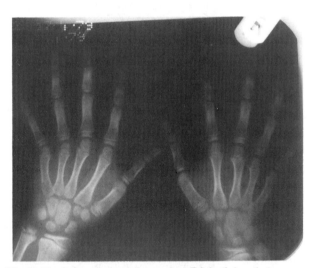

Fig. 10.46   S.G.—11 years 2 months—PA both hands (same
patient as in Fig. 10.44)
Normal.

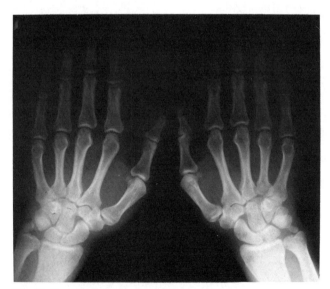

Fig. 10.47   G.B.—28 years—PA both hands
Normal, apart from separation of both ulnar styloid processes.
Same patient: Figs 10.6, 10.38, 10.39, 10.43

Chapter Eleven
METAPHYSEAL CHONDRODYSPLASIA—
Type McKusick

Less common than Type Schmid, the metaphyseal lesions are more widespread, and the skeletal disorder is accompanied by fine, sparse hair and excessive joint laxity.

Inheritance

Autosomal recessive.

Frequency

Probable prevalence is less than 3 per million population.

Clinical features

Facial appearance—normal apart from fine sparse hair, including that of the eyebrows and eyelashes.

Intelligence—normal.

Stature—reduced, due to limb shortening, and of a greater severity than Type Schmid (is reported as ranging from 105 to 157 cm—3′5″ to 5′2″).

Presenting feature/age—can be recognised at birth, with sparse hair and short limbs, but more probably in infancy or early childhood, with shortness of stature.

Deformities—Probably none, although there may be some lateral bowing in the lower limbs.

Associated anomalies—there is an association with intestinal malabsorption, with Hirschsprung's disease, and with chronic neutropenia and lymphopenia, giving an abnormality of cellular immunity—smallpox vaccination, for example, has been fatal. Excessive joint laxity may be a feature.

Radiographic features

Skull—slight involvement with large fontanelles.

Spine—in early childhood there are oval-shaped vertebral bodies on the lateral view and sometimes odontoid hypoplasia.

Thorax—widening of the ribs anteriorly, with cupping of osteochondral junctions. There may be anterior angulation of the sternum.

Limb epiphyses—normal.

Limb metaphyses—sclerosis and scalloping of the margin. This is usually particularly obvious around the knee joint.

Hands—marked involvement of the hands is one of the main characteristics of MCD Type McKusick. There are cup-shaped deformities of the phalangeal metaphyses, with complementary delta-shaped epiphyses.

Bone maturation

Probably normal, but no definite data.

Biochemistry

No Ca/P imbalance.

Differential diagnosis

Other metaphyseal chondrodysplasias, in particular Type Schmid, but here it is principally the upper ends of the femora which are affected; it is of autosomal dominant inheritance and there are no associated malabsorption or immunity problems.

There are other radiographically-distinct metaphyseal chondrodysplasias in which malabsorption and immunity problems occur, but in these the hands are unaffected.

Progress/complications

Shortness of stature, but skeletal complications are not usually a feature. The immunity problems can be serious.

REFERENCES

Kaufmann H J (ed) 1973 Intrinsic diseases of bones. Progress in Pediatric Radiology 4. Philadelphia
Kozlowski K 1976 Metaphyseal and spondylometaphyseal chondrodysplasias. Clinical Orthopedics and Related Research 114:83–93
Virolainen M, Savilahti E, Kaitila I, Perheentupa J 1978 Cellular and humoral immunity in cartilage-hair hypoplasia. Pediatric Research 12/10:961–966

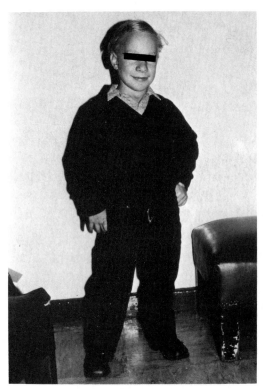

Fig. 11.1   Some disproportionate limb shortening, and the fine, sparse hair is obvious (age 5½ years).

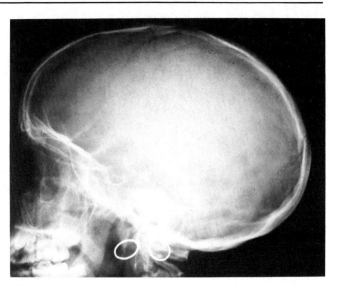

Fig. 11.2   K.Y.—8 years—Lateral skull
Normal.
Same patient: Figs 11.3, 11.4, 11.7, 11.10, 11.11, 11.15

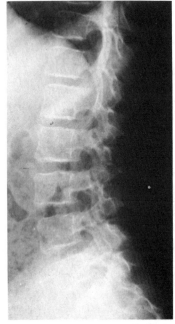

Fig. 11.3   K.Y.—8 years—Lateral spine (same patient as in Fig. 11.2)
Normal.

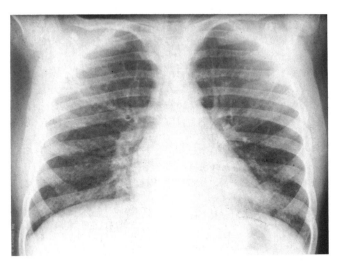

Fig. 11.4   K.Y.—8 years—AP chest (same patient as in Fig. 11.2)
Normal.

Pelvis and hip: 2–8 years

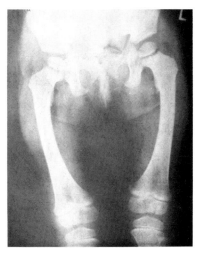

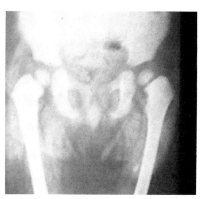

Fig. 11.6   Unknown—2 years—AP hips
There is mild coxa vara. The epiphyseal plates appear wide but there is no other abnormality.
Same patient: Fig. 11.8

Fig. 11.5   C.S.—2 years—AP hips and femora
There is lateral bowing of both femoral shafts, with minimal widening of the proximal femoral epiphyseal plates. The distal femoral metaphyses are within normal limits at this age.

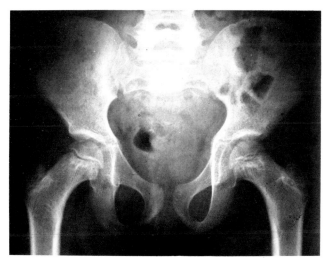

Fig. 11.7   K.Y.—8 years—AP pelvis
The pelvis and capital femoral epiphyses are normal but there has been some irregular growth of the upper femoral metaphysis.
Same patient: Figs 11.2–11.4, 11.10, 11.11, 11.15

Lower limb: 2–8 years

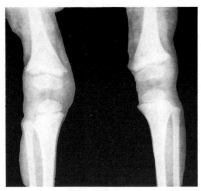

Fig. 11.8   Unknown—2 years—AP both knees
There is overgrowth of the medial femoral condyles with some metaphyseal flaring and irregularity. The tibial metaphyses are also irregular and deficient laterally.
Same patient : Fig. 11.6

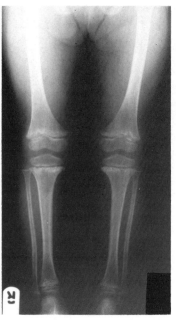

Fig. 11.9   A.S.—5 years 7 months—AP both lower limbs
There is mild metaphyseal flaring and a 'scalloped' outline.
Same patient : Figs 11.14, 11.16

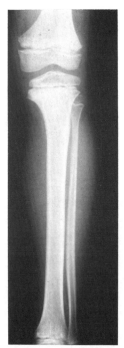

Fig. 11.10   K.Y.—8 years—AP tibia and fibula
There is mild metaphyseal flaring and 'scalloping' at both knee and ankle.
Same patient : Figs 11.2–11.4, 11.7, 11.11. 11.15

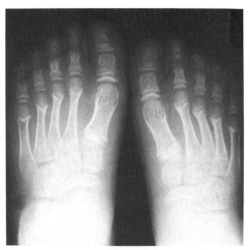

Fig. 11.11   K.Y.—8 years—DP both feet (same patient as in Fig. 11.10)
Cone-shaped epiphyses are present at the proximal phalanges of the second, third and fourth toes and the lateral four metatarsals show cupping of the metaphyses.

Upper limb: 2–5 years

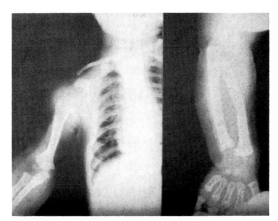

Fig. 11.12   D.S.—2 years—AP left upper limb (brother of child in Fig. 11.5)
There is generalised shortening of the long bones with metaphyseal flaring, and prominence of the deltoid tuberosity. The radius and ulna show mild bowing and there is slight metaphyseal irregularity of the distal radius and ulna.
Same patient: Fig. 11.13

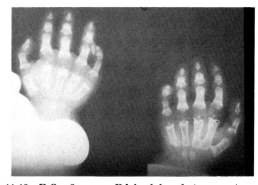

Fig. 11.13   D.S.—2 years—PA both hands (same patient as in Fig. 11.12)
Short phalanges and metacarpals with cone-shaped and delta-shaped epiphyses. Small, pointed terminal phalanges.

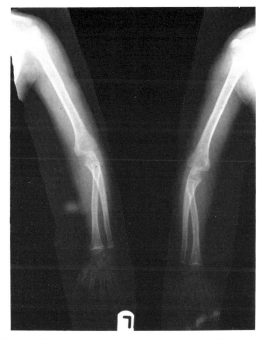

Fig. 11.14   A.S.—5 years 7 months—Both upper limbs
Modelling deformities are minimal. There is mild metaphyseal irregularity involving only the distal radial metaphysis. There are streaks of increased density extending towards the diaphyses of the radii.
Same patient: Figs 11.9, 11.16

Hand : 8–10 years

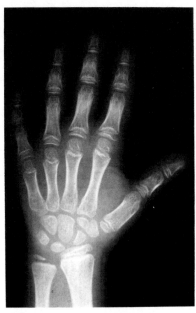

Fig. 11.15   K.Y.—8 years—PA left hand
There are delta-shaped epiphyses of the proximal and middle
phalanges and mild metaphyseal irregularity throughout.
Same patient : Figs 11.2–11.4, 11.7, 11.10, 11.11

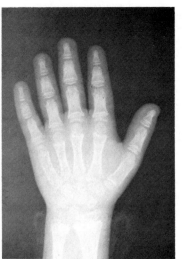

Fig. 11.16   A.S.—9 years 6 months—PA left hand
Characteristic delta-shaped epiphyses are present in the proximal and
middle rows of the phalanges. Very mild metaphyseal irregularity is
present in the distal radius and ulna and the distal phalanges. The
metaphyses of the middle and proximal phalanges show indentations
corresponding to the epiphyseal deformity.
Same patient : Figs 11.9, 11.14

Fig. 11.17   P.D.—10 years—PA left hand (brother of child in
Fig. 11.18)
Very mild metaphyseal irregularities are present; there is Kirner's
deformity of the little finger. Delta-shaped epiphyses are seen in the
proximal phalanges. There is apparent widening of the epiphyseal
plates of the middle and distal phalanges and all the hand bones show
some shortening.

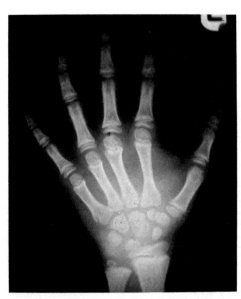

Fig. 11.18   J.D.—10 years—PA left hand (brother of child in Fig.
11.17)
Delta-shaped epiphyses are present in all the phalanges. Modelling of
the long bones is normal.

Chapter Twelve
METAPHYSEAL CHONDRODYSPLASIA—Type Jansen

Short stature, principally of the limbs and of their distal segments; striking bulbous expansion of metaphyseal areas with irregular ossification.

Inheritance

Most are sporadic, but possibly autosomal dominant.

Frequency

Extremely rare—under 0.1 per million prevalence.

Clinical features

Facial appearance—hypertelorism, perhaps exophthalmos and hyperplasia of the fronto-nasal and superciliary areas.

Intelligence—probably normal.

Stature—retarded growth and short stature, around 125 cm (4′ 1″), reported.

Presenting feature/age—infancy, with short stature and deformed limbs.

Deformities—the greater part of the shafts of the long bones appear normal, but there is a tendency to bend at the diaphyseal-metaphyseal junction, particularly in the lower limbs.

Associated anomalies—deafness has been reported, also joint contractures and clubfoot.

Radiographic features

Skull—probably affected, with brachycephaly, short base, small air sinuses and some sclerosis.

Spine—normal (but very little information available).

Limb epiphyses—normal outline, but may be late in appearing.

Limb metaphyses—striking appearance of expanded bulbous area, irregularly mottled. In early years a thin outline of bone surrounds the expanded area; some, particularly the metacarpals, show marked cupping. On the diaphyseal side of the metaphysis there is dense, sclerotic bone, and the epiphyseal plate is widened.

Bone maturation

Delayed.

Biochemistry

Not known.

Differential diagnosis

This condition is unlike any other, being the most severe of the metaphyseal chondrodysplasias.

Progress/complications

Little is known, but as is usual in metaphyseal lesions, there is a tendency to heal, but leaving residual limb deformity.

REFERENCES

Gordon S L, Varano L A, Alandete A, Maisels M J 1976 Jansen's metaphyseal dysostosis. Pediatrics 58/4:556–560

Kaufman H J (ed) 1973 Intrinsic diseases of bones. Progress in Pediatric Radiology 4. Philadelphia

Kozlowski K 1976 Metaphyseal and spondylometaphyseal chondrodysplasias. Clinical Orthopaedics and Related Research 114:83–93

Nazara Z, Hernandez A, Corona-Rivera E 1981 Further clinical and radiological features in metaphyseal chondrodysplasia Jansen type. Radiology 140/3:697–700

Skull and trunk : 2 years

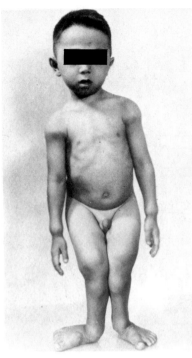

Fig. 12.1   The face and head appear normal in this child, but there is limb deformity due to bending at the diaphyseal-metaphyseal junctions.

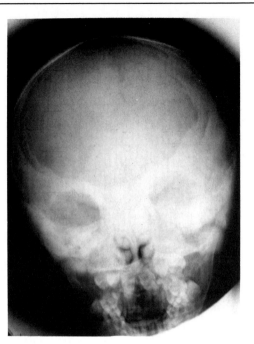

Fig. 12.2   R.D.—2 years—AP skull
There is sclerosis of the base with obliteration of the nasal sinuses.
Same patient : Figs 12.3–12.5, 12.8, 12.10, 12.13

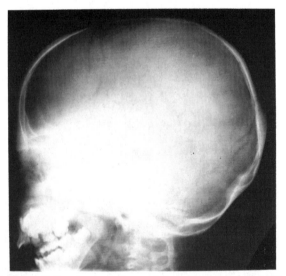

Fig. 12.3   R.D.—2 years—Lateral skull (same patient as in Fig. 12.2)
Similar to Figure 12.2.

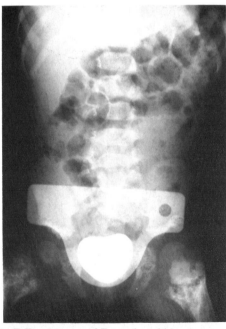

Fig. 12.4   R.D.—2 years—AP trunk and hips (same patient as in Fig. 12.2)
The end-plates are a little irregular, but the striking feature is the increase of interpedicular distances. The upper ends of the femora show irregular ossification of the metaphyses.

Pelvis and femora : 2–7 years

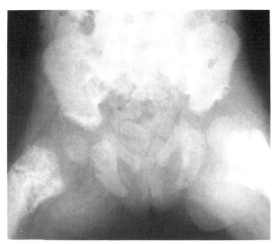

Fig. 12.5   R.D.—2 years—AP pelvis
There is a little irregularity of the acetabular roofs, but the main
defect is of the metaphyseal region, with defective ossification
extending down into the diaphyses.
Same patient; Figs 12.2–12.4, 12.8, 12.10, 12.13

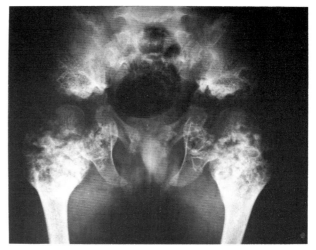

Fig. 12.6   T.H.L.—7 years—Pelvis
The iliac wings are small and square with widening of the symphysis
pubis and patchy ossification in the adjacent pubic bones. The
acetabular roofs are irregular and there is bulbous enlargement of the
upper ends of the femora with patchy sclerotic ossification extending
into the femoral diaphyses. The capital femoral epiphyses are normal.
Same patient: Figs 12.7, 12.9, 12.11, 12.14

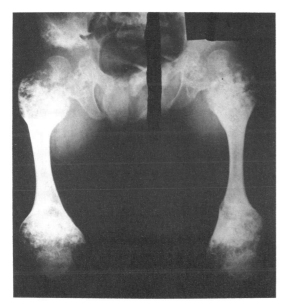

Fig. 12.7   T.H.L.—7 years—AP femora (same patient as in Fig.
12.6)
There is probably shortening of the femora. The diaphyses are of
normal calibre but there is pronounced metaphyseal enlargement
with patchy irregular sclerotic ossification. The distal femoral
epiphyses are abnormally modelled, but have smooth margins and are
of normal density.

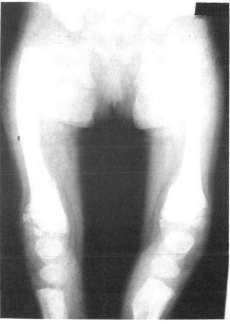

Fig. 12.8   R.D.—7 years—AP femora (same patient as in Fig.
12.5)
This child is less severely affected than that in Figure 12.7, but the
changes are similar with the addition of some lateral femoral bowing.

Lower limb: 7 years

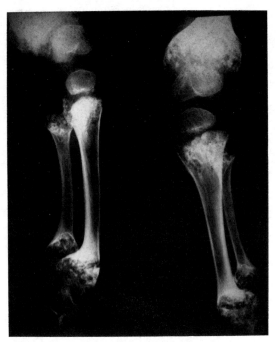

Fig. 12.9   T.H.L.—7 years—Tibiae and fibulae
Both fibulae are short. All metaphyses are enlarged and show patchy
sclerotic ossification, and the bone appears to be soft since there is a
bowing deformity at the lower end. The epiphyses are unusually
rounded.
Same patient: Figs 12.6, 12.7, 12.11, 12.14

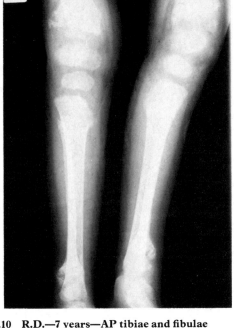

Fig. 12.10   R.D.—7 years—AP tibiae and fibulae
The changes are similar but less marked than in Figure 12.9.
Same patient: Figs 12.2–12.5, 12.8, 12.13

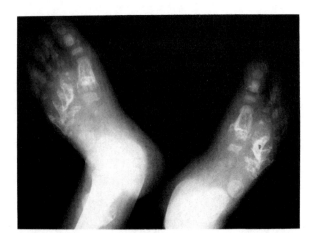

Fig. 12.11   T.H.L.—7 years—PA feet (same patient as in Fig.
12.9)
There is irregular and delayed ossification, particularly of the tarsals
and metatarsals. The phalangeal epiphyses are large and round.

Upper limb: 1–7 years

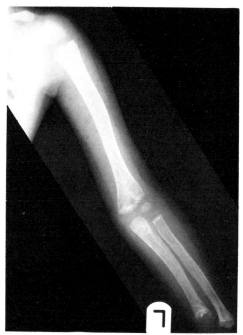

Fig. 12.12   G.S.—1 year 6 months—AP upper limb
The principal metaphyseal changes are in the lower humerus and
lower radius and ulna.

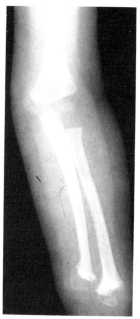

Fig. 12.13   R.D.—2 years—AP forearm
There is irregular ossification, particularly at the lower metaphyses,
and a faint rim of ossification (seen best at the distal ulna) which
appears to outline the poorly ossified bulbous metaphysis. Elsewhere
bone modelling appears normal and there is only minor irregularity of
the distal humeral metaphysis.
Same patient: Figs 12.2–12.5. 12.8, 12.10

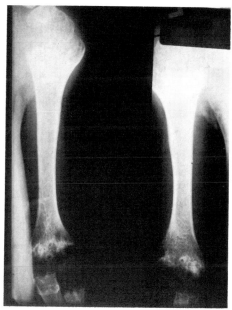

Fig. 12.14   T.H.L.—7 years—AP humeri
There is probably overall shortening and widened metaphyses with
irregular ossification extending towards the diaphyses. There is also
irregularity of the metaphyseal margins, and the epiphyses are
relatively large, but otherwise normal.
Same patient; Figs 12.6, 12.7, 12.9, 12.11

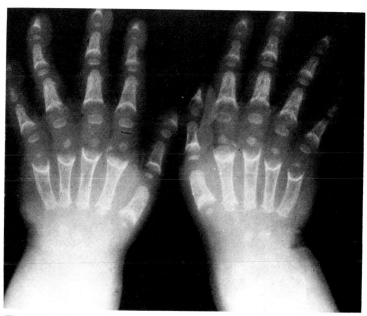

Fig. 12.15   Unknown—2 years 6 months—PA hands
The expanded and cupped metaphyses are well shown, as in the wide
gap between epiphyses and metaphyses. Some of the epiphyses are
strikingly round rather than flat.

Chapter Thirteen
METAPHYSEAL CHONDRODYSPLASIA
with malabsorption and neutropenia

The presenting feature is congenital pancreatic insufficiency and the bone lesions (usually of the upper femur) are discovered only incidentally. Neutropenia, anaemia and thrombocytopenia may be serious.

Inheritance

Likely to be autosomal recessive.

Frequency

Not known, but rare. Likely prevalence less than 3 per million population.

Clinical features

Facial appearance—normal.
Intelligence—normal.
Stature—probably somewhat reduced in view of the general failure to thrive.
Presenting feature/age—failure to thrive in infancy, with diarrhoea and bulky offensive stools. Neutropenia may be present at this age, but may be cyclic.
An alternative presentation in the neonatal period is respiratory distress associated with shortened ribs and reduced size of thorax.
It is probable that later onset of the pancreatic/blood disease will be associated with later onset of the skeletal disorder.
Deformities—coxa vara.
Associated anomalies—no further anomalies known.

Radiographic features

Skull—normal.
Spine—may show irregularity of the upper vertebral plates.
Thorax—shortened ribs with expanded anterior ends.
Pelvis—normal.
Long bones—it is chiefly the metaphyseal region of the upper femur which is affected, although there may be sclerosis and irregularity of metaphyses elsewhere. Coxa vara develops and the disturbance of bone formation is in the shape of an inverted triangle extending down the femoral neck: the epiphyseal plate surface of the epiphysis becomes wedge-shaped, growing into the triangular defect in the metaphysis. In time the epiphysis becomes flattened and the femoral head may not be contained within the acetabulum.

Bone maturation

Retarded, by 1–2 years.

Biochemistry and haematology

Low concentrations of both ductular and acinar secretions, sometimes low chymotrypsin in the stools.
Neutropenia is either permanent or cyclic, and associated with pancytopenia.

Differential diagnosis

Other metaphyseal chondrodysplasias, but the pancreatic and haematological findings distinguish this disorder.

Progress/complications

This is related to the pancreatic and haematological problems. Spontaneous remission can occur, but, without treatment, death by the age of 5 years is likely.

REFERENCES

Danks D M, Haslam R, Mayne V, Kaufmann H J, Holtzapple P G 1976 Metaphyseal chondrodysplasia, neutropenia and pancreatic insufficiency presenting with respiratory distress in the neonatal period. Archives of Diseases of Childhood 51:697–702

Shmerling D H, Prader A, Hitzig W H, Giedion A, Hadorn B, Kuhni M 1969 The syndrome of exocrine pancreatic insufficiency, neutropenia, metaphyseal dysostosis and dwarfism. Helv. paediat. Acta 24:547–575

Sutcliffe J, Stanley P 1973 Metaphyseal Chondrodysplasias. In Kaufman H J (ed) Intrinsic diseases of bones. Progress in Pediatric Radiology 4. Karger, Basel, 250–269

Chest and spine: 7 months–1 year

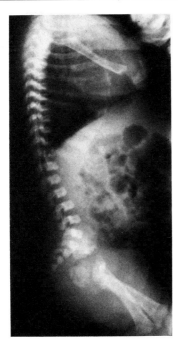

Fig. 13.1 Unknown—1 year—lateral spine
Normal.

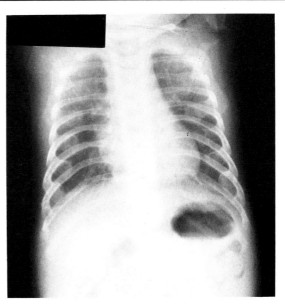

Fig. 13.2 H.M.—7 months—AP chest
The ribs are short, giving rise to a small-volumed thorax. Expanded anterior ends of ribs.
Same patient: Figs 13.3, 13.7

Pelvis: 7 months–12 years

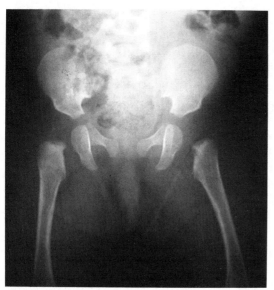

Fig. 13.3   H.M.—7 months—AP pelvis
The pelvis is normal. There is delayed ossification of capital femoral
epiphyses with irregularity of the adjacent metaphyses. (Metaphyseal
changes are not usually seen at such a young age.)
Same patient: Figs 13.2, 13.7

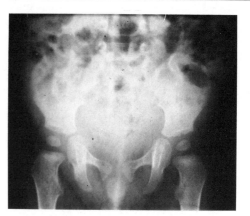

Fig. 13.4   M.C.—2 years—AP pelvis
Normal.
Same patient: Figs 13.8, 13.10

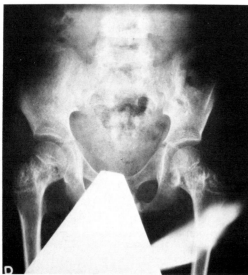

Fig. 13.5   A.R.—12 years—AP pelvis
There is irregular ossification of the broad femoral necks and also
around the greater trochanters; these and the capital femoral
epiphyses are unusually large.
Same patient: Figs 13.6, 13.11

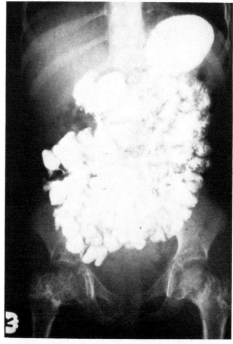

**Fig. 13.6   A.R.—12 years—AP abdomen, with part of pelvis
and barium follow-through examination** (same patient as in
Fig. 13.5)
Barium study was normal, but the femoral necks are short with
irregular ossification.

Lower limb : 7 months–2 years

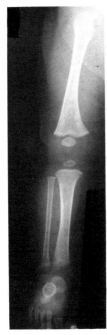

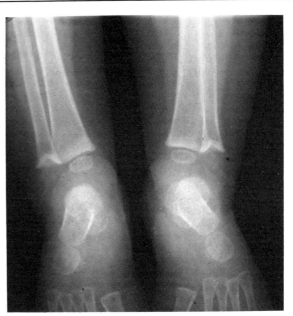

Fig. 13.8   M.C.—2 years—AP lower limbs
There is irregularity and indentation of the distal fibular metaphyses and some disorder of modelling of the distal tibial metaphysis.
Same patient : Figs 13.4, 13.10

Fig. 13.7   H.M.—7 months—AP lower limb
There is irregularity of the proximal femoral metaphysis (as already noted in Fig. 13.3), but metaphyses at the knee and ankle are normal.
Same patient : Figs 13.2, 13.3

Hand: 7 months–13 years

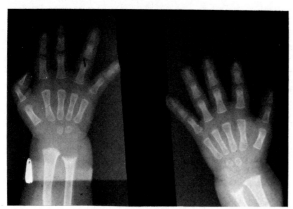

Fig. 13.9   M.C.—2 years—PA both hands
There is some irregularity of the distal radial and ulnar metaphyses,
but the metacarpals and phalanges are normal. Bone maturation
retarded—corresponding to 1 year 3 months.
Same patient: Figs 13.4, 13.8

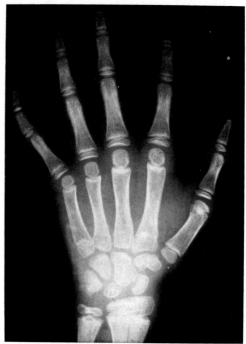

Fig. 13.10   A.R.—13 years—PA left hand
There is widening of epiphyseal plates and metaphyseal irregularity,
especially of the distal radius and ulna. There are additional epiphyses
of the first, second and fifth metacarpals and rather poor modelling of
these bones. Clinodactyly is present, and an unusual square
appearance of the second to fifth terminal phalanges. Bone maturation
is retarded, corresponding to 11 years.
Same patient: Figs 13.5, 13.6

Chapter Fourteen
METAPHYSEAL CHONDRODYSPLASIA—Type Peña

Irregular metaphyseal ossification with some lateral expansion, the lesion extending well into the diaphysis. Accompanied by dwarfing.

Inheritance

Probably autosomal recessive.

Frequency

Extremely rare.

Clinical features

Facial appearance—normal.
Intelligence—normal.
Stature—reduced.
Presenting feature/age—short stature in infancy.
Deformities—severe scoliosis, without congenital segmental vertebral anomalies. One patient had unilateral coxa vara.
Associated anomalies—none known.

Radiographic features

Skull—normal.
Spine—disorder of the cervical spine has been reported, with platyspondyly here, and ossification defects of the vertebral bodies. Scoliosis occurs in the thoracic region.
Limb epiphyses—normal, apart from delayed appearance.
Limb metaphyses—these are unlike those in the other metaphyseal chondrodysplasias. The area of irregular, though somewhat striated metaphyseal growth extends well into the diaphysis. In some cases there is lateral expansion also.

Bone maturation

Retarded by about 3 years.

Biochemistry

Not known.

Progress/complications

Short stature and progressive scoliosis appear to be the main problems.

REFERENCES

Kaufman H. J (ed) 1973 Intrinsic diseases of bone. Progress in Pediatric Radiology 4. Philadelphia
Peña J 1965 Disostosis metafisaria. Una revision. Con aportación de una observación familiar. Una forma mieva de la enfermedad. Radiologia 47:3

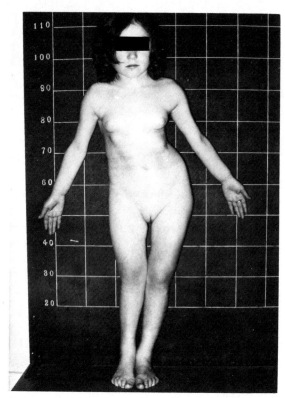

Fig. 14.1 14-year-old girl with short stature and scoliosis.

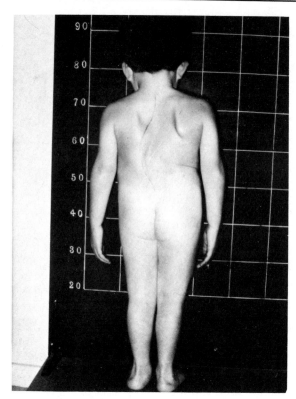

Fig. 14.2 12-year-old boy, also with short stature, but rather less severe scoliosis.

Hip and wrist: 10–14 years

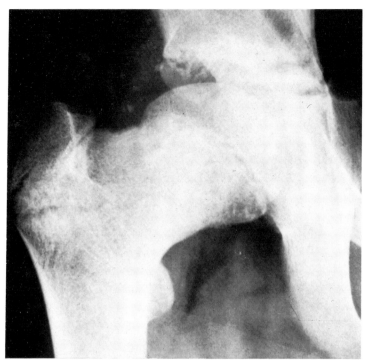

Fig. 14.3 Unknown female—14 years—AP right hip
There is irregularity of the acetabular roof and to a lesser extent of the proximal femoral metaphysis.
Same patient: Figs 14.4, 14.6

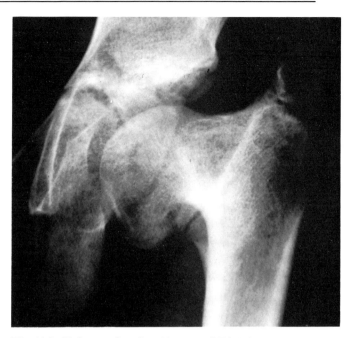

Fig. 14.4 Unknown female—14 years—AP left hip (same patient as in Fig. 14.3)
The acetabular roof is shallow and slopes steeply. Marked coxa vara is present. The capital femoral epiphysis appears to have slipped but there is no lateral view to confirm this.

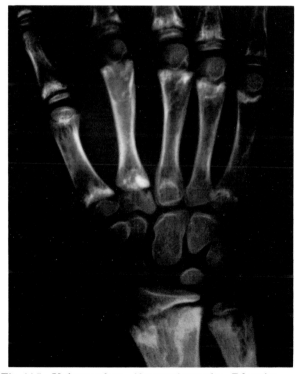

Fig. 14.5 Unknown boy—10 years 4 months—PA wrist
Irregularly ossified defects extend up into the diaphysis. Additional epiphyses are present for the first and second metacarpals. Bone maturation corresponds to 7 years.

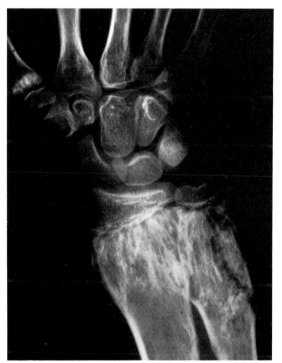

Fig. 14.6 Unknown female—14 years—PA wrist
The epiphyses are normal. There is expansion of the metaphyses extending well into the diaphyses. Irregular ossification with linear striations is present here. Bone maturation corresponds to 11 years.
Same patient: Figs 14.3, 14.4

Chapter Fifteen
METAPHYSEAL CHONDRODYSPLASIA
with retinitis pigmentosa and brachydactyly

One brother and sister with the above combination of disorders have been described.

Inheritance

Presumed autosomal recessive; consanguinity of parents is possible, but not proven.

Frequency

One kindred only is known.

Clinical features

Facial appearance—normal, apart from near blindness.
Intelligence—normal.
Stature—somewhat short, the girl on the 3rd centile and the boy on the 10th.
Presenting feature/age—short stature and stubby hands in childhood. Onset of blindness at 12–15 years.
Deformities—none.
Associated anomalies—apart from a divergent squint in each child, none.

Radiographic features

Skull—normal.
Spine—normal.
Long bones—metaphyseal irregularity particularly of the neck of the femur, and to a lesser extent at the knee, ankle and wrist.
Hands—short metacarpals and terminal phalanges sparing the index finger.

Bone maturation

Probably normal.

Biochemistry

Not known.

Differential diagnosis

From other metaphyseal chondrodysplasias, but the associated features distinguish this disorder.

Progress/complications

The metaphyseal lesions heal, leaving the limbs a little short. The problems are those of the retinitis pigmentosa and near-blindness.

REFERENCE

Phillips C I, Wynne-Davies R, Stokoe N L, Newton M 1981 Retinitis pigmentosa, metaphyseal chondrodysplasia and brachydactyly: an affected brother and sister. Journal of Medical Genetics 18:46–49

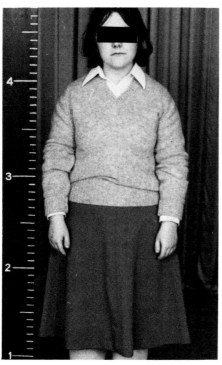

Fig. 15.1 Short stature, but normal facial appearance.

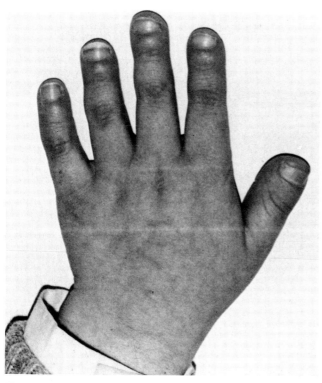
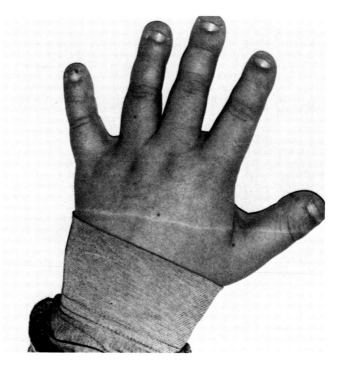

Figs 15.2 and 15.3 Hand of patient in Figure 15.1, and of her brother. There is marked brachydactyly, principally of the metacarpals and terminal phalanges.

Hip : 3–17 years

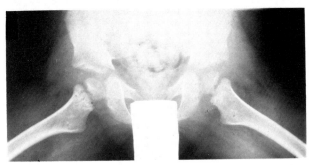

Fig. 15.4   K.E.—3 years 2 months—Lateral hips (brother of child in Fig. 15.5)
There are irregular areas of cyst-like bone in the metaphyses, with cupping, and the acetabula are also abnormal.
Same patient : Figs 15.6, 15.8, 15.9

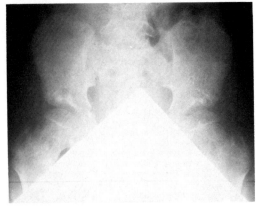

Fig. 15.5   E.E.—9 years 6 months—AP hips (sister of child in Fig. 15.4)
Similar to her brother (Fig. 15.4), but changes are less marked.
Same patient : Figs 15.7, 15.10

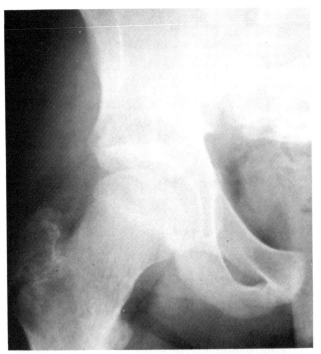

Fig. 15.6   K.E.—14 years 5 months—AP right hip (same patient as in Fig. 15.4)
Apart from a little sclerosis of the acetabulum, the hip is normal.

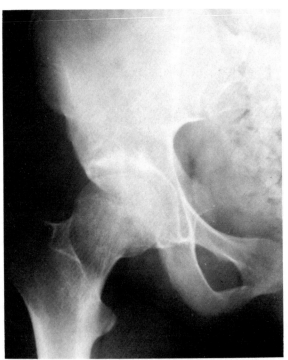

Fig. 15.7   E.E.—17 years—AP right hip (same patient as in Fig. 15.5)
Slight sclerosis of the acetabulum, otherwise normal.

Limbs: 3–17 years

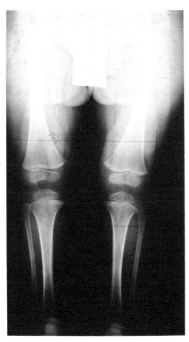

Fig. 15.8   K.E.—3 years 2 months—AP lower limbs
There is mild flaring of the metaphyses with some irregularity.
Same patient: Figs 15.4, 15.6, 15.9

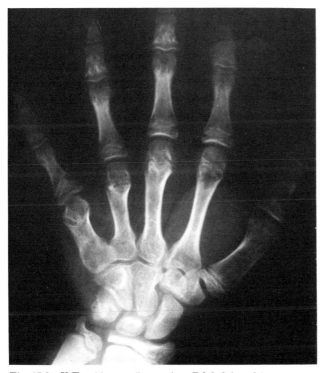

Fig. 15.9   K.E.—14 years 5 months—PA left hand (same
patient as in Fig. 15.8)
The metacarpals are short (except the second) and there are some
cone epiphyses. There is also brachydactyly, due to short middle
phalanges.

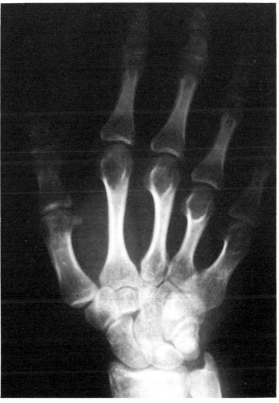

Fig. 15.10   E.E.—17 years 2 months—PA right hand
The epiphyses are fused, but the short metacarpals are obvious.
Same patient: Figs 15.5, 15.7

Section Five
SPONDYLOMETAPHYSEAL DYSPLASIA

Chapters Sixteen and Seventeen
THE SPONDYLOMETAPHYSEAL DYSPLASIAS
(including Types Kozlowski and Sutcliffe)

There are probably several varieties, but the characteristics are platyspondyly with metaphyseal changes in the long bones.

Inheritance

Autosomal dominant is usually described; probably also an X-linked recessive type.

Frequency

Less common than the spondyloepiphyseal disorders, perhaps a prevalence of 1 per million or less.

Clinical features

Facial appearance—normal.
Intelligence—normal.
Stature—short, with a disproportionately short trunk.
Presenting feature/age—usually in infancy or childhood, not at birth. Short stature/short trunk are noticed first, perhaps later abnormal gait associated with coxa vara.
Deformities—limb malalignment, often coxa vara, scoliosis.
Associated anomalies—probably none.

Radiographic features

Skull—normal.
Spine—universal platyspondyly, with only mild irregularity of the end plates.
Pelvis—the base of the ilium is broad, as are the femoral necks. Marked coxa vara may develop.

Long bones—metaphyseal irregularities, with almost normal epiphyses.

Bone maturation

Retarded.

Biochemistry

Not known.

Differential diagnosis

The disorders within the group of spondylometaphyseal dysplasias need to be distinguished, but the spinal as well as limb involvement delineates them from the metaphyseal chondrodysplasias, and their relatively normal epiphyses from more serious forms of dwarfism such as metatropic, Morquio's etc.

Progress/complications

Related to the deformities of coxa vara and scoliosis—there is probably a normal life span.

REFERENCES

Kozlowski K 1976 Metaphyseal and spondylometaphyseal chondrodysplasias. Clinical Orthopaedics and Related Research 114:83–93
Kozlowski K, Barylak A, Middleton R W 1976 Spondylometaphyseal dysplasias. (Report of a case of common type and three pairs of siblings of new varieties). Radiology (Australia) 20/2:154–164
Sutcliffe J 1965 Metaphyseal dysostosis. Annals of Radiology 9:215–223

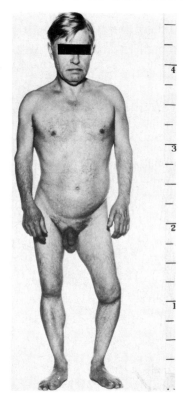

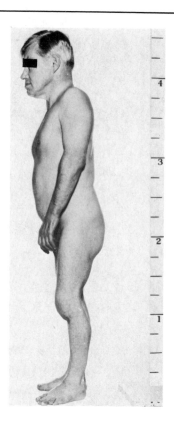

Figs 16.1 and 16.2   Adult with short stature malalignment of the left lower limb.

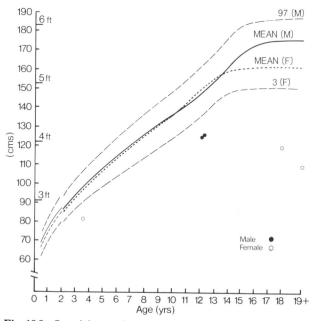

Fig. 16.3   Spondylometaphyseal dysplasias—percentile height.

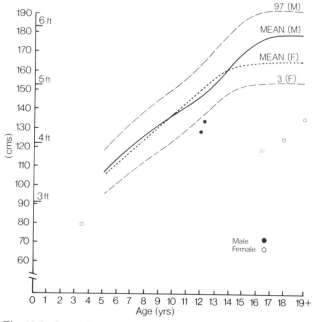

Fig. 16.4   Spondylometaphyseal dysplasias—percentile span.

Spine (Type Kozlowski) : 10–13 years

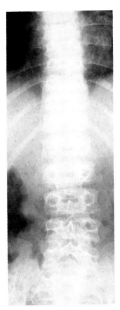

Fig. 16.5   L.H.—10 years—AP thoracic and lumbar spine
Mild platyspondyly is present and there is a progressive decrease in the interpedicular distances from L1 to L5.
Same patient : Figs 16.6, 16.11, 16.14, 16.19

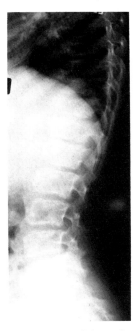

Fig. 16.6   L.H.—10 years—Lateral thoracic and lumbar spine
(same patient as in Fig. 16.5)
There is mild platyspondyly with an increase in the posterior diameters of the vertebral bodies and irregularity of the end-plates.

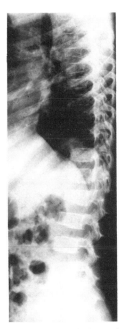

Fig. 16.7   A.H.—10 years—Lateral thoracic and lumbar spine
There is platyspondyly with anterior wedging throughout.
Same patient : Figs 16.10, 16.13, 16.16, 16.18

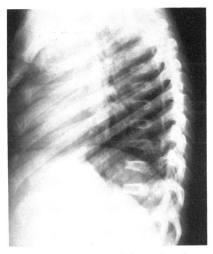

Fig. 16.8   S.H.—13 years—Lateral thoracic spine
The platyspondyly is pronounced the disc spaces being greater than the height of the vertebral bodies.
Same patient : Figs 16.9, 16.17

Pelvis (Type Kozlowski) : 7–14 years

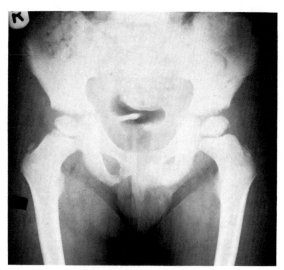

Fig. 16.9   S.H.—7 years—Pelvis
Mild irregularity of the upper femoral metaphyses is present.
Same patient : Figs 16.8, 16.17

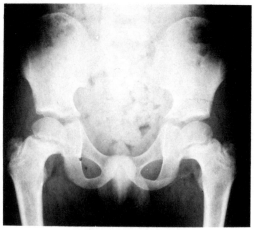

Fig. 16.10   A.H.—9 years—Pelvis
The iliac wings are narrow. The capital femoral epiphyses are large
and the femoral metaphyses flared, with apparent widening of the
epiphyseal plates.
Same patient : Figs 16.7, 16.13, 16.16, 16.18

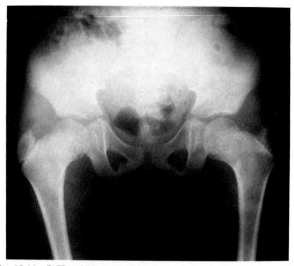

Fig. 16.11   L.H.—10 years—Pelvis
There is bizarre ossification of the broad femoral necks, with coxa
vara.
Same patient : Figs 16.5, 16.6, 16.14, 16.19

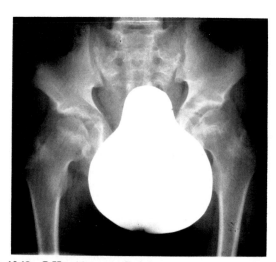

Fig. 16.12   G.H.—14 years—Pelvis
There is mild irregularity of the upper femoral metaphyses and of the
acetabular roofs.
Same patient : Figs 16.15, 16.20

Lower limb (Type Kozlowski): 9–14 years

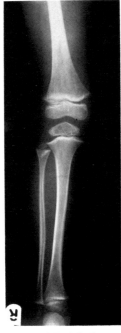

Fig. 16.13   A.H.—9 years—AP right leg
All metaphyses are flared, irregular and poorly ossified.
Same patient: Figs 16.7, 16.10, 16.16, 16.18

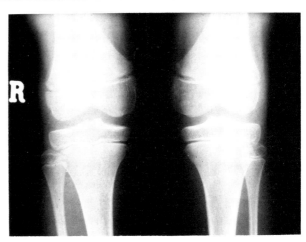

Fig. 16.14   L.H.—10 years—AP knees
The metaphyses show only minor irregularity.
Same patient: Figs 16.5, 16.6, 16.11, 16.19

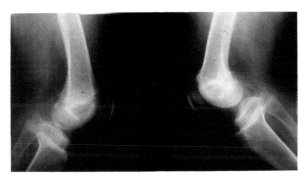

Fig. 16.15   G.H.—14 years—Lateral knees
All metaphyses are flared and irregular.
Same patient: Figs 16.12, 16.20

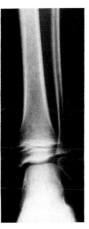

Fig. 16.16   A.H.—9 years—AP ankle (same patient as in Fig. 16.13)
There is apparent widening of the epiphyseal plates with a little metaphyseal irregularity.

Upper limb (Type Kozlowski): 6–14 years

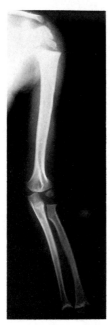

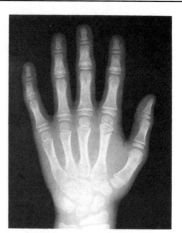

Fig. 16.18   A.H.—9 years—PA left hand
The short middle phalanges are associated with metaphyseal irregularity and cupping. Bone maturation is normal.
Same patient : Figs 16.7, 16.10, 16.13, 16.16

Fig. 16.17   S.H.—6 years—AP upper limb
The shoulder and wrist metaphyses show irregularity, mild flaring and cupping.
Same patient : Figs 16.8, 16.9

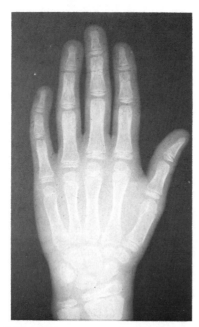

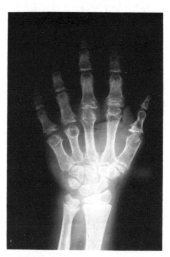

Fig. 16.20   G.H.—14 years—PA left hand
All the phalanges are short, and have flared, cupped metaphyses. The fourth and fifth metacarpals are short. Bone maturation is retarded by approximately 2 years.
Same patient : Figs 16.12, 16.15

Fig. 16.19   L.H.—10 years—PA left hand
Normal.
Same patient : Figs 16.5, 16.6, 16.11, 16.14

Type Sutcliffe: 7 years

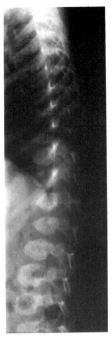

Fig. 17.1   C.S.—7 years—Lateral thoracic and lumbar spine
The vertebral bodies are oval with an irregular outline.
Same patient: Figs 17.2–17.4

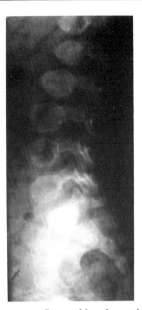

Fig. 17.2   C.S.—7 years—Lateral lumbar spine (same patient as in Fig. 17.1)
The vertical bodies have an oval configuration with anterior pointing and there is a 'bone-in-a-bone' appearance.

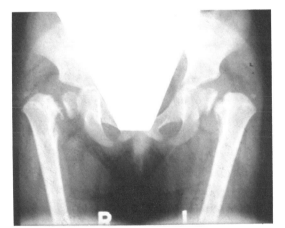

Fig. 17.3   C.S.—7 years—Pelvis (same patient as in Fig. 17.1)
There is marked coxa vara with widening and irregularity of the zones of provisional calcification at the upper femoral metaphyses.

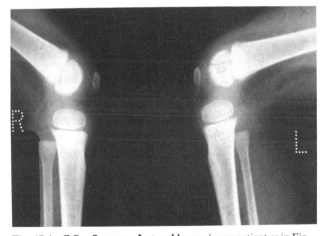

Fig. 17.4   C.S.—7 years—Lateral knees (same patient as in Fig. 17.1)
There is metaphyseal flaring and irregularity with a little sclerosis.

Section Six
SHORT LIMBS— NORMAL TRUNK

Chapter Eighteen
ACHONDROPLASIA

Dwarfism characterised by short limbs, lumbar lordosis, a bulging forehead and low nasal bridge.

Inheritance

Autosomal dominant, but nearly all cases are new mutants. Rarely, autosomal recessive inheritance has been reported.

Frequency

Population incidence of mutant achondroplasia is about 3 per million.

Clinical features

Facial appearance—small maxillary area, low nasal bridge and bulging forehead.

Intelligence—normal.

Stature/body proportions—see Figures 17.5–17.9. Trunk is of almost normal length. Proximal segment of limbs is shorter than distal. Soft tissues in excess causing skin folds.

Presenting feature/age—short limbs at birth. Occasionally hydrocephalus.

Deformities

—Lumbar lordosis (all)
—Limitation of elbow extension (all)
—Lumbar kyphos in infancy (about half)
—Persistent lumbar kyphos (after 10 years: 19%)
—Occasional scoliosis
—Genu varum (15%)
—Long fibula with occasionally disruption at the ankle mortice

Associated anomalies—none.

Radiographic features

Skull—small area of facial bones in comparison with vault (see also hypochondroplasia). Sometimes hydro-cephalus, associated with small foramen magnum and local obstruction. Short base of skull.

Spine—very characteristic, with short pedicles throughout its length and scalloped posterior border of vertebral bodies. The interpedicular distances usually (two-thirds of cases) narrow progressively from L1 to L5, the remainder are parallel, never widening as do 40% of normal individuals. Persistence of lumbar kyphos is associated with wedging of vertebral bodies at this site.

Thorax—shortened ribs, some flattening of antero-posterior diameter.

Pelvis and hip joints—the iliac wings are squared with horizontal acetabular roofs. There is a sharp spur at the medial edge of the triradiate cartilage and the greater sciatic notch is narrowed. The upper femoral metaphysis is splayed. Epiphysis unaffected.

Limbs—widening of metaphyses with normal epiphyses.

Hands—in about half the cases 'starfish' or 'trident' hands present, with metacarpals of almost equal length. Others normal.

Bone maturation

Little information is available but it does seem that the pattern is not completely normal, although not grossly delayed. Some early delay is followed by an excessive spurt of bone growth.

Biochemistry

Not known.

Differential diagnosis

From *other forms of short limbed dwarfism*. Classical achondroplasia is identified by the appearance of the skull and facial bones, by the spine with its short pedicles and scalloped posterior border of the vertebrae and by the typical pelvis. Epiphyses are not affected in achondroplasia.

Progress/complications

About 3% have hydrocephalus requiring drainage. During childhood there are usually recurrent upper respiratory problems due to poor development of nasal sinuses. Adults not affected. Most individuals are fit and strong. Main problems relate to narrowed spinal canal and symptoms of spinal stenosis—paralysis has been reported in 11%. This complication is particularly likely in the presence of a persistent kyphos. Tetraplegia has rarely been reported.

Lower limb problems occur due to disproportionate growth between tibia and fibula—the latter can appear too long at either knee or ankle, or both, although it is chiefly at the lower end that mechanical problems develop. Since epiphyses are normal, osteoarthritis is not a feature of this disease—provided normal limb alignment is maintained.

REFERENCES

Gardner R J M 1977 A new estimate of the achondroplasia mutation rate. Clinical Genetics 11:31–38

Maynard J A, Ippolito E G, Ponseti I V, Mickelson M R 1981 Histochemistry and ultrastructure of the growth plate in achondroplasia. Journal of Bone and Joint Surgery (Series A) 63/6:969–979

Morgan D F, Young R F 1980 Spinal neurological complications of achondroplasia. Results of surgical treatment. Journal of Neurosurgery 52/4:463–472

Nehme A-M E, Riseborough E J, Tredwell S J 1976 Skeletal growth and development of the achondroplastic dwarf. Clinical Orthopaedics and Related Research 116:8–23

Oberklaid F, Danks D M, Jensen F, Stace L, Rosshandler S 1979 Achondroplasia and hypochondroplasia. Comments on frequency, mutation rate, and radiological features in skull and spine. Journal of Medical Genetics 16:140–146

Wynne-Davies R, Gormley J, Walsh K 1981 Clinical variation and spinal stenosis in achondroplasia and hypochondroplasia. Journal of Bone and Joint Surgery 63B:508–515

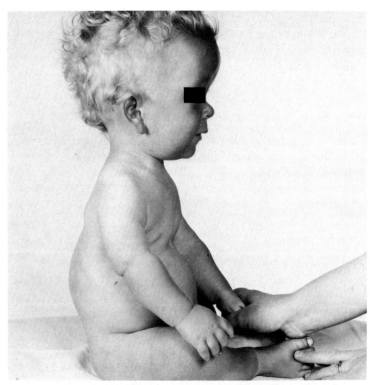

Fig. 18.1   The bulging forehead and depressed bridge of nose are typical of this condition. The patient still has a lumbo-dorsal kyphos—an ominous sign if persisting after the age of 3 or 4 years.

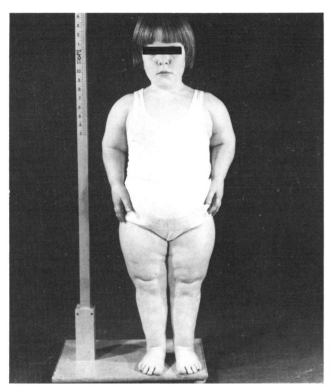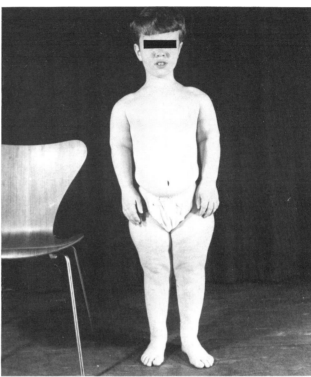

Figs 18.2 and 18.3 The body disproportion is obvious—an almost normal length of trunk, with short limbs, particularly in the proximal segment.

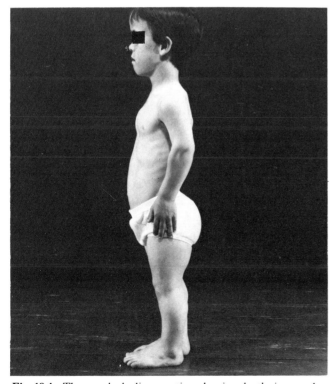

Fig. 18.4 The same body disproportion, showing also the increased lumbar lordosis.

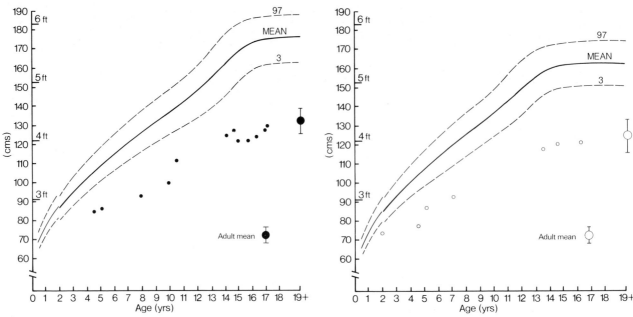

Fig. 18.5 Achondroplasia—percentile height (males).

Fig. 18.6 Achondroplasia—percentile height (females).

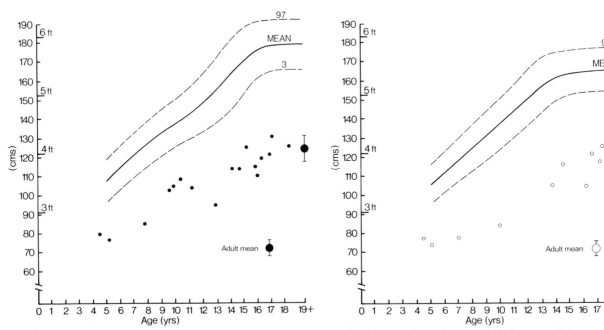

Fig. 18.7 Achondroplasia—percentile span (males).

Fig. 18.8 Achondroplasia—percentile span (females).

Whole body : Neonate

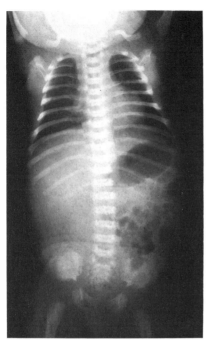

Fig. 18.9   C.M.—Neonate—AP whole body
Short ribs. Iliac wings small and squared. Horizontal acetabular roofs with medial beak.

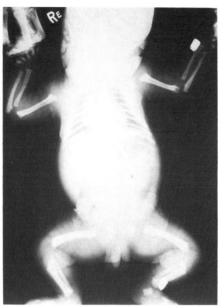

Fig. 18.10   B.B.—Neonate—AP whole body
Iliac wings small and squared. Limbs are short and metaphyses sloping, particularly well seen at the proximal end of the tibiae. Fibulae are long in relation to the tibiae. Folds of redundant skin are obvious.
Same patient : Fig. 18.11

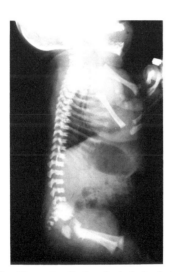

Fig. 18.11   B.B.—Neonate—Lateral whole body (same patient as in Fig. 18.10)
The normal biconvexity of the vertebra is absent. Limbs are short, and femora have spurs at the proximal metaphyses.

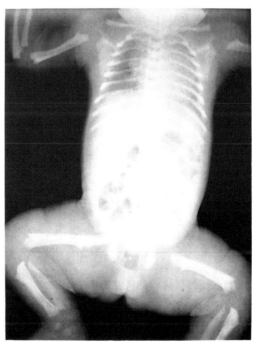

Fig. 18.12   Neonate—AP whole body
Short ribs. Short long bones with sloping metaphyses with spurs. Bowed fibulae. Horizontal acetabular roofs with medial beaks.

Skull: Birth–1½ years

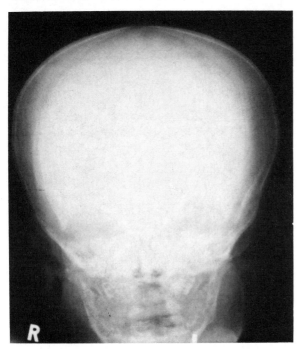

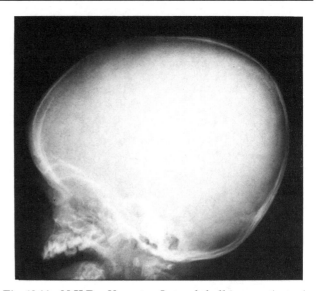

Fig. 18.14   M.H.D.—Neonate—Lateral skull (same patient as in Fig. 18.13)
The vault is large in comparison with the area of facial bones. The base is short.

Fig. 18.13   M.H.D.—Neonate—AP skull
The vault is large with prominence of the parietal bones.
Same patient: Fig. 18.14

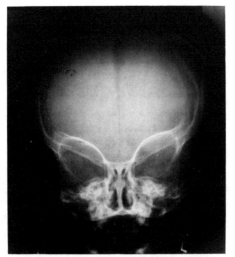

Fig. 18.15   C.S.—1½ years—AP skull
Cranial sutures are wide for this age. Wide bi-parietal diameter. A metopic suture is present (mid-line frontal).

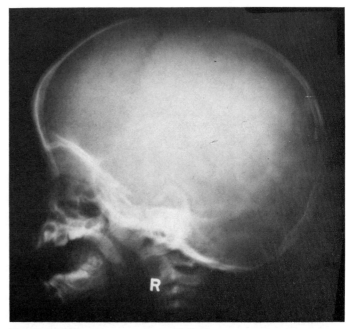

Fig. 18.16   S.S.—1½ years—Lateral skull
As in Figure 18.14, the vault is large in comparison with the facial bones, and the base short.

Skull: 5 years

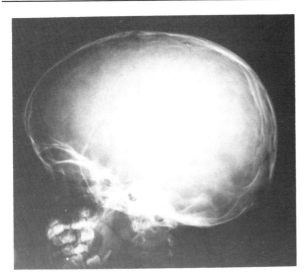

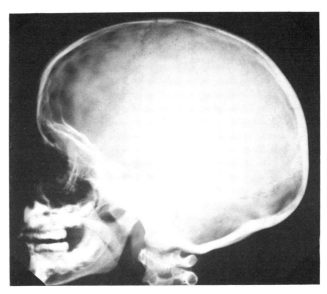

Fig. 18.17   M. McK.—5 years—Lateral skull
Large vault in comparison to facial bones. In the normal child by this age, the maxillae have grown and the disproportion is less marked (see Fig. 18.18).
Same patient: Figs 18.33, 18.48

Fig. 18.18   Normal skull—5 years—for comparison

Lumbar spine : 1–11 years

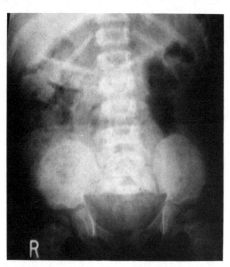

Fig. 18.19   S.P.—1½ years—AP lumbar spine
Pronounced backward angulation of the sacrum (associated with lumbar lordosis). Failure of the normal widening between pedicles from the first to last lumbar vertebrae. L5 is low-set between the iliac wings.

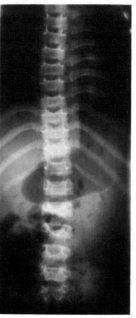

Fig. 18.20   J.I.—1½ years—AP dorso-lumbar spine
Progressive narrowing of interpedicular distances between L1 and L5.
Same patient : Fig. 18.40.

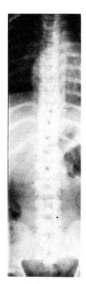

Fig. 18.21   K.G.—7 years—AP dorso-lumbar spine
Progressive narrowing of interpedicular distances from L1 to L5. Backward angulation of the sacrum and low-set L5. Spina bifida occulta of S1.
Same patient : Figs 18,22, 18.44

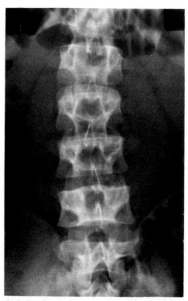

Fig. 18.22   K.G.—11½ years—AP lumbar spine (same patient as in Fig. 18.21)
Progressive narrowing of interpedicular distances caudally. L5 is low-set between the iliac wings. Spina bifida occulta of S1.

Lumbar spine: 13 years—Adult

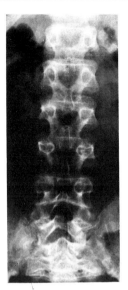

Fig. 18.23   P.D.—13 years—AP lumbar spine
Progressive narrowing of interpedicular distances from L1 to L5. L5 is low-set.

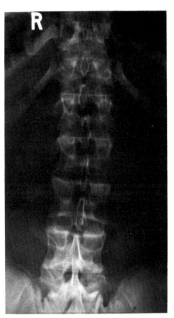

Fig. 18.24   J.E.—15 years 9 months—AP lumbar spine
Progressive caudal narrowing of interpedicular distances. L5 is low-set.
Same patient: Figs 18.27, 18.53, 18.57

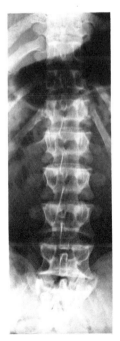

Fig. 18.25   A.G.—21 years—AP lumbar spine
As Figures 18.23 and 18.24.

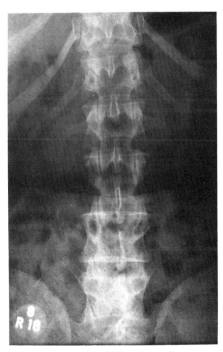

Fig. 18.26   M.S.—29 years—AP lumbar spine
Interpedicular distances appear parallel, not narrowing—this being a feature in about one-third of achondroplastic patients.

Lateral spine : 5–18 months

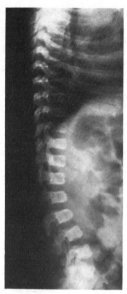

Fig. 18.27   J.E.—5 months—Lateral spine
There is no lumbar lordosis, but no kyphos either.
Same patient: Figs 18.24, 18.53, 18.57

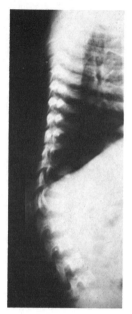

Fig. 18.28   D.H.—8 months—Lateral spine
There is a dorso-lumbar kyphos. Pedicles are short and posterior
scalloping is now apparent.

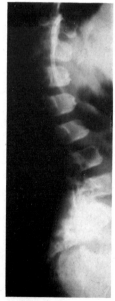

Fig. 18.29   T.S.—1 year—Lateral spine
There is a dorso-lumbar kyphos and some anterior wedging already
present. The vertebral bodies show posterior scalloping. Increased
backward angulation of the sacrum is obvious.
Same patient: Fig. 18.39

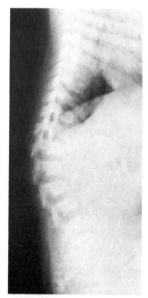

Fig. 18.30   G.M.—1½ years—Lateral spine
Dorso-lumbar kyphos with anterior wedging. Short pedicles with
narrow spinal canal.

Lateral spine : 2–14 years

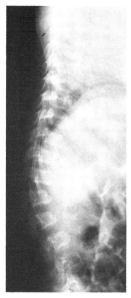

Fig. 18.31   J.W.—2 years—Lateral spine
Dorsal kyphos with anterior wedging. Posterior scalloping of lumbar
vertebral bodies.

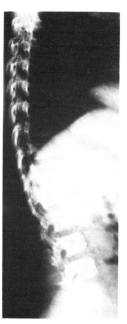

Fig. 18.32   J.C.—5 years—Lateral spine
Pedicles are short with posterior scalloping of lumbar vertebral
bodies. The lumbar kyphos is minimal, and will probably disappear.
Same patient : Fig. 18.64

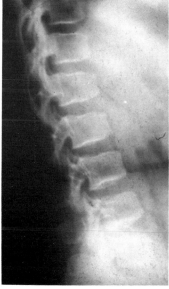

Fig. 18.33   M.McK.—11 years—Lateral spine
Very mild kyphos with wedging. Pedicles short.
Same patient : Figs 18.18, 18.42

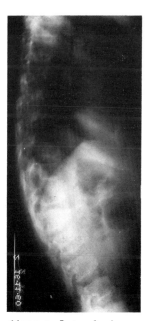

Fig. 18.34   A.K.—14 years—Lateral spine
Mild kyphos with pronounced anterior wedging of lumbar vertebrae.
Pedicles short and spinal canal narrowed.

Lateral spine : 14 years—Adult

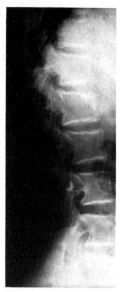

Fig. 18.35   B.W.—14 years—Lateral spine
Pedicles are short and there is posterior scalloping. No kyphos.
Irregularity of the anterior end of the vertebral end plates of T12 and
L1.
Same patient : Figs 18.46, 18.59, 18.63

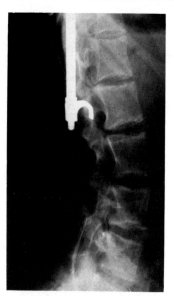

Fig. 18.36   A.B.—16½ years—Lateral spine
Short pedicles and posterior scalloping. Rather tall vertebral bodies.
The lower end of a Harrington rod can be seen—operative treatment
for a developing scoliosis.

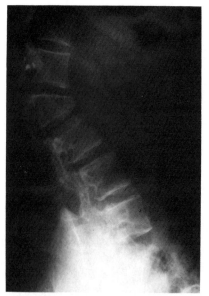

Fig. 18.37   J.B.—19 years—Lateral spine
Pronounced kyphos and wedging of L2. This patient became
paraplegic at 18 years.

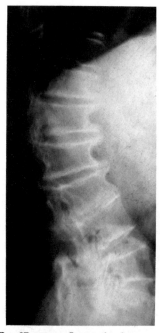

Fig. 18.38   M.B.—67 years—Lateral spine
Lumbar kyphos present, and marked spondylosis has now developed.

Pelvis : Birth–5 years

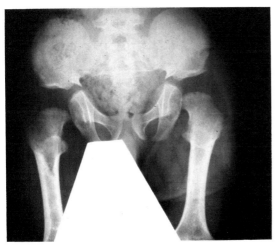

Fig. 18.39   T.S.—Neonate—AP pelvis
Iliac wings are small and squared. Acetabular roofs are horizontal with pronounced medial beaking and small sacro-sciatic notches.
Same patient : Fig. 18.29

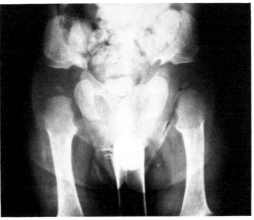

Fig. 18.40   J.I.—6 months—AP pelvis
As Figure 18.39, but in addition there are radiolucent oval areas seen at the upper ends of the femora (associated with abnormal development of the metaphyses).
Same patient : Fig. 18.20

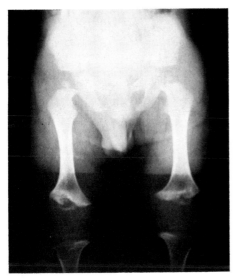

Fig. 18.41   P.R.—3 years—AP pelvis
As Figure 18.39. Capital femoral epiphyses are unusually small for this age. Metaphyseal flaring of the femora.
Same patient : Fig. 18.60

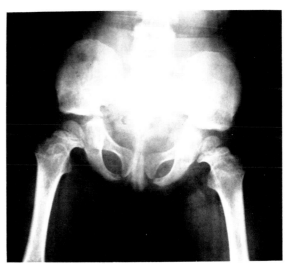

Fig. 18.42   M.McK.—5 years—AP pelvis
As Figure 18.39. Femoral capital epiphyses are of normal size for this age.
Same patient : Figs 18.17, 18.33

Pelvis : 7–14 years

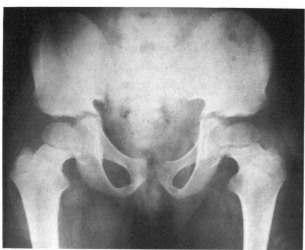

Fig. 18.43   M.R.—7 years 9 months—AP pelvis
Iliac wings are small and squared. Acetabular roofs are horizontal and
the sacro-sciatic notches remain small. Medial beaking no longer
present. Femoral capital epiphyses are large, metaphyses splayed and
necks short.
Same patient : Fig. 18.65

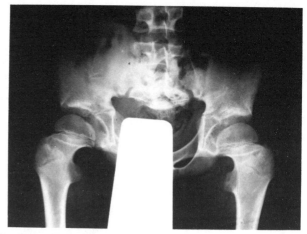

Fig. 18.44   K.G.—9 years—AP pelvis
As Figure 18.43. Posterior sacral angulation is well shown.
Same patients : Figs 18.21, 18.22

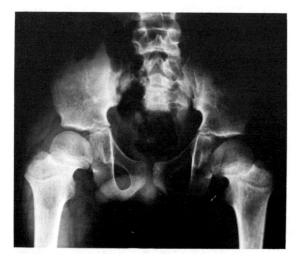

Fig. 18.45   Q.S.—11 years—AP pelvis
As Figure 18.43.

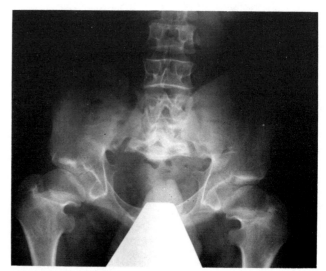

Fig. 18.46   B.W.—14 years—AP pelvis
As Figure 18.43. Short femoral necks.
Same patient : Fig. 18.35, 18.59, 18.63

Pelvis : 15 years–Adult

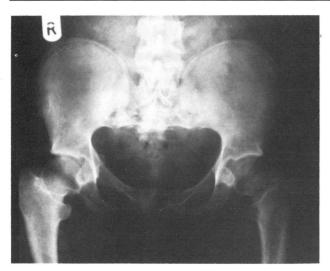

Fig. 18.47   B.C.—15 years 4 months—AP pelvis
Iliac wings small and squared. Horizontal acetabular roofs. Hip joints normal, apart from slightly short femoral necks.

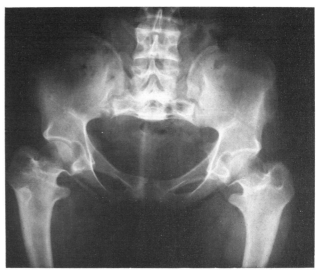

Fig. 18.48   C. McL.—17 years—AP pelvis
As Figure 18.47. Pronounced posterior angulation of sacrum.

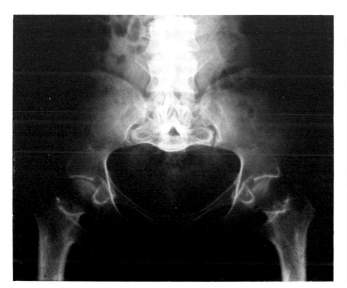

Fig. 18.49   M.A.—26 years—AP pelvis
As Figure 18.47.
Same patient : Fig. 18.54

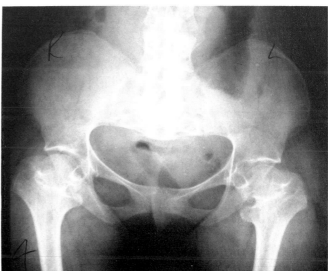

Fig. 18.50   E.H.—33 years—AP pelvis
As Figure 18.47.

Lower limb: Birth–1 year

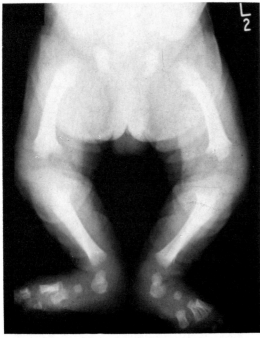

Fig. 18.51   W.C.—Neonate—AP lower limbs
Femora and tibiae are short. The fibulae are long proximally in relation to the tibiae. Flared metaphyses. Bone maturation corresponds to less than 30 weeks gestation. Excessive skin folds are seen.

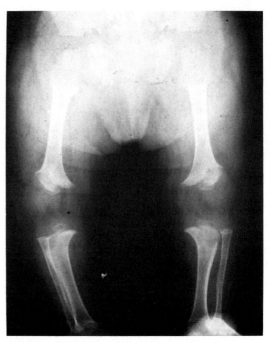

Fig. 18.52   T.G.—Under 1 year—AP lower limbs
As Figure 18.51. There is a pronounced slope of the metaphyses around the knee joint.

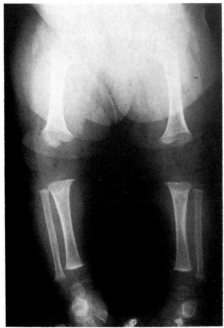

Fig. 18.53   J.E.—5 months—AP lower limbs
As Figure 18.51 but the fibulae are long *distally*.
Same patient: Figs 18.24, 18.27, 18.57

Lower limb: 2–15 years

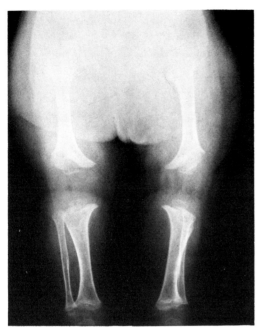

Fig. 18.54   M.A.—2 years 8 months—AP lower limbs
Shortening of the long bones. Flared metaphyses, with abnormal slope. The fibula is not disproportionately long in comparison with the tibia. Early chevron deformity of the distal femur. Epiphyses around the knee are small for this age.
Same patient: Fig. 18.49

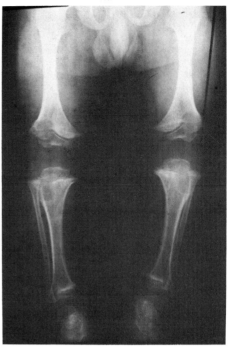

Fig. 18.55   G.W.—4 years—AP lower limbs
As Figure 18.54, but the fibulae are long in relation to the tibiae. The oval radiolucencies around the knee joint are well demonstrated, and are due to the abnormal metaphyseal slope.

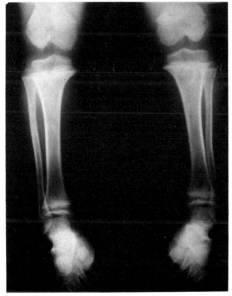

Fig. 18.56   A.B.—11 years—AP tibiae and fibulae
Metaphyses flared and the fibulae are long distally, disrupting the ankle mortice.

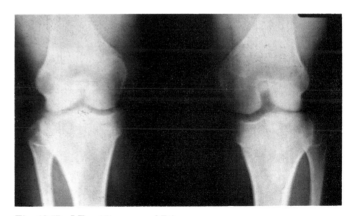

Fig. 18.57   J.E.—15½ years—AP knees
Deep intercondylar notches of femora.
Same patient: Figs 18.24, 18.27, 18.53

Upper limb : Birth–6 years

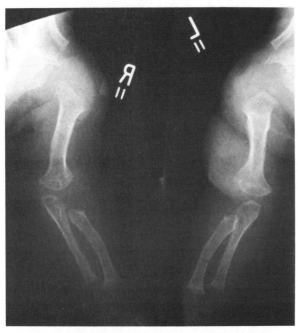

Fig. 18.58   P.E.—Neonate—AP upper limbs
Shortened long-bones with metaphyseal flaring. Deltoid tuberosities
are pronounced. Skin folds are obvious.

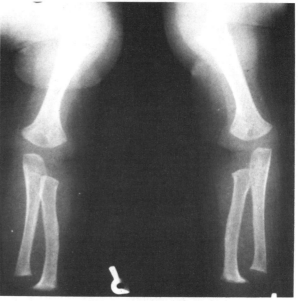

Fig. 18.59   B.W.—9 months—AP upper limbs
As Figure 18.58. Distal ulna appears short.
Same patient : Figs 18.35, 18.46, 18.63

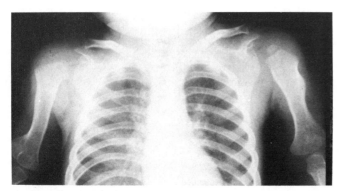

Fig. 18.60   P.R.—3 years—AP shoulders
Humeri short and broad with pronounced muscle markings. Clavicle
and ribs normal.
Same patient : Fig. 18.41

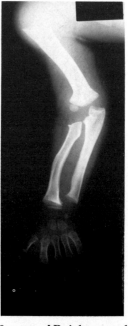

Fig. 18.61   H.E.—6 years—AP right upper limb
As Figure 18.58. Distal end of ulna appears short in relation to the
radius.

Hand: Birth–7 years

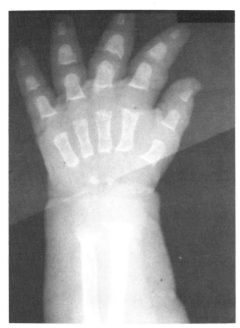

Fig. 18.62   J.L.—Neonate—PA left hand
Long bones short and broad, particularly at the base of the phalanges.
The 'starfish' or 'trident' appearance of the hand is not marked, but
this is only found in about half the patients with achondroplasia.

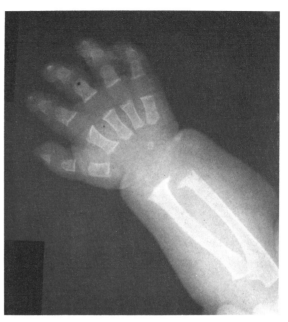

Fig. 18.63   B.W.—1½ years—AP forearm and hand
As Figure 18.62. In addition there is flaring of the distal radial and
ulnar metaphyses, and the distal ulna is short in relation to the radius.
Same patient: Figs 18.35, 18.46, 18.59

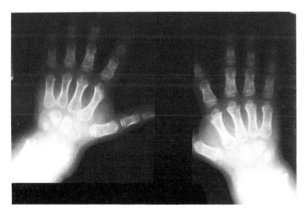

Fig. 18.64   J.C.—5½ years—PA both hands
Although the long bones are short, there is less modelling deformity
than in Figure 18.62.
Same patient: Fig. 18.32

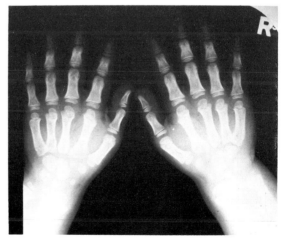

Fig. 18.65   M.R.—7 years 9 months—PA hands
Apart from some shortening of the long bones, these hands are almost
normal.
Same patient: Fig. 18.43

Chapter Nineteen
HYPOCHONDROPLASIA

At its most severe, hypochondroplasia is indistinguishable from achondroplasia, and at its least severe, from normality. The usual clinical picture is of short-limbed short stature and an almost normal skull and facial appearance.

Inheritance

Autosomal dominant, distinct from achondroplasia.

Frequency

Not possible to ascertain the true incidence since there must be many unidentified cases in the population. Likely prevalence is 3–4 per million, similar to achondroplasia.

Clinical features

Facial appearance—may be normal, or with slight reduction in facial area in comparison with the vault of the skull.

Intelligence—normal.

Stature—see graphs. Almost normal length of trunk with shortened arm span and increased head-pubis/pubis-heel ratio.

Presenting feature/age—short stature at 2–3 years.

Deformities—Lumbar lordosis. Kyphos is not a feature. Sometimes limitation of elbow extension. Genu varum (8%).

Associated anomalies—none.

Radiographic features

Skull—normal, or somewhat decreased size of facial bones in comparison with size of vault.

Spine—normal, or with some reduction in diameter of spinal canal associated with short pedicles. Posterior scalloping of vertebral bodies is unusual.

Pelvis and hip joint—large capital femoral epiphyses and trochanters.

Limbs—little abnormality apart from some metaphyseal flaring, prominent femoral condyles and often increased length of fibula.

Bone maturation

Probably normal.

Biochemistry

Not known.

Differential diagnosis

Achondroplasia is usually more severe, with shorter stature, large cranial vault and greater disproportion of limb length. Milder cases are clinically and radiologically indistinguishable, the only firm evidence for separate delineation is genetic (no overlap of the two conditions in families).

Metaphyseal chondrodysplasia Type Schmid—here the lower limbs are principally involved and the upper of normal length. Coxa vara and bowed femora are usual, and the skull and vertebrae normal.

Dyschondrosteosis typically shows failure of development of the medial part of the lower radial epiphysis, with a Madelung type deformity. However, this may be minimal and differentiation is not easy. Both disorders have shortened leg bones, but the fibula in dyschondrosteosis may be too short or too long. Hypochondroplasia may show a long fibula.

Progress/complications

Normal life span and few complications apart from genu (tibia?) varum. Spinal stenosis problems have been reported, but are rare. In general these patients are strong and muscular.

REFERENCES

Glasgow J F T, Nevin N C, Thomas P S 1978 Hypochondroplasia. Archives of Disease in Childhood 53/11:868–872

Heselson N G, Cremin B J, Beighton P 1979 The radiographic manifestations of hypochondroplasia. Clinical Radiology 30/1 : 79–85

Oberklaid F, Danks D M, Jensen F, Stace L, Rosshandler S 1979 Achondroplasia and hypochondroplasia. Comments on frequency, mutation rate, and radiological features in skull and spine. Journal of Medical Genetics 16 : 140–146

Wynne-Davis R, Gormley J, Walsh K 1981 Clinical variation and spinal stenosis in achondroplasia and hypochondroplasia. Journal of Bone and Joint Surgery 63B : 508–515

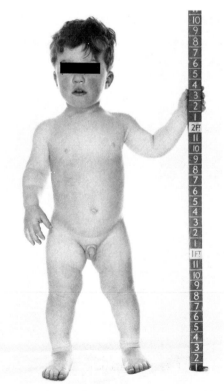
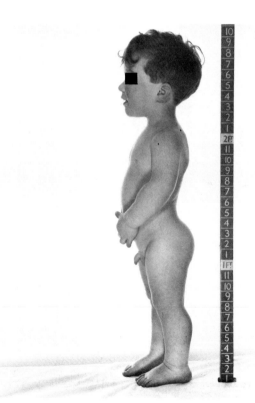

Figs 19.1 and 19.2   Some shortness of stature with a normal length of trunk and short limbs, but not so marked as in achondroplasia. The head and face are clinically normal.

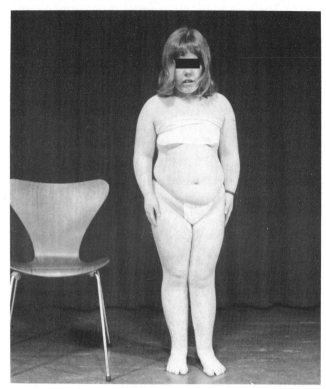

Fig. 19.3   An older girl, clearly with short upper limbs, but she has had tibial lengthening and the lower segment is more nearly normal.

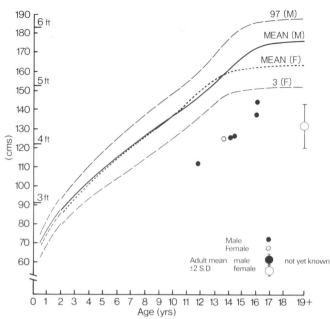

Fig. 19.4   Hypochondroplasia—percentile height.

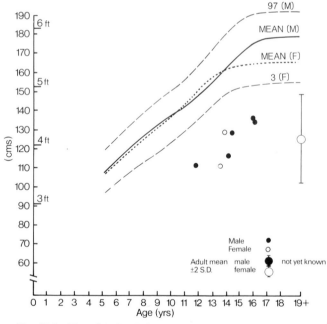

Fig. 19.5   Hypochondroplasia—percentile span.

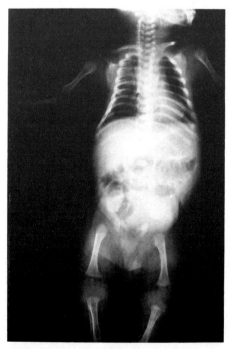

Fig. 19.6   R.W.—1 month—AP whole body
There is some shortening of the long bones in relationship to the trunk. Bone maturation at the knees is delayed. The thoracic cage is small.
Same patient: Figs 19.18, 19.27, 19.33

Skull: 1–13 years

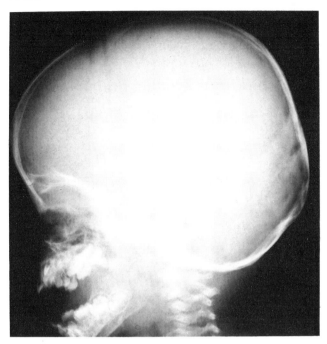

Fig. 19.7   D.K.—1 year 10 months—Lateral skull
The vault is large in relation to the facial bones—the discrepancy being greater than is usual at this age, although not so pronounced as in achondroplasia.

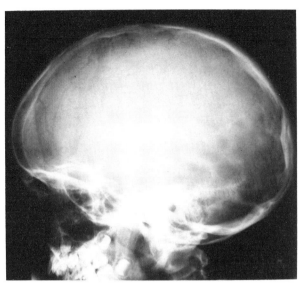

Fig. 19.8   D.C.—3 years—Lateral skull
Moderate vault enlargement with slight prominence of the frontal bones.
Same patient: Figs 19.26, 19.32, 19.36

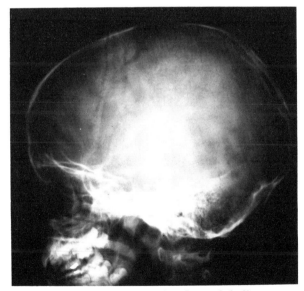

Fig. 19.9   S.M.—8 years 11 months—Lateral skull
The facial bones are small in comparison with the vault of the skull. Normally, by this age, the maxillae have grown considerably.
Same patient: Fig. 19.39

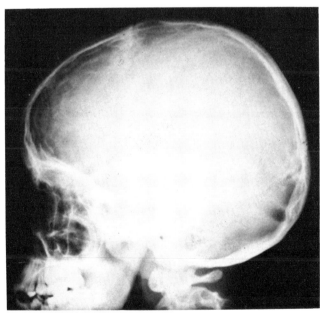

Fig. 19.10   G.M.—13 years—Lateral skull
Normal skull vault, odontoid peg, sinuses and facial proportions.
Same patient: Fig. 19.35

Spine: 3–17 years

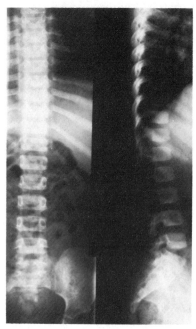

Fig. 19.11   T.H.—3 years—AP and lateral spine
Minor spina bifida of S1. There is no progressive caudal widening of
the interpedicular distances in the lumbar spine.
Same patient: Figs 18.23, 18.37

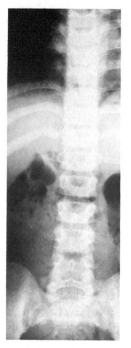

Fig. 19.12   M.R.—7 years—AP spine
Normal spine apart from spina bifida of S1.
Same patient: Fig. 19.19

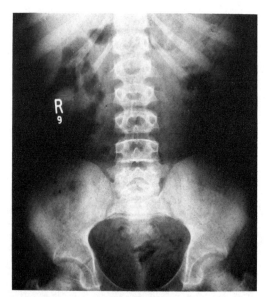

Fig. 19.13   C.M.—7½ years—AP pelvis and lumbar spine
The femoral heads are rather large, otherwise a normal radiograph
(spina bifida S1).

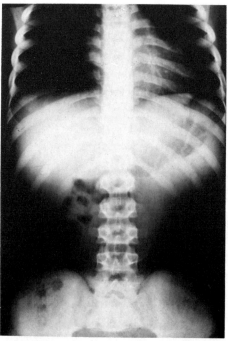

Fig. 19.14   J.C.—17 years 4 months—AP spine
There is some narrowing of the interpedicular distances in the lower
lumbar spine. L5 is rather low-set between the iliac wings, indicating
probable lumbar lordosis and pelvic tilt.

Spine: Adult

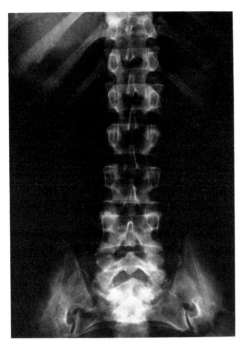

Fig. 19.15   O.H.—32 years—AP lumbar spine
L5 is low-set. There is no progressive widening of the interpedicular distances.
Same patient: Fig. 19.22

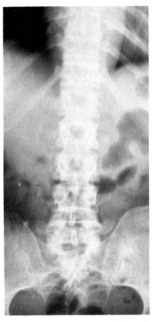

Fig. 19.16   E.M.—51 years 9 months—AP spine
L5 is low-set. The interpedicular distances narrow progressively in the lumbar spine. (There is spina bifida of S1.)

Lateral spine : 4–11 years

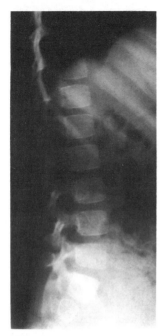

Fig. 19.17   A.W.—4½ years—Lateral spine
Normal.

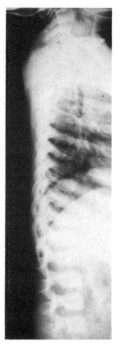

Fig. 19.18   R.W.—4 years 9 months—Lateral spine
Rather oval configuration indicating some immaturity. No other
abnormality.
Same patient : Figs 19.6, 19.27, 19.33

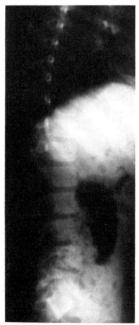

Fig. 19.19   M.R.—7 years—Lateral spine
Normal.
Same patient : Fig. 19.12

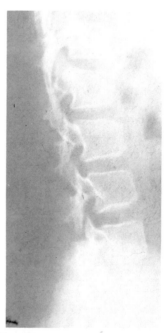

Fig. 19.20   H.W.—11 years 3 months—Lateral lumbar spine
The pedicles are very short. There is posterior scalloping of the
lumbar vertebral bodies. (This is indistinguishable from classical
achondroplasia.)
Same patient : Fig. 19.29

Lateral spine : 11 years–Adult

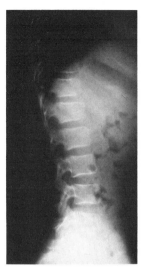

Fig. 19.21   G.S.—11 years 9 months—Lateral spine
The pedicles are rather short, but there is no scalloping.

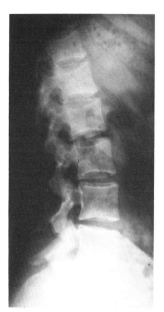

Fig. 19.22   O.H.—32 years—Lateral spine
The pedicles are very short causing a reduction in the AP diameter of the spinal canal.
Same patient : Fig. 19.15

Pelvis: 3–16 years

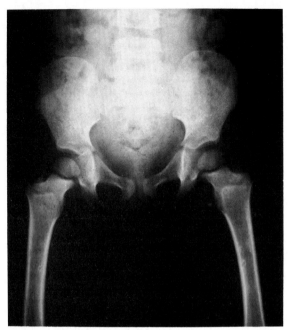

Fig. 19.23   T.H.—3 years—AP pelvis
Normal pelvic configuration. The femoral heads are large and the
femoral necks appear slightly short.
Same patient: Figs 19.11, 19.37

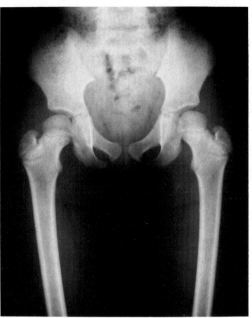

Fig. 19.24   J.C.—11 years—AP pelvis
Large femoral heads incompletely covered by the acetabular roofs.
Large greater trochanters.
Same patient: Fig. 19.28

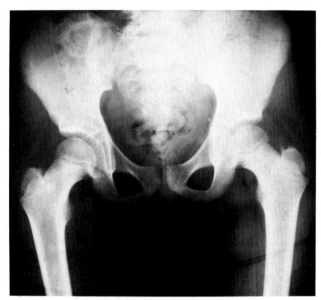

Fig. 19.25   F.G.—16 years—AP pelvis
Probably normal.
Same patient: Fig. 19.31

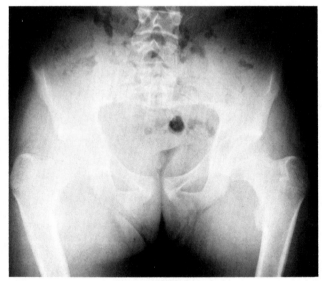

Fig. 19.26   D.C.—16 years—AP pelvis
L5 and the sacrum are low-set. The femoral heads are large.
Same patient: Figs 19.8, 19.32, 19.36

Lower limb: 5–13 years

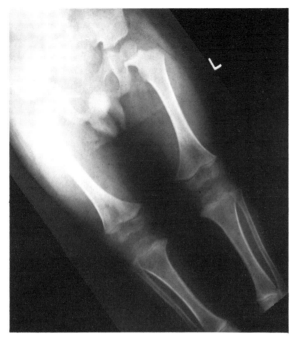

Fig. 19.27   R.W.—5 years 9 months—AP both lower limbs
There is shortening of the long bones with over-modelling and slight
metaphyseal flare. There is mild coxa vara with short femoral necks.
The capital femoral epiphyses appear unusually round. The proximal
fibulae are long.
Same patient: Figs 19.6, 19.18, 19.33

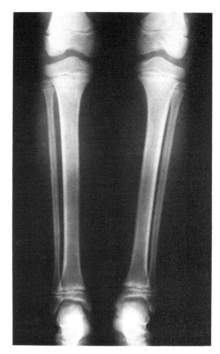

Fig. 19.28   J.C.—11 years—AP tibiae and fibulae
There is slight prominence of both medial femoral epicondyles.
Otherwise normal.
Same patient: Fig. 19.24

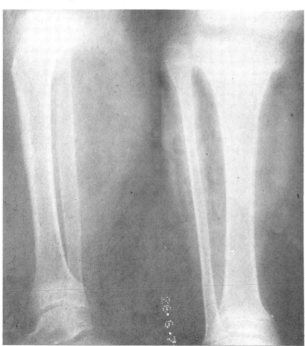

**Fig. 19.29   H.W.—11 years 9 months—AP and lateral tibia
and fibula**
The fibula is long in relation to the tibia, both proximally and distally.
The tibia is short with over-modelling and metaphyseal flaring. The
epiphyses are large.
Same patient: Fig. 19.20

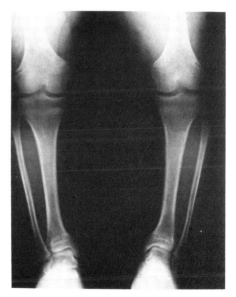

Fig. 19.30   J.K.—13 years—AP tibiae and fibulae
The medial femoral condyles are large. The intercondylar notches are
narrow and there are chevron deformities of the distal femoral
epiphyses. These changes are approaching those found in classical
achondroplasia. The tibiae are short with over-modelling. The fibulae
are long with some lateral bowing and deformity at the ankle joints.

Lower limb: 16 years

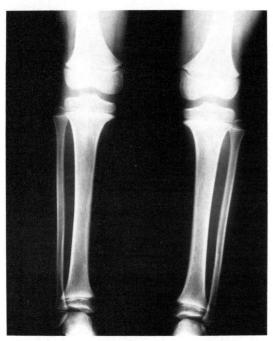

Fig. 19.31   F.G.—16 years—AP both tibiae and fibulae
There is mild shortening of the tibiae with some metaphyseal flaring.
The epiphyses are rather large. The distal ends of the fibulae are long.
Same patient: Fig. 19.25

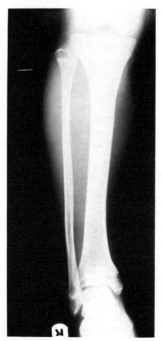

Fig. 19.32   D.C.—16 years—AP right tibia and fibula
The distal fibula is long with a tilt of the talus and lateral bowing of
the fibular shaft.
Same patient: Figs 19.8, 19.26, 19.36

Upper limb: 2–16 years

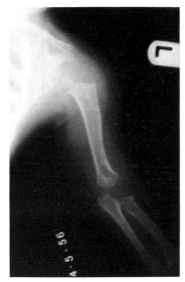

Fig. 19.33   R.W.—2 years 4 months—AP upper limb
There is shortening of the long bones which are rather broad. The deltoid tuberosity is pronounced.
Same patient: Figs 19.6, 19.18, 19.27

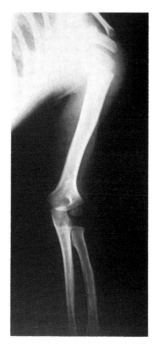

Fig. 19.34   M.P.—4 years 4 months—AP upper limb
The appearances are essentially normal.

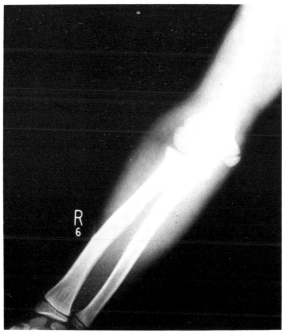

Fig. 19.35   G.M.—13 years—AP right forearm
Normal bone modelling.
Same patient: Fig. 19.10

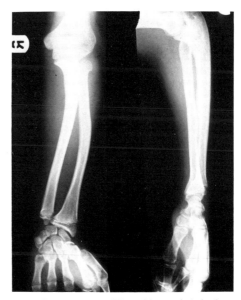

Fig. 19.36   D.C.—16 years—AP and lateral right forearm
The distal ulna is short and there is lateral bowing of the radius and ulna.
Same patient: Figs 19.8, 19.26, 19.32

Hand : 3–14 years

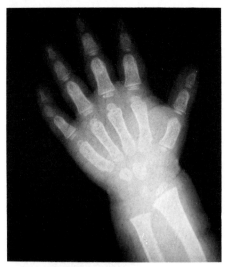

Fig. 19.37   T.H.—3 years—PA left hand
Normal appearances, apart from clinodactyly.
Same patient : Figs 19.11, 19.23

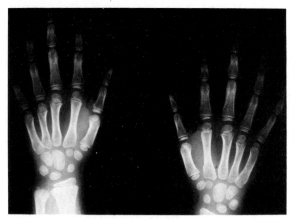

Fig. 19.38   J.H.—7 years—PA hands (daughter of patient in
Figure 19.15)
Normal appearances.

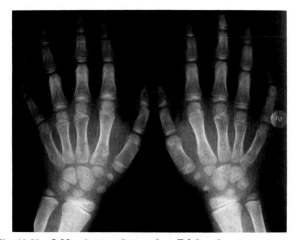

Fig. 19.39   S.M.—9 years 5 months—PA hands
There is bilateral clinodactyly. Otherwise normal.
Same patient : Fig. 19.9

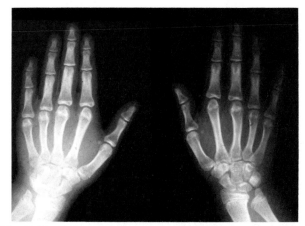

Fig. 19.40   M.R.—14 years—PA both hands
Both ulnae are short distally and their epiphyseal plates are angled
towards the radii.

Chapter Twenty
DYSCHONDROSTEOSIS

A disorder affecting the forearm and leg bones only, with disproportionate growth between the radius/ulna and tibia/fibula, leading sometimes to disruption at the wrist, knee or ankle joints.

Inheritance

Autosomal dominant, probably only 50% penetrance.

Frequency

At least 3 per million index patients, 7 per million including affected relatives. There may be more, since patients may not attend hospital and may remain undiagnosed.

Clinical features

Facial appearance—normal.
Intelligence—normal.
Stature—short-limbed short stature, associated with forearm and tibia/fibula shortening.
Presenting feature/age—in childhood, with short stature. Occasionally later, with wrist pain.

Deformities

Forearm—dorsal dislocation of lower end of ulna, with a short bowed radius.
Leg—usually less severely affected, but there may be disruption of the ankle mortice due to the fibula being 'too long or too short', or tibia varum with disruption at the knee joint.
Range of severity—wide. Some individuals have only minimal shortening of both bones, and no mechanical disruption at wrist, knee or ankle.
Associated anomalies—none.

Radiographic features

Skull
Spine
Thorax } all normal.
Shoulder girdle/humerus
Pelvic girdle/femur

Radius—varies from near normality, apart from reduction in overall length, to marked shortening and bowing associated with failure of development of the medial half of the lower epiphysis.
Ulna—varies from normality apart from reduction in overall length to 'Madelung's deformity' with dorsal dislocation at the lower end.
Wrist—pyramidal appearance of the carpus with the lunate at the apex, fitting in between the radius and ulna.
Tibia and fibula—occasional medial beaking at the upper end of tibia. There may be equal shortening of both bones. The fibula may be 'too short' at upper or lower ends, or both, with consequent disruption at the knee or ankle mortice.

Bone maturation

Apparently normal, apart from the lower radial epiphysis.

Biochemistry

Not known.

Differential diagnosis

Trauma/infection may cause similar (secondary) effects on the growth plates, but this is unlikely to be bilateral.
Diaphyseal aclasis—here the ulna is short, not the radius, and exostoses are present elsewhere in the skeleton.
Hypochondroplasia—this may be difficult to distinguish, but Madelung's deformity does not develop and metaphyses may be splayed.
Turner's syndrome may have a Madelung-type wrist deformity but is easily distinguished on other clinical features.

Progress/complications

Symptoms are insignificant in most cases, in spite of ulnar dislocation. During the growing period problems may arise from disproportionate growth of the tibia and fibula, either at the knee or ankle joints.

REFERENCES

Beals R K, Lovrien E W 1976 Dyschondrosteosis and Madelung's
    deformity: report of three kindreds and review of the literature.
    Clinical Orthopaedics and Related Research 116:24–28
Dawe C, Wynne-Davies R 1982 Clinical variation in
    dyschondrosteosis. A report of 13 individuals in 8 kindreds.
    Journal of Bone and Joint Surgery 64B:377–381
Golding J S R, Blackburne J S 1976 Madelung's disease of the wrist
    and dyschondrosteosis. Journal of Bone and Joint Surgery
    58B:350–352

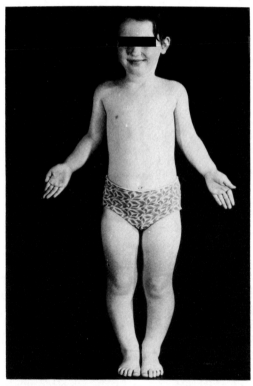

Fig. 20.1   Shortening of limbs, particularly the upper in this child.
Disproportionate growth between the tibia and fibula is leading to
tibia varum, more marked on the left than the right.

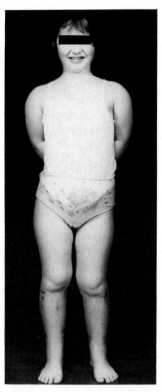

Fig. 20.2   Three years later, a postoperative picture following fibular
osteotomy and tibial lengthening.

Upper limb: 4–6 years

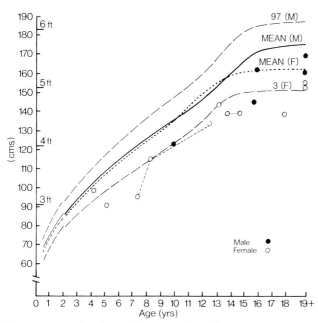

Fig. 20.3 Dyschondrosteosis—percentile height.

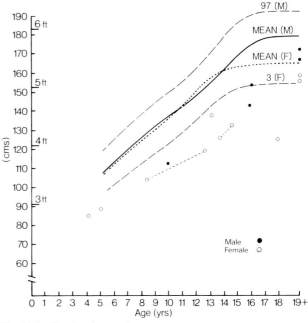

Fig. 20.4 Dyschondrosteosis—percentile span.

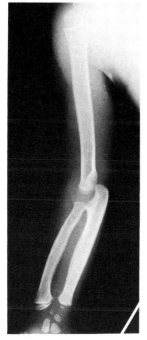

Fig. 20.5 L.C.—4 years 3 months—Right upper limb (sister of child in Figure 20.6)
Even at this age the failure of development of the medial side of the distal radial epiphysis is apparent. There is also slight lateral bowing of the radial diaphysis.

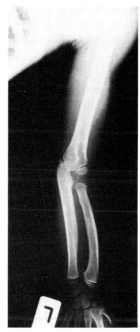

Fig. 20.6 N.C.—6 years 8 months—Left upper limb (sister of child in Figure 19.5)
The sister, here, is 2 years older and the radius has a similar appearance, now slightly more obvious.

Wrist: 5–12 years

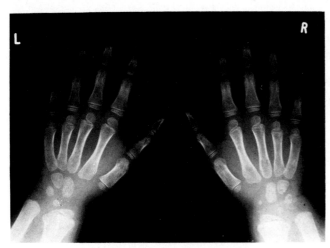

Fig. 20.7   G.C.—5 years 3 months—AP both wrists
The failure of the medial aspect of the distal radial epiphysis is well shown. The lunate is abnormally situated being more proximal and lateral than usual.

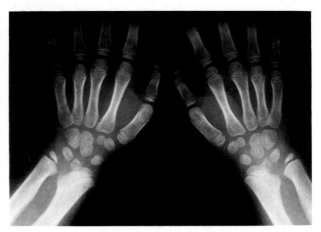

Fig. 20.8   A.M.—5 years 9 months—AP both wrists
The distal radius is abnormal on its medial side. The epiphyseal plate has already fused here but the lateral radius continues to grow giving an abnormal slope. This radial defect has altered the configuration of the carpus, and the lunate lies in the angle between the radius and ulna.

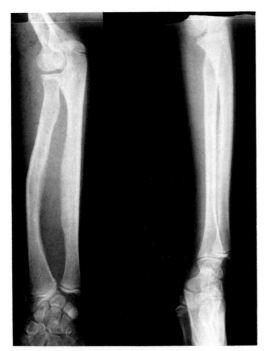

Fig. 20.9   K.J.—12 years 10 months—AP and lateral right forearm and wrist
The medial aspect of the distal radial epiphysis is less well developed than normal. The lateral side has continued to grow giving rise to lateral bowing of the shaft. The lunate lies in the angle between the radius and ulna. The ulna shows an unusually pronounced intraosseous ridge.
Same patient: Fig. 20.17

Wrist: 13–15 years

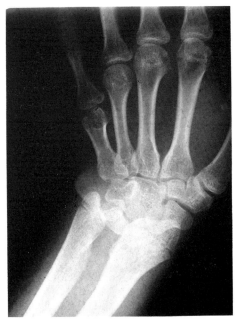

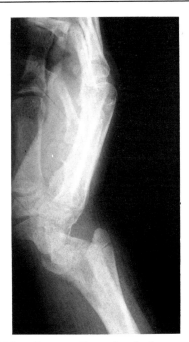

Fig. 20.10   H.L.—13 years—AP wrist
The lower ends of the radius and ulna are abnormal, being of equal length, widely separated, and with disruption at the wrist joint. There is an abnormally short metacarpal and middle phalanx of the little finger.

Fig. 20.11   H.L.—13 years—Lateral wrist (same patient as in Figure 20.10)
    The dorsal dislocation of the ulna is apparent.

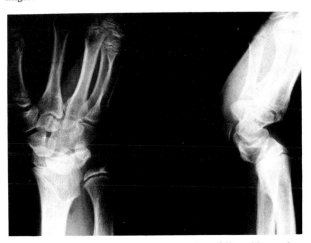

Fig. 20.12   C.MacG.—15 years 7 months—AP and lateral wrist
There is a similar Madelung-type deformity here, with failure of development of the medial side of the distal radial epiphysis.

Wrist: 19 years–Adult

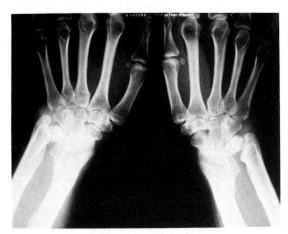

Fig. 20.13   D.S.—19 years—AP both wrists
The radius and ulna are of equal length and the disruption at the
wrist is obvious. The ulnar styloid process is unusually pronounced.
Same patient: Figs 20.14, 20.18

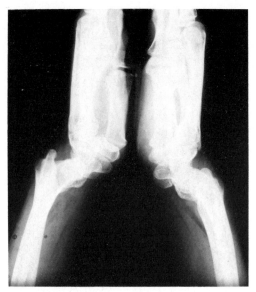

Fig. 20.14   D.S.—19 years—Lateral both wrists (same patient as
in Figure 20.13)
Lateral view.

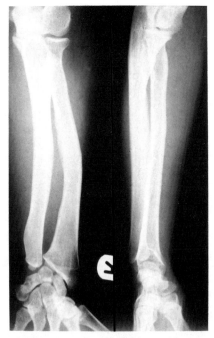

Fig. 20.15   A.H.—32 years—AP and lateral forearm and wrist
This patient is less severely affected than the one in Figures 20.13 and
20.14, but the disorder at the lower radial epiphysis and bowing are
obvious, as well as the pyramidal appearance of the carpus with the
lunate at the apex. There is only a minor degree of subluxation of the
distal ulna. The interosseous ridge on the radius is unusually
pronounced.

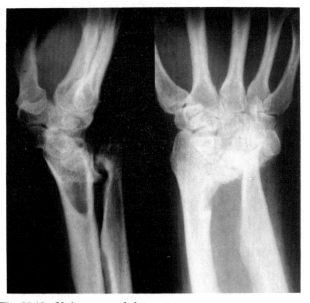

Fig. 20.16   Unknown—adult
There is a bizarre deformity of the neck of the ulna and adjacent part
of the radius, possibly caused by an unusual degree of proximal
displacement of the lunate.

Lower limb: 13 years–Adult

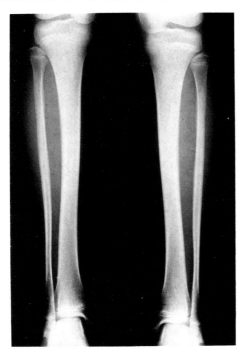

Fig. 20.17 K.J.—13 years 1 month—AP tibiae and fibulae
There is little abnormality to be seen here other than slight shortening of the fibulae proximally.
Same patient: Fig. 20.9

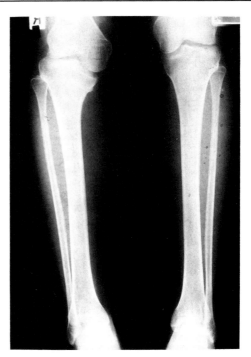

Fig. 20.18—D.S.—19 years—AP tibiae and fibulae
The right leg is the more severely affected. There has been disordered epiphyseal growth at the upper tibia medially.
Same patient: Figs 20.13, 20.14

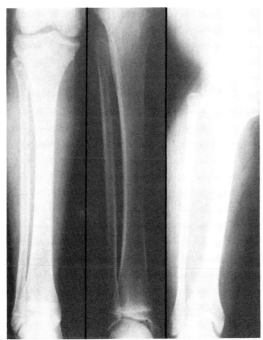

Fig. 20.19 Composite of three patients
Three different examples of disproportionate tibio-fibular growth.
Left: The fibula is short at its upper end. **Centre:** The fibula is too short at both ends. **Right:** The fibula is too long at both ends.

Chapter Twenty-one
FETAL FACE SYNDROME (Robinow)

A characteristic facies with bulging forehead and hypertelorism, associated with forearm/leg shortening and genital hypoplasia.

Inheritance

Autosomal dominant.

Frequency

Extremely rare.

Clinical features

Facial appearance—so-called 'fetal face' because of the disproportionately large neurocranium, hypertelorism and short up-turned nose.

Intelligence—normal.

Stature—short, principally due to the short limbs.

Presenting feature/age—at birth, with peculiar facies, short limbs and genital anomalies.

Deformities—variable skeletal deformities including congenital scoliosis and dislocation of the radial head.

Associated anomalies—genital hypoplasia with small or absent penis/clitoris.

Radiological features

Skull—normal.

Spine and thorax—congenital vertebral anomalies, rib fusions and premature calcification of costal cartilages.

Long bones—the shortening is more marked in the radius and ulna than in the tibia and fibula, but rarely may be rhizomelic.

Hands/feet—normal.

Bone maturation

Not known.

Biochemistry

Not known.

Differential diagnosis

From other *mesomelic forms of dwarfism*, but the normal hands, together with genital hypoplasia should differentiate the Robinow syndrome.

Progress/complications

No complications other than noted above. The life span is probably normal.

REFERENCES

Kaitila I I, Leisti J T, Rimoin D L 1976 Mesomelic skeletal dysplasias. Clinical Orthopaedics and Related Research 114:94–106

Vera-Roman J M 1973 Robinow dwarfing syndrome accompanied by penile agenesis and hemivertebrae. American Journal of Diseases of Children 126:206

Wadlington W B, Tucker V L, Schimke R N 1973 Mesomelic dwarfism with hemivertebrae and small genitalia (the Robinow syndrome). American Journal of Diseases of Children 126:202–205

Skull and spine: 7 years

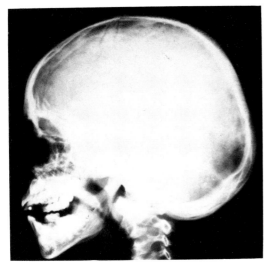

Fig. 21.1   A.K.—7 years—Lateral skull
Normal.
Same patient: Figs 21.2–21.8

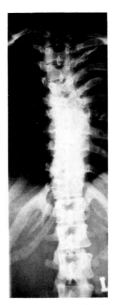

Fig. 21.2   A.K.—7 years—AP dorsal spine (same patient as in Fig. 21.1)
In addition to a left sided mid-dorsal hemi-vertebra there are anterior cleft bodies and vertebral fusions. Rib deformities are present and there is abnormal costal cartilage calcification.

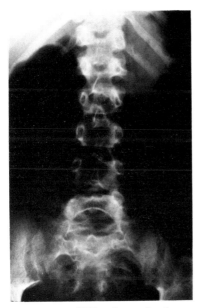

Fig. 21.3   A.K.—7 years—AP lumbar spine (same patient as in Fig. 21.1)
Disc space narrowing is present at L2 to L3 and L4 to L5.

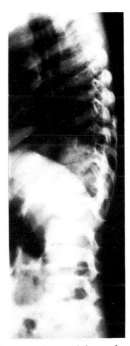

Fig. 21.4   A.K.—7 years—Lateral dorso-lumbar spine (same patient as in Fig. 21.1)
The bodies of D11 and D12 are fused. Disc space narrowing is present in the lumbar region.

Chest and limbs: 7 years

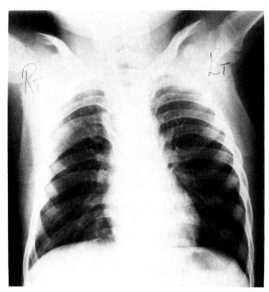

Fig. 21.5   A.K.—7 years—AP chest
The right third and fourth ribs are fused and all the ribs are unusually broad.
Same patient: Figs 21.2–21.4, 21.6–21.8

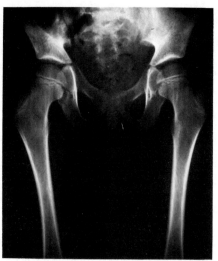

Fig. 21.6   A.K.—7 years—AP hips and femora (same patient as in Fig. 21.5)
The hip joints appear normal.

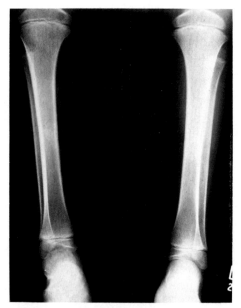

Fig. 21.7   A.K.—7 years—AP tibiae and fibulae (same patient as in Fig. 21.5)
Both fibulae are short, especially at their upper ends.

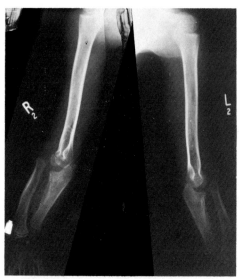

Fig. 21.8   A.K.—7 years—AP both upper limbs (same patient as in Fig. 21.5)
Pronounced shortening with some bowing of both ulnae and radii is present and the radial heads are dislocated.

Chapter Twenty-two
MESOMELIC DYSPLASIA
(Werner type: absent/dysplastic tibiae with five-fingered hand)

Absent or dysplastic tibiae are associated with intact fibulae, five-fingered hands, sometimes pre-axial polydactyly in addition and up to seven digits in the foot.

Inheritance

Autosomal dominant.

Frequency

Extremely rare.

Clinical features

Facial appearance—normal.
Intelligence—normal.
Stature—reduced on account of the tibial deformity or absence, otherwise normal.
Presenting feature/age—birth, with limb defects.
Deformities—in addition to the obvious structural deformities, there is secondary 'club-foot' (equinovarus). Knees and ankles may be dislocated.
Associated anomalies—congenital heart defects have been reported.

Radiographic features

Tibia—this varies from normality in mildly affected individuals to total absence. If a shortened tibia is present it becomes grossly thickened during growth.
Fibula—this is characteristically intact, but if it is the main leg bone, it becomes grossly thickened.
Hands—the five-fingered hand (i.e. with no thumb) is characteristic. The first digit has no thenar eminence, a metacarpal of approximately equal length to the remaining digits, and a distal metacarpal epiphysis. In addition there may be a sixth digit (pre-axial polydactyly) having the characteristics of a thumb, with a short metacarpal and proximal epiphysis.
Feet—pre-axial polydactyly may be present, with up to seven toes.

Bone maturation

Normal.

Biochemistry

Not known.

Differential diagnosis

From other limb 'absence' defects, but the association of abnormalities here is specific.

Progress/complications

The complications are related to the severity of the leg and foot deformities. The hands are improved functionally by pollicisation of first digit.

REFERENCES

Hall C M 1981 Werner's mesomelic dysplasia with ventricular septal defect and Hirschsprung's disease. Pediatric Radiology 10:247–249
Lamb D W, Wynne-Davies R, Whitmore J M 1983 Five fingered hand associated with partial or complete tibial absence and pre-axial polydactyly. Journal of Bone and Joint Surgery 65B:60–63

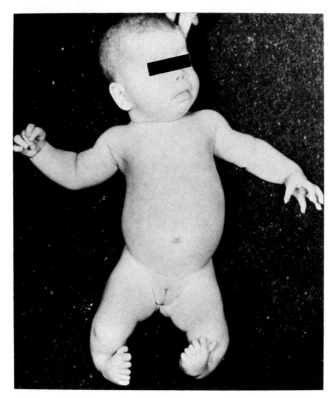

Fig. 22.1   Absence of tibiae, pre-axial polydactyly of the toes and a five-fingered hand.

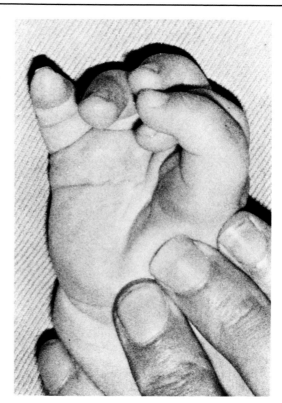

Fig. 22.2   Five digits, but all are 'finger-like'; there is no thumb rotation and the thenar eminence is absent.

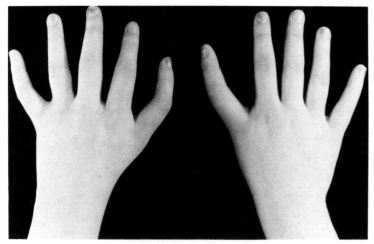

Fig. 22.3   An older patient (a cousin of the patients in Figures 22.1 and 22.2), showing a typical five-fingered hand.

Lower limb: Birth–Adult

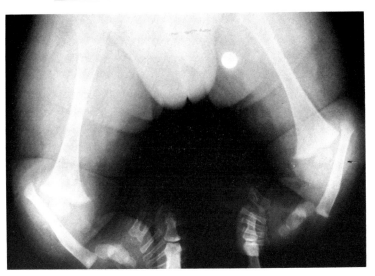

Fig. 22.4   J.D.—Neonate—AP lower limb
There is total absence of both tibiae and pre-axial polydactyly of the toes.

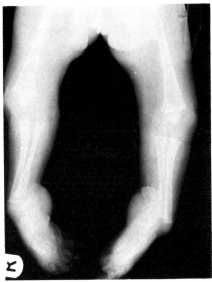

Fig. 22.5   J.A.—5 months—AP lower limb
Both tibiae are present, but hypoplastic, and there is dislocation at the knee and ankle.
Same patient: Figs 22.6, 22.8, 22.9

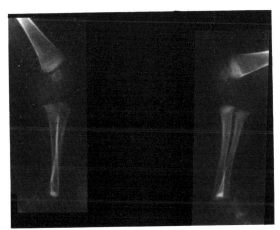

Fig. 22.6   J.A.—5 months—Lateral tibia and fibula (same patient as in Fig. 22.5)
Lateral view.

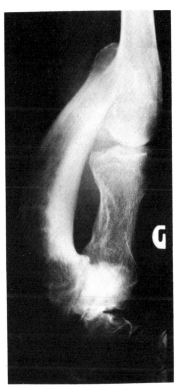

Fig. 22.7   A.D.—56 years—AP left tibia and fibula
(grandfather of child in Fig. 22.4)
There is massive thickening of both leg bones particularly the fibula, but the tibia is disproportionately short.

Hands and feet: 5 months—Adult

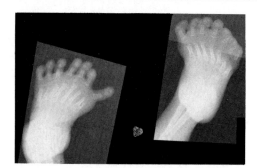

Fig. 22.8   J.A.—5 months—PA feet
There are seven or eight digits on each foot and duplication of tarsal bones.
Same patient: Figs 22.5, 22.6, 22.9

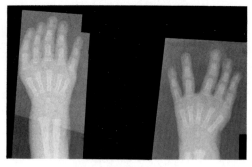

Fig. 22.9   J.A.—5 months—PA hands (same patient as in Fig. 22.8)
This is a 'five-fingered' hand—the first digit being un-rotated, not having a shorter metacarpal than digits 2–5, and possessing three phalanges.

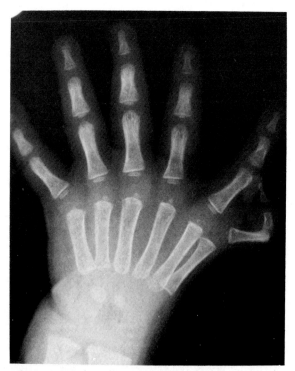

Fig. 22.10   B.M.—2 years 6 months—PA hands (cousin of patient in Fig. 22.11)
Similar to Figure 22.9, but there is also pre-axial polydactyly. The true first digit here has a distal metacarpal epiphysis, unlike a normal thumb metacarpal.

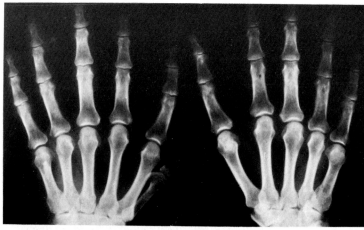

Fig. 22.11   A.D.—Adult—PA hands (cousin of child in Fig. 22.11; father of child in Fig. 22.4)
A five-fingered hand similar to Figures 22.9 and 22.10, with pre-axial polydactyly on one side only.

Chapter Twenty-three
MESOMELIC DYSPLASIA
(severe dominant form)

Gross shortening of distal limb bones with partial absence of the fibula.

Inheritance

Autosomal dominant.

Clinical features

These are confined to the forearms and legs. The fibulae are partially or completely absent, causing disruption at the ankle joint. The ulna is shorter than the radius, which is bowed, and the radial head is malformed and dislocated.

REFERENCES

McKusick V A 1972 Heritable disorders of connective tissue. C V Mosby, St Louis, p 782
Leroy J G, De Vos J, Timmermans J 1975 Dominant mesomelic dwarfism of the hypoplastic tibia, radius type. Clinical Genetics 7 (4):280
Silverman F N 1973 In: Kaufmann H J (ed) Intrinsic diseases of bone. Progress in Pediatric Radiology 4. S Karger, Basel, pp 546–562

Limbs: Birth–Adult

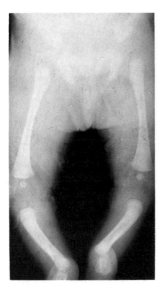

Fig. 23.1   B.S. Neonate—AP lower limbs
Both tibiae are short with absent epiphyses at the knees and the
fibulae are almost entirely absent. There is dislocation at the ankles.

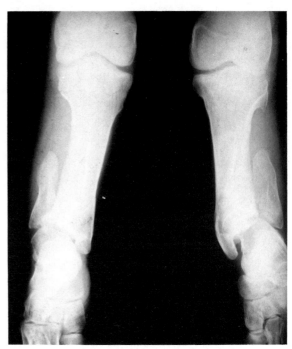

Fig. 23.2   M.S.—Adult—AP both legs (mother of child in Fig.
23.1)
The tibiae are short and very broad and only the distal halves of the
fibulae are present. Both ankle joints are disorganised.

Fig. 23.3   M.S.—Adult—AP forearm (same patient as in Fig.
22.2)
In addition to ulnar shortening the radius is short and bowed with
dislocation of its head.

Chapter Twenty-four
ACROMESOMELIC DWARFISM

Short stature due predominantly to shortening of the forearm and leg, with marked brachydactyly.

Inheritance

Autosomal recessive.

Frequency

Not known: extremely rare.

Clinical features

Facial appearance—may be normal or with slight nasal hypoplasia.

Intelligence—normal.

Stature—short, due to shortening of limb bones. Adult height about 120 cm.

Presenting feature/age—short limbs at birth, with stubby hands and feet.

Deformities—lumbar kyphos, bowing of forearm and leg bones, dislocated head of radius.

Associated anomalies—none known.

Radiological features

Skull—normal, or with some facial hypoplasia.

Spine—interpedicular narrowing from L1 to L5 and short pedicles on the lateral view. Lumbar kyphos.

Pelvis/hips—large capital femoral epiphyses with acetabular dysplasia.

Long bones—short, particularly in the distal part of the limb, with flared metaphyseal areas.

Hands/feet—all bones are short and rather broad.

Bone maturation

Not known.

Biochemistry

Not known.

Differential diagnosis

From *achondroplasia* but here the limb shortening is predominantly of the proximal segment, and the hands do not show the same degree of brachydactyly.

From other disorders in which *brachydactyly* is a feature, but in acro-mesomelic dwarfism there is also marked shortening of the forearm and leg bones.

Progress/complications

No complications are known, apart from kyphosis or scoliosis, and possibly secondary osteoarthritis of the hips.

REFERENCES

Hall C M, Stoker D J, Robinson D C, Wilkinson D J 1980 A case report on acromesomelic dysplasia. British Journal of Radiology 53:999–1003

Langer L O Jr, Beals R K, Solomon I L et al 1977 Acromesomelic dwarfism: manifestations in childhood. American Journal of Medical Genetics 1:87–100

Langer L O, Garrett R T 1980 Acromesomelic dysplasia. Radiology 137:349

Skull, spine and chest: 8 years–Adult

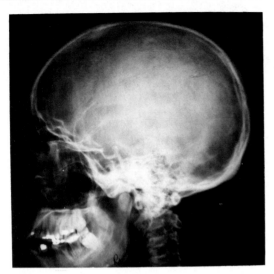

Fig. 24.1   S.H.—Adult—Lateral skull
The mandible is large in relation to the vault and maxilla.
Same patient: Figs 24.4, 24.6–24.8, 24.10

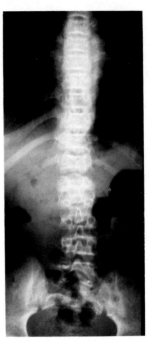

Fig. 24.2   G.H.—8 years—AP spine
There is progressive narrowing of the interpedicular distances from L1 to L5 and the sacrum is low-set between the ilia.
Same patient: Figs 24.3, 24.5, 24.9

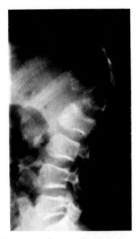

Fig. 24.3   G.H.—8 years—Lateral spine (same patient as in Fig. 24.2)
The pedicles are short. There is platyspondyly with some posterior wedging and a kyphos at L1.

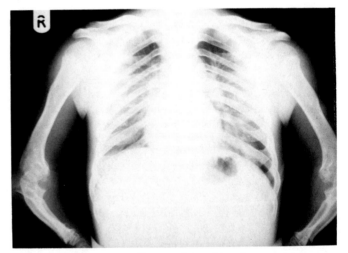

Fig. 24.4   S.H.—Adult—Chest and shoulders (same patient as in Fig. 24.1)
The proximal humeral epiphyses are abnormally modelled and the deltoid tuberosities unusually prominent.

Pelvis : 8 years–Adult

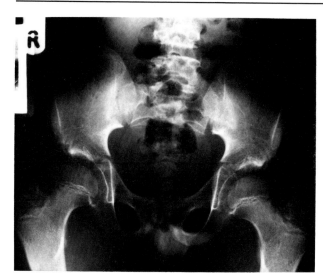

Fig. 24.5   G.H.—8 years—AP pelvis
The acetabular roofs are poorly developed and the large, rounded
capital femoral epiphyses incompletely covered.
Same patient : Figs 24.2, 24.3, 24.9

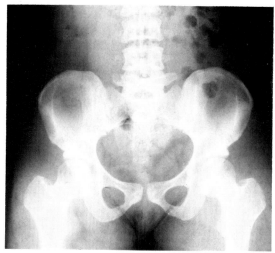

Fig. 24.6   S.H.—Adult—AP pelvis
The iliac wings are small and the femoral heads large with narrow
femoral necks.
Same patient : Figs 24.1, 24.4, 24.7, 24.8, 24.10

Limbs: 8 years–Adult

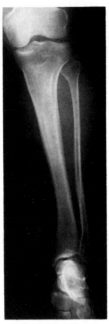

Fig. 24.7   S.H.—Adult—AP tibia and fibula
Both the tibia and fibula are short.
Same patient: Figs 24.1, 24.4, 24.6, 24.8, 24.10

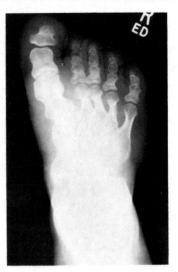

Fig. 24.8   S.H.—Adult—PA foot (same patient as in Fig. 24.7)
All long bones are short and 'squat'.

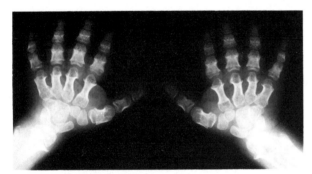

Fig. 24.9   G.H.—8 years—PA hands
The distal radial and ulnar epiphyses are unusually large and
rounded, the adjacent metaphyses flared, and the ulnae short. All the
metacarpals and phalanges are extremely short and cone epiphyses are
present.
Same patient: Figs 24.2, 24.3, 24.5

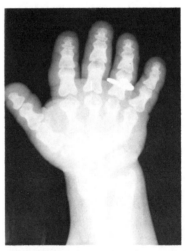

Fig. 24.10   S.H.—Adult—PA right hand (same patient as in Fig.
24.7)
The interosseous ridge on the ulnar shaft is unusually prominent, and
the lower end of the radius enlarged. There is marked shortening of
all metacarpals and phalanges.

Chapter Twenty-five
ACRODYSOSTOSIS
(Peripheral dysostosis)

Short stature with peripheral limb shortening, stubby hands and feet and nasal hypoplasia.

Inheritance

Autosomal dominant, but most appear to be sporadic.

Frequency

Not known—very rare.

Clinical features

Facial appearance—sometimes 'pug' nose with anteverted nostrils and nasal hypoplasia. There may be hypertelorism, maxillary hypoplasia and malocclusion.

Intelligence—often, but not invariably retarded.

Stature—short, principally mesomelic (of radius/ulna and tibia/fibula).

Presenting feature/age—odd facies at birth, later short stature, mental retardation and short, stubby hands and feet.

Deformities—none other than noted above.

Associated anomalies—none known.

Radiographic features

Skull—sometimes brachycephaly and thickening of the vault and base.

Spine—minor changes similar to Scheuermann's disease.

Pelvis—possibly some acetabular dysplasia.

Hands/feet—shortening of all metacarpals and phalanges with cone epiphyses and premature fusion.

Bone maturation

Early fusion of hand bones.

Biochemistry

Not known.

Differential diagnosis

From many other disorders which feature *brachydactyly*, but the facial appearance and short stature of acrodysostosis should differentiate it. The short stature is not so severe as in *acromesomelic dwarfism*.

Progress/complications

There appear to be no complications, except possibly acetabular dysplasia with secondary osteoarthritis.

REFERENCES

Giedion A 1973 Acrodysplasias. In: Kaufmann H J (ed) Intrinsic diseases of bone. Progress in Pediatric Radiology 4. S Karger, Basel, pp 325–345

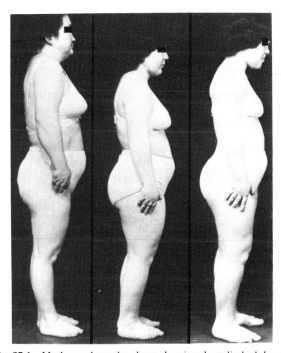

Fig. 25.1 Mother and two daughters showing short-limbed short stature and markedly stubby hands.

Skull and chest: 8 years–Adult

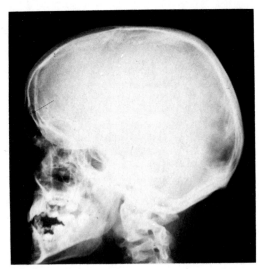

Fig. 25.2   F.M.—8 years—Lateral skull
The frontal bone is rather prominent and this is associated with some mid-face flattening.
Same patient: Figs 25.5–25.10

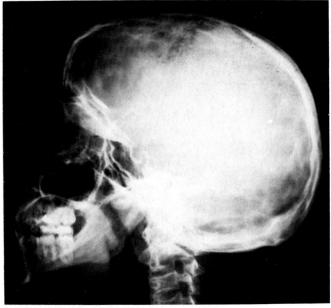

Fig. 25.3   D.M.—14 years—Lateral skull
There is patchy increased density of the vault and base.
Same patient: Fig. 25.11.

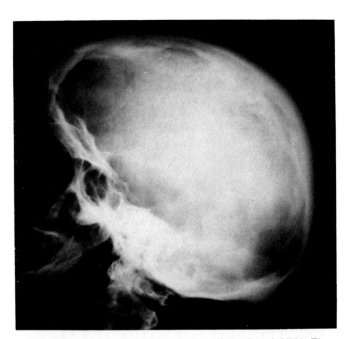

Fig. 25.4   M.M.—Adult—Lateral skull (mother of child in Fig. 25.3)
The vault and base are sclerotic and there is some maxillary hypoplasia.
Same patient: Fig. 25.12

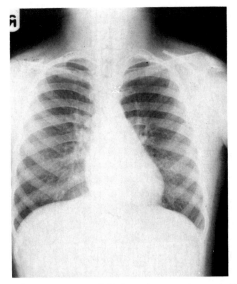

Fig. 25.5   F.M.—8 years—AP chest (same patient as in Fig. 25.2)
The ribs are broad and the clavicles rather long.

Spine and lower limb : 8 years

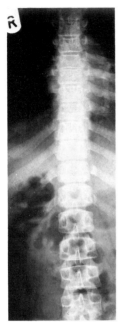

Fig. 25.6 F.M.—8 years—AP thoraco-lumbar spine
The disc spaces are a little narrowed and the ribs rather broad.
Same patient: Figs 25.2, 25.5, 25.7–25.10

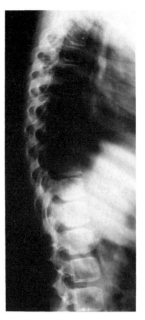

Fig. 25.7 F.M.—8 years—Lateral thoraco-lumbar spine
(same patient as in Fig. 25.6)
The vertebral bodies are reduced in their AP diameters, and the end-plates have an oval configuration. Some disc space narrowing is present.

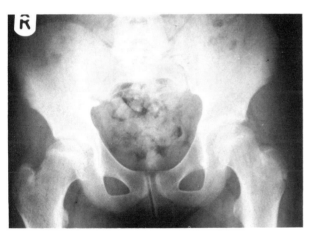

Fig. 25.8 F.M.—8 years—Pelvis (same patient as in Fig. 25.6)
The capital femoral epiphyses are slightly flattened and incompletely covered by the acetabular roofs.

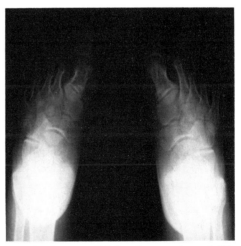

Fig. 25.9 F.M.—8 years—PA feet (same patient as in Fig. 25.6)
There is marked shortening of the metatarsals and phalanges.

Hand: 8 years–Adult

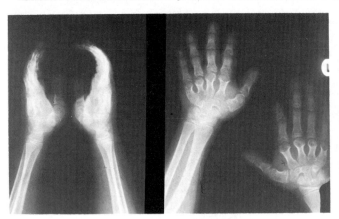

Fig. 25.10   F.M.—8 years—Lateral and PA hands and wrists
Pronounced shortening of all metacarpals is present, associated with premature fusion of their epiphyses. The short phalanges (with some sparing of the proximal row) have cone shaped epiphyses.
Same patient: Figs 25.2, 25.5–25.9

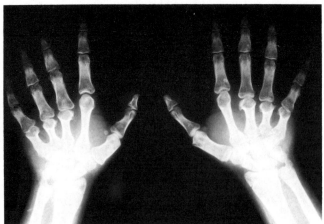

Fig. 25.11   D.M.—14 years—PA hands
The third to fifth metacarpals and the middle and terminal phalanges are short.
Same patient: Fig. 25.3.

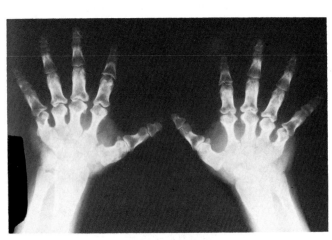

Fig. 25.12   M.M.—Adult—PA hands
The long bones of the hands are all short, but this shortening is most pronounced in the metacarpals and terminal phalanges.
Same patient: Fig. 25.4

Section Seven
SHORT LIMBS AND TRUNK

Chapter Twenty-six
PSEUDOACHONDROPLASIA

Short-limbed dwarfism, sometimes very severe, and (unlike achondroplasia) with marked epiphyseal as well as metaphyseal involvement, leading to premature osteoarthritis in most joints.

Inheritance

Autosomal dominant and recessive forms both occur, although most cases are sporadic. Radiological delineation is not possible.

Frequency

Possible prevalence is about 4 per million population—about half as frequent as classical achondroplasia.

Clinical features

Facial appearance—normal.
Intelligence—normal.
Stature—see graphs. Adult height between 82 and 130 cm, with marked limb shortening.
Presenting feature/age—frequently not diagnosed at birth, the short stature and disproportionately short limbs not being apparent until 2–3 years of age.
Deformities—those of secondary osteoarthritis. Some patients have a gross degree of joint laxity with genu valgum, varum or recurvatum. Scoliosis may occur.
Associated anomalies—some with marked joint laxity and instability: perhaps commoner in the autosomal recessive form.

Radiological features

Skull—normal.
Spine—there are variable deformities of the vertebrae on the lateral view: persistent oval shape in early childhood, anterior 'beaking', platyspondyly, triangular outline, odontoid dysplasia; but some are relatively normal.
Long bones—markedly shortened with flared metaphyses (as in achondroplasia), but here the epiphyses are also grossly abnormal: delayed in time of appearance,

small, irregular and fragmented. These changes are very obvious in the hands and feet as well as more proximally.
Pelvis and hips—the triradiate cartilage is strikingly wider than normal and the acetabulum poorly formed. The capital femoral epiphyses, as all other epiphyses, are late in appearing and remain small. The femoral neck usually shows medial beaking. After epiphyseal fusion, the appearance of the hip joint is similar to that in many other epiphyseal dysplasias.

Bone maturation

Much delay in pelvic maturation (that is, in closure of the triradiate cartilage and ischiopubic ramus) and in development of the greater trochanteric epiphyses. The capital femoral epiphyses remain at a bone age of 3 years or under. Delayed maturation at the hand and wrist is less marked, but still grossly retarded compared with normal.

Biochemistry

Not known.

Differential diagnosis

Achondroplasia. Here the head is usually large, with a prominent frontal region and depressed bridge of nose. Radiographically the epiphyses are normal in appearance, and the pelvis characteristically square, with small sciatic notches.
Severe cases of multiple epiphyseal dysplasia may be difficult to differentiate, since these patients are also of short stature with grossly abnormal epiphyses and premature osteoarthritis. The shape of the pelvis should differentiate the two, that of MED being essentially normal apart from some scalloping of the acetabular margin.
Spondyloepiphyseal dysplasia congenita. Characteristically the periphery of the limbs are nearly normal, whereas the hip joints are more severely disorganised than in pseudoachondroplasia.
Other rarer spondylo–epi–metaphyseal disorders may be difficult to differentiate.

Diastrophic dwarfism is associated with severe joint contractures and usually scoliosis from birth or early infancy.

Metatropic dwarfism has 'dumb-bell-shaped' long bones and less epiphyseal involvement than pseudoachondroplasia; also the vertebrae are more severely involved with paper-thin platyspondyly in infancy.

Progress/complications

Life expectancy is normal, but there is early and increasing disability from osteoarthritis.

REFERENCES

Bergsma D 1969 The First Conference on the Clinical Delineation of Birth Defects. Part IV. Skeletal dysplasias. Birth Defects: Original Article Series 5/4: 242–259

Cranley R E, Williams B R, Kopits S E, Dorst J P 1975 Pseudoachondroplastic dysplasia: five cases representing clinical, roentgenographic and histologic heterogeneity. Birth Defects: Original Article Series 11/6: 205–215

Hall J G 1975 Pseudoachondroplasia. Birth Defects: Original Article Series 11/6: 187–202

Kozlowski K 1976 Pseudoachondroplasia (Maroteaux Lamy). A critical analysis. Australian Radiology 20/3: 255–269

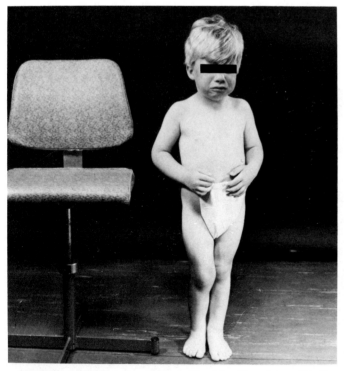
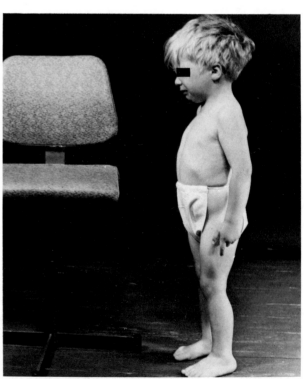

Figs 26.1 and 26.2   A normal face and head, but short stature with disproportionately short limbs.

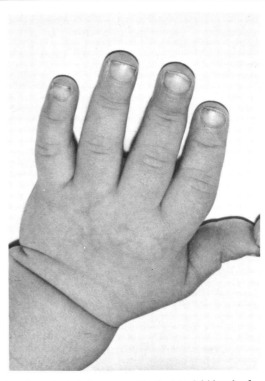

Figs 26.3 and 26.4 Short, stubby hands, particularly affecting the metacarpals and terminal phalanges—unlike the 'starfish' hands of classical achondroplasia.

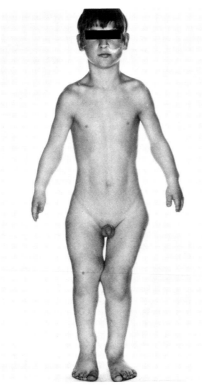

Figs 26.5 and 26.6 In addition to the obvious limb shortening, contractures are now developing and there is bony swelling associated with several joints.

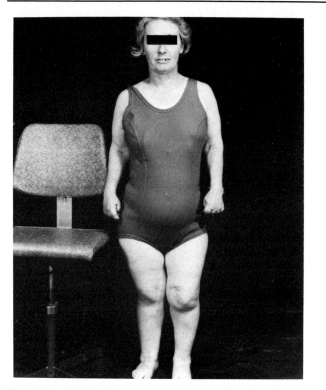
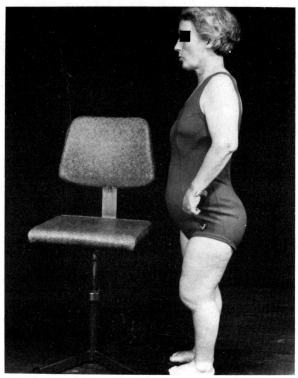

Figs 26.7 and 26.8   Short stature, short limbs and obvious arthritic problems in the knees.

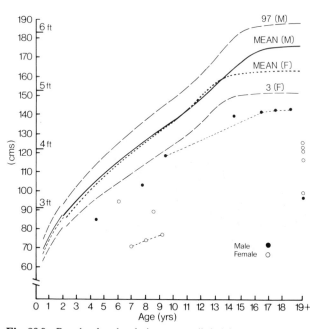

Fig. 26.9   Pseudoachondroplasia—percentile height.

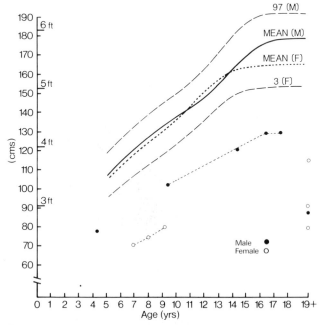

Fig. 26.10   Pseudoachondroplasia—percentile span.

Skull : 4 months–Adult

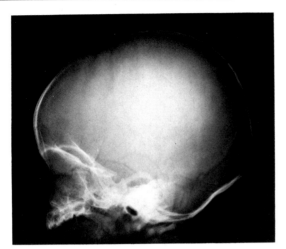

Fig. 26.11   F.B.—4 months—Lateral skull
Normal.
Same patient : Figs 26.15, 26.23, 26.29, 26.38, 26.64

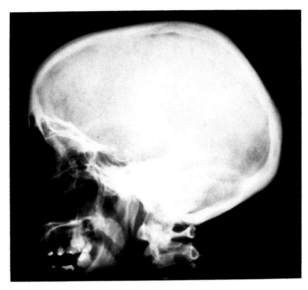

Fig. 26.12   E.T.—53 years—Lateral skull
Normal.
Same patient : Figs 26.48, 26.49, 26.65

Spine: 4 years

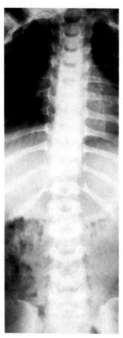

Fig. 26.13   W.G.—4 years—AP spine (one of four affected sibs, autosomal recessive type; sister of children in Figs 26.26, 26.28 and 26.59)
There is widening of the costovertebral joint spaces, with some cupping of the posterior ends of the ribs. The vertebrae appear normal in this projection.
Same patient: Figs 26.14, 26.52

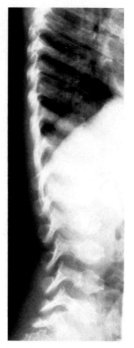

Fig. 26.14   W.G.—4 years 1 month—Lateral spine (same patient as in Fig. 26.13)
There is anterior beaking of the vertebral bodies with rounded posterior halves of the vertebral end plates giving almost a diamond shape. The pedicles are long (unlike achondroplasia).

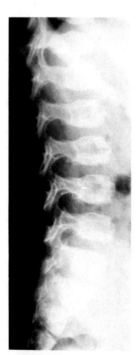

Fig. 26.15   F.B.—4 years 9 months—Lateral lumbar spine
There is anterior beaking and minor irregularity of the vertebral end plates. The pedicles are long.
Same patient: Figs 26.11, 26.23, 26.29, 26.38, 26.64

Spine: 7–9 years

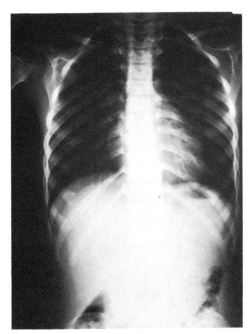

Fig. 26.16   P.H.—7 years—AP spine
The spine appears normal, but there is some widening of the costovertebral joint spaces.
Same patient: Figs 26.17, 26.43, 26.53, 26.61

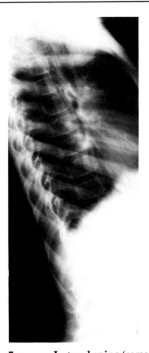

Fig. 26.17   P.H.—7 years—Lateral spine (same patient as in Fig. 26.16)
The vertebral bodies are elongated in their AP diameters, but vertebral end-plates are smooth. The bodies have an immature oval configuration.

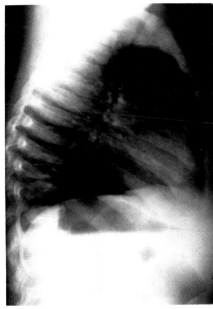

Fig. 26.18   N.B.—7 years 6 months—Lateral spine
There is generalised platyspondyly and the vertebral bodies are increased in their AP diameters. The vertebral end-plates are irregular in the lower dorsal and upper lumbar regions and here there are also antero-superior defects. Increase of the normal dorsal kyphosis.
Same patient: Fig. 26.24

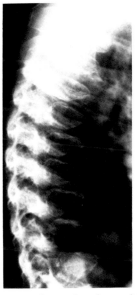

Fig. 26.19   A.E.—9 years 2 months—Lateral dorsal spine
There is platyspondyly with anterior pointing (triangular appearance) and irregularity of the vertebral end-plates.
Same patient: Figs 26.27, 26.30

Spine: 17 years–Adult

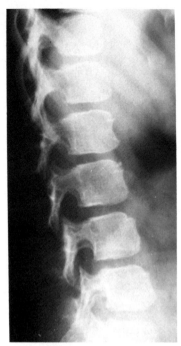

Fig. 26.20   E.M.—17 years—Lateral lumbar spine (autosomal dominant type)
Essentially normal appearance. The ring apophyses are just beginning to ossify.
Same patient: Figs 26.39, 26.47, 26.55, 26.56, 26.63

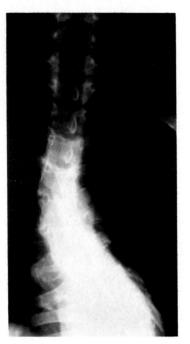

Fig. 26.21   V.W.—21 years 2 months—AP dorsal spine
There is a marked scoliosis convex to the left in the lower dorsal region, but apart from this the vertebrae appear normal.
Same patient: Figs 26.32, 26.40, 26.54

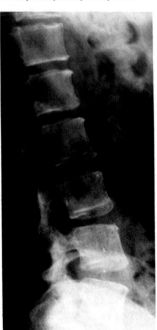

Fig. 26.22   T.M.—39 years—Lateral lumbar spine (mother of patient in Fig. 26.20)
There is disc space narrowing at D12/L1 with vertebral end-plate irregularity here, and mild anterior osteophyte formation at several levels, indicating degenerative changes. The overall configuration is normal.
Same patient: Fig. 26.41

Pelvis: 1–7 years

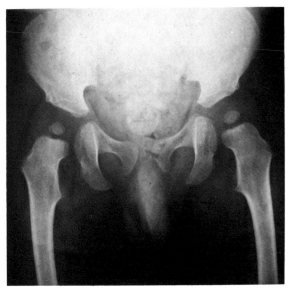

Fig. 26.23 F.B.—1 year 11 months—Pelvis
The capital femoral epiphyses are small and there is coxa valga. The pelvis appears to be normal.
Same patient: Figs 26.11, 26.15, 26.29, 26.38, 26.64

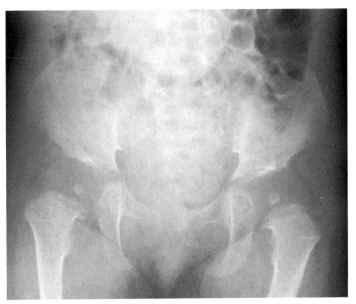

Fig. 26.24 N.B.—3 years—Pelvis
The iliac wings are flared. The Y cartilage is large and unossified. Both capital femoral epiphyses are small and the femoral necks are short and irregular, with some medial breaking.

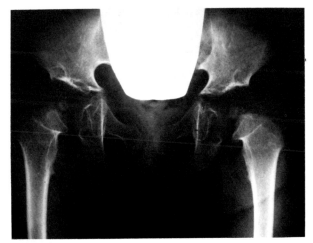

Fig. 26.25 A.B.—7 years—Pelvis
The acetabular area is clearly abnormal, with a wide triradiate cartilage and V-shaped defect here. The femoral necks are short and the metaphyses irregular and domed. The capital femoral epiphyses are very small, corresponding to those of a 6-month-old infant.
Same patient: Figs 26.35, 26.37, 26.58

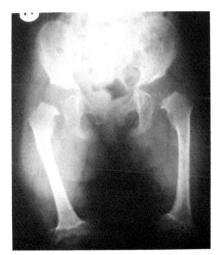

Fig. 26.26 D.G.—7 years—Pelvis (sister of children in Figs 26.13, 26.28 and 26.59)
There is virtually no ossification of the femoral capital epiphyses, and the acetabular roofs are similar to Figure 26.25, with a V-shaped defect. The ischio-pubic ramus is not yet fused.
Same patient: Figs 26.51, 26.60

Pelvis : 9–12 years

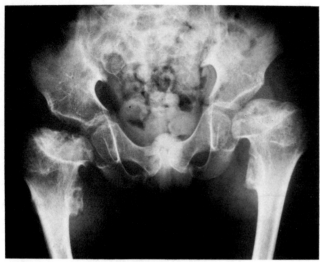

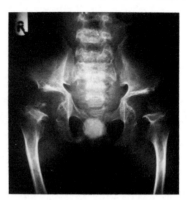

Fig. 26.28   J.G.—11 years—Pelvis (brother of children in Figs 26.13, 26.26 and 26.59)
There are huge defects in the acetabular region, and minute capital femoral epiphyses.

Fig. 26.27   A.E.—9 years 2 months—Pelvis
The acetabula are very shallow and the triradiate cartilage unusually wide for this age. The capital femoral epiphyses are not visible within the disordered metaphyseal area.
Same patient : Figs 26.19, 26.30

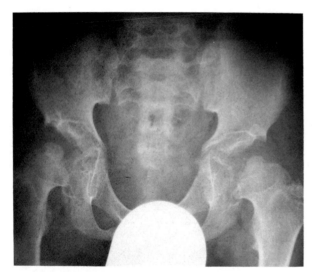

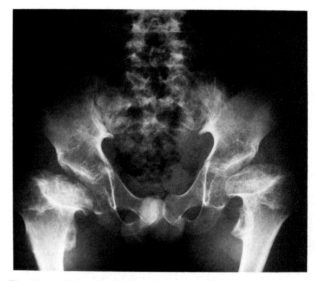

Fig. 25.29   F.B.—11 years 1 month—Pelvis
Defective and irregular acetabula; the capital femoral epiphyses have hardly changed in size over nine years.
Same patient : Figs 26.11, 26.15, 26.23, 26.38, 26.64

Fig. 26.30   A.E.—12 years 4 months—Pelvis (same patient as in Fig. 26.27)
Very poorly formed acetabula. No capital femoral epiphyses visible, and the metaphyses are enlarged.

Pelvis: 17 years–Adult

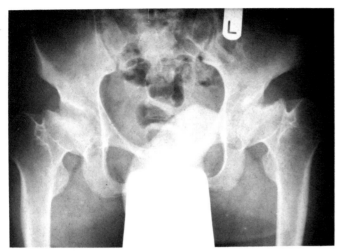

Fig. 26.31   R.M.—17 years 7 months—Pelvis
The femoral heads are flattened and very irregular and the acetabula shallow. The inferior part of the ilium is narrowed.
Same patient: Figs 26.44–26.46

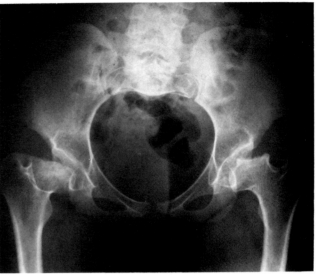

Fig. 26.32   V.W.—20 years 6 months—Pelvis
Brim view of pelvis with pronounced sacral tilt. Acetabula are shallow and femoral heads ill-formed.
Same patient: Figs 26.21, 26.40, 26.54

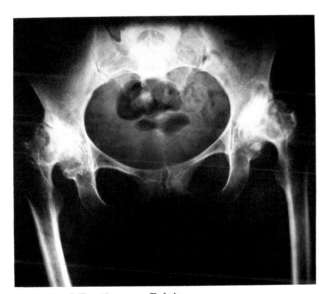

Fig. 26.33   O.F.—48 years—Pelvis
Virtually no hip joint space can be identified and there is subarticular sclerosis and cyst formation, indicating severe secondary osteoarthritis.

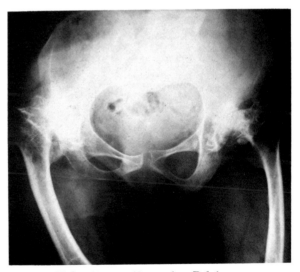

Fig. 26.34   M.G.—61 years 10 months—Pelvis
Severe osteoarthritis of the hips and also lateral bowing of the femora.
Same patient: Figs 26.42, 26.57

Lower limb: 3–11 years

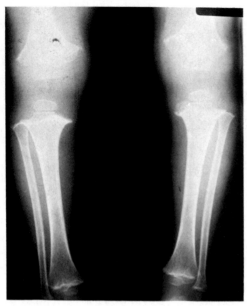

Fig. 26.35   A.B.—3 years 7 months—AP knees and ankles
There is flaring and irregularity of the metaphyses and the epiphyses
are small. The distal end of the fibula is long in relation to the tibia.
Same patient: Figs 26.25, 26.37, 26.58

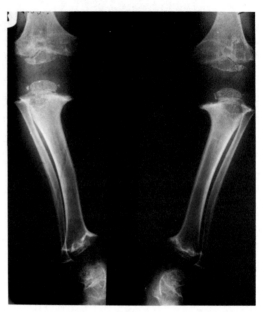

Fig. 26.36   C.D.—5 years—AP knees and ankles
The metaphyses are splayed and irregular and epiphyses small,
irregular and fragmented. The distal ends of the fibulae are long in
relation to the tibiae, associated with a marked varus tilt here. The
upper ends of the tibiae are also abnormal, with medial beaking.
Same patient: Fig. 26.50

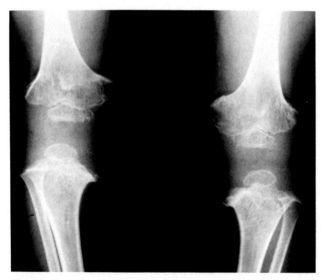

Fig. 26.37   A.B.—7 years—AP knees (same patient as in Fig.
26.35)
The metaphyses are flared and irregular, and epiphyses small with
marginal fragmentation.

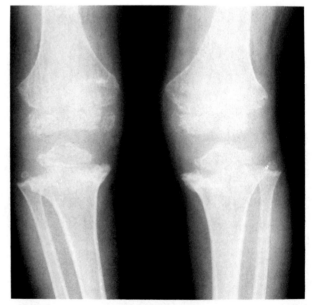

Fig. 26.38   F.B.—11 years 1 month—AP knees
The metaphyses are flared and irregular, and epiphyses small with
marginal fragmentation.
Same patient: Figs 26.11, 26.15, 26.23, 26.29, 26.64

Lower limb : 15 years–Adult

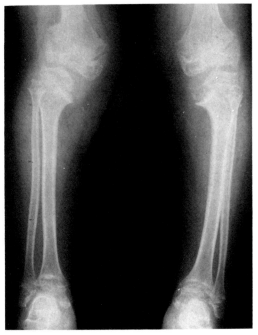

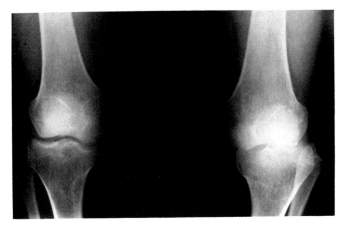

Fig. 26.40   V.W.—23 years 6 months—AP knees
The fibulae are abnormally long in relation to the tibiae and the knee joints are not horizontal.
Same patient : Figs 26.21, 26.32, 26.54

Fig. 26.39   E.M.—15 years—AP knees and ankles
There is a pronounced medial tilt of the knee joints with medial subluxation of the femora. The metaphyses are flared and irregular, and epiphyses fragmented. The fibula is long in relation to the tibia both proximally and distally.
Same patient : Figs 26.20, 26.47, 26.55, 26.56, 26.63

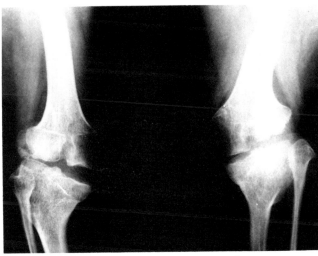

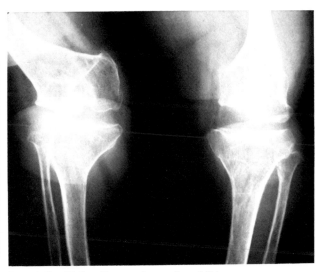

Fig. 26.41   T.M.—39 years—AP knees
The fibulae are disproportionately long, and the knee joints grossly disorganised with a pronounced medial slope and osteoarthritic changes. (On the left there is a healed osteotomy of the lower femoral shaft.)
Same patient : Fig. 26.22

Fig. 26.42   M.G.—61 years 9 months—AP knees
Gross disorganisation of both knee joints with osteoarthritic changes.
Same patient : Figs 26.34, 26.57

Feet: 7–12 years

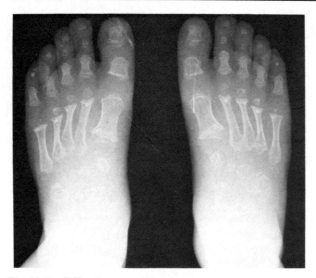

Fig. 26.43   P.H.—7 years—PA feet
There is generalised shortening of the long bones with metaphyseal irregularity. The tarsal centres are small and irregular, and all other epiphyses retarded.
Same patient: Figs 26.16, 26.17, 26.53, 26.61

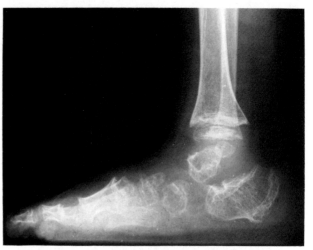

Fig. 26.44   R.M.—8 years—Standing lateral foot
All metaphyses are flared and irregular. The tarsal centres are small with irregular margins. The distal fibula is not unduly long in this patient.
Same patient: Figs 26.31, 26.45, 26.46

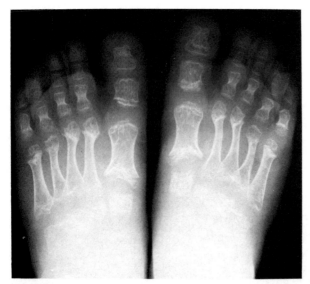

Fig. 26.45   R.M.—11 years 9 months—PA feet (same patient as in Fig. 26.44)
The long bones are all shortened. The metaphyses are irregular and epiphyses show defects on the articular side. There are chevron deformities at the bases of the first metatarsals and the tarsal centres are small and irregular.

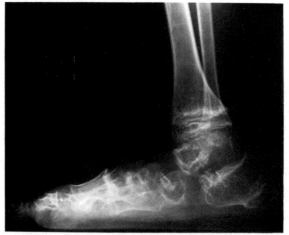

Fig. 26.46   R.M.—12 years—Standing lateral foot (same patient as in Fig. 26.44)
There is narrowing of the ankle joint space and tarsal centres have irregular margins. Metaphyseal flaring is present.

Feet: 14 years–Adult

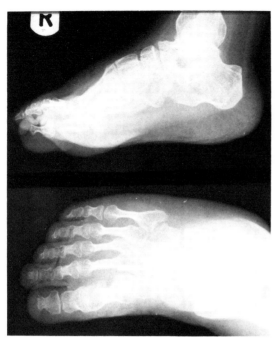

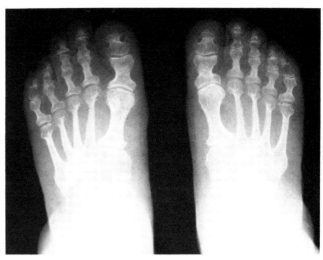

Fig. 26.48 E.T.—53 years—PA feet
There is shortening of the long bones and rather abrupt metaphyseal flaring.
Same patient: Figs 26.12, 26.49, 26.65

Fig. 26.47 E.M.—14 years—Lateral and PA right foot
The epiphyses are already fused and there is severe shortening of the long bones with widened metaphyses and, by contrast, narrow diaphyses.
Same patient: Figs 26.20, 26.39, 26.55, 26.56, 26.63

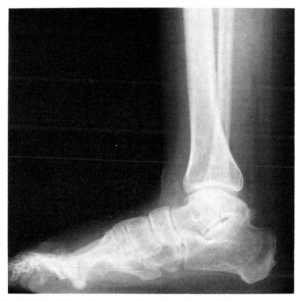

Fig. 26.49 E.T.—53 years—Standing lateral foot (same patient as in Fig. 26.48)
There is narrowing of the ankle joint space and flattening of the trochlear surface of the talus. The flaring of the distal tibia is not pronounced and the distal fibula is not unduly elongated.

Upper limb: 1–7 years

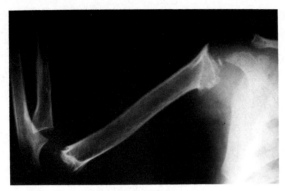

Fig. 26.50   C.D.—1 year—AP right arm
The long bones are short. There is metaphyseal splaying and
irregularity and epiphyses are small.
Same patient: Fig. 26.36

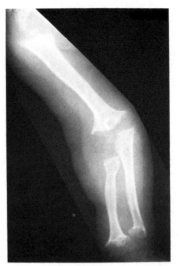

Fig. 26.51   D.G.—2 years—AP left arm (sister of child in Fig.
26.52)
The long bones are short with metaphyseal flaring, irregularity and
small epiphyses. There is a varus deformity of the proximal humerus.
Same patient: Figs 26.26, 26.60

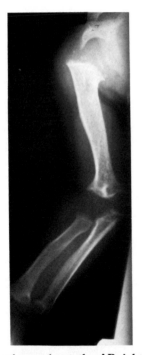

Fig. 26.52   W.G.—4 years 1 month—AP right arm (sister of
child in Fig. 26.51)
The long bones are short with metaphyseal flaring, irregularity and
with small, fragmented epiphyses. (Identical with her sister, Fig.
26.51.) There is a varus deformity of the proximal humerus.
Same patient: Figs 26.13, 26.14

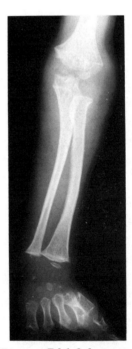

Fig. 26.53   P.H.—7 years—PA left forearm
The long bones are short with metaphyseal flaring and irregularity.
The epiphyses are small and irregular and bone maturation is delayed.
Same patient: Figs 26.16, 26.17, 26.43, 26.61

Upper limb: 8 years–Adult

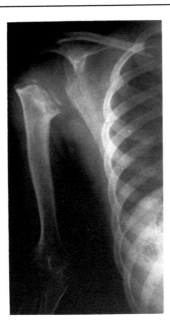

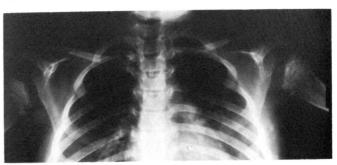

Fig. 26.55   E.M.—12 years—AP shoulders
The proximal humeral epiphyses are small and fragmented and the adjacent metaphyses are irregular. There is widening of the costovertebral joint spaces.
Same patient: Figs 26.20, 26.39, 26.47, 26.56, 26.63

Fig. 26.54   V.W.—8 years 4 months—AP right shoulder
There is a varus deformity of the proximal humerus with metaphyseal splaying and irregularity. The epiphyses are small and fragmented.
Same patient: Figs 26.21, 26.32, 26.40

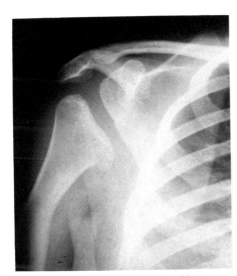

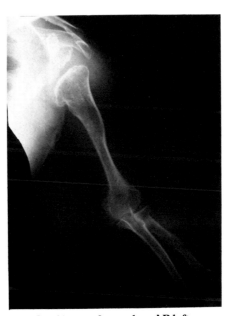

Fig. 26.56   E.M.—19 years—AP right shoulder (same patient as in Fig. 26.55)
There is a marked varus deformity of the proximal humerus and a flared metaphysis. Articular surfaces are irregular and flattened.

Fig. 26.57   M.G.—61 years 9 months—AP left arm
The long bones are short with metaphyseal flaring. The deltoid tuberosity is pronounced and there is a varus deformity of the proximal humerus. At the shoulder the articular margins show sclerosis and there is some marginal osteophyte formation, indicating secondary osteoarthritis.
Same patient: Figs 26.34, 26.42

Hand: 4–7 years

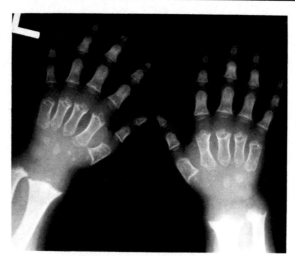

Fig. 26.58   A.B.—4 years 5 months—PA hands
The long bones are shortened and metaphyses splayed and irregular.
The epiphyses are small and fragmented. There is proximal pointing
of the 2nd–5th metacarpals (as in the mucopolysaccharide disorders)
possibly related to pseudoepiphyses at these sites.
Same patient: Figs 26.25, 26.35, 26.37

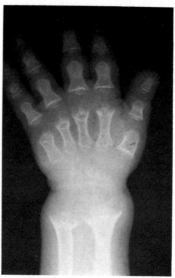

Fig. 26.59   S.G.—6 years—PA hand (sister of children in Figs
26.13, 26.26 and 26.28)
The long bones are short with flared irregular metaphyses. The
epiphyses and carpal centres are small and irregular. Bone maturation
is delayed.

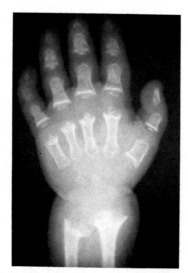

Fig. 26.60   D.G.—7 years—PA hands (sister of child in Fig.
26.59)
There is shortening of the long bones and the metaphyses are flared
and irregular. Epiphyses are small and fragmented and bone
maturation delayed.
Same patient: Figs 26.26, 26.51

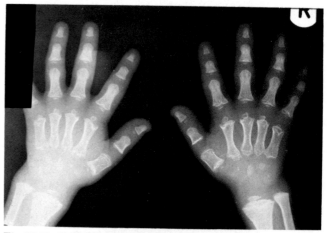

Fig. 26.61   P.H.—7 years—PA hands
The long bones are short and metaphyses irregular. There is proximal
pointing of the second to fifth metacarpals. The carpal centres are
small and irregular and bone maturation is retarded.
Same patient: Figs 26.16, 26.17, 26.43, 26.53

Hand : 8 years–Adult

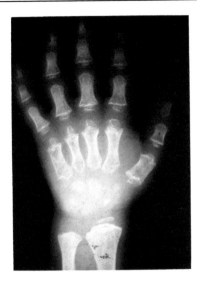

Fig. 26.62   G.M.—8 years—Hand
Short long bones with irregular small epiphyses and proximal pointing of the metacarpals (not as sharp as in the MPS disorders). Delayed bone maturation.

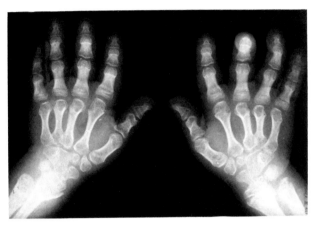

Fig. 26.63   E.M.—12 years—Hands
Short long bones and unusual flaring of the distal radial and ulnar metaphyses with central radiolucent defects.
Same patient : Figs 26.20, 26.39, 26.47, 26.55, 26.56

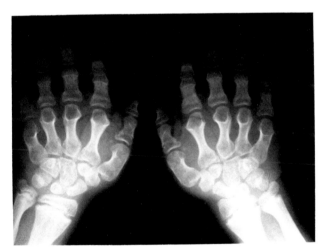

Fig. 26.64   F.B.—14 years 5 months—Hands
Short but well-modelled long bones, with premature fusion of the epiphyses in the digits, although the radial and ulnar epiphyses are still wide open.
Same patient : Figs 26.11, 26.15, 26.23, 26.29, 26.38

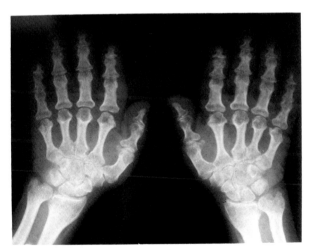

Fig. 26.65   E.T.—53 years—Hands
The long bones are short with flared metaphyses and diaphyseal constriction.
Same patient : Figs 26.12, 26.48, 26.49

Chapter Twenty-seven
DIASTROPHIC DYSPLASIA (Dwarfism)

Short-limbed dwarfism accompanied by contractures of many joints, foot deformities and scoliosis.

Inheritance

Autosomal recessive.

Frequency

Very rare. Possible prevalence of 1 per million population.

Clinical features

Facial appearance—normal.
Intelligence—normal.
Stature—markedly reduced with disproportionately short limbs. Adult height between 80 and 140 cm (2′ 7″ to 4′ 7″).
Presenting feature/age—birth, with short limbs, talipes equinus or equinovarus, contractures of many joints and sometimes a cystic swelling of the ear (a haematoma, which should be aspirated).
Deformities—severe progressive contractures, including clubfoot, scoliosis and dislocation of various joints; secondary osteoarthritis later, with associated deformities.
Associated anomalies—cleft palate in about one-quarter of cases.

Radiographic features

Skull—probably normal, apart from cleft palate.
Spine—dysplastic odontoid process. Appearance of vertebrae is very variable, with irregular bodies, some flattened and some unusually tall. The pedicles may be short and the bodies show posterior scalloping (as in achondroplasia). In some cases the vertebrae are normal.
Pelvis and hips—variable. Probably flared iliac wings, but some appear less than usually flared. The capital femoral epiphyses are late in appearing, but may be subsequently well-formed, or flat and fragmented. There may be coxa vara or valga.
Limb bones—short, and a general lack of modelling; 'chevron' deformities at the distal ends of femora and tibiae; irregular epiphyses and metaphyses.
Hands—irregular length of metacarpals and bizarre ossification. A frequent feature is marked shortening of the first metacarpal, which is more proximally placed than is normal ('hitch-hiker's thumb').

Bone maturation

Probably generally delayed, but carpal ossification has been reported as advanced.

Differential diagnosis

Achondroplasia. Here the skull is affected, and joint contractures are not a feature (except for limitation of elbow extension). The limb bones in achondroplasia have flared metaphyses, with smooth ends, and epiphyses are normal.

No other rarer form of short-limbed dwarfism is associated with the severe contractures so typical of diastrophic dysplasia.

Progress/complications

Some patients die in infancy from respiratory failure, but many appear to have a normal life span. Severe scoliosis leads to cardio-respiratory problems; atlanto-axial instability may lead to cord compression; the clubfoot and scoliosis are particularly intractable and secondary osteoarthritis inevitably a feature.

REFERENCES

Lachman R, Sillence D, Rimoin D 1981 Diastrophic dysplasia: the death of a variant. Radiology 140/1:79–86
Walker B A, Scott C I, Hall J G, Murdoch J L, McKusick V A 1972 Diastrophic dwarfism. Medicine 51:41–59

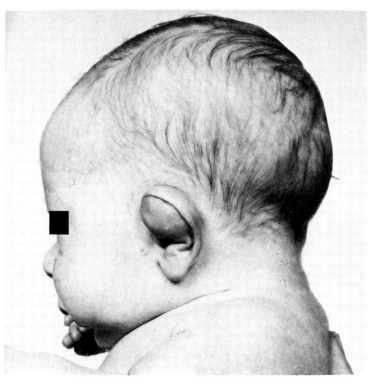

Fig. 27.1 A haemorrhagic cyst of the ear, sometimes seen in newborn infants. Easily treated by aspiration.

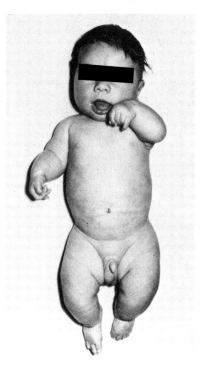

Fig. 27.2 Disproportionate shortening of both upper and lower limbs, in comparison with the trunk.

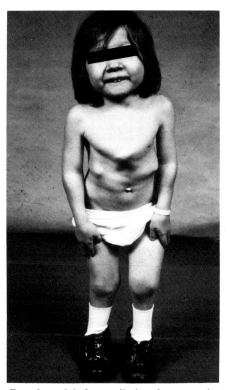

Fig. 27.3 Dwarfism, club-feet, scoliosis and pectus carinatum, as well as disproportionate limb shortening.

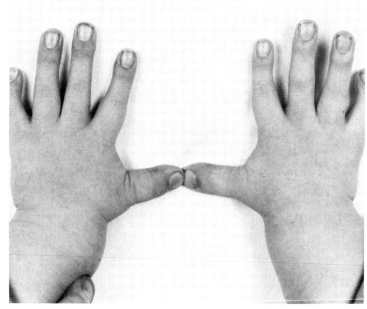

Fig. 27.4 Under-development of the first metacarpal leads to the thumbs being more proximally placed than usual ('hitch-hiker's thumb').

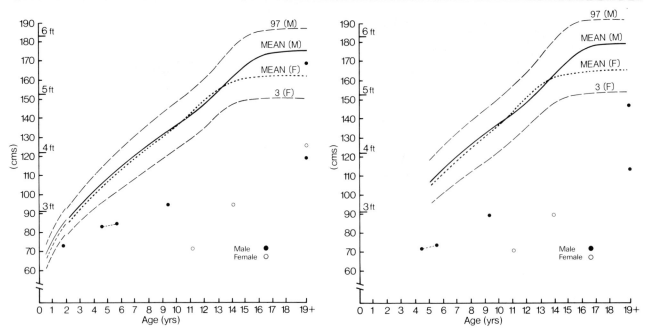

Fig. 27.5 Diastrophic dysplasia—percentile height.

Fig. 27.6 Diastrophic dysplasia—percentile span.

Whole body: Birth

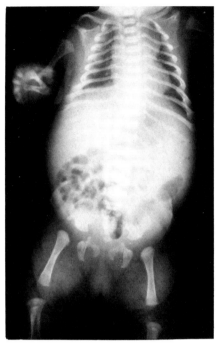

Fig. 27.7   Neonate—AP whole body
Both humeri and femora are disproportionately short, the former being pointed at the lower end. The thoracic cage is small, and the clavicles relatively long.
Same patient: Figs 27.8, 27.45, 27.46

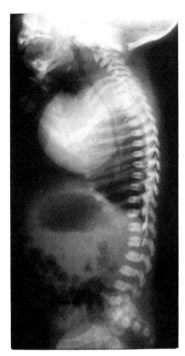

Fig. 27.8   Neonate—Lateral whole body (same patient as in Fig. 27.7)
Short upper limbs. Normal spine.

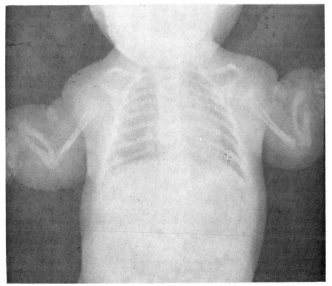

Fig. 27.9   T.M.—Neonate—AP whole body
The thoracic cage is small. The clavicles are relatively long with symmetrical lateral hooks. The radius on both sides is short and bowed and there is dislocation at the elbow. The ulna is short, with distal pointing and ulnar deviation at the wrist.
Same patient: Figs 27.17, 27.19, 27.47

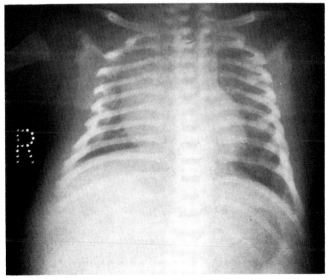

Fig. 27.10   R.M.—Neonate—Chest
The thoracic cage is small and the ribs short.
Same patient: Figs 27.11, 27.52

Skull and cervical spine : Birth–Adult

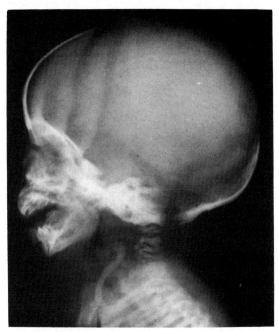

Fig. 27.11   R.M.—Neonate—Skull
The sutures are abnormally wide and the floor of the anterior fossa slopes steeply.
Same patient : Figs 27.10, 27.52

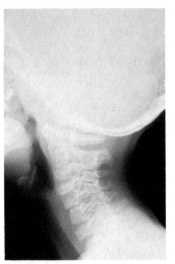

Fig. 27.12   A.D.—4 years—Lateral cervical spine
The vertebral bodies are irregular in size and shape ; some are flattened and elongated, others tall with a narrow antero-posterior diameter.
Same patient : Figs 27.18, 27.25, 27.34, 27.54

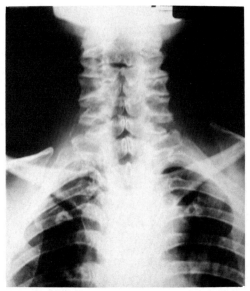

Fig. 27.13   M.S.—48 years—AP cervical spine
There is spinal dysraphism extending from C5 to D1 and disc space narrowing with some platyspondyly and marginal osteophytes.
Same patient : Figs 27.14, 27.31, 27.39, 27.51, 27.55

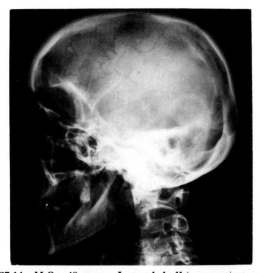

Fig. 27.14   M.S.—48 years—Lateral skull (same patient as in Fig. 27.13)
No abnormality in the skull, but the odontoid process is virtually absent and there is cervical spondylosis.

Spine : Birth–6 years

Fig. 27.15   R.W.—Neonate—Lateral spine
There is mild dorsal wedging of vertebral bodies and an increase in
the AP diameter. The sacrum is almost horizontal.
Same patient : Figs 27.27, 27.28

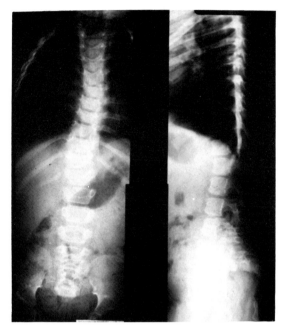

Fig. 27.16   S.C.—1 year 3 months—AP and lateral spine
There is minor scoliosis with a triple curve. There is pronounced
sacral angulation; L.4 appears reduced in its AP diameter and there is
a posterior scalloping. There are transitional vertebrae both at the
thoraco-lumbar and lumbo-sacral junctions.
Same patient : Figs 27.24, 27.40, 27.53

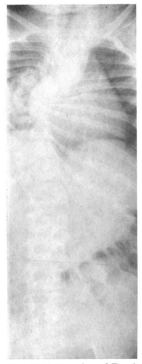

Fig. 27.17   T.M.—2 years 11 months—AP spine
There is scoliosis (or more probably kyphoscoliosis) convex to the
right in the mid-dorsal region, but the vertebrae appear otherwise
normal.
Same patient : Figs 27.9, 27.19, 27.47

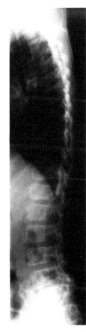

Fig. 27.18   A.D.—6 years 10 months—Lateral spine
Normal.
Same patient : Figs 27.12, 27.25, 27.34, 27.54

Spine: 9 years–Adult

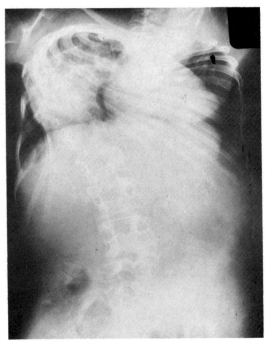

Fig. 27.19   T.M.—9 years 3 months—AP spine
The kyphoscoliosis has markedly deteriorated since the age of 3 years
(Fig. 27.17). (These curves are extremely rigid.)
Same patient: Figs 27.9, 27.17, 27.47

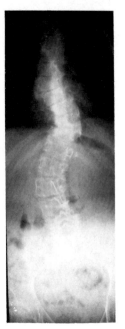

Fig. 27.20   A.B.—12 years 6 months—AP spine
There is a triple curve and mild vertebral irregularity is present.
Same patient: Fig. 27.37

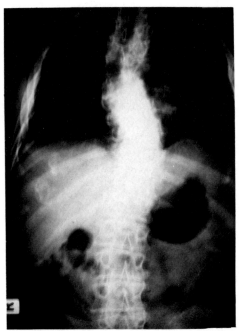

Fig. 27.21   S.H.—25 years—AP spine
A triple curve is present but is not severe. There are marginal
osteophytes indicating early degenerative changes.
Same patient: Figs 27.22, 27.26, 27.30, 27.36, 27.38, 27.41, 27.48, 27.49

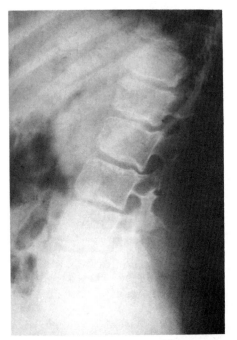

Fig. 27.22   S.H.—25 years—Lateral lumbar spine (same patient
as in Fig. 27.21)
There is a pronounced lumbar lordosis. The disc spaces are narrow;
Schmorl's nodes are present at L2–L3 and L3–L4 and there is mild
platyspondyly with some wedging at D12 and L1. The pedicles are
short and the spinal canal is narrow.

Pelvis: Birth–7 years

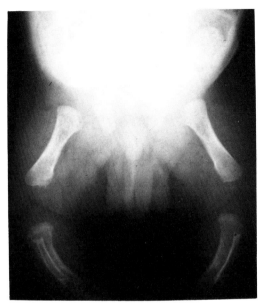

Fig. 27.23   M.Sc.—Neonate—AP pelvis and lower limbs
There is delayed ossification of the pubic rami. The limbs are short and bone maturation is delayed—only one small epiphysis at the lower end of the left femur. There is lateral bowing of the leg bones, and a general lack of modelling.
Same patient: Fig. 27.44

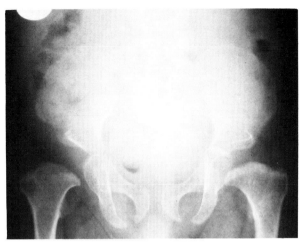

Fig. 27.24   S.C.—1 year 6 months—Pelvis
The iliac wings are flared and there is a pronounced notch above the acetabula. The capital femoral epiphyses have not yet ossified and the femoral necks are broad, with some coxa vara.
Same patient: Figs 27.16, 27.40, 27.53

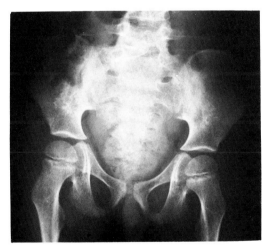

Fig. 27.25   A.D.—6 years 10 months—Pelvis
The acetabular roofs are horizontal. There is coxa valga and the femoral heads are large. There is spinal dysraphism of S1.
Same patient: Figs 26.12, 26.18, 26.34, 26.54

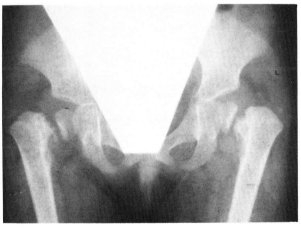

Fig. 27.26   S.H.—7 years—Pelvis
The iliac wings are somewhat flared. The capital femoral epiphyses are well formed and maintained within the acetabula, but both epiphyseal plates are wide and there is marked coxa vara. The appearances suggest a slip or fracture through the epiphyseal plates.
Same patient: Figs 26.21, 26.22, 26.30, 26.36, 26.38, 26.41, 26.48, 26.49

Pelvis and hip : 7–12 years

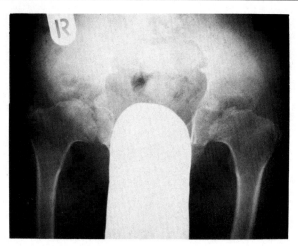

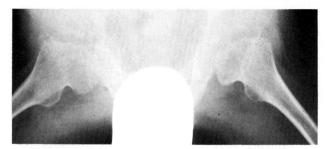

Fig. 27.28   R.W.—7 years 3 months—Lateral hips (same patient as in Fig. 27.27)
The femoral heads are flat and fragmented, and the necks short and broad.

Fig. 27.27   R.W.—7 years 3 months—AP hips
There are flattened and fragmented capital femoral epiphyses with short broad femoral necks and coxa vara.
Same patient : Figs 27.15, 27.28

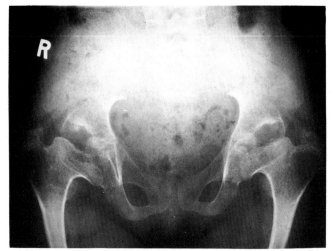

Fig. 27.29   R.G.—12 years—Pelvis
There is bilateral coxa vara with very poorly ossified femoral heads. The iliac wings are small in relation to the upper femora.
Same patient : Figs 27.33, 27.35

Pelvis: Adult

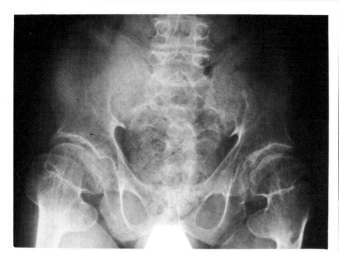

Fig. 27.30   S.H.—25 years—Pelvis
There is loss of the normal flare of the iliac wings, and the bases of the ilia are unusually broad. The femoral heads are irregular and there is early osteoarthritis. The hips are rotated, but there is also unusual prominence of both greater and lesser trochanters.
Same patient: Figs 27.21, 27.22, 27.36, 27.38, 27.41, 27.48, 27.49

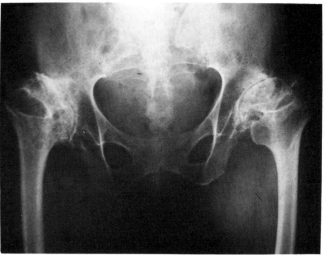

Fig. 27.31   M.S.—48 years—Pelvis
There is severe osteoarthritis of both hips. The femoral heads are eroded and flattened and the femoral necks are short. The iliac wings are small and the iliac bases broad.
Same patient: Figs. 27.13, 27.14, 27.39, 27.51, 27.55

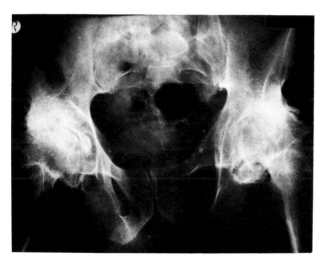

Fig. 27.32   R.B.—48 years—Pelvis
There is severe joint space narrowing with osteoarthritis. The bases of the ilia are broader than normal.
Same patient: Figs 27.42, 27.43, 27.50

Lower limb: 3–8 years

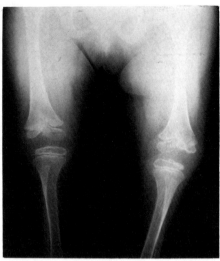

Fig. 27.33   R.G.—3 years 7 months—AP lower limbs
There is mild flaring at the metaphyses. A chevron deformity is
present at the distal femoral metaphysis, involving the adjacent
epiphysis. The epiphyses in general are flat with abnormal modelling.
Same patient: Figs 27.29, 27.35

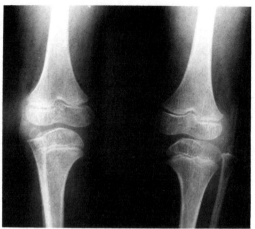

Fig. 27.34   A.D.—6 years 10 months—AP knees
Mild metaphyseal flaring is present and the epiphyses are abnormally
rounded. Very mild chevron deformities are present at the distal
femoral metaphyses. The proximal ends of the fibulae are
exceptionally long.
Same patient: Figs. 27.12, 27.18, 27.25, 27.54

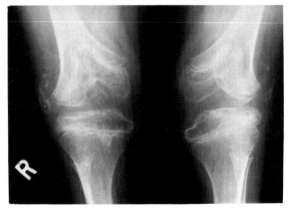

Fig. 27.35   R.G.—8 years 7 months—AP knees (same patient as
in Fig. 27.33)
There is genu valgum. The epiphyses are flattened and irregular,
especially laterally. Marked chevron deformities are present at the
distal femora. There is lateral subluxation of the patellae which
appear fragmented.

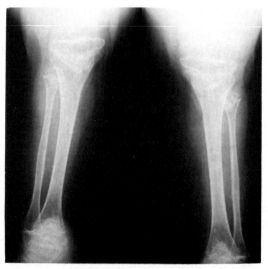

Fig. 27.36   S.H.—8 years 7 months—AP tibiae and fibulae
There is a bilateral genu valgum. The epiphyses at the proximal tibiae
are flat. There are chevron deformities of all metaphyses (both ends of
the tibiae and fibulae) and there is also patchy sclerosis extending into
the diaphyses. In this patient, the fibulae are disproportionately
shorter than usual.
Same patient: Figs 27.21, 27.22, 27.26, 27.30, 27.38, 27.41, 27.48, 27.49

Lower limb: 12 years–Adult

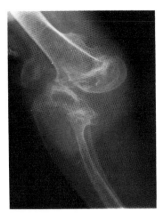

Fig. 27.37 A.B.—12 years 6 months—Lateral right knee
There is anterior dislocation of the knee. The epiphyses are
abnormally modelled and irregular.
Same patient: Fig. 27.20

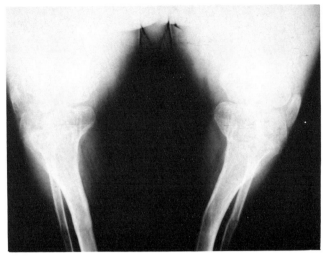

Fig. 27.38 S.H.—25 years—AP knees
These knee joints are grossly disorganised, with the patellae shown on
the 'lateral' aspect. The tibial shafts are not straight, and there is
marked flaring at their upper ends.
Same patient: Figs 27.21, 27.22, 27.26, 27.30, 27.36, 27.41, 27.48, 27.49

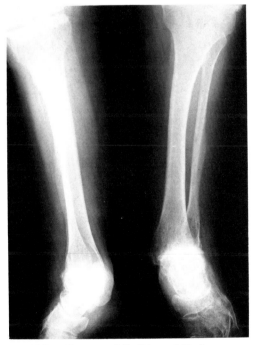

Fig. 27.39 M.S.—48 years—AP tibia and fibula
There is joint-space narrowing at both knee and ankle joints, with
gross osteoarthritis.
Same patient: Figs 27.13, 27.14, 27.31, 27.51, 27.55

Ankle and foot : 4 years–Adult

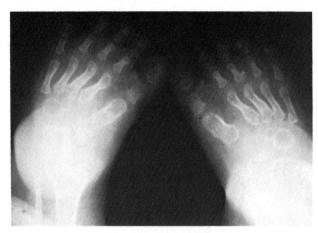

Fig. 27.40   S.C.—4 years 6 months—PA feet
All the long bones are short, but the first metatarsal is particularly short and broad.
Same patient : Figs 27.16, 27.24, 27.53

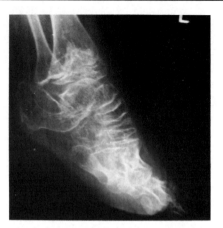

Fig. 27.41   S.H.—8 years 8 months—Lateral ankle
The tarsal centres are small and irregular and there is a chevron deformity of the distal tibial epiphysis.
Same patient : Figs 27.21, 27.22, 27.26, 27.30, 27.36, 27.38, 27.48, 27.49

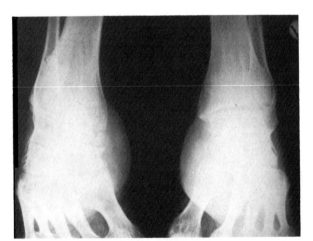

Fig. 27.42   R.B.—41 years—AP ankles
There is osteoarthritis and also fragmented ossicles of the right medial malleolus.
Same patient : Figs 27.32, 27.43, 27.50

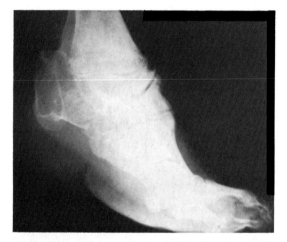

Fig. 27.43   R.B.—41 years—Lateral right foot (same patient as in Fig. 27.42)
There is severe joint space narrowing around all the tarsal bones and at the ankle joint. Articular sclerosis is present and talonavicular osteophyte formation.

Upper limb: Birth–5 years

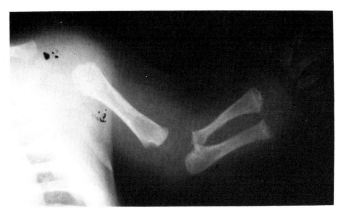

Fig. 27.44 M.Sc.—Neonate—AP left upper limb
There are marked modelling defects and shortening of all the long bones. The metaphyses are irregular with mild flaring.
Same patient: Fig. 27.23

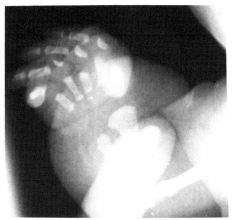

Fig. 27.45 Neonate—Forearm and hand
There is dislocation at the elbow and shortened long bones. In the hand, there is unusual irregular shortening of the metacarpals, and bizarre ossification of the proximal phalanges.
Same patient: Figs 27.7, 27.8, 27.46

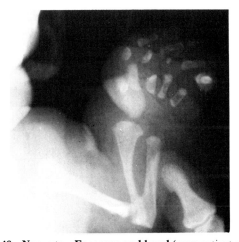

Fig. 27.46 Neonate—Forearm and hand (same patient as in Fig. 27.45).
Similar to the contralateral limb (Fig. 27.45).

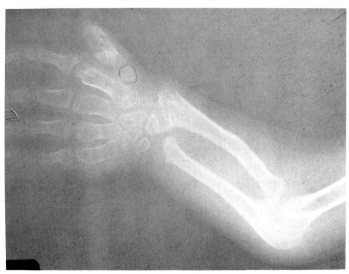

Fig. 27.47 T.M.—5 years 9 months—Forearm
The long bones are short and metaphyses flared and irregular. The first metacarpal is very short and poorly ossified (marked on the original film).
Same patient: Figs 27.9, 27.17, 27.19

Upper limb: Adult

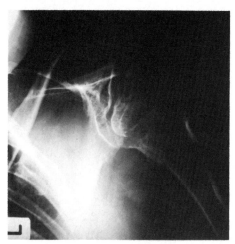

Fig. 27.48   S.H.—25 years—AP left shoulder
The joint space is narrow, and the glenoid fossa and humeral head are irregular. The humeral shaft is short with pronounced muscle attachments.
Same patient: Figs 27.21, 27.22, 27.26, 27.30, 27.36, 27.38, 27.41, 27.49

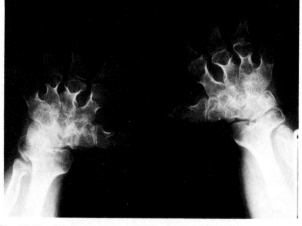

Fig. 27.49   S.H.—25 years—PA wrists (same patient as in Fig. 27.48)
There is marked shortening of the long bones. The first metacarpals are proximally placed and particularly short. The carpal centres are crowded and the articular spaces narrow. There is distal flaring of the radii, and the ulnae are short. A pronounced tubercle is present on the medial side of the radial shafts.

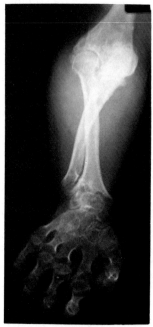

Fig. 27.50   R.B.—47 years—forearm
The long bones are short with flared metaphyses. The carpal bones appear to be fused.
Same patient: Figs 27.32, 27.42, 27.43

Fig. 27.51   M.S.—48 years—AP right shoulder
There is flattening of the humeral head and joint space narrowing. The articular surfaces are sclerotic and marginal osteophytes present. The humerus is short with a prominent deltoid tuberosity.
Same patient: Figs 27.13, 27.14, 27.31, 27.39, 27.55

Hand : Birth–Adult

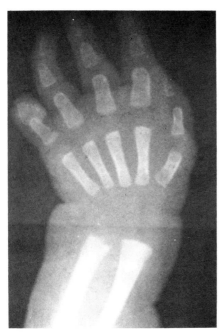

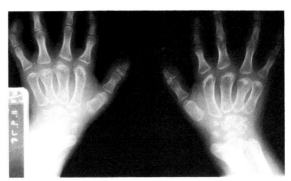

Fig. 27.53   S.C.—4 years 6 months—PA hands
The carpal centres are advanced, irregular in outline and have a horizontal orientation. The long bones are irregularly shortened with flared metaphyses and sloping articular margins.
Same patient : Figs 27.16, 27.24, 27.40

Fig. 27.52   R.M.—Neonate—PA hand
Little abnormality apart from some shortening of the first metacarpal.
Same patient : Figs 27.10, 27.11

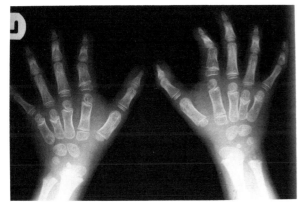

Fig. 27.54   A.D.—6 years 10 months—PA hands
Carpal centres are here retarded. There is irregular shortening of the long bones of the hand, the fourth metacarpals being particularly short, also with cone epiphyses. There is bilateral clinodactyly and flexion deformities of the index fingers. Large pseudoepiphyses are present at the bases of the second metacarpals. The epiphyses at the bases of the proximal phalanges of the index fingers are long with a triangular configuration, the apex of the triangle being on the ulnar side.
Same patient : Figs 27.12, 27.18, 27.25, 27.34

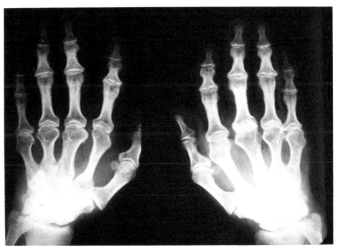

Fig. 27.55   M.S.—48 years—PA hands
All the hand bones are shortened, with joint space narrowing both of the digits and carpus.
Same patient : Figs 27.13, 27.14, 27.31, 27.39, 27.51

Chapter Twenty-eight
METATROPIC DWARFISM

Markedly short limbs at birth but subsequently the disproportion may become less obvious as spinal deformity and trunk shortening develop. A striking characteristic is the 'dumb-bell' appearance of long bones on radiography.

Inheritance

Little is known, both autosomal dominant and recessive forms have been reported.

Frequency

Very rare. Possible prevalence is under 1 per million population.

Clinical features

Facial appearance—normal.
Intelligence—normal.
Stature—markedly reduced, with disproportionately short limbs. Adult height between 110 and 120 cm (3′ 7″ and 3′ 11″).
Presenting feature/age—birth, with short limbs and long trunk, narrow chest, sometimes with respiratory difficulties. A small tail-like appendage over the sacrum has been reported.
Deformities—scoliosis or kyphosis is usual and potentially severe. Secondary osteoarthritis with associated deformities develops from the disordered epiphyseal growth.
Associated anomalies—none regularly reported.

Radiographic features

Skull—normal.
Cervical spine—platyspondyly and sometimes absence or hypoplasia of the odontoid peg.
Thoracic and lumbar spine—in the neonate and in early infancy platyspondyly is very striking ('paper thin' vertebrae); there are coronal clefts and an apparent increase in the depth of the disc spaces. Subsequently

the appearance is variable, with diamond shaped vertebrae, or with a hump superiorly only, and irregular ossification of the end-plates. Finally there may be only mild platyspondyly, although there is elongation of the antero-posterior diameter of the vertebrae.

Scoliosis or kyphosis is usual and may be already present at birth.

Thorax—in the neonate, the ribs are short with widening of the costochondral junctions, and the thoracic cage is small.

Pelvis—small square ilia with the anterior superior iliac spines apparently more inferiorly placed than is normal. The acetabular roofs are horizontal with a sharpened lateral beak in infancy. The sciatic notches are small and the pubic symphysis may be widened. The most striking feature is the great expansion of the intertrochanteric region and the delayed appearance of the capital femoral epiphyses.

Long bones—the characteristic appearance of all long bones, including those of the hand, is of diaphyseal constriction and widely flaring metaphyses. Ossification is delayed and epiphyses small and irregular—but this is not an invariable feature.

Knee joint—intercondylar notches may be unusually deep. The fibulae may be long in relation to the tibiae.

Bone maturation

Usually delayed.

Biochemistry

Not known.

Differential diagnosis

Achondroplasia. Here the skull is affected, metaphyseal flaring is not so marked and the epiphyses are normal.

Morquio's disease. This is not characterised by short-limbed dwarfism, pectus carinatum is usual and the radiological signs of a mucopolysaccharide disorder are present.

Kniest disease is radiologically similar, but patients

have a characteristic facies and do not have the narrow thorax of metatropic dwarfism. Associated defects such as cleft palate and mental retardation are often present.

Fibrochondrogenesis has recently been described in the English literature: a chondrodysplasia, lethal at birth, with 'dumb-bell' appearance of the long bones, posterior hypoplasia of vertebrae with clefting, short ribs and a hypoplastic pelvis (see Figs 28.2, 28.3, 28.4). It is of autosomal recessive inheritance.

Progress/complications

Death may occur in early infancy from respiratory insufficiency, and at a later age from cardio-respiratory problems associated with the spinal and thoracic deform-ity. Secondary osteoarthritis develops following the disordered epiphyseal growth, and there may be mechanical problems at the knee and ankle associated with disproportionate growth of the tibiae and fibulae.

REFERENCES

Etesan D J, Adomian G E, Ornoy A, Koide T, Sugiura Y, Calabro A, Lungarotti S, Mastroiacovo P, Lachman R S, Rimoin D L 1984 Fibrochondrogenesis. Radiologic and histologic studies. American Journal of Medical Genetics 19:277–290

Jenkins P, Smith M B, McKinnell J S 1970 Metatropic dwarfism. British Journal of Radiology 43:561

Rimoin D L, Siggers D C, Lachman R S, Silberberg R 1976 Metatropic dwarfism, the Kniest syndrome and the pseudoachondroplastic dysplasias. Clinical Orthopaedics 114:70–82

Fig. 28.1 A woman in her twenties with severe dwarfing. The limbs are disproportionately short and bony swelling around several joints indicates premature arthritis.

Whole body: Birth

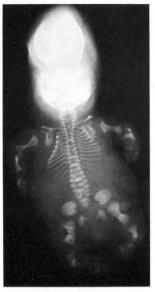

Fig. 28.2   Stillbirth—AP whole body★
There is severe shortening of all the long bones with pronounced diaphyseal constriction, metaphyseal flaring and some bowing deformities. The ribs are short and the thoracic cage small. In the spine there is irregular and abnormal ossification of the vertebral bodies. Platyspondyly is seen in the dorsal region. In the lumbar region there are several butterfly vertebrae. The iliac wings are small and square and acetabular roofs horizontal. The pubic and ischial bones are squat and heavy and there is widening of the symphysis pubis. The skull has a 'clover-leaf' appearance, as sometimes seen associated with thanatophoric dwarfism. This is not a usual feature of metatropic dwarfism.

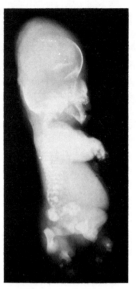

Fig. 28.3   Stillbirth—Lateral whole body★
There is irregular platyspondyly and coronal clefts are present in the lumbar region. Pronounced shortening of the long bones is present, with expanded metaphyses.

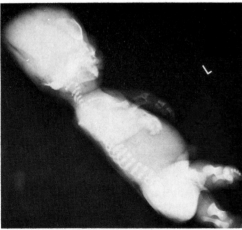

Fig. 28.4   Stillbirth—Lateral whole body★
There is absence of the neural arch of C2 and forward subluxation of the body. Coronal clefts are present in the lumbar spine, and platyspondyly. There is also marked limb shortening.

★ These three figures illustrate the recently described Fibrochondrogenesis, not metatropic dwarfism (see Differential diagnosis).

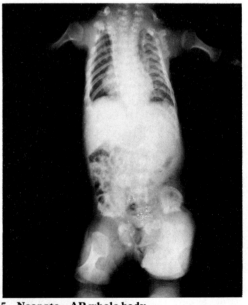

Fig. 28.5   Neonate—AP whole body
There is marked limb shortening but, apart from some shortening of ribs with widening of the costochondral junctions, the trunk appears to be normal. There is a left thoraco-lumbar scoliosis and 'butterfly' vertebrae in the cervico-dorsal area. Interpedicular distances in the lumbar region are very narrow. The long bones show the characteristic 'dumb-bell' deformity and bone maturation is retarded.

Skull and cervical spine: Birth–15 years

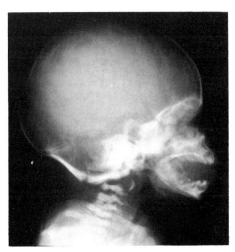

Fig. 28.6 V.M.—Neonate—Lateral skull
The vault, base and sutures appear normal, but there is platyspondyly of the cervical spine.
Same patient: Figs. 28.9, 28.10, 28.16, 28.21, 28.23, 28.27, 28.30, 28.34, 28.39, 28.42, 28.43

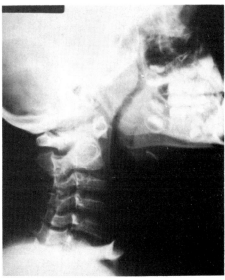

Fig. 28.7 K.P.—10 years—Lateral cervical spine
There is platyspondyly throughout the cervical spine and the odontoid peg is virtually absent. The clivus is short. Dentition appears normal.
Same patient: Figs 28.11, 28.20, 28.24, 28.26, 28.31

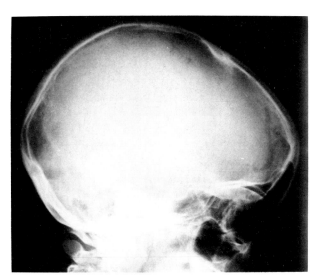

Fig. 28.8 F.G.—15 years—Lateral skull
Normal appearances of the vault and base.
Same patient: Figs. 28.17–28.19, 28.22, 28.28, 28.29, 28.35–28.38, 28.44, 28.45

Spine: Birth–1 year

Fig. 28.9   V.M.—Neonate—AP spine
There is very severe platyspondyly with increase in the depth of the intervertebral disc spaces. There is also scoliosis convex to the left in the thoraco-lumbar region.
Same patient: Figs 28.6, 28.10, 28.16, 28.21, 28.23, 28.27, 28.30, 28.34, 28.39, 28.42, 28.43

Fig. 28.10   V.M.—Neonate—Lateral spine (same patient as in Fig. 28.9)
The platyspondyly is very striking, only the anterior part of the lumbar vertebral bodies being ossified at this stage. The spinous processes are relatively normal.

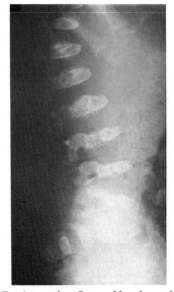

Fig. 28.11   K.P.—9 months—Lateral lumbar spine
Platyspondyly is still present at this age but L3, L4 and L5 show coronal clefts, and L1 and L2 have a central superior hump.
Same patient: Figs 28.7, 28.20, 28.24, 28.26, 28.31

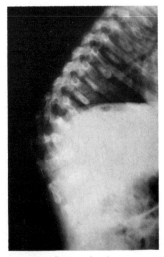

Fig. 28.12   S.R.—1 year—Lateral spine
There is pronounced platyspondyly with posterior wedging of the vertebral bodies, although L1 is anteriorly wedged with a kyphos at this level.
Same patient: Figs 28.13, 28.32, 28.40

Spine: 1–5 years

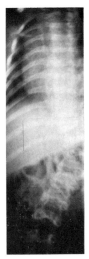

Fig. 28.13 S.R.—1 year—AP spine
There is flattening of the vertebral bodies with an increase in the size of the disc spaces, and left thoraco-lumbar scoliosis.
Same patient: Figs 28.12, 28.32, 28.40

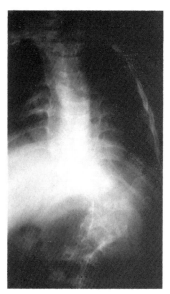

Fig. 28.14 R.B.—4 years—AP spine
There is irregular ossification of the end-plates and a left lumbar scoliosis.
Same patient: Figs 28.15, 28.25, 28.33, 28.41

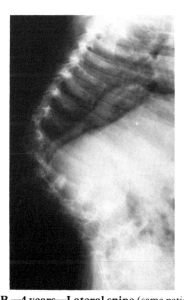

Fig. 28.15 R.B.—4 years—Lateral spine (same patient as in Fig. 28.14)
There is a kyphos in the lower dorsal region with anterior wedging. In the mid-dorsal region there is a 'diamond' configuration of the vertebral bodies, platyspondyly and the disc spaces are widened.

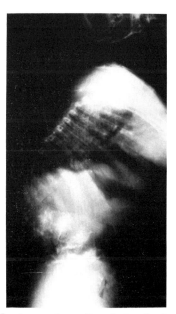

Fig. 28.16 V.M.—5 years 4 months—Lateral spine
There is a kyphos in the dorso-lumbar region and the vertebral bodies are flattened, with irregular ossification of the end-plates.
Same patient: Figs 28.6, 28.9, 28.10, 28.21, 28.23, 28.27, 28.30, 28.34, 28.39, 28.42, 28.43

Spine: 14 years–Adult

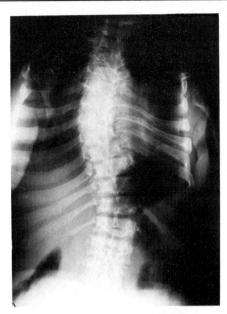

Fig. 28.17   F.G.—14 years 6 months—AP spine
There is mild end-plate irregularity. Scoliosis is present (right thoracic/left lumbar).
Same patient: Figs 28.8, 28.18, 28.19, 28.22, 28.28, 28.29, 28.35–28.38, 28.44, 28.45

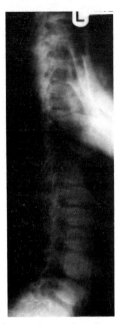

Fig. 28.18   F.G.—14 years 6 months—Lateral spine (same patient as in Figure 28.17)
Mild anterior wedging is present in the lower dorsal region, and there is humping in the middle portions of the vertebral bodies with some posterior constriction. The vertebral bodies appear to be of good height (cf. Fig 28.19).

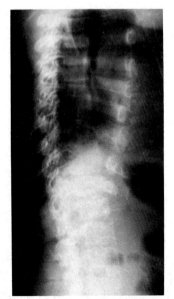

Fig. 28.19   F.G.—23 years—Lateral spine (same patient as in Fig. 28.17)
There is some elongation in the AP diameter of the vertebral bodies. Mild platyspondyly is present involving particularly the anterior parts of the vertebral bodies.

Chest: 4 months–15 years

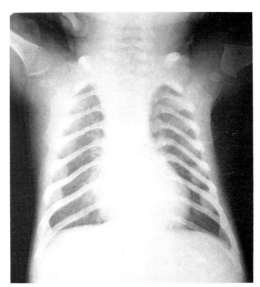

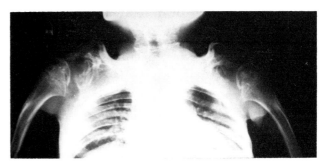

Fig. 28.21   V.M.—8 years—AP both shoulder joints
There is some lateral bowing of the humeral shafts and slight
irregularity of metaphyseal ossification at the proximal humeri and
glenoid fossae. The clavicles are elevated.
*Same patient: Figs 28.6, 28.9, 28.10, 28.16, 28.23, 28.27, 28.30, 28.34,
28.39, 28.42, 28.43*

Fig. 28.20   K.P.—4 months—PA chest
The bony thorax is small, and the ribs are short with expanded
anterior ends.
Same patient: Figs 28.7, 28.11, 28.24, 28.26, 28.31

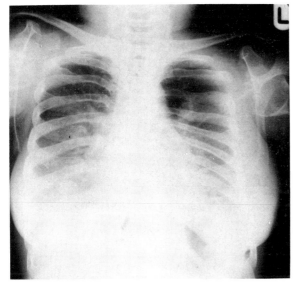

Fig. 28.22   F.G.—15 years—PA chest
There is only minimal reduction in the size of the thoracic cage. The
clavicles are slender and long.
*Same patient: Figs 28.8, 28.17–28.19, 28.28, 28.29, 28.35–28.38, 28.44,
28.45*

Pelvis : Birth–6 years

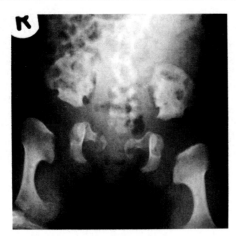

Fig. 28.23   V.M.—Neonate—Pelvis
The iliac wings are small and square, the acetabular roofs horizontal and the anterior superior spines low. The sacro-sciatic notches are small. There is marked enlargement of the proximal femora in the intertrochanteric regions, and flaring of the metaphyses at the lower end.
Same patient : Figs 28.6, 28.9, 28.10, 28.16, 28.21, 28.27, 28.30, 28.34, 28.39, 28.42, 28.43

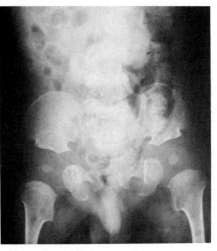

Fig. 28.24   K.P.—8 months—Pelvis
The iliac bones are small and square with low-set anterior superior iliac spines. The sacro-sciatic notches are small and the acetabular roofs horizontal. The pubic and ischial bones are rather broad and squat and there is slight widening of the symphysis. A 'butterfly' vertebra is present at L4. The capital femoral epiphyses are well placed within the acetabular fossae but the epiphyseal plates are unusually wide. There is expansion of the intertrochanteric regions, with short femoral necks.

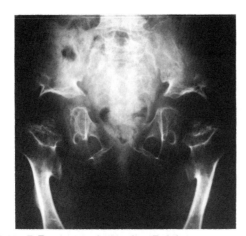

Fig. 28.25   R.B.—3 years 8 months—Pelvis
The acetabular roofs are horizontal with a sharpened lateral edge and the sacro-sciatic notches are small. The capital femoral epiphyses are still very small and the femoral necks irregularly ossified. There is broadening of the intertrochanteric regions.
Same patient : Figs 28.14, 28.15, 28.33, 28.41

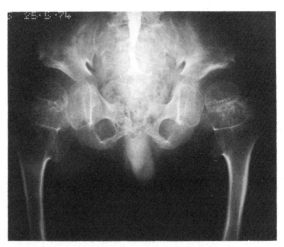

Fig. 28.26   K.P.—6 years 1 month—Pelvis (same patient as in Fig. 28.24)
The acetabular roofs are horizontal with some lateral beaking. The sacro-sciatic notches are small. The femoral heads are well ossified and well placed in the acetabula. There is irregular patchy ossification of the femoral necks and flaring in the intertrochanteric regions.

Pelvis: 9 years–Adult

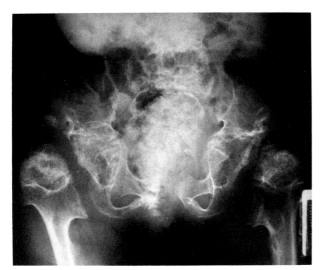

Fig. 28.27 V.M.—9 years 2 months—Pelvis
The pubic symphysis is wide; the iliac bones are small, acetabular roofs horizontal and the anterior superior iliac spines low-set. There is some protrusio acetabuli but the capital femoral epiphyses are poorly ossified and rather fragmented. The femoral necks are extremely short but there is enlargement of the intertrochanteric regions.
Same patient: Figs 28.6, 28.9, 28.10, 28.16, 28.21, 28.23, 28.30, 28.34, 28.39, 28.42, 28.43

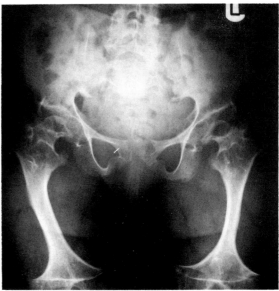

Fig. 28.28 F.G.—14 years 6 months—Pelvis
The iliac bones are small and the acetabular roofs horizontal. The femora are short with lateral bowing of the diaphyses and pronounced metaphyseal flaring. There is constriction of the femoral necks and the femoral heads are small.
Same patient: Figs 28.8, 28.17–28.19, 28.22, 28.29, 28.35–28.38, 28.44, 28.45

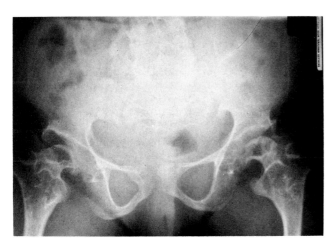

Fig. 28.29 F.G.—23 years—Pelvis (same patient as in Fig. 28.28)
The femoral heads are small and the necks short. The intertrochanteric regions are enlarged and the greater trochanters are at a higher level than normal.

Lower limb : Birth–2 years

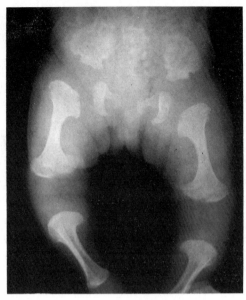

Fig. 28.30   V.M.—Neonate—Lower limbs
The long bones are shortened, with pronounced diaphyseal
constriction and metaphyseal flaring and the metaphyses have smooth
rounded ends ('dumb-bell' shaped). The iliac bones are small and
square, acetabular roofs horizontal and the inferior portions of the
iliac bones underdeveloped. Knee ossification centres have not
appeared, indicating delayed bone maturation compared with the
normal neonate.
*Same patient : Figs 28.6, 28.9, 28.10, 28.16, 28.21, 28.23, 28.27, 28.34,
28.39, 28.42, 28.43*

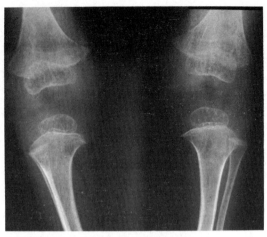

Fig. 28.31   K.P.—1 year 6 months—AP knees
There is osteoporosis and metaphyseal flaring is present, causing
pronounced medial beaking of the proximal tibial and the distal
femoral metaphyses.
Same patient : Figs 28.7, 28.11, 28.20, 28.24, 28.26

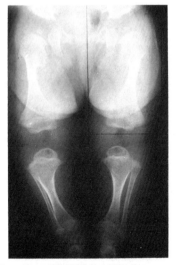

Fig. 28.32   S.R.—2 years—AP lower limbs
The long bones are short with diaphyseal constriction and
metaphyseal flaring. The capital femoral epiphyses are present but
small. The fibulae are long in relationship to the tibiae and there is
some lateral tilt at the ankle joints.
Same patient : Figs 28.12, 28.13, 28.40

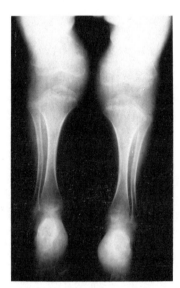

Figs 28.33   R.B.—2 years 9 months—AP tibiae and fibulae
Some metaphyseal flaring is present. The epiphyses are large and
rather flat, conforming to the increased size of the metaphyses. There
is diaphyseal constriction and medial bowing of both fibulae, which
are long in relation to the tibiae.
Same patient : Figs 28.14, 28.15, 28.25, 28.41

Knees: 8 years–Adult

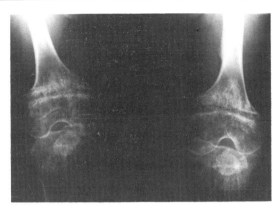

Fig. 28.34   V.M.—8 years—AP knees
There is metaphyseal flaring and irregular ossification involving the metaphyses and extending in a ∧-shape towards the diaphysis. The epiphyses are large and poorly modelled.
Same patient: Figs 28.6, 28.9, 28.10, 28.16, 28.21, 28.23, 28.27, 28.30, 28.39, 28.42, 28.43

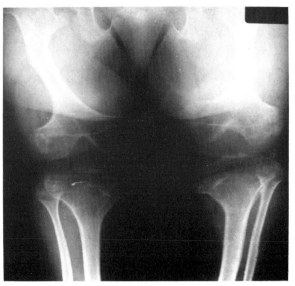

Fig. 28.35   F.G.—14 years 6 months—AP knees
Pronounced metaphyseal flaring is present and the fibulae are long in relation to the tibiae.
Same patient: Figs 28.8, 28.17–28.19, 28.22, 28.28, 28.29, 28.36–28.38, 28.44, 28.45

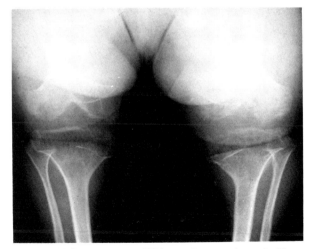

Fig. 28.36   F.G.—23 years—AP knees (same patient as in Fig. 28.35)
Pronounced flaring is present at the upper ends of the tibiae, and there is abnormal modelling of the joint surfaces with some irregularity. The intercondylar notches are unusually deep. The proximal ends of the fibulae are long.

Feet: 15 years

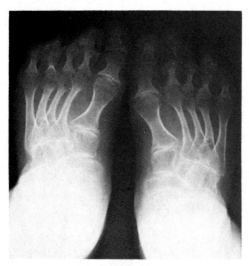

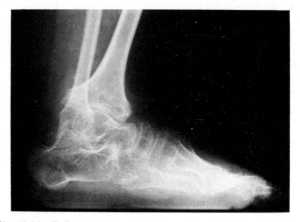

Fig. 28.38   F.G.—15 years—Standing lateral foot (same patient as in Fig. 28.37)
The distal end of the fibula is long in relation to the tibia, the talus is flattened, and lower end of the tibia flared.

Fig. 28.37   F.G.—15 years—Standing PA feet
There is bilateral metatarsus varus. The long bones show diaphyseal constriction and shortening.
Same patient: Figs 28.8, 28.17–28.19, 28.22, 28.28, 28.29, 28.35, 28.36, 28.38, 28.44, 28.45

Upper limb: Birth–2 years

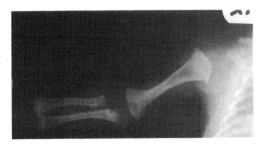

Fig. 28.39 V.M.—Neonate—Right upper limb
The long bones are short and there is pronounced metaphyseal flaring with irregularity.
Same patient: Figs 28.6, 28.9, 28.10, 28.16, 28.21, 28.23, 28.27, 28.30, 28.34, 28.42, 28.43

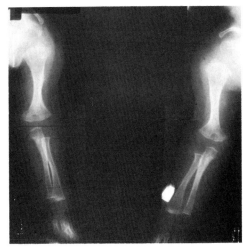

Fig. 28.40 S.R.—1 year—AP both upper limbs
The long bones are short with pronounced metaphyseal flaring and the deltoid tuberosities are prominent.
Same patient: Figs 28.12, 28.13, 28.32

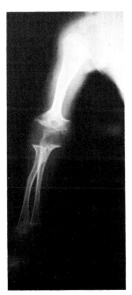

Fig. 28.41 R.B.—2 years 9 months—AP left upper limb
The long bones are short, particularly the humerus (rhizomelic appearance). Metaphyseal flaring and irregularity is pronounced, especially around the elbow joint. The proximal humeral epiphyses are present; they are flat and broad, conforming to the overall increased size of the epiphyseal plate.
Same patient: Figs 28.14, 28.15, 28.25, 28.33

Upper limb: 8 years–adult

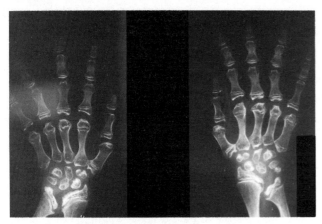

Fig. 28.42   V.M.—8 years—PA hands
The phalanges and metacarpals show shortening and mild
metaphyseal flaring. The carpal centres are small and rather angular
in outline. There is a pronounced V-shaped deformity at the wrist
with medial sloping of the distal ulnar epiphysis and to a lesser extent,
radial. The distal ulna is shortened.
Same patient : Figs 28.6, 28.9, 28.10, 28.16, 28.21, 28.23, 28.27, 28.30,
28.34, 28.39, 28.43

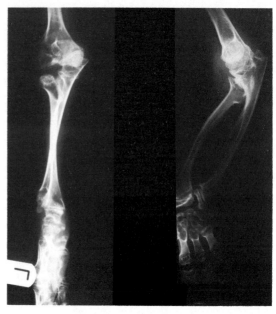

Fig. 28.43   V.M.—9 years—AP and lateral left forearm (same
patient as in Fig. 28.42)
There is bowing of the radius and ulna, metaphyseal flaring and
medial angulation of the distal epiphyses with some subluxation of the
distal ulna.

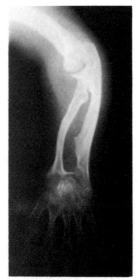

Fig. 28.44   F.G.—15 years—Left forearm
There is pronounced metaphyseal flaring of the distal end of the
radius with irregularity of the surface adjacent to the epiphyseal plate.
The ulna is relatively long and the distal end shows abnormal
modelling. The interosseous ridge of the ulna is pronounced and there
appears to be ossification extending into the interosseous membrane.
Same patient : Figs 28.8, 28.17–28.19, 28.22, 28.28, 28.29, 28.35–28.38,
28.45

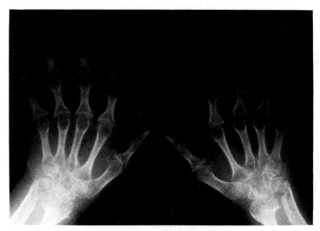

Fig. 28.45   F.G.—23 years—PA both hands (same patient as in
Fig. 28.44)
There is mild ulnar deviation of the fingers. The middle and distal
phalanges are especially short. The size of the carpus is reduced and
there is medial angulation of the distal end of the radius. The ulna is
relatively long and shows abnormal modelling with subluxation. The
appearances of the distal radius and ulna are essentially those of a
Madelung's deformity.

Chapter Twenty-nine
KNIEST DISEASE

Similar to metatropic dwarfism, but with hypertelorism, depressed nasal bridge, deafness and myopia as well as short stature with a short trunk and broad thorax.

Inheritance

Autosomal dominant.

Frequency

Extremely rare, probably under 0.1 per million prevalence.

Clinical features

Facial appearance—rather flat, with depressed nasal bridge and hypertelorism.

Intelligence—some patients are normal and rarely some retarded—but visual and auditory problems may account for this in part.

Stature—markedly reduced, and spinal deformity may reduce still further the length of the trunk.

Presenting feature/age—signs are present at birth of short stature, joint stiffness and enlargement and facial abnormalities.

Deformities—scoliosis, joint contractures, malalignment of lower limbs.

Associated anomalies—myopia and detached retina, conductive and neural hearing loss, cleft palate, cataract.

Radiographic features

Skull—normal.

Spine—irregular platyspondyly. Thoraco-lumbar kyphos.

Pelvis—broad (rectangular) ilia which are hypoplastic inferiorly.

Long bones—very broad metaphyses and irregular epiphyses particularly around the knee and trochanteric area.

Hands—extra ossification centres at distal ends of middle phalanges.

Bone maturation

This is unusually advanced.

Biochemistry

Not known.

Differential diagnosis

From *metatropic dwarfism* but here the face is normal and the thorax narrow.

Spondylo-epiphyseal dysplasia congenita is similar, but does not have the facial changes or so great an expansion of the trochanteric area of femora.

Progress/complications

Marked dwarfing (final height between 106 and 145 cm) and secondary osteoarthritis associated with the epiphyseal dysplasia. Chronic otitis and deafness, myopia and possible retinal detachment are also likely.

REFERENCES

Kozlowski K, Barylak A, Kobielowa Z 1977 Kniest syndrome. Report of two cases. Australasian Radiology 21/1:60–67

Langer L O, Gonzales-Ramos M, Chen H, Espiritu C E, Courtney N W, Opitz J M 1976 A severe infantile micromelic chondrodysplasia which resembles Kniest disease. European Journal of Pediatrics 123:29–38

Maroteaux P, Spranger J 1973 La maladie de Kniest. Archives françaises de pédiatrie 30:735

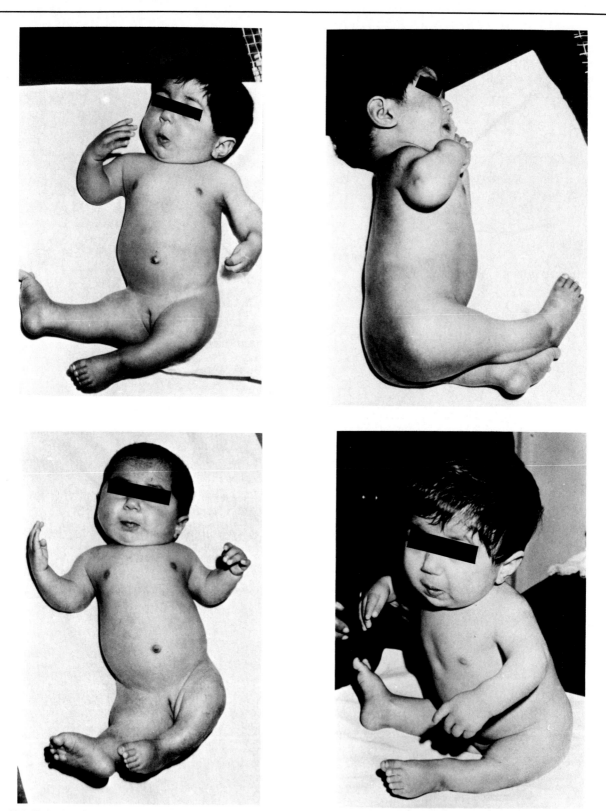

Figs 29.1–29.4   A rather flat, squashed face, with hypertelorism. There is a 'windswept' deformity of the lower limbs.

Skull and cervical spine: Birth–1 year

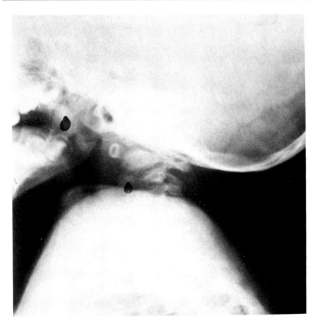

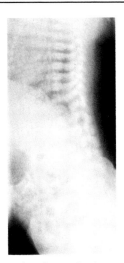

Fig. 29.6 S.R.—3 weeks—Lateral spine
There are coronal clefts present, otherwise no abnormality.
Same patient: Figs 29.7, 29.10, 29.12–29.14, 29.16, 29.17

Fig. 29.5 S.W.—1 year 9 months—Lateral skull and cervical spine
The base of the skull is normal, but the cervical spine shows an absent odontoid peg and platyspondyly.
Same patient: Figs 29.8, 29.9, 29.11, 29.15, 29.18, 29.19

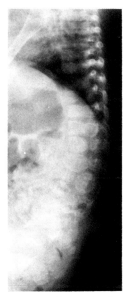

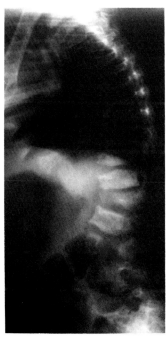

Fig. 29.7 S.R.—5 months—Lateral spine (same patient as in Fig. 29.6)
Coronal clefts are present involving the whole of the thoracic and lumbar spine.

Fig. 29.8 S.W.—1 year 9 months—Lateral spine (same patient as in Fig. 29.5)
There is platyspondyly with irregular vertebral end-plates, also a kyphos associated with a hypoplastic first lumbar vertebra. L.3 has a coronal cleft.

Spine and thorax : 1–2 years

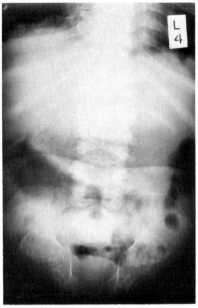

Fig. 29.9   S.W.—1 year 9 months—AP spine
There is platyspondyly and spina bifida occulta of L5 and the sacrum.
Same patient : Figs 29.5, 29.8, 29.11, 29.15, 29.18, 29.19

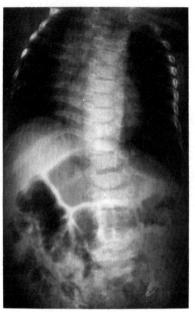

Fig. 29.10   S.R.—2 years—AP spine
The pedicles are wide apart—beyond the outline of the vertebral bodies.
Same patient : Figs 29.6, 29.7, 29.12–29.14, 29.16, 29.17

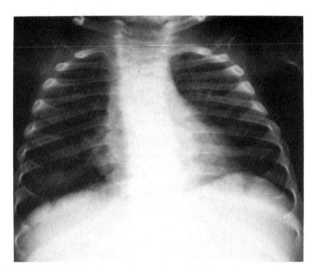

Fig. 29.11   S.W.—1 year 9 months—AP chest (same patient as in Fig. 29.9)
The ribs and thorax appear normal, but platyspondyly can be seen.

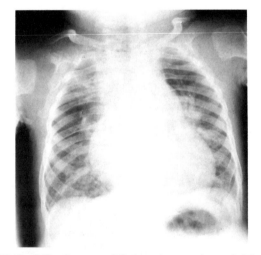

Fig. 29.12   S.R.—2 years—AP chest (same patient as in Fig. 29.10)
The heart is enlarged, but the thorax is normal. The proximal humeral metaphysis is flared and rounded; there are no epiphyses.

Pelvis and lower limb: Birth–21 months

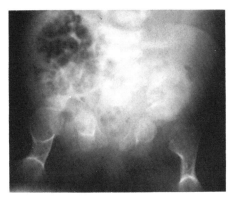

Fig. 29.13   S.R.—Neonate—Pelvis and femora
The ilia are rather small and rounded. The femora are short with
rounded ends, but not so clearly 'dumb-bell' shaped as in metatropic
dwarfism. The hips are subluxed.
Same patient: Figs 29.6, 29.7, 29.10, 29.12, 29.14, 29.16, 29.17

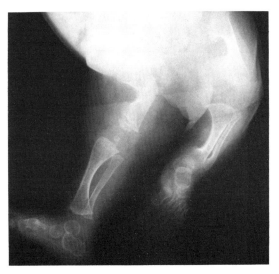

Fig. 29.14   S.R.—Neonate—AP lower limbs (same patient as in
Fig. 29.13)
All long bones are short with flared metaphyses. Bone maturation is
normal at the knees.

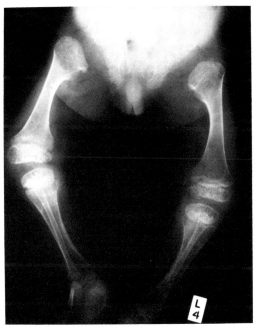

Fig. 29.15   S.W.—1 year 9 months—AP lower limbs
All long bones have expanded heads. The capital femoral epiphyses
have not appeared, and those around the knee show irregularity and
patchy sclerosis.
Same patient: Figs 29.5, 29.9, 29.11, 29.18, 29.19

Upper limb: Birth–21 months

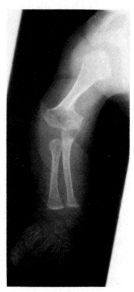

Fig. 29.17   S.R.—Neonate—AP and lateral right elbow (same patient as in Fig. 29.16)
There are flared metaphyses and a dislocated head of radius.

Fig. 29.16   S.R.—Neonate—AP upper limb
As in the lower limbs, all long bones are short with flared metaphyses.
Carpal maturation is advanced by about 2½ years.
Same patient: Figs 29.6, 29.7, 29.10, 29.12–29.14, 29.17

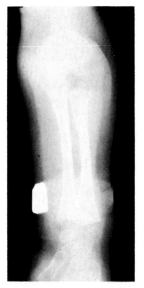

Fig. 29.18   S.W.—1 year 9 months—AP forearm
All metaphyses are flared.
Same patient: Figs 29.5, 29.8, 29.9, 29.11, 29.15, 29.19

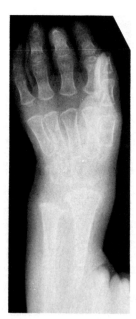

Fig. 29.19   S.W.—1 year 9 months—PA hand (same patient as in Fig. 29.18)
The ends of all long bones are rounded and smooth. The phalangeal bone age is normal, but the carpal centres show maturation advanced by about 2½ years.

Chapter Thirty
DYGGVE-MELCHIOR-CLAUSEN DISEASE

Short-trunk dwarfism together with mental retardation and characteristic radiological signs in the spine and ilia.

Inheritance

Autosomal recessive.

Frequency

Extremely rare, less than 0.1 per million prevalence.

Clinical features

Facial appearance—normal.
Intelligence—usually mental retardation.
Stature—short trunk dwarfism.
Presenting feature/age—the short stature becomes apparent in infancy or early childhood as does the mental retardation.
Deformities—sternal protrusion (similar to Morquio's disease); scoliosis; limb malalignment.
Associated anomalies—stiff joints.

Radiographic features

Skull—normal, or microcephalic.
Spine—marked platyspondyly with a characteristic notched appearance on the lateral view.
Pelvis—disordered ossification of the iliac crest, which has a bizarre lace-like appearance.

Long bones—slightly irregular epiphyseal and metaphyseal development.
Hands—sometimes coned epiphyses.

Bone maturation

Delayed.

Biochemistry

Not known.

Differential diagnosis

From *Morquio's disease*, but here intelligence is normal, corneal clouding is present and the vertebral shape and iliac development is different. Biochemical tests will confirm the diagnosis.

Progress/complications

Little is known but premature osteoarthritis is likely and perhaps scoliosis.

REFERENCES

Schorr S, Legum C, Ochshorn M 1977 The Dyggve-Melchior-Clausen syndrome. American Journal of Roentgenology 128/1:107–113
Toledo S P A, Saldanha P H, Lamego C, Mourao P A S, Dietrich C P, Mattar E 1979 Dyggve-Melchior-Clausen syndrome: genetic studies and report of affected sibs. American Journal of Medical Genetics 4:255–261

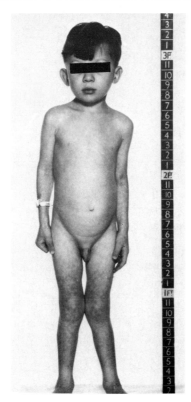
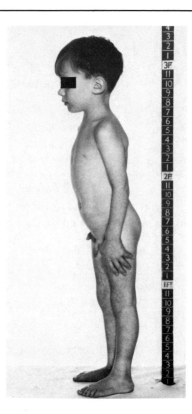

Figs 30.1 and 30.2   Little abnormality apart from genu valgum and some shortness of stature—although trunk and limbs are in proportion.

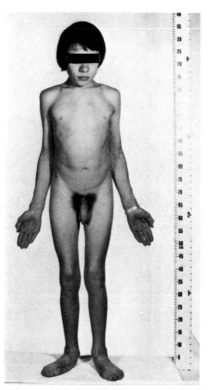
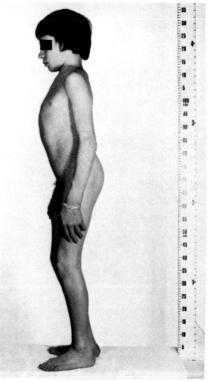

Figs 30.3 and 30.4   Same patient, aged 16 years, now shows a disproportionately short trunk, sternal protrusion and contractures of the hips and knees.

Skull and thorax: 10–14 years

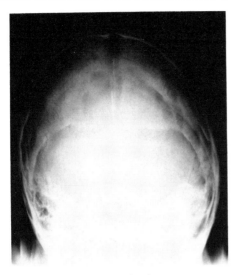

Fig. 30.5 S.F.—10 years—Towne's view
There is calcification in the falx cerebri, not normally seen at this age.
Same patient: Figs 30.7, 30.11, 30.13, 30.15, 30.16, 30.18

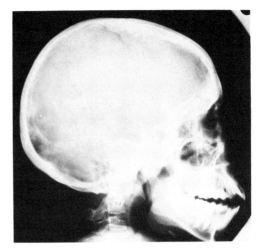

Fig. 30.6 B.B.—14 years—Lateral skull
There is microcephaly, but no other abnormality.
Same patient: Figs 30.8–30.10, 30.12, 30.14, 30.17

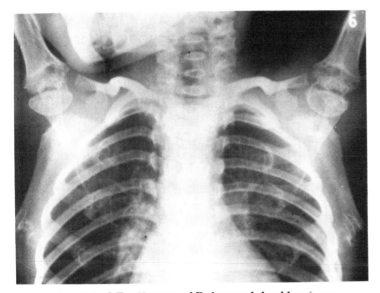

Fig. 30.7 S.F.—10 years—AP chest and shoulders (same patient as in Fig. 30.5)
The ribs are normal apart from deep anterior cupping. The scapulae here have a lace-like appearance at the lower angles, and the humeral metaphyses are irregular, flared and translucent. There is platyspondyly of the cervical vertebrae.

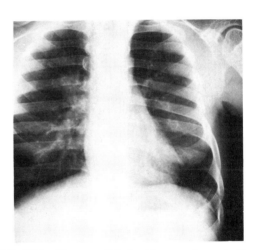

Fig. 30.8 B.B.—14 years—PA chest (same patient as in Fig. 30.6)
The ribs are broad and lacking the normal posterior constriction.

Spine : 6–15 years

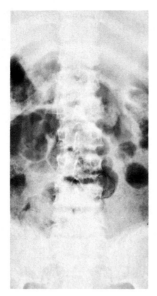

Fig. 30.9   B.B.—15 years—AP lumbar spine
There is platyspondyly with minor irregularities of the vertebral end-plates.
Same patient : Figs 30.6, 30.8, 30.10, 30.12, 30.14, 30.17

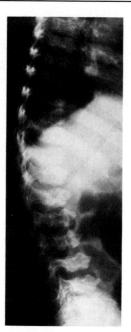

Fig. 30.10   B.B.—6 years—Lateral thoracic and lumbar spine
(same patient as in Fig. 30.9)
The upper lumbar vertebrae have an anterior tongue. The characteristic feature of this disease is shown in the lower lumbar vertebrae, with the notching of the superior and inferior surfaces (persistence of coronal clefts). There is also a thoraco-lumbar kyphos.

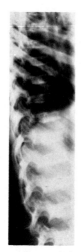

Fig. 30.11   S.F.—10 years—Lateral thoracic and lumbar spine
There is marked platyspondyly, but no persisting coronal clefts.
Same patient : Figs 30.5, 30.7, 30.13, 30.15, 30.16, 30.18

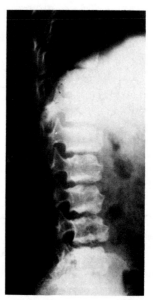

Fig. 30.12   B.B.—14 years—Lateral lumbar spine (same patient as in Fig. 30.9)
There is no platyspondyly at this age, but the vertebral end-plates are very irregular.

Pelvis and lower limb: 6–15 years

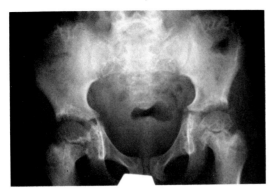

Fig. 30.13 S.F.—6 years—Pelvis
The main characteristic feature is the lace-like ossification of the iliac crests. The capital femoral epiphyses are large, and the adjacent metaphyses irregular.
Same patient: Figs 30.5, 30.7, 30.11, 30.15, 30.16, 30.18

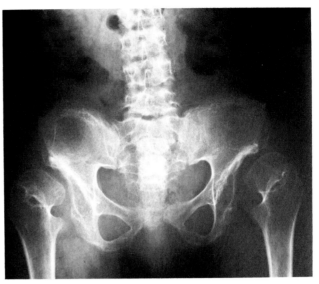

Fig. 30.14 B.B.—15 years—Pelvis and hips
The iliac crest is irregular, and the whole ilium poorly developed, with marked acetabular dysplasia and subluxation of the hips.
Same patient: Figs 30.8–30.10, 30.12, 30.17

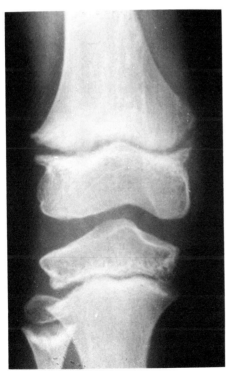

Fig. 30.15 S.F.—9 years—AP knee (same patient as in Fig. 30.13)
There are widened epiphyseal plates, with adjacent sclerosis. The metaphyses are flared (almost spur-like) as are the epiphyses.

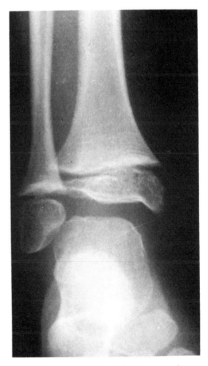

Fig. 30.16 S.F.—9 years—AP ankle (same patient as in Fig. 30.13)
Some sclerosis and cupping of the metaphyses, is present, also incomplete development of the lower tibial epiphysis.

Hand: 10–12 years

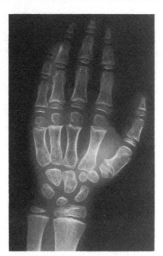

Fig. 30.17 B.B.—10 years—PA hand
The lower radius and ulna are normal, but the metacarpals are short and the first has epiphyses at each end. There are also several coned epiphyses in the phalanges.
Same patient: Figs 30.6, 30.8–30.10, 30.12, 30.14

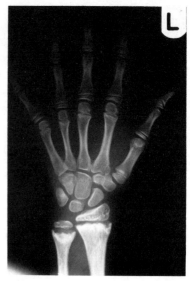

Fig. 30.18 S.F.—12 years—PA hand
In this patient there are marked metaphyseal changes at the wrist, the hand being relatively normal. The bone age is retarded by about 2 years.
Same patient: Figs 30.5, 30.7, 30.11, 30.13, 30.15, 30.16

Chapter Thirty-one
CHONDRO-ECTODERMAL DYSPLASIA
(Ellis-van Creveld syndrome)

Short-limbed dwarfism associated with dysplastic nails, hair and teeth accompanied by post-axial polydactyly and congenital heart disease.

Inheritance

Autosomal recessive.

Frequency

Extremely rare, under 0.1 per million prevalence.

Clinical features

Facial appearance—may be normal, apart from thin sparse hair and disordered eruption or absence of teeth. Sometimes micrognathos, and there may be a small central cleft of the upper lip, tied to the alveolar ridge.

Intelligence—normal.

Stature—may be markedly reduced, and with disproportionately short limbs, particularly in the distal segment. Less severe cases may reach 160 cm final height.

Presenting feature/age—the short stature and polydactyly is apparent at birth and there may be respiratory difficulties from the short ribs and long narrow thorax.

Deformities—there may be none, but some epiphyses are disordered and limb malalignment can occur. The head of radius is sometimes dislocated.

Associated anomalies—congenital heart disease (usually a septal defect) may be present. Hair and teeth anomalies are noted above.

Radiographic features

Skull—normal.

Spine—normal.

Thorax—short ribs and a long narrow thorax in infancy, but this tends to become normal as the child grows.

Pelvis—small iliac bones, and sometimes a downwardly-directed spike in the region of the triradiate cartilage.

Long bones—markedly short, particularly the distal segments. The femora and humeri may be rather thick and bowed, and the terminal phalanges hypoplastic (associated with the absent or hypoplastic nails).

Knee joints—the upper tibial epiphyses appear to be placed too far medially, leading to sometimes severe genu valgum.

Wrists/hands—post-axial polydactyly; fusion of capitate and hamate. Later, coned epiphyses may be seen.

Bone maturation

Not known.

Biochemistry

Not known.

Differential diagnosis

In the neonatal period, from *asphyxiating thoracic dystrophy* and other forms of lethal dwarfism. These are not accompanied by hypoplastic nails or a 'tied' upper lip. Later, from other forms of short-limbed dwarfism, but the short forearm and legs, upper tibial deformity and fused capitate and hamate should distinguish the Ellis-van Creveld syndrome.

Progress/complications

The tendency is towards improvement if the infant survives initial respiratory difficulties: about half die in infancy from this cause. The prognosis is related also to the congenital heart defect.

REFERENCES

Ellis R W B, Andrew J D 1962 Chondro-ectodermal dysplasia. Journal of Bone and Joint Surgery 44B:626

Da Silva E O, Janovitz D, Cavalcanti de Albuquerque S 1980 Ellis-van Creveld syndrome: report of 15 cases in an inbred kindred. Journal of Medical Genetics 17:349–356

Trunk and limbs: 2–8 years

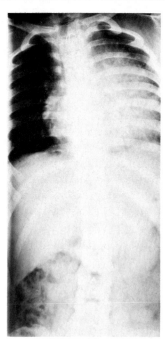

Fig. 31.1   E.H.—8 years—AP spine
The heart is enlarged (and there are sternal sutures indicating previous cardiac surgery).
Same patient: Fig. 31.4

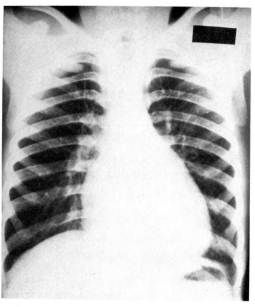

Fig. 31.2   J.M.—2 years 6 months—AP chest
The heart is enlarged; the lung fields are clear and there are normal pulmonary vessels, with normal-sized thorax.
Same patient: Figs 31.3, 31.5, 31.7

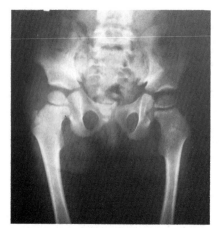

Fig. 31.3   J.M.—3 years—Pelvis (same patient as in Fig. 31.2)
The sciatic notches are small and the femoral necks broadened but by this age the distinctive signs of the Ellis–van Creveld syndrome have already improved.

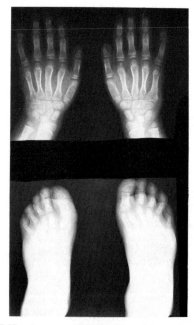

Fig. 31.4   E.H.—8 years—AP hands and feet (same patient as in Fig. 31.1)
The right hand shows a thickened fifth metacarpal (associated with an excised sixth digit). All the terminal phalanges are hypoplastic and there is fusion of the capitate and hamate. The right foot shows six metatarsals, two being associated with the hallux.

Hands and feet : 3 years–Adult

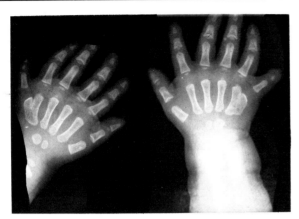

Fig. 31.5 J.M.—3 years—PA hands
There is post-axial polydactyly with partial fusion of the fifth and
sixth metacarpals. The middle and distal phalanges are short and the
cone epiphyses are developing in the middle row.
Same patient : Figs 31.2, 31.3, 31.7

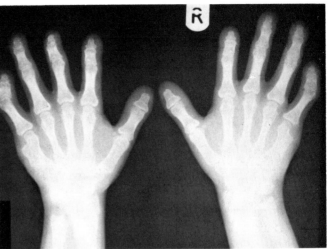

Fig. 31.6 E.W.—37 years—PA hands
There is an attempt at duplication of both fifth metacarpal bones and
disproportionate shortening of the middle and terminal phalanges.
Both distal ulnae are short.
Same patient : Fig. 31.8

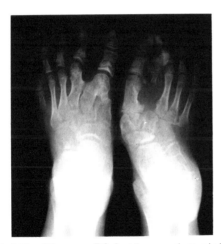

Fig. 31.7 J.M.—13 years—PA feet (same patient as in Fig. 31.5)
Five toes are present on each side, but the metatarsals are
disorganised : on the left the fifth is absent, the first is duplicated and
partially fused to the second; on the right the second metatarsal is
absent. There is disorganisation of the tarsal centres, with absence of
the lateral cuneiforms.

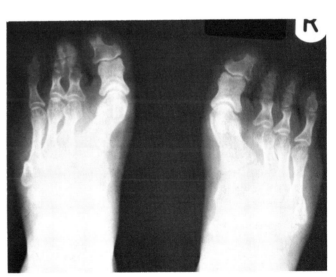

Fig. 31.8 E.W.—37 years—PA feet (same patient as in Fig. 31.6)
Four toes are present on each foot. There appears to have been
amputation on the post-axial sides, causing deformities of the
proximal fourth metatarsals. All metatarsals are short.

Chapter Thirty-two
ASPHYXIATING THORACIC DYSPLASIA (Jeune's disease)

Congenital narrowing of the rib cage, with respiratory difficulties in infancy, short limbs and sometimes post-axial polydactyly.

Inheritance

Autosomal recessive.

Frequency

Not known. Severe cases die in the neonatal period, but more mildly affected patients live, and may not be diagnosed.

Clinical features

Facial appearance—normal.
Intelligence—normal.
Stature—short, with disproportionately short limbs.
Presenting feature/age—usually respiratory difficulties in the neonatal period. Post-axial polydactyly may be present.
Deformities—not a feature, apart from the long, narrow chest.
Associated anomalies—some develop chronic nephritis in later childhood.

Radiographic features

Skull—normal.
Spine—normal.
Thorax—small, with short, horizontal ribs, becoming more normal as the child grows.

Pelvis—sometimes a small ilium with a downward projecting spur at the triradiate cartilage.
Limbs—short, with some metaphyseal irregularities.
Hands/feet—some cone epiphyses with premature fusion of epiphyses; sometimes post-axial polydactyly.

Bone maturation

No data.

Biochemistry

Not known.

Differential diagnosis

Chondro-ectodermal dysplasia—but Jeune's disease does not have the hair and teeth anomalies.
In severe cases, other forms of lethal dwarfism.

Progress/complications

If the infant survives the initial respiratory difficulties, then there may be no further complications. Renal failure may develop in later childhood or adult life.

REFERENCES

Kozlowski K, Masel J 1976 Asphyxiating thoracic dystrophy without respiratory disease: report of two cases of the latent form. Pediatric Radiology 5/1: 30–33
Oberklaid F, Danks D M, Mayne V, Campbell P 1977 Asphyxiating thoracic dysplasia. Archives of Disease in Childhood 52: 758–765

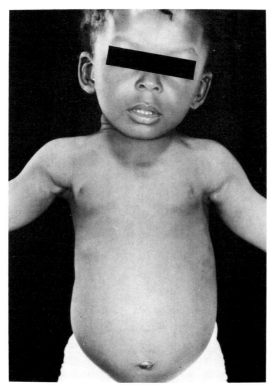

Fig. 32.1 No abnormality, apart from a long, narrow thorax.

Trunk and spine : Birth

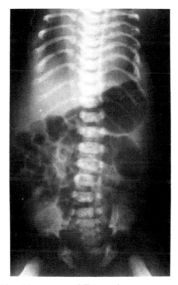

Fig. 32.2 O.H.—Neonate—AP trunk
There is a narrow thorax with short ribs. The iliac wings are small with medial downward projecting spikes.
Same patient : Fig. 32.6

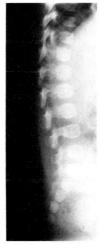

Fig. 32.3 A.H.—Neonate—Lateral lumbar spine
Normal.
Same patient : Figs 32.4, 32.8, 32.10, 32.12, 32.13

Chest: Birth–18 months

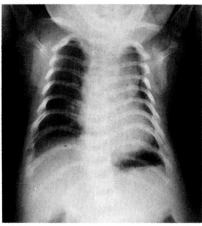

Fig. 32.4   A.H.—Neonate—AP chest
The ribs are short, the anterior ends extending only to the mid-axillary line. There is metaphyseal irregularity of the upper humeri.
Same patient: Figs 32.3, 32.8, 32.10, 32.12, 32.13

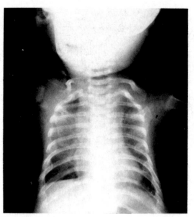

Fig. 32.5   R.N.—Neonate—AP chest
Similar to Figure 32.4; the heart appears large in relation to the small thoracic cavity.
Same patient: Fig. 32.9

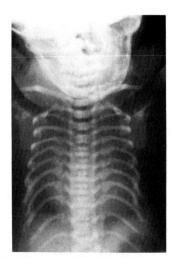

Fig. 32.6   O.H.—Neonate—AP chest
There is a long narrow thorax with short horizontal ribs.
Same patient: Fig. 32.2

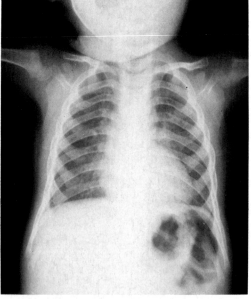

Fig. 32.7   A.N.—18 months—AP chest
The thorax is normal by this age.
Same patient: Fig. 32.14

Pelvis: Birth–1 year

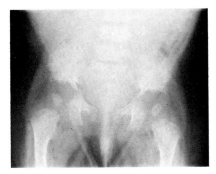

Fig. 32.8 A.H.—Neonate—Pelvis
The ilia are small and there are downward projecting spikes at the tri-radiate cartilage. The capital epiphyses are prematurely ossified.
Same patient: Figs 32.3, 32.4, 32.10, 32.12, 32.13

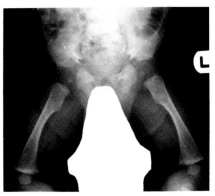

Fig. 32.9 R.N.—Neonate—AP hips and femora
Similar to Figure 32.8, but the capital femoral epiphyses are not unduly premature.
Same patient: Fig. 32.5

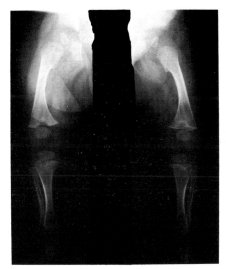

Fig. 32.10 A.H.—Neonate—AP lower limbs (same patient as in Fig. 32.8)
The long bones are short and flared, with metaphyseal irregularity.

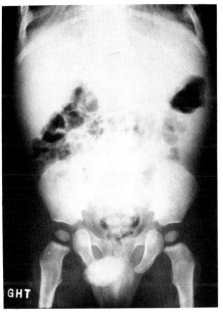

Fig. 32.11 B.C.—1 year—AP pelvis
The configuration is more normal, but small spurs at the triradiate cartilage are still visible.

Upper limb : Birth–6 years

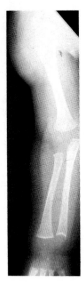

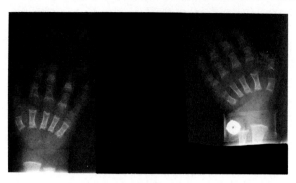

Fig. 32.13   A.H.—Neonate—PA hands (same patient as in Fig. 32.12)
Very unusually there are cone epiphyses of all the phalanges, and all metaphyses are irregular.

Fig. 32.12   A.H.—Neonate—AP right upper limb
The long bones are normal, apart from cupping of the distal ulna.
Same patient : Figs 32.3, 32.4, 32.8, 32.10, 32.13

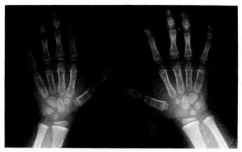

Fig. 32.14   A.N.—5 years—PA hands
The middle and terminal phalanges are short with cone epiphyses.
Same patient : Fig. 32.7

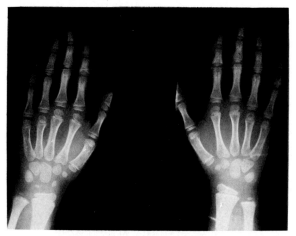

Fig. 32.15   B.N.—6 years—PA hands (brother of children in Figs 32.5 and 32.7)
Some of the middle and terminal phalanges are short, with cone epiphyses. Bone maturation is advanced by some 1–2 years.

Section Eight
LETHAL FORMS OF SHORT-LIMBED DWARFISM

Chapters Thirty-three to Thirty-eight LETHAL FORMS OF SHORT-LIMBED DWARFISM

There are many types, all lethal from respiratory failure at, or shortly after birth. Only the commoner, well-established conditions are described here:

1. Thanatophoric
2. Short rib/polydactyly syndromes
   Type I (Saldino–Noonan)
   Type II (Majewski)
   Type III (Naumoff)
3. Achondrogenesis (Types I and II)

Inheritance

Thanatophoric—sporadic, probably non-genetic.
Saldino-Noonan—autosomal recessive.
Majewski—autosomal recessive.
Naumoff—autosomal recessive.
Achondrogenesis—autosomal recessive.

Frequency

A possible incidence for thanatophoric dwarfism is 4–5 per million births; the other types are less common.

Clinical and radiographic features

The principal differentiating clinical and radiographic features are noted separately with each condition.

Bone maturation

Thanatophoric—retarded throughout.
All short rib/polydactyly syndromes—retarded at the knees, but advanced at the hips and shoulders.
Achondrogenesis—retarded throughout.

Biochemistry

Not known.

Differential diagnosis

From heterozygous *achondroplasia*, but in all lethal forms of dwarfism the limbs are very much shorter and the chest narrower.

The short-rib polydactyly syndromes are similar to *chondro-ectodermal dysplasia* but the additional ectodermal (hair, nails and teeth) anomalies differentiate the latter. *Asphyxiating thoracic dystrophy* may be indistinguishable.

Achondrogenesis needs to be differentiated from the severe lethal form of *hypophosphatasia* but the ossified skull and extremely short limbs of the former, with the blood chemistry of the latter, will distinguish them.

REFERENCES

General
Maroteaux P, Stanescu V, Stanescu R 1976 The lethal chondrodysplasias. Clinical Orthopaedics 114:31–45

Thanatophoric dwarfism (Chapter 33)
Fruchter Z 1973 Thanatophoric dwarfism. In: Kaufmann H J (ed.) Progress in pediatric radiology 4. S Karger, Basel, p 125
Horton W A, Rimoin D L, Hollister D W, Lachman R S 1979 Further heterogeneity within lethal neonatal short-limbed dwarfism: the platyspondylic types. Journal of Pediatrics 94: 736–742
Moir D H, Kozlowski K 1976 Long survival in thanatophoric dwarfism. Pediatric Radiology 5;123–125
Moore Q S et al 1980 Ultrasound scanning in a case of thanatophoric dwarfism with clover-leaf skull. British Journal of Radiology 53:241

Saldino–Noonan (Chapter 34)
Saldino R M, Noonan C D 1972 Severe thoracic dystrophy with striking micromelia, abnormal osseous development, including the spine, and multiple visceral anomalies. American Journal of Roentgenology 114:257–263
Spranger J, Grimm B, Weller M, Weissenbacher G, Hermann J, Gilbert E, Krepler R 1974 Short rib polydactyly (SRP) syndromes type Majewski and Saldino–Noonan. Zeitschrift für Kinderheilkunde 116:73–94

Majewski (Chapter 35)
Chen H, Yong S S, Gonzales E, Fowler M, Al Saddi A 1980 Short rib-polydactyly syndrome, Majewski type. American Journal of Medical Genetics 7:215–222

Cooper C P, Hall C M 1982 Lethal short-rib polydactyly syndrome of the Majewski type: a report of three cases. Radiology 144:513

Thomson G S M, Reynolds C P, Cruickshank J 1982 Antenatal detection of recurrence of Majewski dwarf (short rib-polydactyly syndrome type II Majewski). Clinical Radiology 33:509–517

Naumoff (Chapter 36)

Belloni C, Beluffi G 1981 Short rib-polydactyly syndrome, type Verma–Naumoff. Fortschritte auf dem Gebiete der Röntgenstrahlen und der Nukliarmedizin 134:431–435

Naumoff P, Young L W, Mazer J, Armartegui A 1977 Short rib polydactyly syndrome type e. Radiology 122:443–447

Achondrogenesis (Chapter 38)

Anderson P E Jr 1981 Achondrogenesis type II in twins. British Journal of Radiology 54:61

Saldino R M 1971 Lethal short-limbed dwarfism: achondrogenesis and thanatophoric dwarfism. American Journal of Roentgenology 112:185–197

Sillence D O, Lachman R, Rimoin D L 1978 Neonatal dwarfism. Pediatric Clinics of North America 25 (3):453–483

33. THANATOPHORIC DWARFISM

Clinical features

Large head, very short limbs and a narrow thorax.

Radiological features

Skull—there may be a clover leaf deformity.

Spine—platyspondyly with notched end-plates (H-shaped on AP view).

Thorax—short horizontal ribs.

Long bones—shortening and bowing—'telephone-receiver-shaped' femora.

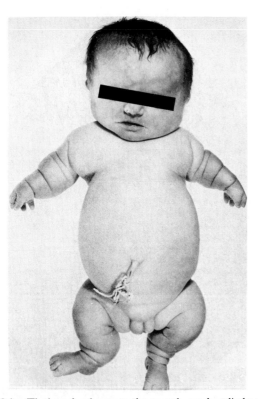

Fig. 33.1. The large head, narrow thorax and very short limbs are quite dissimilar to achondroplasia.

Thanatophoric dwarfism

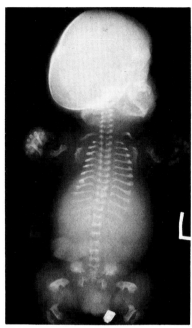

Fig. 33.2 Stillbirth—AP whole body
The ribs are short and horizontal. All long bones are short with bowing deformities and the femora have 'telephone receiver' appearance characteristic of thanatophoric dwarfism.

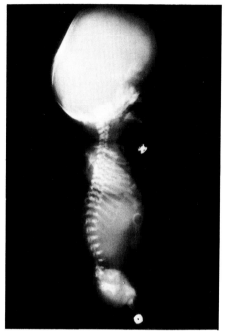

Fig. 33.3 Stillbirth—Lateral whole body
Pronounced platyspondyly is present, the skull vault is large and has the 'clover leaf' deformity sometimes associated with thanatophoric dwarfism.

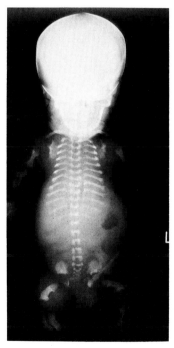

Fig. 33.4 Stillbirth—AP whole body
Similar to Figure 33.1 with short horizontal ribs and short, bowed long bones. The platyspondyly gives rise to an H-shaped appearance of the vertebrae.
Same case: Fig. 33.5

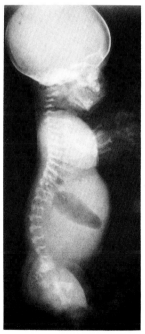

Fig. 33.5 Stillbirth—Lateral whole body (same case as in Fig. 33.4)
The kyphos and platyspondyly are clearly seen, with notched end-plates

Saldino–Noonan syndrome

34. SHORT RIB/POLYDACTYLY SYNDROME TYPE I (Saldino–Noonan)

Clinical features

Oedematous appearance, very short limbs, narrow chest and polydactyly (it is not always possible to tell whether this is pre-axial or postaxial).

Radiographic features

Skull—poor mineralisation, but otherwise normal.
Thorax—short horizontal ribs.
Pelvis—small iliac bones with medial spurring at the triradiate cartilage.
Long bones—short with rather jagged metaphyses.
Hands/feet—polydactyly.

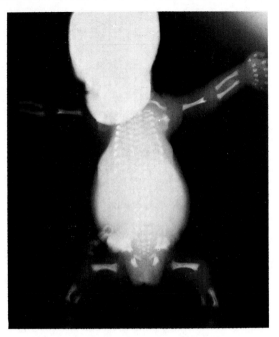

Fig. 34.1 B.B.—Stillborn—AP whole body
The ribs are short and horizontal and the long bones have irregular metaphyses. Post-axial polydactyly is present in the hands.

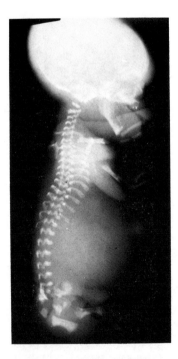

Fig. 34.2 B.B.—Stillborn—Lateral whole body
The vertebral bodies are small and have an angular configuration. The teeth are poorly formed.

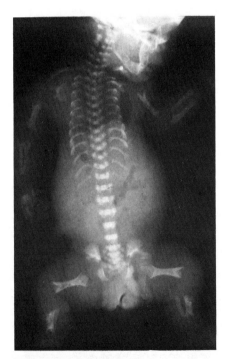

Fig. 34.3 Stillborn—AP whole body
The ribs are short and the long bones short with irregular metaphyses. (The soft tissue shows the appearance following autopsy.)

Majewski/syndrome

35. SHORT RIB/POLYDACTYLY SYNDROME TYPE II (Majewski)

Clinical features

Similar to the Saldino–Noonan type.

Radiographic features

Thorax—short, horizontal ribs.

Pelvis—normal ilia, but very small pubic rami.

Long bones—short, with smooth, rounded metaphyses. The tibia is minute (smaller than the fibula), and oval in shape.

Hands/feet—polydactyly (pre- and postaxial).

Majewski syndrome : whole body

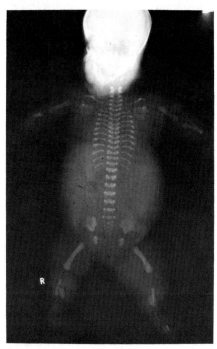

Fig. 35.1   Stillborn—AP whole body
The ribs are extremely short and horizontal. There is shortening of all long bones but this is especially pronounced in the tibiae. All metaphyses are smooth and rounded.

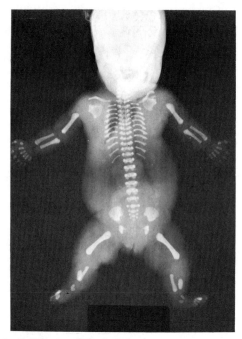

Fig. 35.2   Stillborn—AP whole body
The appearances are similar to Figure 35.1. The pubic rami and ischia are poorly developed. Only 11 pairs of ribs are present. Bone maturation is retarded (no epiphyses at the knees) in this full-term stillbirth corresponding to a bone age of about 29 weeks gestation.

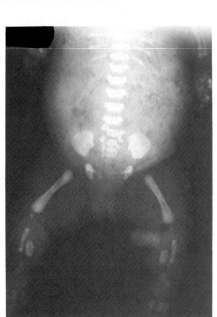

Fig. 35.3   Stillborn—AP trunk and lower limbs (sibling of case in Fig. 35.2)
The characteristic oval tibiae are present. Although only one epiphysis is present at each knee (gestation age about 32 weeks) ossification is just beginning in the capital femoral epiphyses.

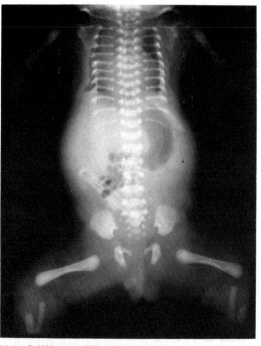

Fig. 35.4   Stillborn—AP whole body
Premature ossification is present in the upper humeral epiphyses and in the left capital femoral epiphysis. A little gas in the proximal small bowel demonstrates the presence of a malrotation, the jejunum lying on the right.
Same case : Figs 35.5–35.8

Majewski syndrome—Limbs

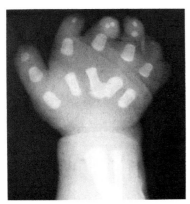

Fig. 35.5   Stillborn—PA right hand
Soft tissue post-axial polydactyly is present and there is fusion of the bases of the third and fourth metacarpals. There is severe shortening or absence of several middle and terminal phalanges.
Same case: Figs 35.4, 35.6–35.8

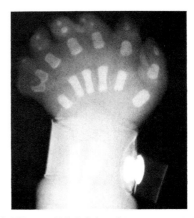

Fig. 35.6   Stillborn—PA left hand (same case as in Fig. 35.5)
Similar to Figure 35.5. but here there is osseous post-axial polydactyly.

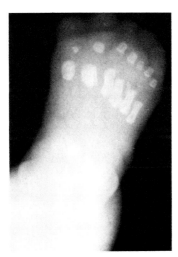

Fig. 35.7   Stillborn—PA right foot (same case as in Fig. 35.5)
Pre- and post-axial polydactyly is present in the feet. There is absent or poor ossification of the middle and distal phalanges.

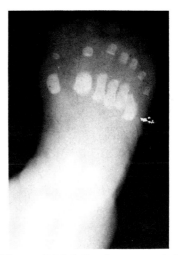

Fig. 35.8   Stillborn—PA left foot (same case as in Fig. 35.5)
Similar to Figure 35.7.

Naumoff syndrome

36. SHORT RIB/POLYDACTYLY SYNDROME TYPE III (Naumoff)

Clinical features

Similar to the Saldino–Noonan type.

Radiographic features

Thorax—short, horizontal ribs.
Pelvis—the iliac bones are small.
Long bones—shorter than in the Saldino–Noonan type, with irregular jagged metaphyses.
Hands/feet—absent ossification of the phalanges.

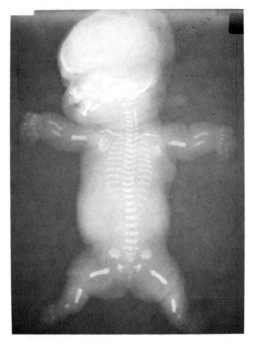

Fig. 36.1   Stillborn—AP whole body
The ribs and long bones are short. The metaphyses are pointed and irregular and there is absent ossification of the fibulae and most of the phalanges. Polydactyly is present in the hands and feet.

Unnamed short-rib polydactyly syndrome

37. A SHORT RIB/POLYDACTYLY SYNDROME (unnamed)

Clinical features

Similar to the Saldino–Noonan type.

Radiographic features

Thorax—short horizontal ribs.
Pelvis—normal.
Long bones—short with smooth, rounded metaphyses which have marginal 'spikes'.
Hands/feet—polydactyly (postaxial in the hands; feet uncertain).

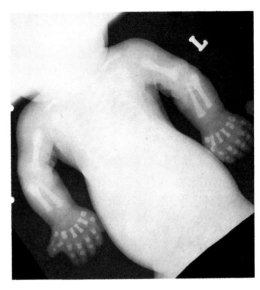

Fig. 37.1 B.P.—Stillbirth—AP trunk
The ribs are short and horizontal. Both hands show post-axial polydactyly.
Same case: Figs 37.2, 37.3

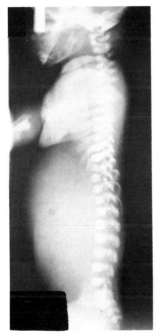

Fig. 37.2 B.P.—Stillbirth—Lateral trunk (same case as in Fig. 37.1)
The vertebral bodies are well-formed.

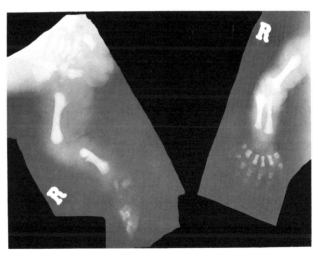

Fig. 37.3 B.P.—Stillbirth—Lateral right lower limb and AP right upper limb (same case as in Fig. 37.1)
The long bones are short with flared, rounded metaphyses and some metaphyseal 'spikes'. No ossification of the epiphyses at the knee is present but the capital femoral and upper humeral epiphyses have appeared. There is post-axial polydactyly in the hand.

38. ACHONDROGENESIS TYPE I

Clinical features

Oedematous appearance, with extremely short limbs and a protuberant abdomen.

Radiographic features

Skull—poorly mineralised (but not so severe as in hypophosphatasia).

Spine—there are areas of total absence of ossification, nearly always involving the sacrum.

Thorax—the ribs are short, horizontal and have a beaded appearance due to intra-uterine fractures.

Pelvis—the only part mineralised is the upper part of the iliac wings.

Long bones—minute, but broadened, with spiky appearance.

(Type II (Fig. 38.4) is less severe, with more complete ossification and no evidence of rib fractures.)

Achondrogenesis Types I and II

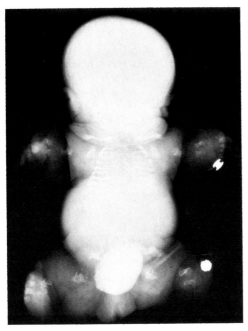

Fig. 38.1 Stillbirth—AP whole body (Achondrogenesis Type I)
The rib shortening is very striking and there is virtually no
ossification of the skull vault, spine or sacrum. The ribs are short and
beaded, and the metaphyses 'spiky'.

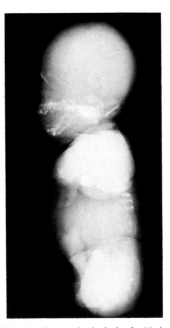

Fig. 38.2 Stillbirth—Lateral whole body (Achondrogenesis
Type I)
The limbs show the characteristic 'flipper' appearance.

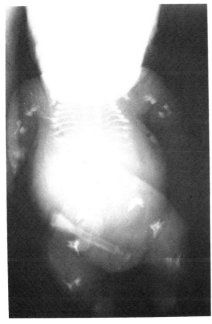

Fig. 38.3 Stillbirth—AP whole body (Achondrogenesis Type I)
Similar to Figure 38.1 but there is rather more ossification in the
spine. The curious appearance of the iliac wings is quite typical of
achondrogenesis Type I.

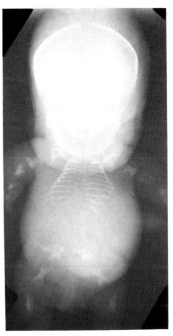

Fig. 38.4 Stillbirth—AP whole body (Achondrogenesis Type
II)
The ossification defects are less severe than in Type I but there is still
total absence of mineralisation in some areas of the spine. (There is no
beading of the ribs in Type II.)

Section Nine
INCREASED LIMB LENGTH

Chapter Thirty-nine
MARFAN SYNDROME

The commoner of variable characteristics are: body disproportion with long limbs, dislocated lens, generalised joint laxity, scoliosis, herniae and later aortic aneurysm.

Inheritance

Autosomal dominant.

Frequency

Difficult to determine since many individuals are only mildly affected, but an incidence of 1.5 per million has been estimated. A prevalence of 11 per million index patients (or 17 per million including affected relatives) is a possible figure for those presenting with orthopaedic complications only.

Clinical features

Facial appearance—normal.
Intelligence—normal.
Stature—increased with an arm span which is greater than the height, and upper segment (head to pubis) less than lower (pubis to heel).
Presenting feature/age—body disproportion with increased height, dislocated lens, scoliosis and joint laxity are the commoner presenting signs, usually during childhood. Less common features are myopia, detached retina, cataract, high-arched palate, hernia, congenital or recurrent dislocation of joints, genu valgum or recurvatum. The diagnosis of Marfan's disease may be made only because there are known affected relatives.
Deformities—scoliosis is the most serious and develops in perhaps half of the patients. Hand contractures are not uncommon, even though excessive joint laxity may be a feature elsewhere. Talipes equino-varus or other foot deformities are sometimes a feature.
Associated anomalies—very variable, and noted above.

Radiographic features

Skull—normal.

Spine—scoliosis frequently develops, vertebrae may be tall on the lateral view, with a decreased anteroposterior diameter.
Thorax—thoracic cage is normal but heart or aortic complications may be noted.
Long bones—surprisingly normal apart from unusual length.
Hands/feet—if arachnodactyly is present, then all the metacarpals (metatarsals) and phalanges are thin and elongated.

Bone maturation

Probably normal.

Biochemistry

It is likely there is a collagen defect but the precise structural abnormality has not yet been defined.

Differential diagnosis

Homocystinuria—here there is the same body disproportion, but patients are frequently mentally retarded, osteoporosis is a feature and the disorder is of autosomal recessive inheritance (enzyme defect: cystathionine synthetase).
Congenital contractural arachnodactyly—also of dominant inheritance, and with the same body disproportion. However, there are contractures of many joints, present at birth but subsequently improving somewhat. The additional features of Marfan's disease do not develop.

Progress/Complications

This is extremely variable. Many individuals are only mildly affected and remain undiagnosed. At its worst, death occurs in early adult life from aortic aneurysm or other cardiovascular problems, perhaps exacerbated by severe scoliosis with its respiratory complications.

REFERENCES

Brenton D P, Dow C J, James J I P, Hay R L, Wynne-Davies R 1972 Homocystinuria and Marfan's syndrome. A comparison. Journal of Bone and Joint Surgery 54B:277

Pyeritz E, Murphy E A, McKusick V A 1979 Clinical variability in the Marfan syndrome(s). Birth defects: Original Article Series XV (5B):155–178
Robins P R, Moe J H, Winter R B 1975 Scoliosis in Marfan's syndrome. Journal of Bone and Joint Surgery 57A:358–368

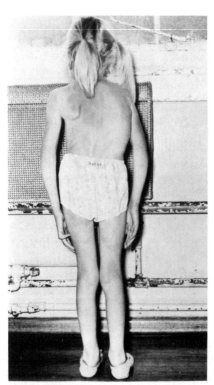

Fig. 39.1   Disproportionately long limbs as well as a right thoracic scoliosis.

Fig. 39.2   An older boy, with severe scoliosis and increased limb length.

Figs 39.3 and 39.4 Brother and sister (twins) with disproportionately long limbs.

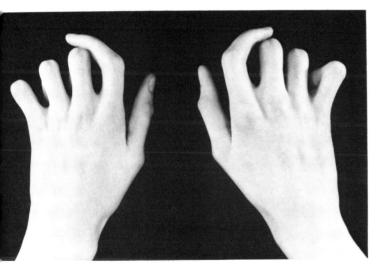
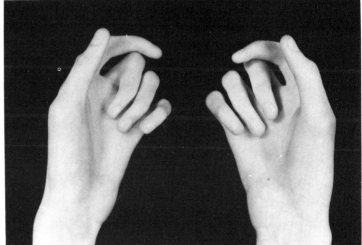

Figs 39.5 and 39.6 Arachnodactyly, this patient having quite severe finger contractures as well.

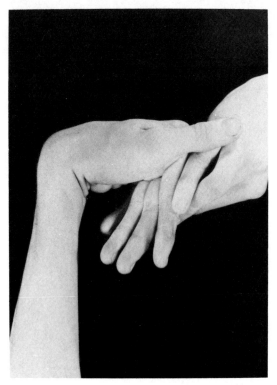

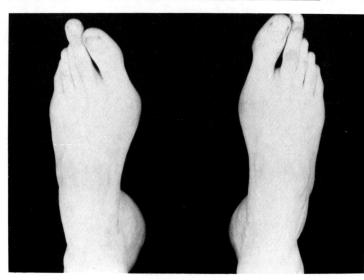

Fig. 39.8   Long, narrow feet and toes.

Fig. 39.7   The more usual excessive joint laxity is shown here.

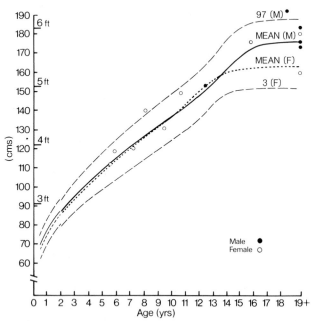

Fig. 39.9   Marfan syndrome—percentile height.

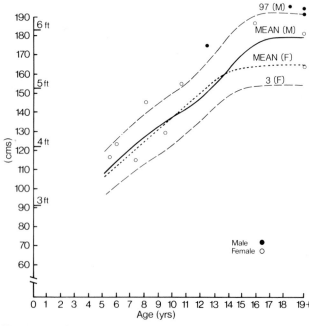

Fig. 39.10   Marfan syndrome—percentile span.

Spine: 2–10 years

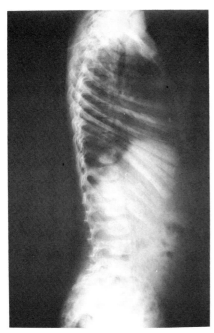

Fig. 39.11   A.S.—2 years—Lateral spine
No abnormality apart from loss of the normal lumbar lordosis.
Same patient: Figs 39.13, 39.16, 39.19

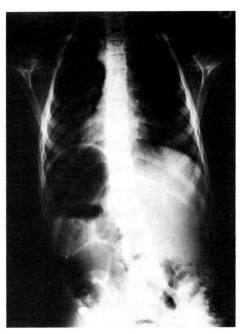

**Fig. 39.12
A.T.—7 years—AP spine**
There is a long C-shaped spinal curve to the left.

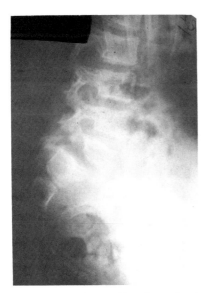

Fig. 39.13   A.S.—9 years—Lateral lumbar spine (same patient
as in Fig. 39.11)
There are bilateral defects of the pars interarticulares, with
spondylolisthesis of L5 on S1.

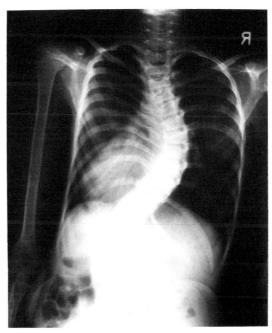

Fig. 39.14   T.B.—10 years—AP spine
There is right thoracic and left lumbar scoliosis.

Spine: 10–15 years

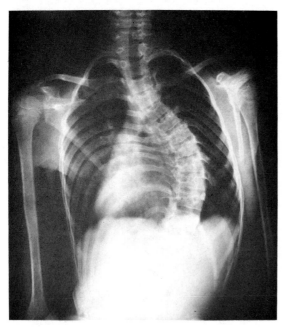

Fig. 39.15   A.W.—10 years—AP spine
A long C-shaped spinal curve is present.
Same patient: Fig. 39.26

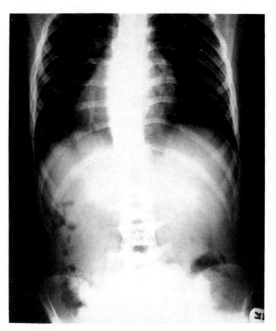

Fig. 39.16   A.S.—12 years—AP spine
There is minor scoliosis only.
Same patient: Figs 39.11, 39.13, 39.19

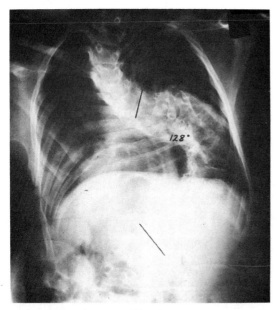

Fig. 39.17   K.T.—13 years—AP spine
There is a very severe (128°) thoracic curve.

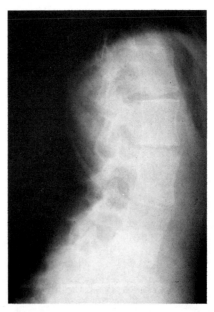

Fig. 39.18   J.H.—15 years—Lateral lumbar spine
The vertebral bodies are unusually tall, with a reduced antero-posterior diameter. The spinal canal is wide.
Same patient: Figs 39.23, 39.30

Spine: 16–19 years

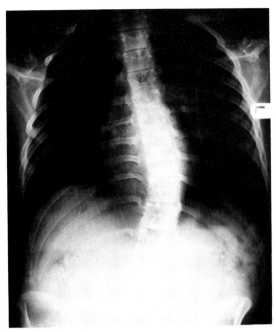

Fig. 39.19 A.S.—16 years—AP spine
Mild scoliosis only.
Same patient: Figs 39.11, 39.13, 39.16

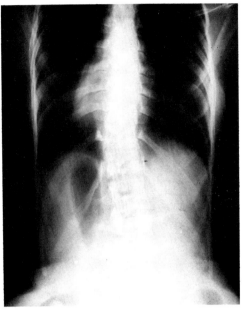

Fig. 39.20 J.W.—16 years—AP spine
Very mild spinal curvature, but it can be seen that the vertebral
bodies are unusually tall.

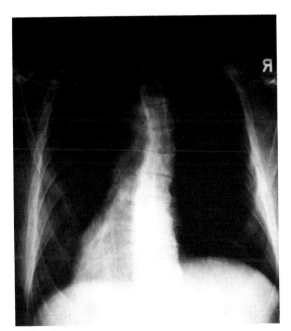

Fig. 39.21 I.G.—19 years—AP spine
Mild scoliosis only.
Same patient: Fig. 39.25

Pelvis: 6–15 years

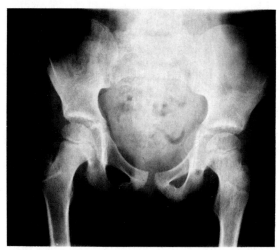

Fig. 39.22   G.O.—6 years—Pelvis
Essentially normal.
Same patient: Figs 39.24, 39.28

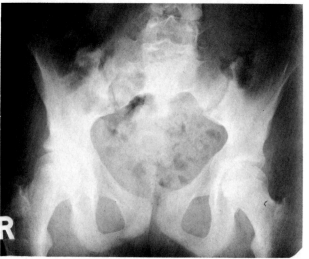

Fig. 39.23   J.H.—15 years—Pelvis
Essentially normal apart from a segmentation defect of the lumbo-sacral region on the right.
Same patient: Figs 39.18, 39.30

Foot : 6 years–Adult

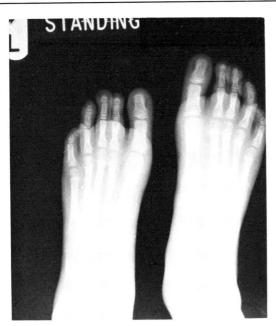

Fig. 39.24   G.O.—6 years—PA feet
All the long bones of the feet are slender and elongated.
Same patient : Figs 39.22, 39.28

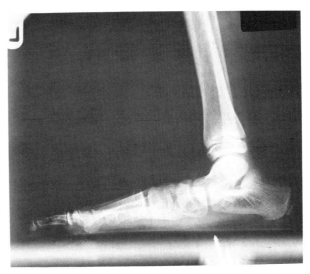

Fig. 39.25   I.G.—8 years—Standing lateral foot
There is pes planus and a vertically inclined talus.
Same patient : Fig. 39.21

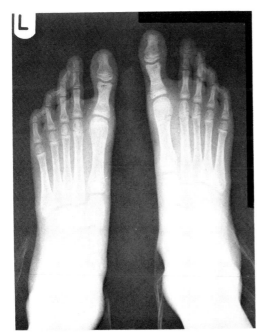

Fig. 39.26   A.W.—13 years—PA feet
Similar to Figure 39.24
Same patient : Fig. 39.15

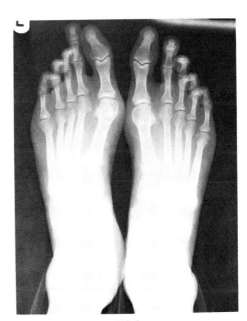

Fig. 39.27   R.M.—23 years—PA feet
Similar to Figure 39.24, and also slight hallux valgus.
Same patient : Fig. 39.3

Hand: 7 years–Adult

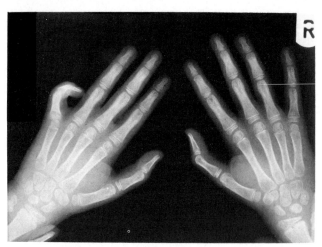

Fig. 39.28   G.O.—7 years—PA hands
There is arachnodactyly with contracture of the left little finger.
Same patient: Figs 39.22, 39.24

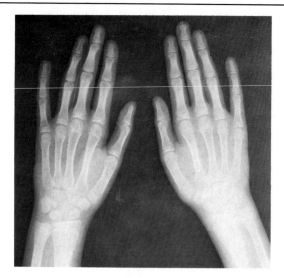

Fig. 39.29   A.T.—9 years—PA hands
Arachnodactyly.

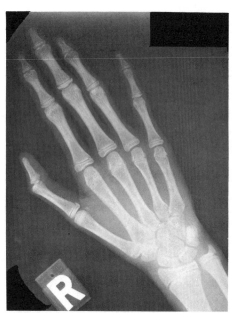

Fig. 39.30   J.H.—15 years—PA right hand
There is arachnodactyly, with some abnormal modelling.
Same patient: Figs 39.18, 39.23

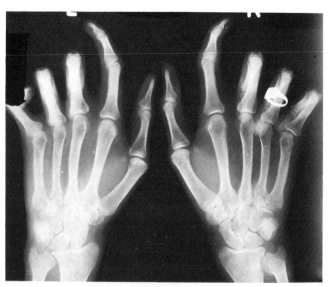

Fig. 39.31   R.M.—23 years—PA hands
Deformed hands, with contractures, ulnar deviation of the index
fingers and arachnodactyly.
Same patient: Fig. 39.27

Chapter Forty
HOMOCYSTINURIA

Body disproportion similar to Marfan's syndrome, accompanied frequently by mental retardation, osteoporosis in childhood and lens dislocation.

Inheritance

Autosomal recessive.

Frequency

Rare. Possible prevalence of 1 per million.

Clinical features

Facial appearance—normal.

Intelligence—over half the patients are mentally retarded.

Stature may be normal or increased and the body proportions are similar to Marfan's disease—the arm span exceeding the height and the head to pubis length less than pubis to heel.

Presenting feature/age—mental retardation, lens dislocation, body disproportion, scoliosis or symptoms related to osteoporosis—but none of these signs may be present.

Deformities—scoliosis, malalignment of lower limbs.

Associated anomalies—sometimes joint laxity, arachnodactyly, pectus excavatum or carinatum, high arched palate.

Radiographic features

Skull—normal.

Spine—structurally normal, but with developing osteoporosis vertebrae become flattened or biconcave on the lateral view.

Thorax—normal, apart from the chest deformity.

Long bones—there may be a striking enlargement of metaphyses, particularly around the knee joint.

Hands—some patients have arachnodactyly, but fewer than in Marfan's syndrome.

Bone maturation

Probably normal.

Biochemistry

The defective enzyme is cystathionine synthetase.

Differential diagnosis

From *Marfan's syndrome*, but here there is no osteoporosis, infrequent mental retardation and it is of autosomal dominant inheritance.

Progress/complications

This is related to the development of osteoporosis and to scoliosis and its complications.

An unusual feature is thrombosis in arteries or veins (perhaps precipitated by surgery), which may be related to abnormal stickiness of platelets caused by the presence of homocystine in the blood. The life span is probably normal.

REFERENCES

Brenton D P 1977 Skeletal abnormalities in homocystinuria. Postgraduate Medical Journal 53/622:488–496

Brenton D P, Dow C J, James J I P, Hay R L, Wynne-Davies R 1972 Homocystinuria and Marfan's syndrome. A comparison. Journal of Bone and Joint Surgery 54B:277

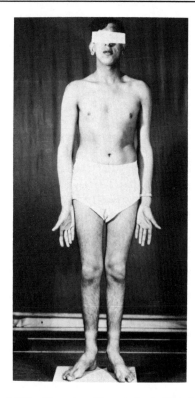 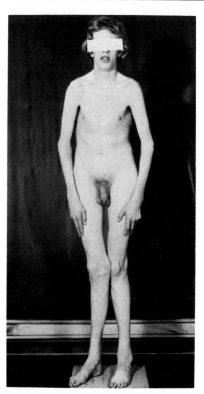

Figs 40.1 and 40.2   Show the wide variation in clinical appearance of these patients. Figure 40.1 is normal apart from some increased limb length, whereas Figure 40.2 has more obvious body disproportion, a narrow thorax and is mentally retarded.

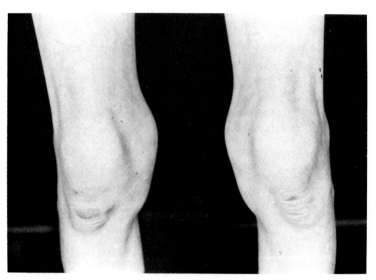

Fig. 40.3   Bony enlargement around the knees.

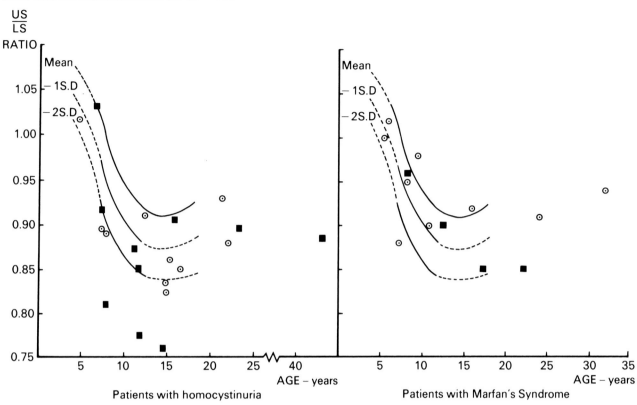

Fig. 40.4 Upper/lower segment ratios in homocystinuria and Marfan's syndrome.

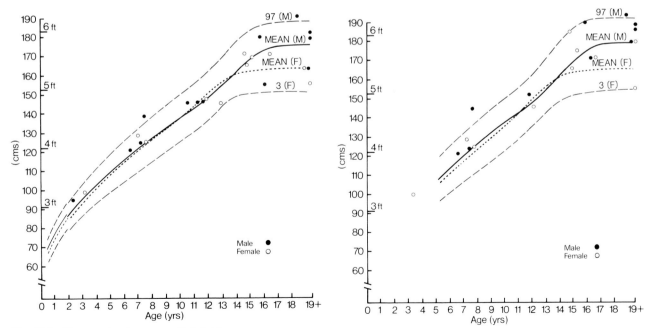

Fig. 40.5 Homocystinuria—percentile height.

Fig. 40.6 Homocystinuria—percentile span.

Spine: 9–14 years

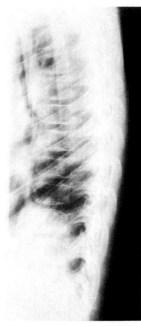

Fig. 40.7 C.S.—9 years—Lateral thoracic spine
The vertebral bodies are reduced in height, but have an increased
anteroposterior diameter.
Same patient: Figs 40.11, 40.12, 40.17

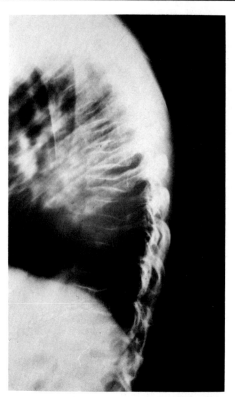

Fig. 40.8 J.C.—12 years—Lateral thoracic spine
There is uniform partial collapse of the vertebral bodies due to
osteoporosis.

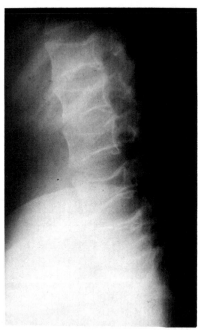

Fig. 40.9 A.C.—14 years 8 months—Lateral lumbar spine
The vertebral bodies are biconcave and flattened.
Same patient: Figs 40.10, 40.14, 40.18

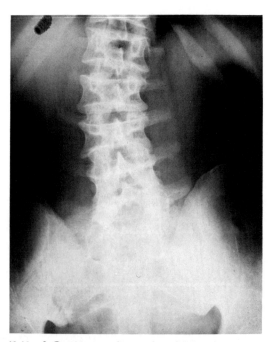

Fig. 40.10 A.C.—14 years 8 months—AP lumbar spine (same
patient as in Fig. 40.9)
There is partial collapse of the vertebral bodies and lumbar scoliosis.

Chest and pelvis: 9–14 years

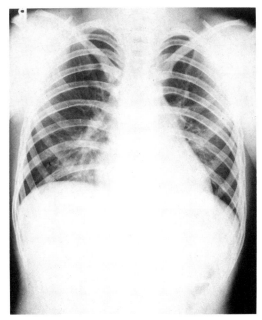

Fig. 40.11   C.S.—9 years—AP Chest
Normal.
Same patient: Figs 40.7, 40.12, 40.17

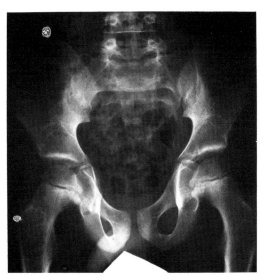

Fig. 40.12   C.S.—9 years—Pelvis (same patient as in Fig. 40.11)
There is irregularity of the articular margin of the left capital femoral epiphysis. This could be a result of early aseptic necrosis following a thrombotic episode.

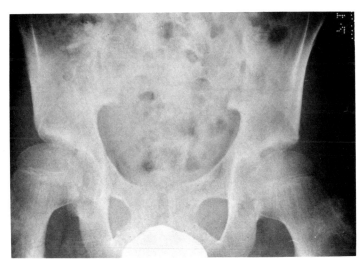

Fig. 40.13   C.G.—12 years—AP hips
The capital femoral epiphyses are large, and incompletely covered.

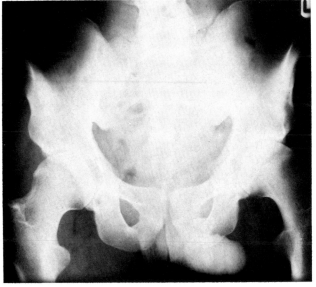

Fig. 40.14   A.C.—14 years 8 months—Pelvis
The femoral heads are large.
Same patient: Figs 40.9, 40.10, 40.18

Limbs : 5–14 years

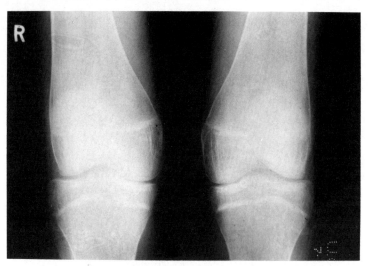

Fig. 40.15   F.M.—13 years—AP knees
The epiphyses are large and the metaphyses poorly modelled.

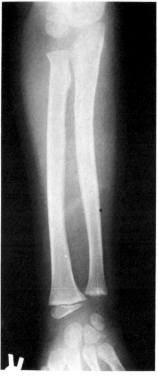

Fig. 40.16   J.B.—5 years—AP right forearm
Small spicules are present at the distal ulnar metaphysis (this may be seen as a normal variant at this age).

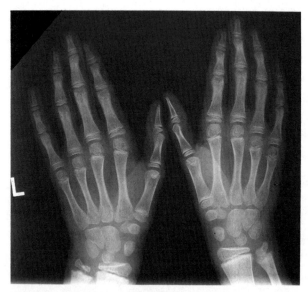

Fig. 40.17   C.S.—9 years—PA hands
There is arachnodactyly and osteoporosis. The capitate and hamate are unduly large in relation to the other carpal centres.
Same patient : Figs 40.7, 40.11, 40.12

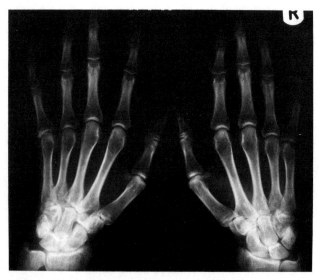

Fig. 40.18   A.C.—14 years 8 months—PA hands
Arachnodactyly is present.
Same patient : Figs 40.9, 40.10

Section Ten
STORAGE DISEASES

THE MUCOPOLYSACCHARIDOSES

This is a group of storage diseases with known biochemical features, differing clinically from each other but with strikingly similar radiological findings—differing only in their degree of severity.

THE MUCOLIPIDOSES

This is a group of storage diseases with some skeletal findings similar to the mucopolysaccharidoses, but without excess MPS in the urine.

Chapter Forty-one
HURLER SYNDROME

The facies is coarse ('gargoylism'), stature short but in proportion and there is progressive mental retardation with enlargement of the liver and spleen as mucopolysaccharides accumulate in the tissues. Joints are stiff and radiological signs characteristic (see below).

Inheritance

Autosomal recessive.

Frequency

The prevalence of the whole group is about 3 per million population. Hurler's is the commonest of them.

Clinical features

Facial appearance—may be normal at birth progressing to the 'gargoyle' appearance with protruding tongue. The head is large with frontal bossing, and due to maldevelopment of nasal passages, there is stertorous breathing and chronic upper respiratory infection with developing deafness.

Intelligence—initially appears normal, but mental retardation is progressive from infancy onwards.

Stature—normal at birth but growth slows during infancy and patients become of markedly short stature.

Presenting feature/age—probably the coarse facies, stiffness of joints, herniae, hepatosplenomegaly, developing short stature and mental retardation. Also corneal opacities. None of these signs may be apparent at birth, but develop gradually over the first few months of life.

Deformities—thoraco-lumbar kyphosis and coxa valga. Stiffness of joints may be associated with contractures. Sometimes congenital dislocation of the hips.

Radiographic features

Skull—normal to 6 months of age, then the sagittal suture closes first, giving a scaphocephalic shape. The skull is enlarged, with thick diploë, base and orbital roofs. Enlarged J-shaped sella turcica. Hydrocephalus may develop due to obstruction with thickened men-inges. The mandible is short and broad with missing condyles and the molar teeth may be sited within the ramus.

Spine—persistence of infantile biconvexity followed by thoraco-lumbar kyphos with hypoplastic vertebrae at the apex. Vertebrae then become 'hooked', being deficient in their anterosuperior parts.

Thorax—clavicles markedly broad at their medial ends. Ribs are 'paddle'-shaped, being broad at their anterior ends, narrowing posteriorly.

Pelvis—flared ilia, narrowing in their inferior parts adjacent to the acetabula, which are sloping. Coxa valga.

Long bones—lack of diaphyseal modelling (thickening) may lead to the shafts being wider than the metaphyseal regions.

Wrists/hands—sloping lower ends of radius and ulna give a ∧-shaped deformity. The second to fifth metacarpals are pointed at their proximal ends, and phalanges short and stubby ('bullet-shaped').

Bone maturation

Delayed and epiphyses are small and irregular when they do appear.

Biochemistry

Increased urinary excretion, of dermatan sulphate and heparan sulphate, the basic enzyme defect being α-L-iduronidase.

Differential diagnosis

From other mucopolysaccharide disorders but the distinction may be impossible on clinical and radiological grounds only.

Progress/complications

A progressive disease, death usually occurring by the age

of 10–15 years from cardio-respiratory problems, including coronary artery and valvular obstruction.

REFERENCES

Grossman H, Dorst J 1973 The mucopolysaccharidoses and mucolipidoses. In: Kaufmann H J (ed) Intrinsic diseases of bones. Progress in Paediatric Radiology 4. S Karger, Basel, pp 495–544

Pennock C A, Barnes I C 1976 The mucopolysaccharidoses. Journal of Medical Genetics 13:169

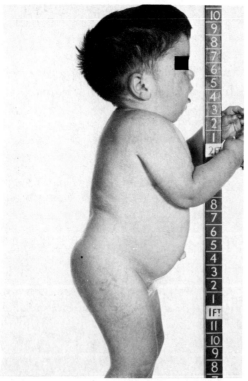

Fig. 41.1 Short stature, coarse facies, an enlarged abdomen (hepato-splenomegaly) and an umbilical hernia.

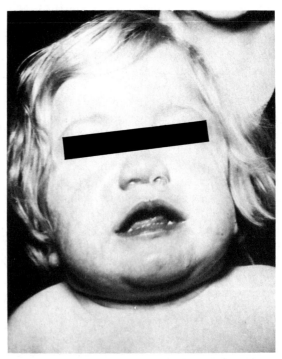

Fig. 41.2 A coarse facies with corneal clouding and obvious 'snuffles'.

Skull: 2–8 years

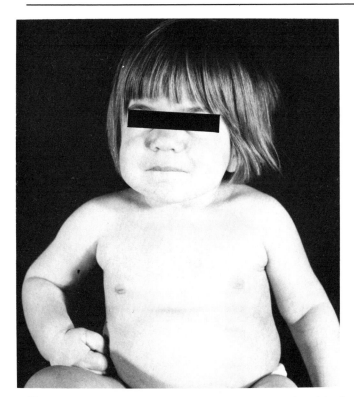 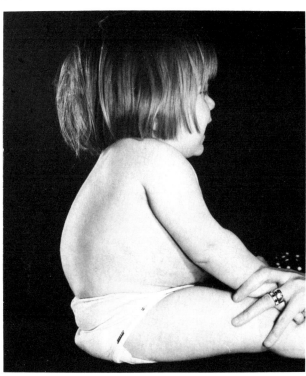

Figs 41.3 and 41.4   This child shows early signs of coarsening of the face and has a lumbar kyphos.

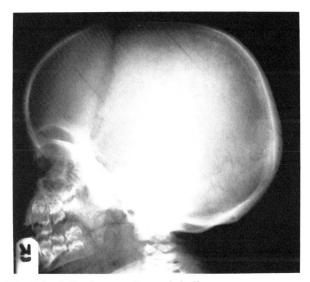

Fig. 41.5   A.C.—2 years—Lateral skull
The vault is large with loss of normal differentiation into inner and outer tables giving a 'ground glass' appearance. The coronal suture is wide indicating raised intracranial pressure. The pituitary fossa is enlarged beneath the anterior clinoid processes (J-shaped). The mandible is normal in this patient, but the odontoid peg is hypoplastic.
Same patient: Figs 41.11, 41.12, 41.16, 41.19

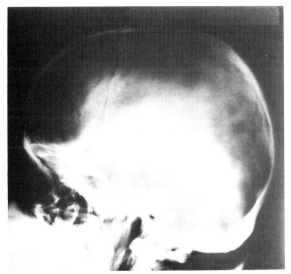

Fig. 41.6   J.W.—8 years—Lateral skull
The pituitary fossa is large and there is some thickening of the occiput.

Spine: 3–6 months

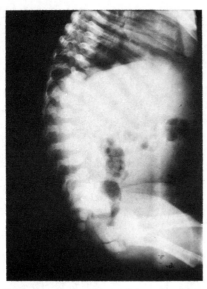

Fig. 41.7   P.B.—3 months—Lateral dorsal and lumbar spine
The vertebral bodies have the infantile oval configuration and there are inferior hooks at L1 and L2 with a dorsolumbar kyphosis.
Same patient: Fig. 41.8

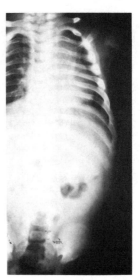

Fig. 41.8   P.B.—3 months—AP spine (same patient as in Fig. 41.7)
The ribs are broad, especially anteriorly, but there is no other abnormality.

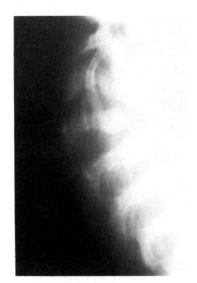

Fig. 41.9   S.D.—6 months—Lateral dorsolumbar region (coned view)
There is marked dorsolumbar kyphosis associated with hypoplasia and posterior displacement of L1.

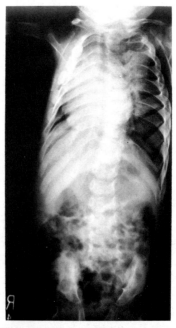

Fig. 41.10   S.W.—6 months—AP whole trunk
There is a mild double scoliosis. The ribs are broad with posterior constrictions. There is hypoplasia of the base of the ilia.

Spine and chest : 2–4 years

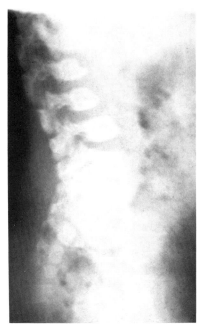

Fig. 41.11   A.C.—2 years—Lateral lumbo-sacral spine
There are inferior hooks on the vertebral bodies in the upper lumbar region, at the site of the kyphos.
Same patient : Figs 41.5, 41.12, 41.16, 41.19

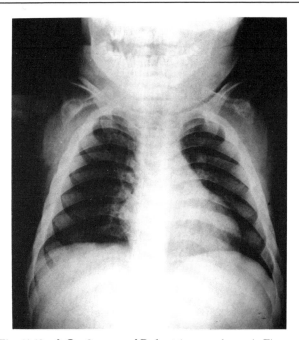

Fig. 41.12   A.C.—2 years—AP chest (same patient as in Fig. 41.11)
The ribs are broad, as are the clavicles at their medial ends. The humeral heads are small and the glenoid fossae poorly formed.

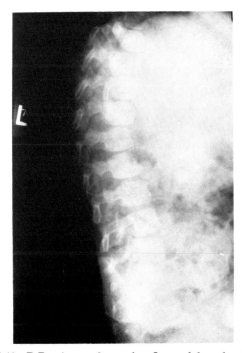

Fig. 41.13   D.R.—4 years 6 months—Lateral dorsal and lumbar spine
There are mild inferior hooks from L1 to L3 and a slight increase in the AP diameters of the vertebral bodies.

Spine and pelvis: 2–13 years

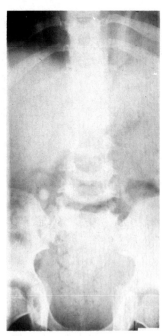

Fig. 41.14   J.B.—13 years—AP spine
The transverse processes of the vertebral bodies are small and
angulated. The anterior ends of the ribs are broad, and the posterior
ends constricted.
Same patient: Figs 41.15, 41.17, 41.18, 41.20

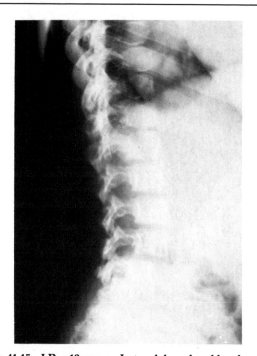

Fig. 41.15   J.B.—13 years—Lateral dorsal and lumbar spine
(same patient as in Fig. 41.14)
There is posterior scalloping of the lumbar vertebral bodies, and the
body of L5 appears to be hypoplastic. There is some persistence of the
infantile biconvexity (T10 and 11) but no inferior 'hook'.

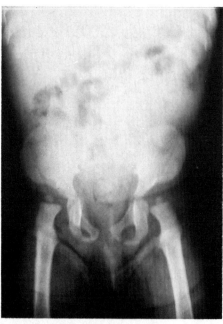

Fig. 41.16   A.C.—2 years—Pelvis
The iliac wings are flared and there is marked hypoplasia of the bases
of the ilia, giving poor demarcation of the acetabular roofs. Bilateral
coxa valga is present and the right hip is dislocating. The capital
femoral epiphyses are small for the chronological age.
Same patient: Figs 41.5, 41.11, 41.12, 41.19

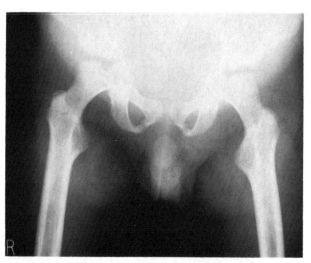

Fig. 41.17   J.B.—13 years—Pelvis (same patient as in Fig. 41.14)
The capital femoral epiphyses are small. There is coxa valga and the
femoral necks appear long. There is mild hypoplasia of the bases of
the ilia.

Upper limb: 2–14 years

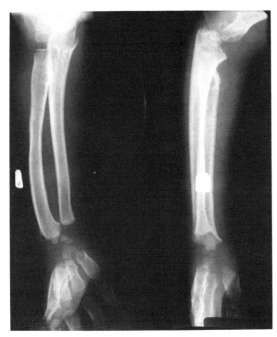

Fig. 41.18   J.B.—13 years—Left forearm
The distal ulna is short and the radial metaphyses curved towards it.
The diaphyses are poorly modelled.
Same patient: Figs 41.14, 41.15, 41.17, 41.20

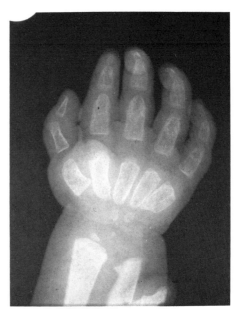

Fig. 41.19   A.C.—2 years—PA right hand
All the long bones are short, thick and poorly modelled, with pointing
of the distal ends of the phalanges ('bullet-shaped'). There is pointing
of the bases of the second to fifth metacarpals and the distal radial and
ulnar metaphyses slope towards each other. Phalangeal bone age is
retarded, but the carpal centres are not.
Same patient: Figs 41.5, 41.11, 41.12, 41.16

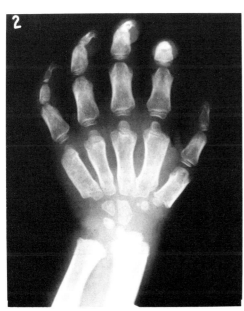

Fig. 41.20   J.B.—13 years—PA left hand (same patient as in Fig.
41.18)
There is flexion deformity of the fingers and some pointing of the 2nd
to 5th metacarpal bases. The carpal centres are retarded and angular.
The distal radial epiphysis is flat and the ulnar one has not ossified.

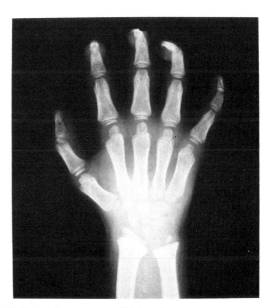

Fig. 41.21   I.M.—14 years—PA hand
There is a flexion deformity of the fingers and the distal radius and
ulna slope towards each other; their epiphyses are small and flat. The
carpal centres are small and angular. The base of the second to fifth
metacarpals is only slightly abnormal.

Chapter Forty-two
HUNTER SYNDROME

Very similar to the Hurler syndrome but less severe physical signs, and all patients are male.

Inheritance

X-linked recessive.

Frequency

Less common than Hurler's.

Clinical features

Facial appearance—similar to Hurler's, but less severe. No corneal clouding.

Intelligence—lesser degree of mental retardation than Hurler's.

Stature—normal at birth, with progressively developing short stature.

Presenting feature/age—of later onset than Hurler's but otherwise similar.

Deformities—similar, but kyphosis is not a usual feature.

Radiographic features

All are similar to Hurler's, but more slowly progressive as well as later in developing.

Biochemistry

Increased urinary excretion of dermatan and heparan sulphate; low enzyme sulphaiduronate sulphatase.

Differential diagnosis

From the other mucopolysaccharide disorders, chiefly Hurler's. Biochemical tests may be needed to distinguish them.

Progress/complications

There is a longer life span than in patients with Hurler's disease, perhaps to the second or third decade, but eventually accumulating mucopolysaccharide in the tissues leads to death from cardio-respiratory disease.

REFERENCES

Grossman H, Dorst J 1973. The mucopolysaccharidoses and mucolipidoses. In: Kaufmann H J (ed) Intrinsic diseases of bones. Progress in Pediatric Radiology 4. S Karger, Basel, pp. 495–544

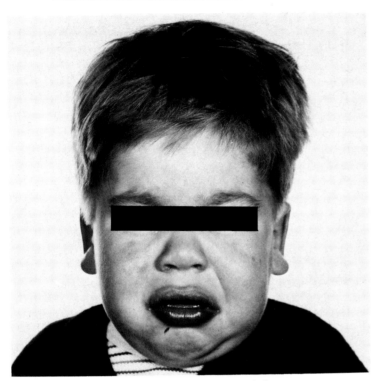

Fig. 42.1   Mental retardation and a coarse facies.

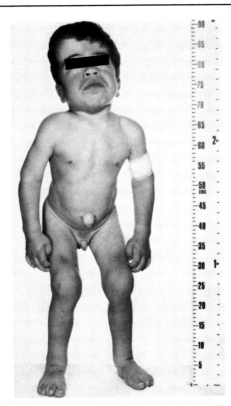

Fig. 42.2   Short stature, probably joint contractures and an umbilical hernia.

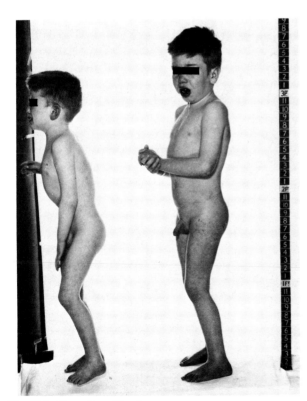

Fig. 42.3   Two affected brothers.

Fig. 42.4   An older child, now with marked hepato-splenomegaly.

Skull: 9 months–14 years

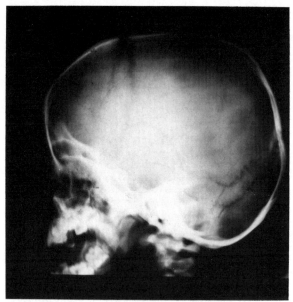

Fig. 42.5   R.C.—9 months—Lateral skull
The vault is large and there is widening of the coronal suture indicating a raised intracranial pressure. The pituitary fossa is elongated under the anterior clinoid processes and there is hypoplasia of the odontoid peg.
Same patient: Fig. 42.17

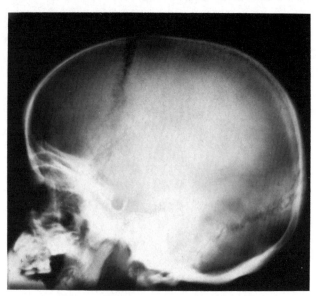

Fig. 42.6   L.F.—3 years 7 months—Lateral skull
The vault is large and has a 'ground-glass' appearance. The pituitary fossa is J-shaped and the odontoid peg hypoplastic.
Same patient: Figs 42.15, 42.18

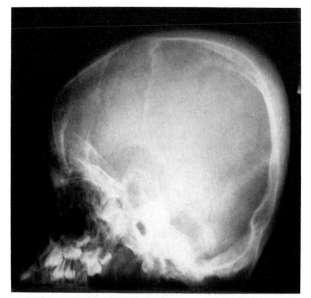

Fig. 42.7   S.N.—9 years 4 months—Lateral skull
There is brachycephaly and thickening of the vault in the parietal and occipital regions, with loss of differentiation into inner and outer tables.
Same patient: Figs 42.9, 42.13

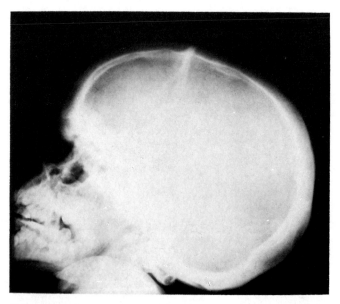

Fig. 42.8   R.C.—14 years—Lateral skull
The vault is large and there is thickening of all the vault bones with loss of differentiation into the tables. (This degree of thickening of the vault could represent a response to treated hydrocephalus.)
Same patient: Fig 42.16

Spine: 11 months–5 years

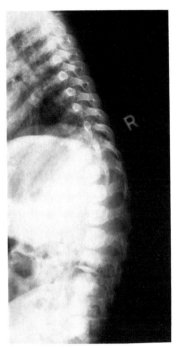

Fig. 42.9   S.N.—11 months—Lateral dorsal and lumbar spine
There is a mild dorso-lumbar kyphos. The vertebral bodies are oval
and rather tall.
Same patient: Figs 42.7, 42.13

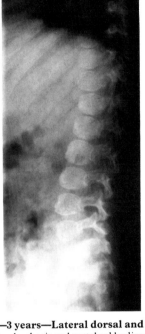

Fig. 42.10   S.M.—3 years—Lateral dorsal and lumbar spine
There is mild lumbar kyphosis and vertebral bodies still have an
immature oval shape.
Same patient: Fig. 42.19

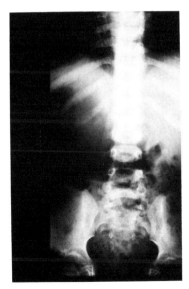

Fig. 42.11   S.R.—5 years—AP spine
There is no abnormality apart from rib expansion anteriorly.
Same patient: Figs 42.12, 42.14, 42.20

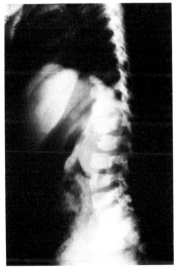

Fig. 42.12   S.R.—5 years—Lateral dorsal and lumbar spine
(same patient as in Fig. 42.11)
There is a mild dorso-lumbar kyphosis and an inferior hook of the
body of L2. The vertebral bodies are increased in their AP diameters.

Trunk and lower limb: 3–13 years

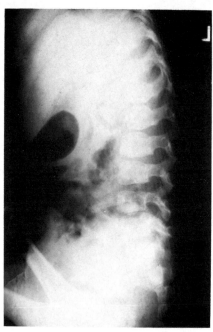

Fig. 42.13   S.N.—6 years 3 months—Lateral dorsal and lumbar spine
The vertebral bodies are elongated in their AP diameters and there is posterior scalloping at L4 and L5. L1 to L3 have vertebral end-plate irregularity and some disc space narrowing. The infantile biconvex shape is still present.
Same patients: Figs 42.7, 42.9

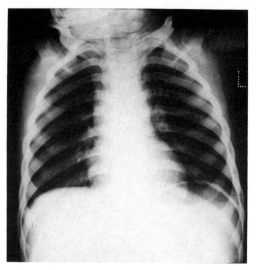

Fig. 42.14   S.R.—5 years—PA chest
The ribs and clavicles are widened and the humeral heads small and with poorly formed glenoid fossae.
Same patient: Figs 42.11, 42.12, 42.20

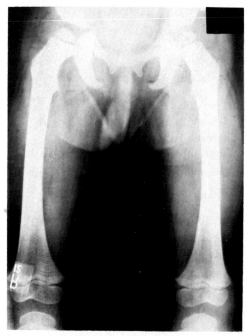

Fig. 42.15   L.F.—3 years 6 months—AP femora
The femoral heads are small and there is coxa valga, but no other abnormality.
Same patient: Figs 42.6, 42.18

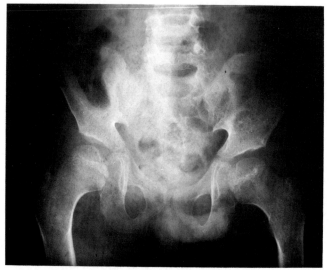

Fig. 42.16   R.C.—13 years 8 months—Pelvis
The femoral heads are small and there is mild hypoplasia at the bases of the ilia—otherwise no abnormality.
Same patient: Fig. 42.8

Upper limb: 3–5 years

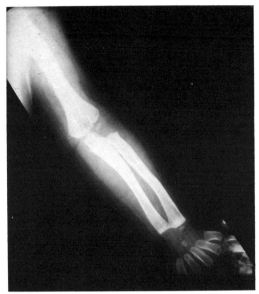

Fig. 42.17   R.C.—3 years 5 months—AP upper limb
Changes are slight with some failure of diaphyseal modelling and
pointing of the second to fifth metacarpal bases.
Same patient: Fig. 42.5

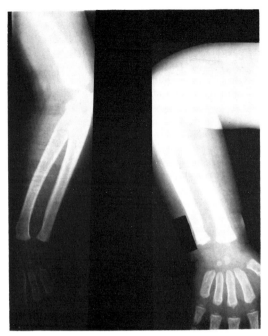

Fig. 42.18   L.F.—3 years 6 months—AP and lateral forearm
There is some failure of modelling of the shafts of the radius and ulna.
The proximal ends of the second to fifth metacarpals are pointed, and
maturation of the carpal bones is delayed.
Same patient: Figs 42.6, 42.15

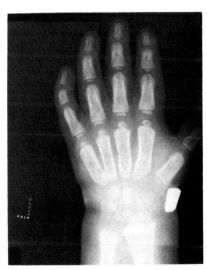

Fig. 42.19   S.M.—3 years 1 month—PA left hand
There is mild pointing of the second to fifth metacarpal bases but the
changes are very mild. Bone maturation is normal.
Same patient: Fig. 42.10

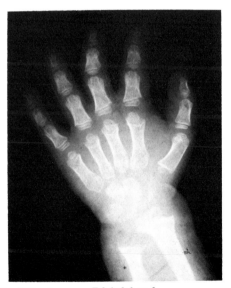

Fig. 42.20   S.R.—5 years—PA left hand
There is mild pointing of the second to fifth metacarpal bases and a
distal epiphysis of the 1st metacarpal. The ulna is short and there is
flaring of both radial and ulnar metaphyses.
Same patient: Figs 42.11, 42.12, 42.14

Chapter Forty-three
SCHEIE SYNDROME

Clinically markedly dissimilar from the Hurler syndrome, although the defective enzyme is the same.

The main feature is joint contractures and corneal clouding, but intelligence, stature and life expectancy are probably normal.

Inheritance

Autosomal recessive.

Frequency

Less common than Hurler's.

Clinical features

Facial appearance—some coarsening, corneal clouding.

Intelligence—probably normal.

Stature—probably normal.

Presenting feature/age—multiple joint contractures, sometimes herniae in infancy or childhood.

Deformities—joint contractures.

Radiographic features

There may be no signs at all, but the most likely are in the *thorax* with widening of the medial end of the clavicles, and hypoplasia of the *small bones* of the hands and feet. An occasional characteristic feature is the later development of small cystic areas in the carpus or digits.

Biochemistry

Increased urinary excretion of mainly dermatan sulphate. Deficient enzyme is α-L-iduronidase.

Differential diagnosis

Dissimilar to the other mucopolysaccharide disorders clinically, needing to be differentiated from numerous conditions in which joint contractures are the main feature. Biochemical tests distinguish this disorder.

Progress/complications

Life span is probably normal, and disability may not be great. Heart disease may develop.

REFERENCES

Grossman H, Dorst J 1973 The mucopolysaccharidoses and mucolipidoses. In : Kaufmann H J (ed) Intrinsic diseases of bones. Progress in Pediatric Radiology 4. S Karger, Basel, pp 495–544

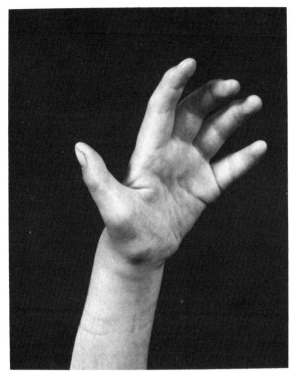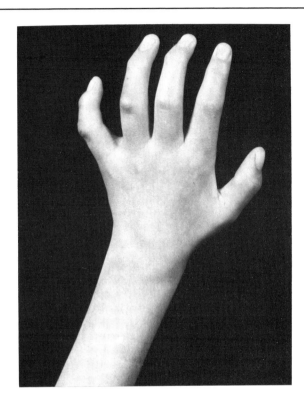

Figs 43.1 and 43.2 Soft-tissue contractures of the fingers and thumb.

Skull, spine and pelvis : 12–14 years

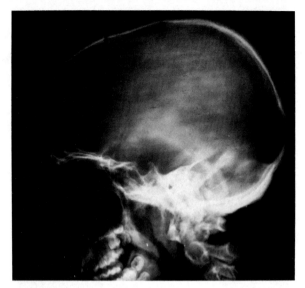

Fig. 43.3　D.H.—14 years 6 months—Lateral skull
Normal, including the odontoid peg.
Same patients: Figs 43.5, 43.8

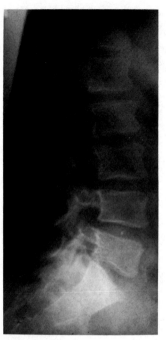

Fig. 43.4　P.H.—12 years 2 months—Lateral lumbar spine
There is a mild increase in the AP diameters of the vertebral bodies
and the pedicles are long.
Same patient: Figs 43.6, 43.7

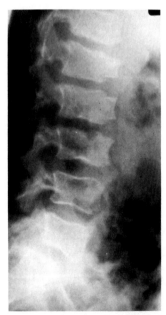

Fig. 43.5　D.H.—14 years 5 months—Lateral lumbar spine
(same patient as in Fig. 43.3)
The vertebral bodies are elongated in their AP diameters. There is
mild anterior wedging of L1 with vertebral end-plate irregularity at
D12/L1. The ring apophyses have not yet ossified, unusually for this
age.

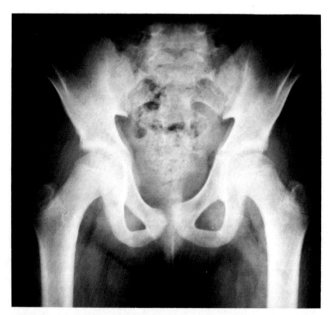

Fig. 43.6　P.H.—12 years 2 months—Pelvis (same patient as in
Fig. 43.4)
There is mild coxa valga and the femoral necks are long. The capital
femoral epiphyses are small. There is mild hypoplasia of the iliac
bases.

Hand: 12 years–Adult

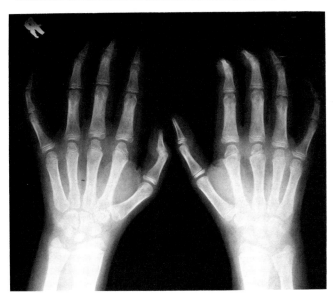

Fig. 43.7 P.H.—12 years 2 months—PA hands
There are flexion deformities of the fingers. All epiphyses and carpal centres are small and there is flattening of the distal radial and ulnar epiphyses. Bone maturation is retarded by some 4 or 5 years.
Same patient: Figs 43.4, 43.6

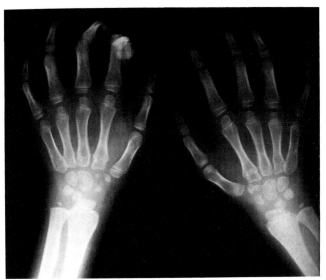

Fig. 43.8 D.H.—13 years 1 month—PA hands
There are flexion deformities of the fingers. All epiphyses are rather small and irregular. Diaphyseal modelling is normal, but the distal radial metaphyses slope towards the ulna. Bone maturation is delayed by about 7 years.
Same patient: Figs 43.3, 43.5

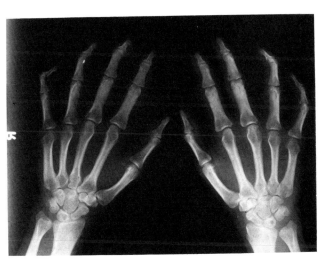

Fig. 43.9 Unknown—Adult—PA hands
There are flexion contractures of the fingers but no other abnormality apart from small bone cysts at the base of the left middle finger, middle phalanx and in both lunates.

Chapter Forty-four
SANFILIPPO SYNDROME

A little similar to the Scheie syndrome but here mental retardation is the main feature.

Inheritance

Autosomal recessive.

Frequency

Less common than Hurler's disease.

Clinical features

Facial appearance—the face is not coarse featured, and there is only minimal corneal clouding.

Intelligence—initially normal, but severe mental retardation develops in infancy and childhood.

Presenting feature/age—mental retardation in infancy, stiff joints and hand contractures.

Deformities—hand or other contractures.

Radiographic features

Skull—the posterior part may be unusually thick.

Spine—persistence of infantile biconvexity but the hook-shape of Hurler's syndrome is not usually present.

Thorax—perhaps normal, but the clavicles may be widened at their medial ends.

Pelvis—similar to Hurler's, but less marked.

Biochemistry

Increased urinary excretion of heparin sulphate. Probably two different enzymes are deficient (clinically indistinguishable); N-heparan sulfatase and α-acetyl-glucosaminidose.

Differential diagnosis

From other mucopolysaccharide disorders, mannosidosis and the mucolipidoses, but the severe mental retardation together with few radiological signs will distinguish it, as well as the biochemical tests.

Progress/complications

The main feature is the severe progressive mental retardation. If radiological signs are present, they tend to regress during childhood and may disappear altogether.

REFERENCES

Grossman H, Dorst J 1973 The mucopolysaccharidoses and mucolipidoses. In: Kaufmann H J (ed) Intrinsic diseases of bones. Progress in Pediatric Radiology 4. S Karger, Basel, pp 495–544

Skull, spine and pelvis : 8 years–Adult

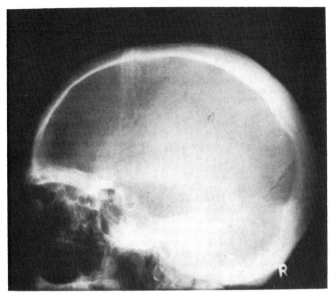

Fig. 44.1   T.B.—8 years—Lateral skull
There is marked thickening of the vault, and generalised loss of
differentiation into inner and outer tables. The sella turcica is normal.
Same patient : Fig. 44.8

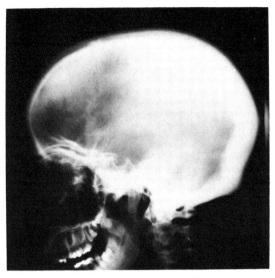

Fig. 44.2   Unknown—Adult—Lateral skull
The main thickening of the vault is over the parietal and occipital
regions. The pituitary fossa, mandible and odontoid peg are normal.

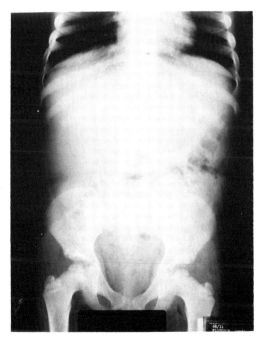

Fig. 44.3   W.F.—10 years—AP spine and pelvis
The ribs are expanded anteriorly and the vertebral bodies are
somewhat flattened. The iliac wings are not flared, but there is some
hypoplasia of the bases. The capital femoral epiphyses are small.
Hepatomegaly is clearly apparent.

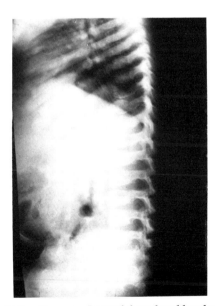

Fig. 44.4   E.B.—12 years—Lateral dorsal and lumbar spine
The vertebral bodies are rather elongated with an immature oval
configuration.
Same patient : Fig. 44.5

Chest, pelvis and hand: 1–12 years

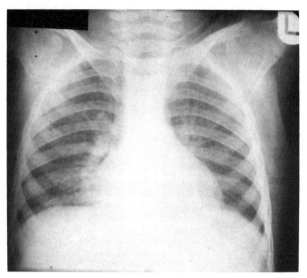

Fig. 44.5　E.B.—12 years—AP chest
The clavicles are short and broad at their medial ends. The ribs are uniformly widened.
Same patient: Fig. 44.4

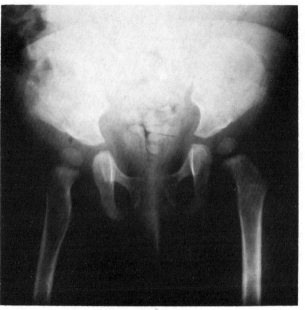

Fig. 44.6　L.M.—1 year 2 months—Pelvis
The iliac wings are flared and there is hypoplasia of their bases, with acetabular roofs which are poorly formed and shallow. These changes are unusually marked for the Sanfilippo type of MPS.
Same patient: Fig. 44.7

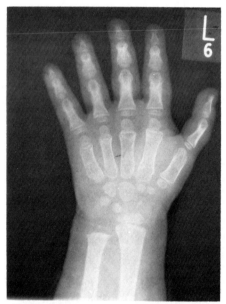

Fig. 44.7　L.M.—2 years 11 months—PA left hand (same patient as in Fig. 44.6)
The metacarpals are poorly modelled and the bases tend towards 'pointing'. The phalanges are unusually 'waisted'. The carpal centres are advanced for this age, but phalangeal epiphyses are normal.

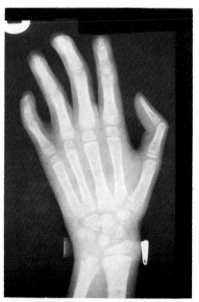

Fig. 44.8　T.B.—8 years—PA left hand
Apart from minor modelling defects and flexion deformities of the fingers, the appearance is normal.
Same patient: Fig. 44.1

Chapter Forty-five
MANNOSIDOSIS

A rare storage disease, with clinical and radiological signs similar to MPS type Sanfilippo.

Inheritance

Autosomal recessive.

Frequency

Extremely rare.

Clinical features

Similar to MPS type Sanfilippo, needing biochemical differentiation.

Biochemistry

Deficiency of enzyme mannosidase.

REFERENCES

Autio S, Norden N E, Ockerman P A, Riek-Kinen P, Rapola J, Louhimo T 1973 Mannosidosis: clinical fine-structural and biochemical findings in three cases. Acta paediatrica Scandinavica 62:555–565

Öckerman P A 1967 A generalised storage disorder resembling Hurler's syndrome. Lancet ii:239–241

Skull and spine : 6–20 years

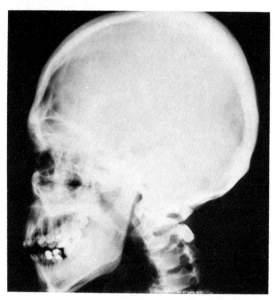

Fig. 45.1   K.Y.—19 years—Lateral skull
The vault is small but the tables considerably thickened—there is
obvious disproportion between the mandible and maxilla compared
with the vault. (The patient is severely mentally retarded.)
Same patient : Figs 45.2, 45.4, 45.6, 45.8

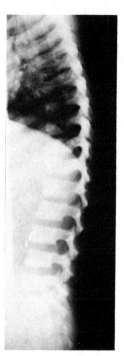

Fig. 45.2   K.Y.—6 years—Lateral thoracic and lumbar spine
(same patient as in Fig. 45.1)
Most of the vertebrae are elongated in their AP diameter. There is
mild dorso-lumbar kyphosis.

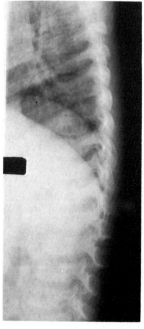

Fig. 45.3   T.C.—7 years—Lateral thoracic and lumbar spine
Similar to Figure 45.2.
Same patient : Figs 45.5, 45.7

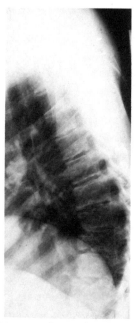

Fig. 45.4   K.Y.—20 years—Lateral thoracic spine (same patient
as in Fig. 45.1)
The vertebral end-plates are irregular; the increase in dorsal kyphosis
is associated with anterior wedging.

Chest and pelvis : 6–20 years

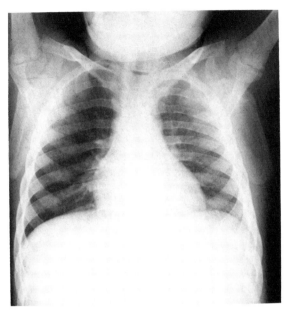

Fig. 45.5   T.C.—7 years—AP chest
The clavicles are similar to those of most MPS disorders, being broadened medially. The ribs are also broad but do not show a posterior constriction.
Same patient : Figs 45.3, 45.7

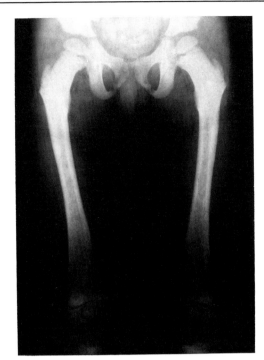

Fig. 45.6   K.Y.—6 years—AP hips and femora
The capital femoral epiphyses are rather small and the adjacent metaphyses broadened and irregular.
Same patient : Figs 45.1, 45.2, 45.4, 45.8

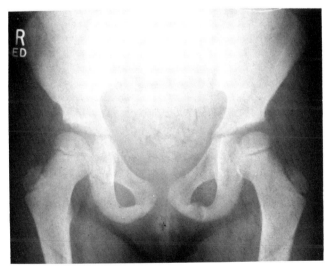

Fig. 45.7   T.C.—7 years—Pelvis (same patient as in Fig. 45.5)
There is mild hypoplasia of the bases of the ilia.

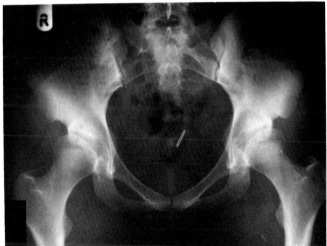

Fig. 45.8   K.Y.—20 years—Pelvis (same patient as in Fig. 45.6)
The hypoplasia at the base of the ilia is associated with acetabular dysplasia, uncovering of the femoral heads with sclerosis and early osteoarthritis.

Chapter Forty-six
MAROTEAUX-LAMY SYNDROME

Characterised by clinical and radiological signs as severe as Hurler's syndrome but these patients are of normal intelligence and some have Perthes-like changes in capital femoral epiphyses.

Inheritance

Autosomal recessive.

Frequency

Extremely rare.

Clinical features

Facial appearance—coarse facies, corneal opacities with poor vision.
Intelligence—normal.
Stature—markedly reduced, after being normal at birth.
Presenting feature/age—poor vision, hepatomegaly (rather than splenomegaly), herniae and sometimes hydrocephalus, all developing during early childhood.
Deformities—dislocation of hips, kyphosis, joint stiffness and contractures.

Radiographic features

Similar to Hurler's syndrome with a wide range of severity.

Pelvis/hips—the feature characteristic of this syndrome is the symmetrical flattening, fragmentation and sclerosis of the capital femoral epiphyses, although not all these changes are necessarily present.

Biochemistry

Increased urinary excretion of dermatan sulphate, with some heparan. Deficient enzyme is N-Ac-gal-4-sulfatase.

Differential diagnosis

Chiefly from the Hurler syndrome, but the normal intelligence and biochemical tests will distinguish it.

Progress/complications

As for Hurler's disease; cardio-vascular complications are common.

REFERENCES

Grossman H, Dorst J 1973 The mucopolysaccharidoses and mucolipidoses. In Kaufmann H J (ed) Intrinsic diseases of bones. Progress in Pediatric Radiology 4. S Karger, Basel, pp 495–544
Pennock C A, Barnes I C 1976. The mucopolysaccharidoses. Journal of Medical Genetics 13:169
Spranger J W, Koch F, McKusick V A, Natzschka J, Wiedemann H-R, Zellweger H 1970 Mucopolysaccharidosis VI (Maroteaux-Lamy's disease). Helvetica paediatrica acta 25:337

Skull: 3–9 years

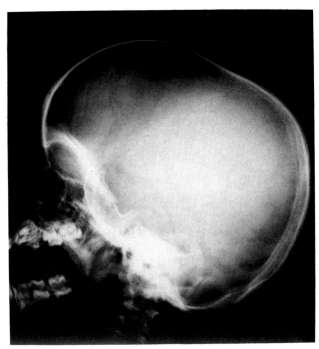

Fig. 46.1   E.B.—3 years 6 months—Lateral skull
The splayed suture lines with pronounced interdigitations indicate hydrocephalus. Other features of an MPS disorder are shown—an elongated pituitary fossa and loss of differentiation of the inner and outer tables.
Same patient: Figs. 46.4–46.6, 46.12

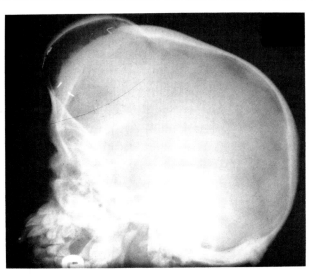

Fig. 46.2   M.W.—9 years 5 months—Lateral skull
There is a bizarre expansion of the frontal area associated with intracranial surgery. Elsewhere the vault shows thickening and a 'ground-glass' appearance. The condyle of the mandible is small and pointed without its normal articular surface.
Same patient: Figs 46.3, 46.7, 46.10, 46.11, 46.15

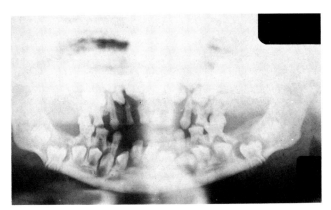

Fig. 46.3   M.W.—7 years 6 months—Orthopantomogram
(same patient as in Fig. 46.2)
This shows well the dental and mandibular anomalies of the MPS disorders: molars within the ramus of the mandible, diminished depth of the body of the mandible, with molar roots extending to its lower surface, mandibular cysts, hypoplastic condylar and coronoid processes.

Myelogram : 3 years

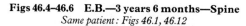

Figs 46.4–46.6   E.B.—3 years 6 months—Spine
Same patient : Figs 46.1, 46.12

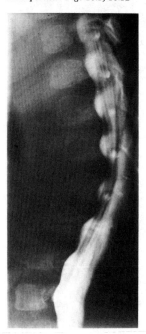

Fig. 46.4   The myelogram shows thickened meninges, causing compression anteriorly at the dorso-lumbar kyphos.

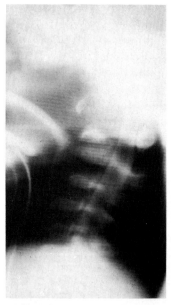

Fig. 46.5   Tomogram of lateral cervical myelogram, neck in extension. The odontoid peg is virtually absent, and the vertebral bodies show the typical biconvex shape.

Fig. 46.6   Similar to Figure 46.5 with neck in flexion. There is now obvious separation between the body of the atlas and the 2nd cervical vertebra, indicating instability here—an unusual finding in the Maroteaux-Lamy syndrome, being more usually found in Morquio's disease.

Spine and chest: 4–12 years

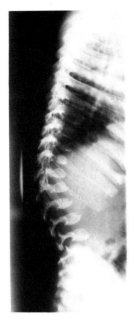

Fig. 46.7   M.W.—4 years—Lateral thoracic and lumbar spine
This shows all the characteristics of a spine in an MPS disorder: some small, biconvex vertebrae; anterior hooks in the lumbar region; dorso-lumbar kyphos; posterior scalloping of bodies; long, spidery pedicles.
Same patient: Figs 46.2, 46.3, 46.10, 46.11, 46.15

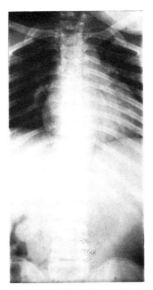

Fig. 46.8   S.M.—12 years—AP spine and ribs
The vertebrae are flattened and sharply angulated. The ribs narrow posteriorly but the clavicles are normal in this patient.
Same patient: Figs 46.9, 46.13, 46.14, 46.17

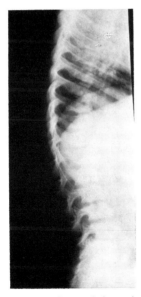

Fig. 46.9   S.M.—12 years—Lateral thoracic and lumbar spine
(same patient as in Fig. 46.8)
This patient is much less severely affected than that in Figure 46.7, but there is some flattening of vertebrae, with a suggestion of anterior hooks and only a mild kyphos.

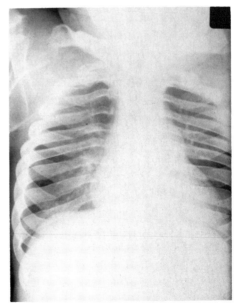

Fig. 46.10   M.W.—9 years 8 months—AP chest (same patient as in Fig. 46.7)
Ribs and both clavicles are broadened.

Pelvis : 4–12 years

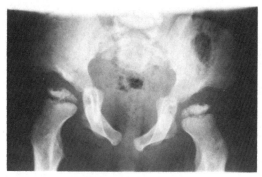

Fig. 46.11   M.W.—4 years—Pelvis
The pelvis shows the typical flared ilia of the MPS disorders,
narrowing at the base ('wine-glass' shape). The capital femoral
epiphyses have features seen only in the Maroteaux–Lamy
syndrome—that is, symmetrical fragmentation, flattening and
sclerosis, somewhat similar to Perthes' disease.
Same patient: Figs 46.2, 46.3, 46.7, 46.10, 46.15

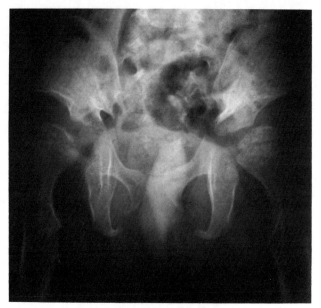

Fig. 46.12   E.B.—4 years 8 months—Pelvis
This patient is severely affected, but his capital femoral epiphyses,
although flattened do not show changes seen in Figure 46.11.
Same patient: Figs 46.1, 46.4–46.7

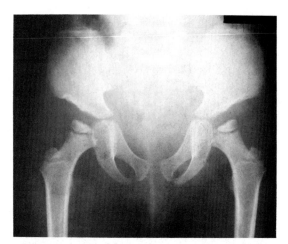

Fig. 46.13   S.M.—12 years—Pelvis
This patient has a less markedly abnormal pelvis than that in Figures
46.11 and 46.12, but the capital femoral epiphyses are sclerotic.
Same patient: Figs 46.8, 46.9, 46.14, 46.17

Upper limb : 4–12 years

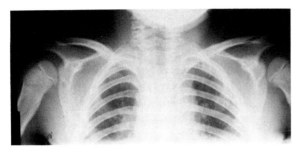

Fig. 46.14   S.M.—12 years—AP shoulders
The clavicles and ribs are poorly modelled and the glenoid fossae flattened. The humeral epiphyses are small.
Same patient : Figs 46.8, 46.9, 46.13, 46.17

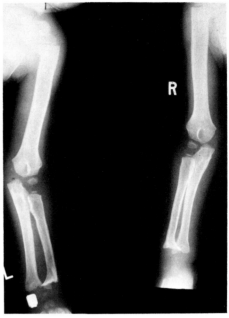

Fig. 46.15   M.W.—4 years—AP both upper limbs
All the long bones are short, broad and poorly modelled. The upper humeral and lower ulnar metaphyses particularly are flared. Epiphyses are small and irregular.
Same patient : Figs 46.2, 46.3, 46.7, 46.10, 46.11

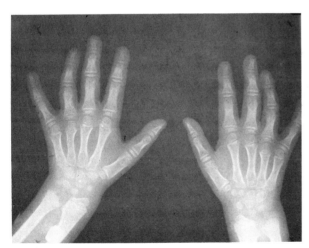

Fig. 46.16   D.T.—7 years—PA hands
The modelling defect is not severe, but there is some pointing at the base of the second to fifth metacarpals and the lower ends of the radius and ulna are irregular. The carpal centres are small and rather angular. Bone maturation corresponds to about 5 years 6 months.

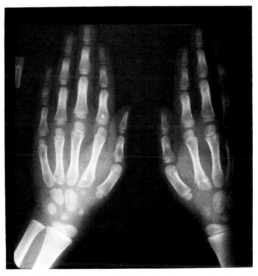

Fig. 46.17   S.M.—12 years—PA hands (same patient as in Fig. 46.14)
There is little abnormality here, but bone maturation is considerably retarded, being equivalent to about 5 years 6 months.

Chapter Forty-seven
MORQUIO SYNDROME

This is the one mucopolysaccharide disorder which can be differentiated on both clinical and radiological grounds from the others. The main features are normal intelligence, developing platyspondyly and increased joint laxity, without some of the long-bone modelling defects of the Hurler syndrome.

Inheritance

Autosomal recessive.

Frequency

The prevalence is probably similar to Hurler's disease.

Clinical features

Facial appearance—no coarsening of the face but the maxilla is prominent and the mouth broad.

Intelligence—normal.

Stature—markedly short stature develops, and (with the platyspondyly) a disproportionately short trunk.

Presenting feature/age—short stature during infancy, together with joint laxity and pectus carinatum. Corneal opacities and hepatosplenomegaly may be present.

Deformities—chiefly genu valgum, associated with the marked joint laxity.

Radiographic features

Skull—may be normal; some flattening of the condyles of the mandible.

Spine—in infancy similar to Hurler's syndrome, with hooked vertebrae and a thoraco-lumbar kyphos. Subsequently the vertebrae gain little in height and platyspondyly with a projecting central tongue is characteristic. The odontoid peg is dysplastic or absent and there may be instability in this region.

Thorax—similar to Hurler's syndrome, but the manubrio-sternal angle may be almost 90°.

Pelvis—flared iliac wings gradually develop during childhood, the inferior part of the ilia, which have been well developed in infancy, becoming absorbed.

Hip joints—the capital femoral epiphyses are initially advanced for their age, but between the ages of 4 and 9 years become smaller and finally disappear altogether.

Knee joints—associated with the severe genu valgum, there is a failure of ossification of the lateral side of the upper tibial epiphyses and metaphyses.

Wrists/hands—as in Hurler's disease, the second to fifth metacarpals are pointed at their proximal ends, but in Morquio's disease the small bones of the hand are well modelled. Carpal bones are initially advanced for their age, subsequently becoming small and irregular.

Long bones—some failure of modelling.

Bone maturation

Initially advanced for the chronological age, subsequently becoming delayed.

Biochemistry

Increased urinary excretion of keratan sulphate, but this seems to be intermittent and also decreases with age; the deficient enzyme is N-Ac-Gal-6-sulfate sulfatase.

Differential diagnosis

From other mucopolysaccharide disorders, but both clinical and radiographic features differentiate it; biochemical tests confirm.

Progress/complications

These patients may live for many decades, although cardio-respiratory disease is common. Atlanto-axial stability may be a problem, and premature osteoarthritis, particularly of the knees, is likely.

REFERENCES

Grossman H, Dorst J 1973 The mucopolysaccharidoses and mucolipidoses. In: Kaufmann H J (ed) Intrinsic diseases of bones. Progress in Pediatric Radiology 4. S Karger, Basel, pp 495–544

Lipson S J 1977 Dysplasia of the odontoid process in Morquio's syndrome causing quadriparesis. Journal of Bone and Joint Surgery 59A; 340–344

Pennock C A, Barnes I C 1976 The mucopolysaccharidoses. Journal of Medical Genetics 13:169

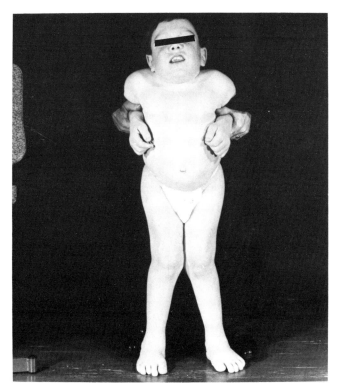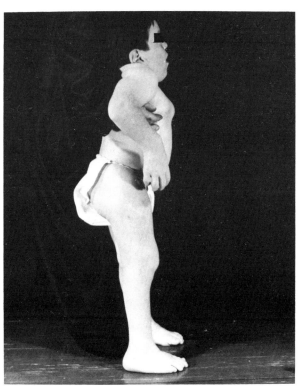

Figs 47.1 and 47.2   A 12-year-old boy with markedly short stature, pectus carinatum and joint laxity with genu valgum.

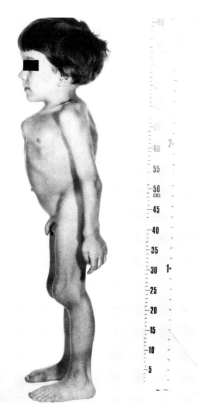

Fig. 47.3   The chest deformity is particularly severe in this child.

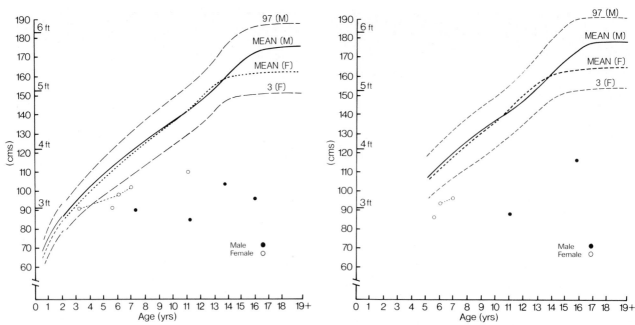

Fig. 47.4   Morquio syndrome—percentile height.

Fig. 47.5   Morquio syndrome—percentile span.

Skull and cervical spine: 5–12 years

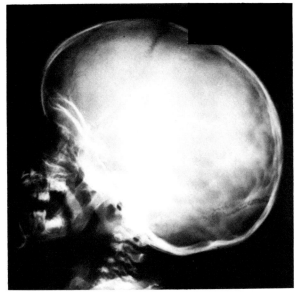

Fig. 47.6   T.N.—5 years 6 months—Lateral skull
The vault is rather large, but there is good differentiation between inner and outer tables. There is anterior elongation of the pituitary fossa under the anterior clinoid processes and the odontoid peg is hypoplastic—typical of an MPS disorder.
Same patient: Figs 47.10, 47.14, 47.30, 47.38

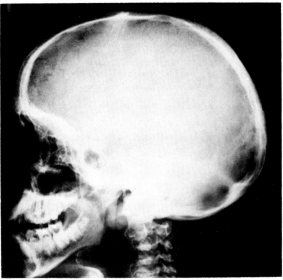

Fig. 47.7   E.H.—11 years—Lateral skull
(sister of patients in Figs 47.12 and 47.13)
Normal appearance of the skull, but the odontoid process is very small.
Same patient: Figs 47.34, 47.35, 47.41, 47.44

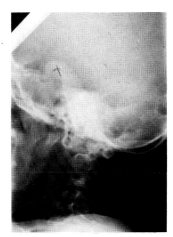

Fig. 47.8   D.W.—7 years 6 months—Lateral cervical spine in flexion
The odontoid peg is absent and there is subluxation of C1 on C2.
Same patient: Figs 47.9, 47.18, 47.19, 47.26–47.29

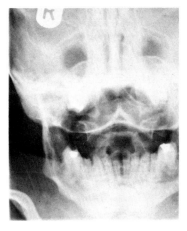

Fig. 47.9   D.W.—12 years 6 months—AP open mouth view of C1 and C2 (same patient as in Fig. 47.8)
There is no odontoid peg.

AP Spine: 5–15 years

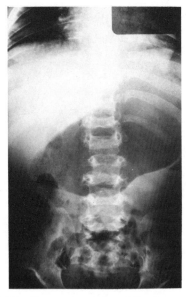

Fig. 47.10   T.N.—5 years 6 months—AP spine
There is mild platyspondyly throughout and the ribs are widened.
Same patient: Figs 47.6, 47.14, 47.30, 47.38

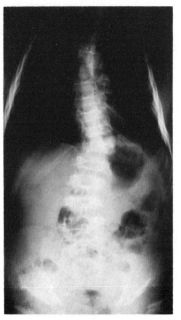

Fig. 47.11   S.K.—6 years—AP spine
There is a scoliosis convex to the left and universal flattening of the
vertebral bodies, with sharpening of the normal rounded angles.
Same patient: Figs 47.23, 47.42

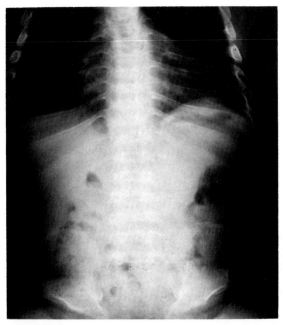

Fig. 47.12   J.H.—14 years—AP spine (brother of patients in Figs
47.7 and 47.13)
There is universal platyspondyly with sharply angled vertebral bodies.
The ribs show the classical appearance of an MPS disorder, with
widening anteriorly.
Same patient: Figs 47.16, 47.21, 47.25, 47.32, 47.36, 47.37, 47.45

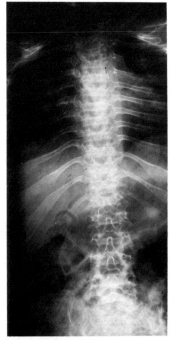

Fig. 47.13   V.H.—15 years—AP spine (sister of patients in Figs
47.7 and 47.12)
There is platyspondyly throughout with small lateral concavities of
the vertebral bodies and some angulation at the corners of the
vertebral end plates. Posteriorly the ribs are of normal size, but they
widen anteriorly.
Same patient: Figs 47.17, 47.31, 47.40

Lateral spine: 5–15 years

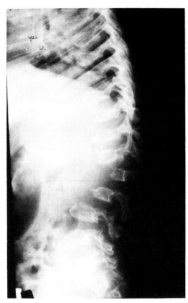

Fig. 47.14   T.N.—5 years 6 months—Lateral thoracic and lumbar spine
There is a kyphos at L2 and the vertebral body is hypoplastic here. Platyspondyly is present throughout and there is an anterior beak at L3 and inferior hooks at L4 and D12, also the vertebral bodies in the lumbar region show posterior scalloping.

(It is only in Morquio's disease that the vertebral bodies develop 'beaking' (an anterior tongue) but frequently there are some with inferior hooks, as in the other MPS disorders).
Same patient: Figs 47.6, 47.10, 47.30, 47.38

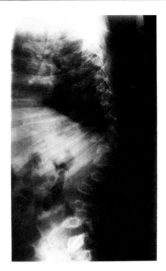

Fig. 47.15   P.S.—6 years 3 months—Lateral thoracic and lumbar spine
There is a kyphos in the lower dorsal region and the vertebral bodies at D12 and L1 show anterior beaking. Above and below these levels the bodies have an antero-inferior hook. All vertebral bodies are flat with an apparent increase in the size of the disc spaces; many of the bodies also show mild posterior wedging.
Same patient: Figs. 47.20, 47.24, 47.39

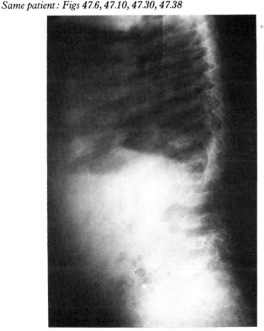

Fig. 47.16   J.H.—12 years—Lateral thoracic and lumbar spine
There is anterior beaking and marked platyspondyly throughout by this age (more severe than is seen in the other MPS disorders).
Same patient: Figs 47.12, 47.21, 47.25, 47.32, 47.36, 47.37, 47.45

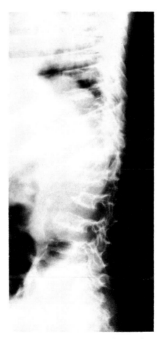

Fig. 47.17   V.H.—15 years—Lateral thoracic and lumbar spine
There is platyspondyly with anterior beaking of the vertebrae, and in the dorsal region also mild posterior wedging. The lumbar vertebrae show posterior scalloping.
Same patient: Figs 47.13, 47.31, 47.40

Chest: 7–12 years

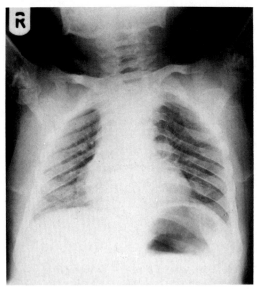

Fig. 47.18   D.W.—7 years 6 months—AP chest
The clavicles are short and broad, especially at their medial ends. The ribs show some anterior widening, and the angles of the scapulae are truncated. Metaphyseal sclerosis can be seen in the upper humeri.
Same patient: Figs 47.9, 47.19, 47.26–47.29

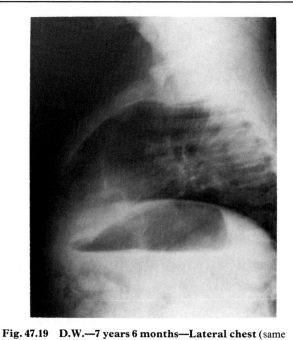

Fig. 47.19   D.W.—7 years 6 months—Lateral chest (same patient as in Fig. 47.18)
The antero-posterior diameter of the chest is markedly increased, but this patient does not show the angled manubrio-sternal joint often seen in Morquio's disease. The sternum is short, having only two segments below the manubrium.

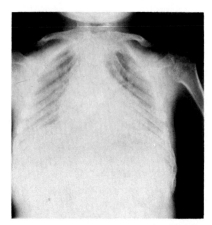

Fig. 47.20   P.S.—8 years 5 months—AP chest
The ribs appear to slope in an unusual manner, but some of this is due to the lordotic projection and the kyphos. The heart appears to be enlarged.
Same patient: Figs 47.15, 47.24, 47.39

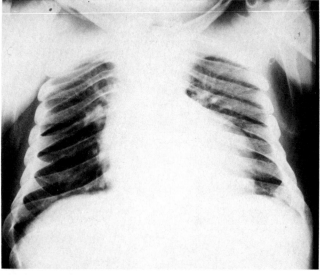

Fig. 47.21   J.H.—12 years—AP chest
The heart is enlarged and the ribs rather broad.
Same patient: Figs 47.12, 47.16, 47.25, 47.32, 47.36, 47.37, 47.45

Pelvis: 5–12 years

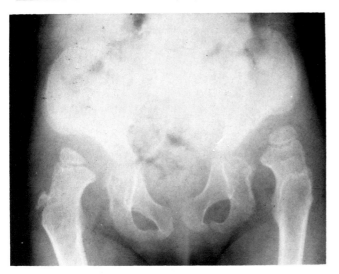

Fig. 47.22   J.G.—5 years 6 months—Pelvis
The iliac wings are flared, the bodies tapered inferiorly, and the acetabular margins not clearly demarcated. The ischia are thickened and point vertically (rather than the normal medially). The capital femoral epiphyses are only a little smaller than normal. There is coxa valga and subluxation of both hips.
Same patient: Fig. 47.43

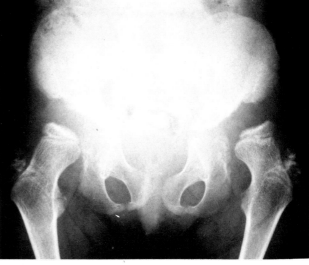

Fig. 47.23   S.K.—6 years—Pelvis
The iliac wings are flared and tapered inferiorly. The ischia are thick and vertically positioned and the capital femoral epiphyses are flattened with a little marginal sclerosis. Both hips are subluxed.
Same patient: Figs 47.11, 47.42

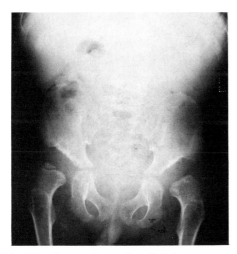

Fig. 47.24   P.S.—8 years 5 months—Pelvis
The acetabular roofs are poorly delineated and slope steeply. The capital femoral epiphyses are by this age very small and flattened.
Same patient: Figs 47.15, 47.20, 47.39

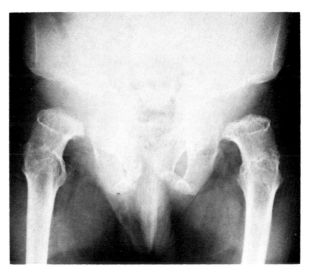

Fig. 47.25   J.H.—12 years—Pelvis
The 'wine-glass' appearance of the pelvis is very obvious, with flared ilia and very narrow bases. The capital femoral epiphyses have become totally reabsorbed.
Same patient: Figs 47.12, 47.16, 47.21, 47.32, 47.36, 47.37, 47.45

Pelvis : 14 months–11 years

Figs 47.26–47.29   D.W.—Pelvis
Progressive absorption of the base of the ilia and disappearance of the capital femoral epiphyses
Same patient : Figs 47.8, 47.9, 47.18, 47.19

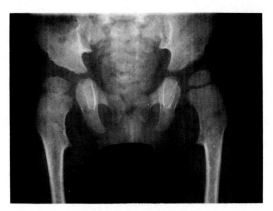

Fig. 47.26   1 year 2 months. The capital femoral epiphysis is unusually large compared with the normal for this age.

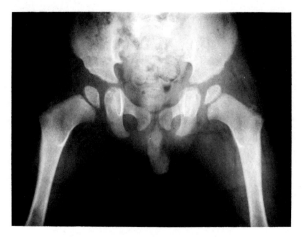

Fig. 47.27   1 year 8 months. Reabsorption is beginning on the medial aspect.

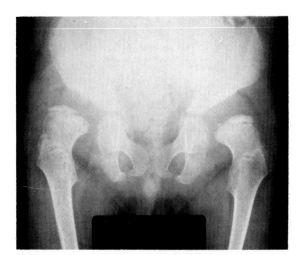

Fig. 47.28   7 years 6 months. The lower iliac absorption is now very marked, and the capital femoral epiphyses are small, irregular, flattened and dense.

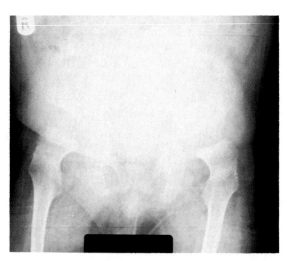

Fig. 47.29   11 years 6 months. There is no sign of either capital femoral epiphysis.

Leg: 5–15 years

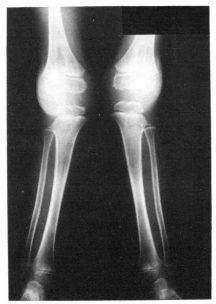

Fig. 47.30   T.N.—5 years 6 months—AP knees and ankles
Long bone modelling is essentially normal apart from the knee joint
area, which shows genu valgum and (probably secondary) defects in
the lateral tibial metaphyses. (Associated also with marked
ligamentous laxity). The fibulae appear short at both ends.
Same patient: Figs 47.6, 47.10, 47.14, 47.38

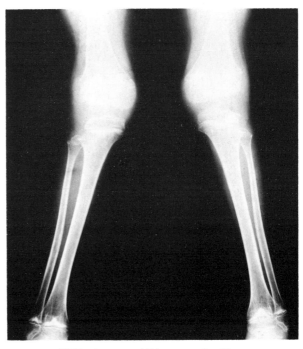

Fig. 47.31   V.H.—9 years—AP knees and ankles
There is genu valgum and the upper tibial metaphyses are markedly
underdeveloped, but the lower femoral ones are less affected;
epiphyses are also flattened laterally. The distal metaphyses show
patchy sclerosis.

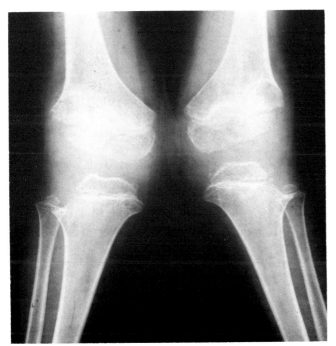

Fig. 47.32   J.H.—12 years—AP knees
There is severe genu valgum with pressure effects seen on the lateral
side of both epiphyses and metaphyses. Modelling of the shafts
appears to be normal.
Same patient: Figs 47.12, 47.16, 47.21, 47.25, 47.36, 47.37, 47.45

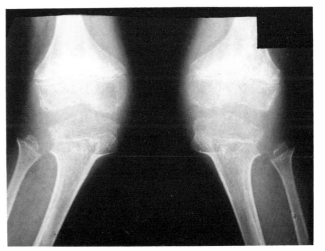

Fig. 47.33   J.P.—15 years—AP knees
There is marked metaphyseal splaying with irregular dense
ossification and angulation, and the adjacent epiphyses have irregular
articular margins. Genu valgum is present, and the modelling defects
are more marked on the lateral sides.

Foot: 11–14 years

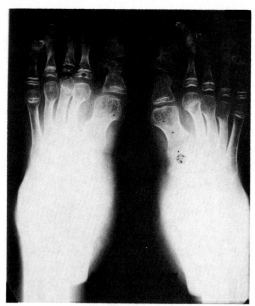

Fig. 47.34   E.H.—11 years—PA feet
There is shortening of both halluces due to premature fusion of the epiphyses of the first metatarsal and of the phalanges, and generalised metatarsal shortening but with normal modelling (unlike Hurler's syndrome).
Same patient: Figs 47.7, 47.35, 47.41, 47.44

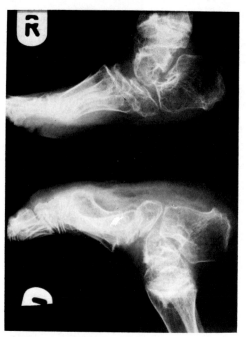

Fig. 47.35   E.H.—11 years—Lateral feet (same patient as in Fig. 47.34)
There is a 'rocker' deformity with a vertical talus—probably associated with the marked ligamentous laxity of this disease. The tarsal bones have sharply angled corners.

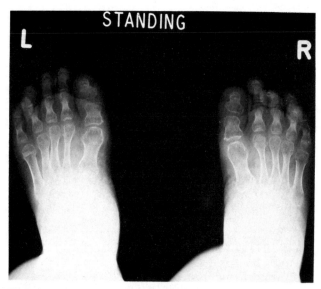

Fig. 47.36   J.H.—14 years—PA feet
There is shortening of the halluces but otherwise modelling is normal.
Same patient: Figs 47.12, 47.16, 47.21, 47.25, 47.32, 47.37, 47.45

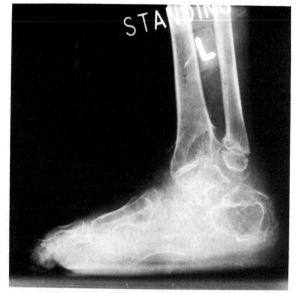

Fig. 47.37   J.H.—14 years—Standing lateral (same patient as in Fig. 47.36)
There is abnormal projection at the ankle joint with a sloping irregular metaphysis. The tarsal bones are all deformed and small, with irregular margins.

Upper limb: 5–20 years

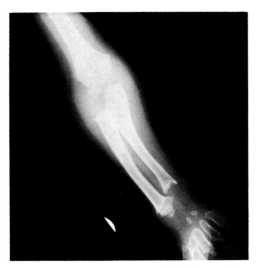

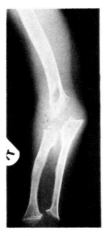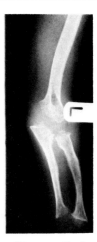

Fig. 47.39   P.S.—8 years 5 months—AP upper limbs
All long bones are short and there is mild metaphyseal flaring with
irregularity. Epiphyses are small, irregular and fragmented.
Same patient : Figs 47.15, 47.20, 47.24

Fig. 47.38   T.N.—5 years 6 months—AP elbow and forearm
All metaphyses are flared and the distal radial and ulnar ones also
cupped. Epiphyses are generally small and irregular.
Same patient : Figs 47.6, 47.10, 47.14, 47.30

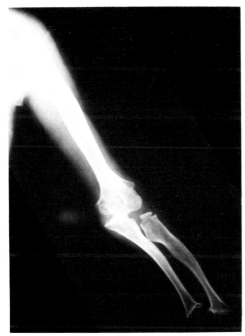

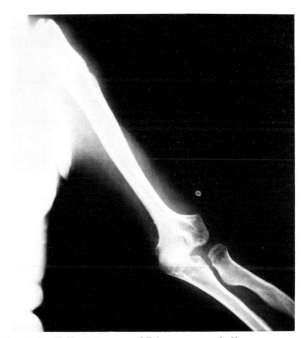

Fig. 47.40   V.H.—9 years—AP upper limb
There is mild metaphyseal flaring with angulation of the distal ulnar
and radial metaphyses.
Same patient : Figs 47.13, 47.17, 47.31

Fig. 47.41   E.H.—20 years—AP humerus and elbow
Modelling of the humerus is irregular and theré is prominence of the
deltoid tuberosity.
Same patient : Figs 47.7, 47.34, 47.35, 47.44

Hand: 3–12 years

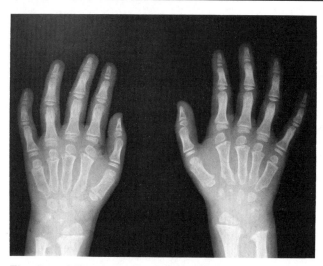

Fig. 47.42   S.K.—3 years 2 months—PA hands
There is mild flaring of the distal radial and ulnar metaphyses with medial sloping of the radius. The carpal centres are small and angular but maturation is *advanced* for the chronological age. The metacarpals show pointing at their proximal ends and there is a pseudo-epiphysis at the distal end of the first metacarpal. The phalanges appear normal (unlike Hurler's syndrome).
Same patient: Figs 47.11, 47.23

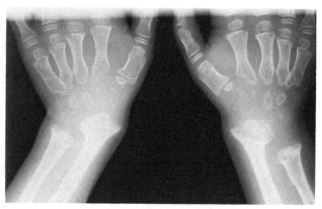

Fig. 47.43   J.G.—5 years 6 months—PA wrists
The metaphyses of the radius and ulna are markedly flared and both ulnae are short. The carpals are small and irregular and there is now *delayed* bone maturation. The bases of the second to fifth metacarpals are pointed, but modelling is otherwise normal.
Same patient: Fig. 47.22

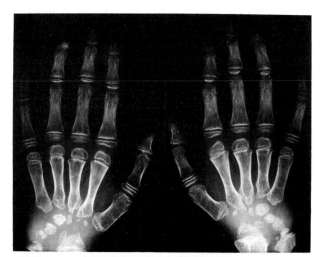

Fig. 47.44   E.H.—11 years—PA hands
The metacarpals are short with proximal pointing of the bases of the second to fifth. The carpal centres are small and angular and the bone age is retarded—especially on the radial side.
Same patient: Figs 47.7, 47.34, 47.35, 47.41

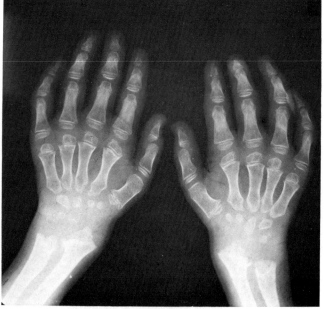

Fig. 47.45   J.H.—12 years—PA hands
The lower radial and ulnar epiphyses and metaphyses are poorly developed for this age, and the carpal bones are small and irregular, also showing retarded maturation—particularly on the radial side.
Same patient: Figs 47.12, 47.16, 47.21, 47.25, 47.32, 47.36, 47.37

Chapters Forty-eight and Forty-nine
THE MUCOLIPIDOSES

This is a group of storage diseases with some skeletal findings similar to the mucopolysaccharidoses, but without excess MPS in the urine.

GM₁ GANGLIOSIDOSIS TYPE 1

Signs are similar to Hurler's disease but apparent at birth. Prognosis is poor and most infants die within 2 years.

Inheritance

Autosomal recessive.

Frequency

Extremely rare.

Clinical features

Facial appearance—coarse facies.
Intelligence—retarded.
Stature—not known since infants do not survive.
Presenting feature/age—failure to thrive from birth; hepatosplenomegaly and stiff joints.

Radiographic features

Skull—similar to Hurler's syndrome, with enlargement and thickening of diploë.
Spine—broad hook-like appearance, although this is not present in neonatal period.
Thorax—clavicles broadened at the medial ends, ribs at their anterior ends.
Long bones—the distinguishing feature is excessive periosteal bone formation and lack of diaphyseal modelling. Subsequently, if the child survives long enough, there is over-constriction of the diaphysis.

Hands—similar to Hurler's disease, with pointed base of the second to fifth metacarpals and 'bullet-shaped' phalanges.

Bone maturation

Not known.

Biochemistry

No MPS in urine. Deficient enzyme is β-galactosidase.

Differential diagnosis

From *Hurler's disease*, but the earlier onset and periosteal bone growth of GM₁ gangliosidosis should differentiate it.

Progress/complications

Death within 2 years from respiratory complications.

MUCOLIPIDOSIS II (I-cell disease)

Clinically similar to GM₁ gangliosodosis type 1, and probably also of autosomal recessive inheritance.

A number of lysosomal enzymes are deficient.

Prognosis is poor, most infants dying within the first year or so from failure to thrive and respiratory complications.

There are other mucolipidoses, but with little skeletal change, and these are not described here.

REFERENCES

Nolte K, Spranger J 1976 Early skeletal changes in mucolipidosis III. Annals of Radiology 19/1:151–159
Leroy J G, Spranger J W, Feingold M, Opitz J M, Crocker A C 1971 I-cell disease. A clinical picture. Journal of Pediatrics 79:360–365
Spranger J W, Wiedemann H R 1970 The genetic mucolipidoses. Diagnosis and differential diagnosis. Humangenetik 9:113–139

GM₁ Gangliosidosis type I

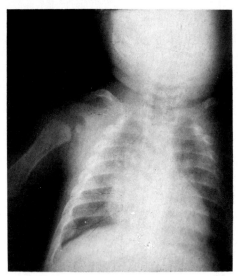

Fig. 48.1   J.T.—Neonate—AP chest
There is generalised osteoporosis. The ribs are broadened. The
humerus is rather short with rounded metaphyses. The heart appears
large.

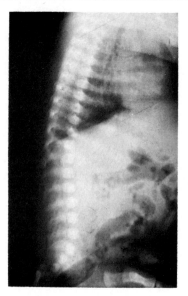

Fig. 48.2   J.T.—Neonate—Lateral thoracic and lumbar spine
(same patient as in Fig. 48.1)
The configuration of the vertebral bodies is normal for this age.

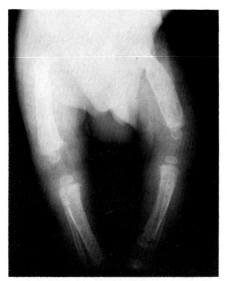

Fig. 48.3   J.T.—Neonate—Lower limbs (same patient as in Fig.
48.1)
The periosteal 'cloaking' of all long bones is clearly shown. The
metaphyses and epiphyses are normal.

Mucolipidosis II : Skull and spine

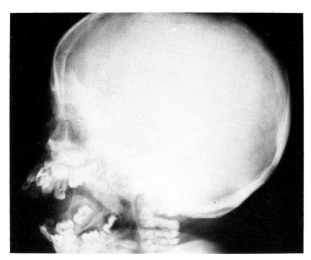

Fig. 49.1   L.P.—1 year 11 months—Lateral skull
The vault is thickened, with a 'ground-glass' appearance. Dentition is abnormal with a narrowed body of mandible, cyst formation and molars developing within the rami.
Same patient : Figs 49.2, 49.4, 49.5, 49.7, 49.9

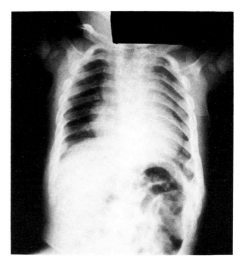

Fig. 49.2   L.P.—Neonate—AP trunk (same patient as in Fig. 49.1)
The thoracic cage is small with short broad ribs.

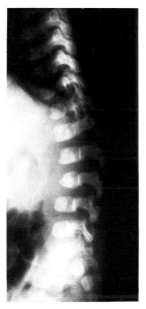

Fig. 49.3   S.H.—5 months—Lateral lumbar spine
The typical anterior hook-like appearance of the vertebrae is now present.
Same patient : Figs 49.6, 49.10

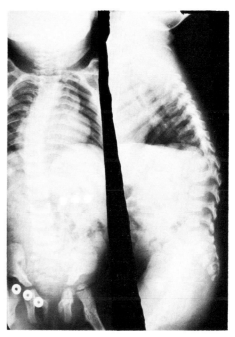

Fig. 49.4   L.P.—1 year 11 months—AP and lateral spine (same patient as in Fig. 49.1)
AP view—the neural arches of a number of dorsal vertebrae have not fused and there is left thoracic scoliosis. The ribs are broadened anteriorly.
   Lateral view—vertebrae are biconvex with posterior scalloping and there is a dorso-lumbar kyphos.

Mucolipidosis II : Pelvis and lower limb

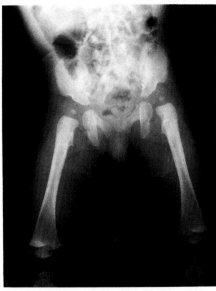

Fig. 49.5   L.P.—Neonate—AP pelvis and femora
Only the lateral part of the base of the ilium is failing to develop. The femora are normal.
Same patient : Figs 49.1, 49.2, 49.4, 49.7–49.9

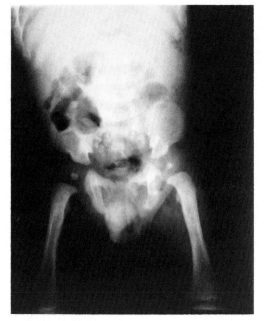

Fig. 49.6   S.H.—3 months—AP pelvis and femora
Similar to Figure 49.5 but femoral modelling is not quite normal.
Same patient : Figs 49.3, 49.10

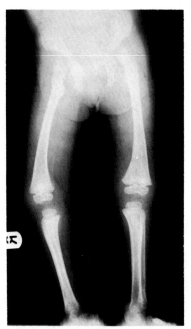

Fig. 49.7   L.P.—1 year 11 months—AP pelvis and lower limbs
(same patient as in Fig. 49.5)
There is now marked failure of development at the bases of the ilia with acetabular dysplasia. The capital femoral epiphyses are small and flattened, as are those around the knee and ankle. The long bones are normal apart from shortening of the fibulae.

Mucolipidosis II : Upper limb

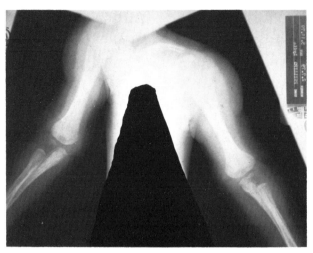

Fig. 49.8 L.P.—Neonate—AP upper limbs
Apart from slightly short and thickened humeri, the long bones are normal.
Same patient : Figs 49.1, 49.2, 49.4, 49.7, 49.9

Fig. 49.9 L.P.—1 year 11 months—AP upper limb and hand
(same patient as in Fig. 49.8)
The long bones are short and broad with widened irregular metaphyses and small, flattened epiphyses. The hand is identical to the MPS disorders with proximal pointing of the second to fifth metacarpals and 'bullet-shaped' phalanges.

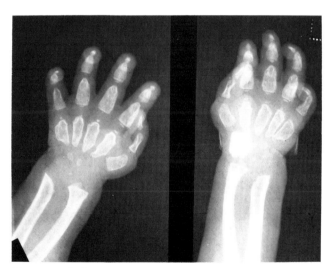

Fig. 49.10 S.H.—2 years 3 months—P.A. hands
The hands and wrists are similar to Figure 49.9 but the changes are now marked. An unusual feature is widening at the inferior radio-ulnar joint.
Same patient : Figs 49.3, 49.6

Section Eleven
METABOLIC BONE DISEASE

Chapter Fifty
HYPOPHOSPHATASIA

A low plasma alkaline phosphatase is accompanied by poor or absent mineralisation in the neonate, resulting in metaphyseal defects then and as the child grows.

Inheritance

Both the severe (lethal) congenital form and the milder form (developing in infancy) are of autosomal recessive inheritance. There is a very rare adult form which is possibly of autosomal dominant inheritance.

Frequency

Not known, but very rare.

Clinical features

Facial appearance—in the severe congenital form the head is 'globular' and soft, with no bone formation.

Intelligence—probably normal in the type which survives.

Stature—may be normal initially, but with persistent failure of bone formation in childhood, stature will be reduced.

Presenting feature/age—in the *severe congenital form* the child is stillborn, with limb deformities. In *infancy* there is failure to thrive, and pathological fractures. The *adult* may have signs of previous rickets and the presenting feature may be a fracture.

Deformities—those of rickets, with bending of bone, frontal bossing of the skull etc.

Associated anomalies—none.

Radiographic features

Skull—in the *severe congenital form* there may be no mineralisation at all of the vault. In the *tarda form* premature fusion of cranial sutures is a feature, with digital markings indicating raised intracranial pressure.

Spine—in the *severe congenital form* the vertebrae appear paper thin, and there may be no mineralisation of the neural arches.

Pectoral and pelvic girdles—small with poor mineralisation, or none in the *severe congenital form*.

Long bones—in the infant, there are large metaphyseal defects, sometimes extending far into the diaphysis. At a later age metaphyseal irregularity remains.

Bone maturation

Probably delayed.

Biochemistry

Alkaline phosphatase is very low. Phosphoethanolamine is present in the urine, and may be detected in heterozygote parents. Some infants also have hypercalcaemia.

Differential diagnosis

The severe congenital form needs to be distinguished from *severe congenital osteogenesis imperfecta*, and from other forms of lethal dwarfism, but the characteristic lack of bone mineralisation and low alkaline phosphatase in hypophosphatasia make the diagnosis clear. The tarda forms are similar to *rickets* but again, low alkaline phosphatase will distinguish hypophosphatasia.

Progress/complications

The severe congenital form is lethal, from respiratory insufficiency. In the milder form the metaphyseal lesions tend to heal, but with bone deformity and there is also premature fusion of epiphyses giving stunted growth. Premature fusion of sutures in the skull gives rise to the problems associated with raised intracranial pressure. In adult life fractures of the long bones occur, similar to the Looser's zones of osteomalacia, but on the convexity of long bones.

REFERENCE

Smith R 1979 Biochemical disorders of the skeleton. Butterworth, London

Whole body : Birth

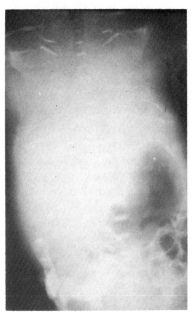

Fig. 50.1   J.S.—Neonate—AP trunk
Mineralisation of the whole skeleton is very poor. Several ribs are
totally unossified and the vertebrae paper-thin (post mortem
radiograph).

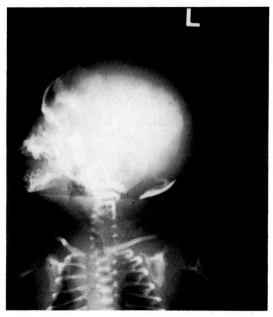

Fig. 50.2   T.C.—Neonate—Lateral skull and upper thorax
There is practically no mineralisation of the skull vault, and parts of
the cervical vertebrae are 'missing'. Dentition is deficient.
Same patient : Figs 50.3, 50.4

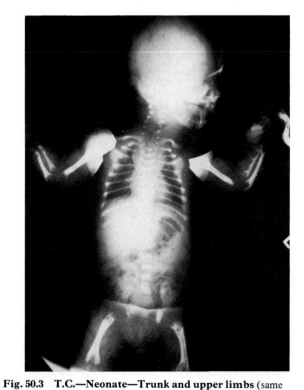

Fig. 50.3   T.C.—Neonate—Trunk and upper limbs (same
patient as in Fig. 50.2)
The ribs and vertebrae are better mineralised than the stillborn infant
(Fig. 50.1). The long bones show jagged metaphyseal defects
extending along the shafts.

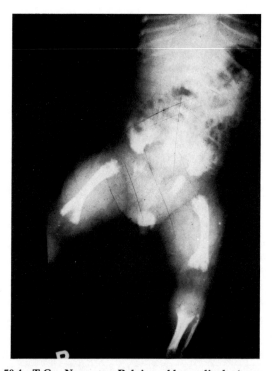

Fig. 50.4   T.C.—Neonate—Pelvis and lower limbs (same
patient as in Fig. 50.2)
Metaphyseal ossification is defective and bone maturation is retarded.

Skull: Birth–5 years

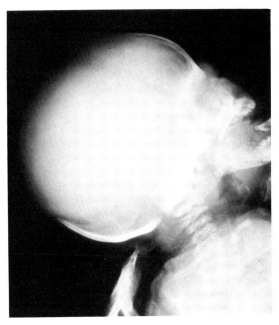

Fig. 50.5   A.T.—Neonate—Lateral skull
There is virtually no mineralisation of the parietal bones.
Same patient: Figs 50.6, 50.7, 50.9, 50.13–50.17, 50.20–50.22

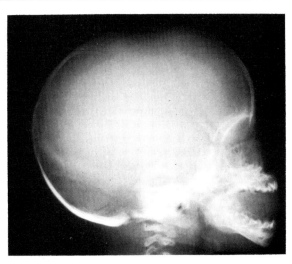

Fig. 50.6   A.T.—2 months—Lateral skull (same patient as in Fig. 50.5)
Mineralisation is progressing but the suture lines and anterior fontanelle are unusually wide for this age.

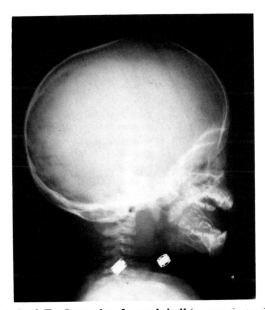

Fig. 50.7   A.T.—7 months—Lateral skull (same patient as in Fig. 50.5)
Further mineralisation has occurred but the straight suture lines (coronal) indicate that premature fusion is occurring.

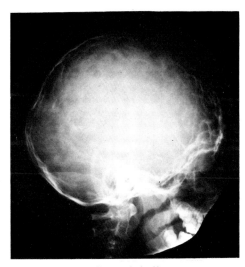

Fig. 50.8   T.L.—5 years—Lateral skull
The skull is unusually round in shape, due to premature fusion of the sutures; the digital markings thus indicate raised intra-cranial pressure.
Same patient: Figs 50.10–50.12, 50.18, 50.19

Spine and chest: 1 month–5 years

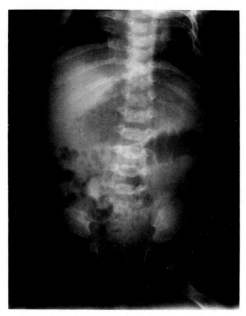

Fig. 50.9   A.T.—2 months—AP spine
There is mild platyspondyly, with apparently laterally placed
pedicles. The ribs are short with an irregular cortical outline.
Same patient: Figs 50.5–50.7, 50.13–50.17, 50.20–50.22

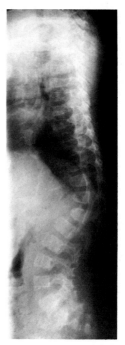

Fig. 50.10   T.L.—1 year 6 months—Lateral spine
There is a thoraco-lumbar kyphos and deep cupping of the anterior
ends of the ribs.
Same patient: Figs 50.8, 50.11, 50.12, 50.18, 50.19

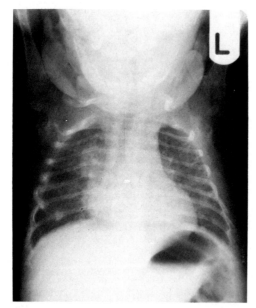

Fig. 50.11   T.L.—1 month—AP chest (same patient as in Fig.
50.10)
The ribs are poorly mineralised, with splayed anterior ends. (There is
a healing fracture of the right eighth rib.)

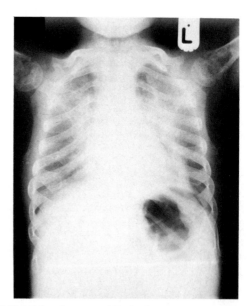

Fig. 50.12   T.L.—5 years—AP chest (same patient as in Fig. 50.8)
The ribs are broad and well mineralised now. Jagged metaphyseal
lesions are seen in the upper humeri.

Pelvis and lower limbs : Birth–5 years

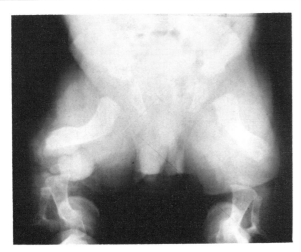

Fig. 50.13   A.T.—Neonate—AP pelvis and lower limbs
The pelvis is virtually normal. The long bones are thickened and
bowed.
Same patient : Figs 50.5–50.7, 50.9, 50.14–50.17, 50.20–50.22

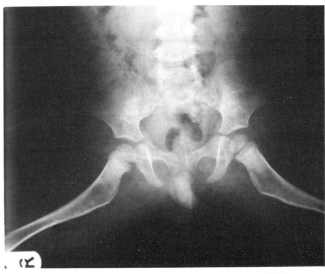

Fig. 50.14   A.T.—3 years—AP pelvis (same patient as in Fig.
50.13)
The pelvis is normal. The femoral shafts are now more than usually
slender with slight metaphyseal irregularity.

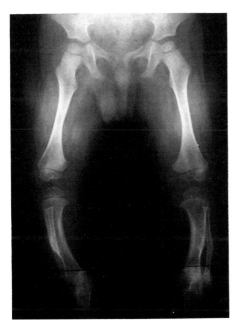

Fig. 50.15   A.T.—5 years 6 months—AP hips and lower limbs
(same patient as in Fig. 50.13)
There are modelling defects of the long bones, with shortening of the
tibia and fibula—there are now no obvious metaphyseal irregularities.

Leg: Birth–12 years

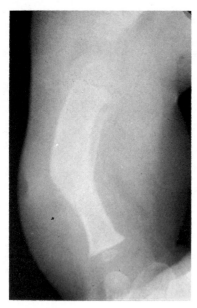

Fig. 50.16   A.T.—Neonate—Lateral tibia and fibula
The tibia is much thickened, with anterior bowing and an overlying cutaneous dimple.
Same patient: Figs 50.5–50.7, 50.9, 50.13–50.15, 50.17, 50.20–50.22

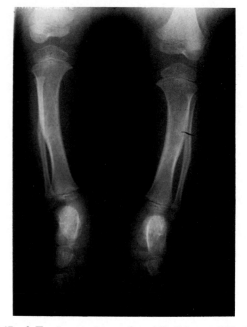

Fig. 50.17   A.T.—5 years 6 months—AP tibiae and fibulae
(same patient as in Fig. 50.16)
There is abnormal modelling of the shafts with persistent bowing, but no metaphyseal defects.

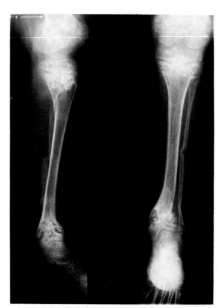

Fig. 50.18   T.L.—10 years—AP and lateral left tibia and fibula
There is obvious metaphyseal sclerosis and irregularity, with a pathological fracture through the distal tibial metaphysis and a mid-fibular pseudarthrosis.
Same patient: Figs. 50.8, 50.10–50.12, 50.19

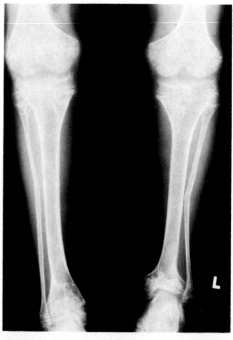

Fig. 50.19   T.L.—12 years—AP tibiae and fibulae (same patient as in Fig. 50.18)
The metaphyseal lesions have improved somewhat, but there appears to be premature fusion of all epiphyses.

Upper limb : Birth–3 years

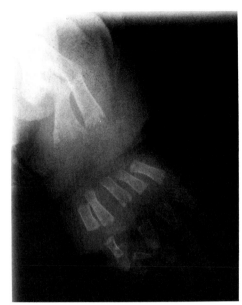

Fig. 50.20 A.T.—Neonate—AP forearm
The radius and ulna are thickened and bowed, and there is irregular mineralisation of the metaphyses.
Same patient: Figs 50.5–50.7, 50.9, 50.13–50.17, 50.21, 50.22

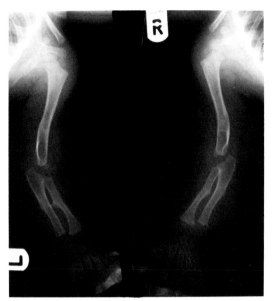

Fig. 50.21 A.T.—3 years—AP upper limbs (same patient as in Fig. 50.20)
There is irregular modelling of all bones, with shortening of the radius and ulna.

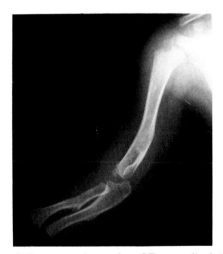

Fig. 50.22 A.T.—3 years 6 months—AP upper limb (same patient as in Fig. 50.20)
Similar to Figure 50.21.

Chapter Fifty-one
IDIOPATHIC HYPERPHOSPHATASIA

Probably several conditions, characterised by high levels of plasma alkaline phosphatase, thickened, fragile bones with patchy sclerosis and severe bowing.

Inheritance

Autosomal recessive is usually reported.

Frequency

Very rare.

Clinical features

Facial appearance—enlargement of the head.
Intelligence—normal.
Stature—if the onset of the disease is early the patient is dwarfed, with later onset there will be some shortness of stature.
Presenting feature/age—the age of onset seems to be variable, from infancy to adolescence. Fracture and long bone deformity are the usual presenting signs.
Deformities—of the soft, bending bone and pathological fractures.
Associated anomalies—none.

Radiographic features

Skull—patchy sclerosis, obliteration of paranasal sinuses, thickened vault and base.
Spine—patchy sclerosis, and there may also be signs of osteoporosis.
Thorax—widened and sclerotic ribs.
Pelvis—patchy sclerosis, with deformities of protrusio acetabuli and coxa vara, the result of bone softening.
Long bones—widened shafts with osteoporosis and loss of differentiation into medulla and cortex; periosteal bone formation in the child and later poor modelling and patchy sclerosis.

Bone maturation

Not known.

Biochemistry

Very high levels of plasma alkaline phosphatase and also urinary hydroxyproline.

Differential diagnosis

Engelmann's disease has rather a similar appearance of the skull and long bones, as do other forms of *craniodiaphyseal dysplasia*, but they are not associated with such a high plasma alkaline phosphatase.

Polyostotic fibrous dysplasia is also characterised by patchy sclerosis, fractures and bending of bone, but again is not associated with a very high plasma alkaline phosphatase, also the skeletal lesions tend to be asymmetrical.

Progress/complications

Little is known but if the child survives there appears to be some improvement, or stabilisation. Obliteration of cranial foramina may occur, with consequent paralysis of cranial nerves.

REFERENCES

Eyring E J, Eisenberg E 1968 Congenital hyperphosphatasia. A clinical, pathological and biochemical study of two cases. Journal of Bone and Joint Surgery 50A : 1099–117
Smith R 1979 Biochemical disorders of the skeleton. Butterworth, London
Thompson R C, Gaull G E, Horwitz S J, Schenk R K 1969 Hereditary hyperphosphatasia. American Journal of Medicine 47 : 209–219

Skull: 6 years–Adult

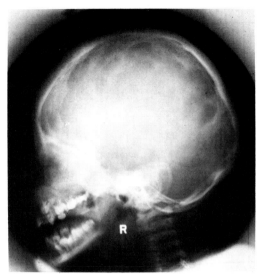

Fig. 51.1 M.S.—6 years—Lateral skull
There is patchy sclerosis and thickening of the vault with obliteration of the paranasal sinuses.
Same patient: Figs 51.4, 51.7, 51.8

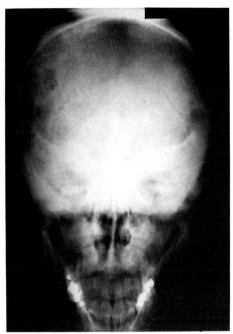

Fig. 51.2 G.M.—22 years—AP skull
There is patchy increased bone density of the skull vault, base and paranasal sinuses, with irregular widening of the diploic space (sclerosis alternating with thinner areas).
Same patient: Figs 51.3, 51.5, 51.6, 51.9, 51.10

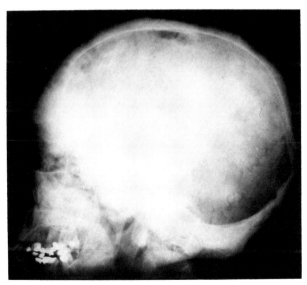

Fig. 51.3 G.M.—22 years—Lateral skull (same patient as Fig. 51.2)
There is gross thickening and sclerosis of the vault, base and paranasal sinuses.

Spine, chest and pelvis : 10 years–Adult

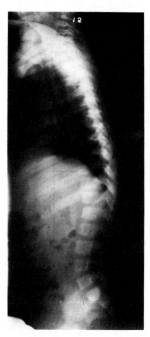

Fig. 51.4   M.S.—10 years—Lateral spine
Apart from some patchy sclerosis, the spine is normal.
Same patient: Figs 51.1, 51.7, 51.8

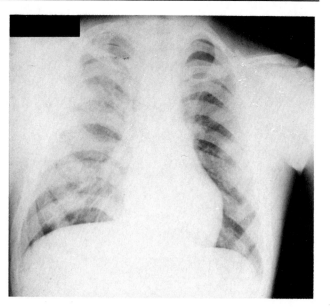

Fig. 51.5   G.M.—17 years 6 months—PA chest
There is obvious widening and sclerosis of the ribs.
Same patient: Figs 51.2, 51.3, 51.6, 51.9, 51.10

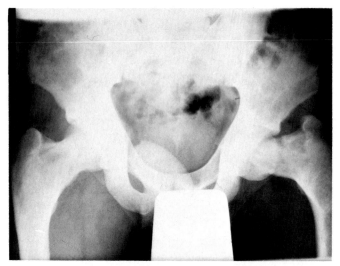

Fig. 51.6   G.M.—22 years—Pelvis (same patient as in Fig. 51.5)
There is patchy sclerosis in all areas, flared ilia and coxa vara, with a
femoral neck fracture on the left.

Limbs: 8–17 years

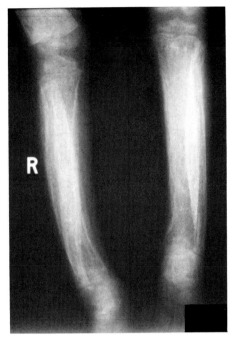

Fig. 51.7   M.S.—8 years—AP tibiae and fibulae
The shafts are all widened, with osteoporosis and loss of differentiation into medulla and cortex.
Same patient: Figs 51.1, 51.4, 51.8

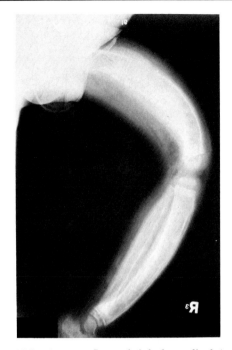

Fig. 51.8   M.S.—8 years—Lateral right lower limb (same patient as in Fig. 51.7)
Periosteal bone formation is obvious in the diaphyses, but is not involving the metaphyses. There is anterior bowing of all bones.

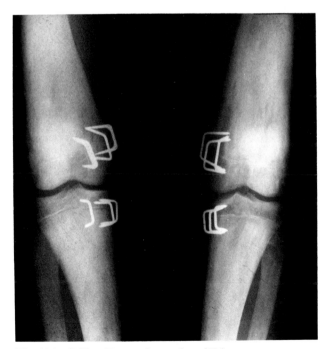

Fig. 51.9   G.M.—17 years 6 months—AP knees
There is poor modelling with widened diaphyses which show patchy sclerosis. (The medial staples were inserted to correct genu valgum.)
Same patient: Figs 51.2, 51.3, 51.5, 51.6, 51.10

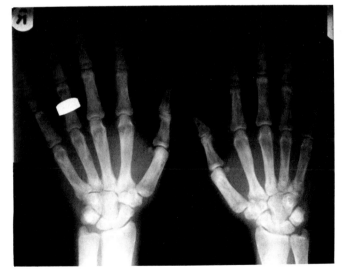

Fig. 51.10   G.M.—17 years 6 months—PA hands (same patient as in Fig. 51.9)
There is poor modelling and patchy sclerosis throughout.

Chapter Fifty-two
HYPOPHOSPHATAEMIC RICKETS
(vitamin D resistant)

This is the commonest of the renal tubular rickets, with an inherited defect in phosphate transport.

Inheritance

X-linked dominant, thus the disease tends to be more severe in males.

Frequency

Not known, but rare.

Clinical features

Facial appearance—prominent forehead.
Intelligence—normal.
Stature—short, principally due to short limbs (bending deformity and premature fusion of epiphyses).
Presenting feature/age—usually limb deformity and short stature in childhood.
Deformities—bending of limb bones; genu valgum or varum is commonest.

Radiographic features

Skull—may be normal, or with mild sclerosis and prominent frontal bones.
Spine—sclerosis of the end-plates. Scoliosis is possible, due to softened bone.
Long bones—in infancy there are wide epiphyseal plates, sometimes indistinguishable from nutritional rickets. Metaphyses may be cupped. At a later age the bones are of increased density with buttressing and enlarged tuberosities at muscle attachments (e.g. the deltoid tuberosity), and Looser's zones may be present.

Bone maturation

Probably premature fusion of epiphyses, contributing to the short stature.

Biochemistry

The only biochemical abnormality is hypophosphataemia. The clearance of phosphate is increased, and the renal tubular reabsorptive capacity for phosphate reduced.

Differential diagnosis

From other forms of '*rickets*' and *osteomalacia* but in vitamin D resistant rickets there are few of the symptoms associated with nutritional rickets—the children have no muscle weakness, only bone deformity and short stature. Bone sclerosis is a feature in adults. The *metaphyseal chondrodysplasias* show a similar appearance of the metaphyses, but have a normal blood chemistry and less tendency for bending of bone.

Progress/complications

In the past the rickets improved with high doses of vitamin D. Current practice includes treatment with 1-α-hydroxy-cholecalciferol together with oral phosphate. The complications are those of long bone malalignment, perhaps protrusio acetabuli and scoliosis. Spinal cord compression can occur from ossification of the ligamenta flava.

REFERENCES

Dent C E, Harris H 1956 Hereditary forms of rickets and osteomalacia. Journal of Bone and Joint Surgery 38B:204–226
Smith R 1979 Biochemical disorders of the skeleton. Butterworth, London

Skull and spine: 3–5 years

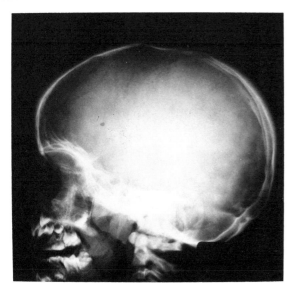

Fig. 52.1   G.B.—5 years—Lateral skull
There is mild sclerosis of the vault and base.
Same patient: Figs 52.3, 52.5

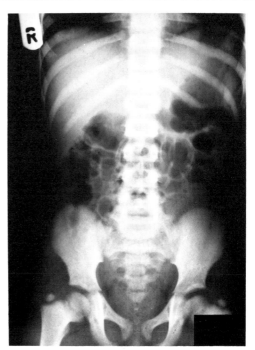

Fig. 52.2   K.G.—3 years—AP lower thoracic and lumbar spine
Modelling is normal but there is sclerosis of all bones and coarsening of the trabecular pattern.
Same patient: Figs 52.4, 52.6, 52.7

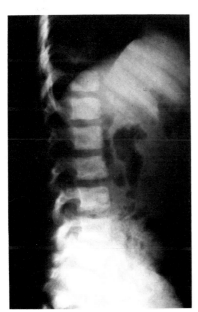

Fig. 52.3   G.B.—5 years—Lateral thoracic and lumbar spine
(same patient as in Fig. 52.1)
Sclerosis involves especially the regions of the vertebral end-plates.

Pelvis and lower limb: 3–12 years

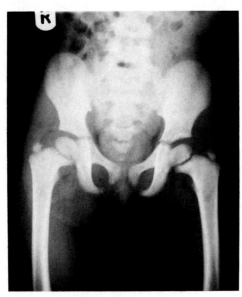

Fig. 52.4   K.G.—3 years—Pelvis
There is an overall increase in bone density but deficient
mineralisation at the epiphyseal plates.
Same patient: Figs 52.2, 52.6, 52.7

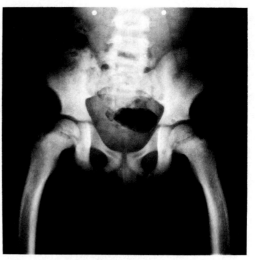

Fig. 52.5   G.B.—5 years—Pelvis
Similar to Figure 52.4, but there is more obvious lateral bowing of the
femora.
Same patient: Figs 52.1, 52.3

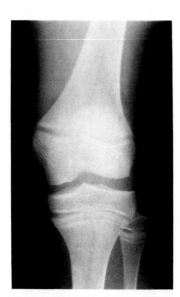

Fig. 52.6   K.G.—12 years—AP knee (same patient as in Fig. 52.4)
There are rachitic changes, with wide epiphyseal plates and splayed
metaphyses (indistinguishable from nutritional or renal rickets here).
The growth arrest lines in the upper tibia and fibula indicate previous
acute episodes.

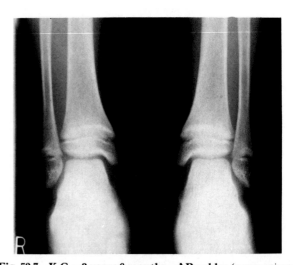

Fig. 52.7   K.G.—3 years 6 months—AP ankles (same patient as
in Fig. 52.4)
Similar to Figure 52.6, but here the metaphyses are also cupped.

Chapter Fifty-three
PSEUDOHYPOPARATHYROIDISM

Characterised by the signs of hypoparathyroidism with the addition of short metacarpals and sometimes mental retardation.

Inheritance

Thought to be of X-linked dominant inheritance.

Frequency

Very rare.

Clinical features

Facial appearance—a round head with short neck is typical.

Intelligence—mental retardation is likely to be a part of the disease, although worsening with the prolonged effects of hypocalcaemia.

Stature—short.

Presenting feature/age—the short metacarpals (metatarsals) are congenital; onset of other signs is variable, and the disorder may be discovered only incidentally.

Deformities—none of note.

Associated anomalies—none other than those noted above.

Radiographic features

Skull—normal. Rarely ectopic calcification of the basal ganglia can be seen.

Hands/feet—short metacarpals, usually the third, fourth and fifth.

Ectopic calcification occurs in the subcutaneous tissues.

Bone maturation

Probably normal.

Biochemistry

As for hypoparathyroidism, but the production of cyclic AMP is defective and there is no response to parathormone.

Differential diagnosis

From *hypoparathyroidism* (see above). From other disorders with metacarpal shortening, but the additional clinical features distinguish pseudohypoparathyroidism.

Progress/complications

The features of hypocalcaemia may develop: epilepsy, cataracts, ectopic calcification, rarely spontaneous tetany.

REFERENCES

Potts J T 1978 Pseudohypoparathyroidism. In: Stanbury J B, Wyngarden J B, Frederickson A S (eds) Metabolic basis of inherited disease. McGraw Hill, New York, pp 1350–1365
Smith R 1979 Biochemical disorders of the skeleton. Butterworth, London

Hand and foot: 9 years–Adult

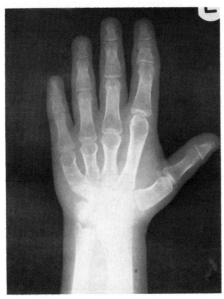

Fig. 53.1   D.H.—9 years—PA left hand
There is shortening of the third to fifth metacarpals, all terminal and the second and fifth middle phalanges.
Same patient: Fig. 53.3

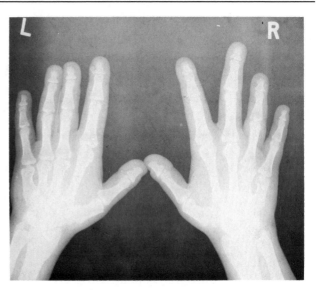

Fig. 53.2   Unknown—Adult—PA hands
The left third to fifth metacarpals are short, and in the right only the fourth and fifth.
All terminal phalanges are short.

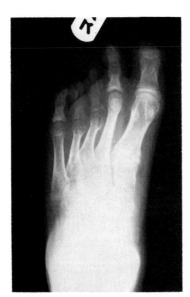

Fig. 53.3   D.H.—9 years—PA right foot (same patient as in Fig. 53.1)
The third and fourth metatarsals are short.

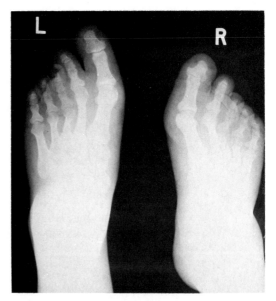

Fig. 53.4   Unknown—Adult—PA feet
The third to fifth metatarsals are short.

Section Twelve
DECREASED BONE DENSITY

Chapter Fifty-four
OSTEOGENESIS IMPERFECTA

A disorder characterised by fragility of bone, but with wide variation in severity. At its worst the infant is stillborn with multiple fractures, and at its mildest there may be no fractures, but only one of the accompanying features of the disease (see below).

Inheritance

Unifactorial, but it is clear there are several different genes. The following are likely (in order of prevalence):

  (i) Mild to moderately severe, with bright blue sclerae: autosomal dominant
 (ii) Very severe, with progressive dwarfing, pale blue or white sclerae: sporadic, possibly new dominant mutant
(iii) Similar to (i) but with white sclerae: autosomal dominant
 (iv) Severe congenital type (perinatal death): autosomal recessive
  (v) Similar to (i) but autosomal recessive inheritance (probably only 2% of all cases)

Frequency

Identifiable at birth: 1 in 20000 total births. Possible prevalence in population is 16 per million index patients (34 per million including affected relatives).

Clinical features

Facial appearance—large head, with temporal bulge and small triangular face.
Intelligence—normal.
Stature—unremarkable at birth, although telescoped fractures may make the limbs appear short. Types (i), (iii) and (v) above tend to be short, although mild cases are of normal stature. In type (ii) above, growth appears to cease altogether around 3–4 years.
Presenting feature/age—at birth and in infancy—fractures, blue sclerae (if a feature), occasional dislocation from joint laxity (congenital dislocation of hip).
Deformities—*of fracture mal-union*, and contractures may occur but are preventable. *Of osteoporosis*: that is,

bowing of long bones, platyspondyly and biconcave vertebrae, trefoil pelvis, basilar invagination. Scoliosis occurs in about one-third of patients.

Associated features indicating disordered connective tissue: generalised joint laxity, herniae, undue ease of bruising, poor scar formation. Also deafness (in second or third decade, usually), dentinogenesis imperfecta.

Radiological features

Skull—wide intertemporal diameter, Wormian bones in suture lines, thin cortex.
Spine—normal at birth. From childhood onwards—osteoporosis, with disc-like or biconcave vertebrae. Mild cases are normal.
Long bones—may be normal apart from fractures, or short and stubby from telescoped fractures at birth, the shafts later becoming thin with narrow cortices. Occasional 'cystic' or 'popcorn' appearance in severe cases. Fractures are commoner in the lower limb than in the upper (unlike normal children), and only in 2–3% of all cases does separation of epiphyses occur (a common injury in normal children).

Bone maturation

Probably normal.

Biochemistry

There are probably genetic collagen defects in all types of the disease.

Differential diagnosis

1. *Other cases of childhood oesteoporosis*, including idiopathic juvenile osteoporosis. Usually Wormian bones, blue sclerae and a family history will distinguish osteogenesis imperfecta.

2. *The 'battered baby'*. Here there may be evidence of other injury. Bruising is an unreliable sign since this occurs as a feature of osteogenesis imperfecta; however when occurring in a 'battered baby' there should be

improvement while he is in a protected environment. A family history, blue sclerae, Wormian bones and the deformities of osteoporosis (see above) usually distinguish osteogenesis imperfecta, but bone histology may be necessary.

3. *Camptomelic dwarfism* is unassociated with fractures. It is characterised by congenital bowing and angulation of long bones.

Progress/complications

Fractures usually cease around puberty. Residual deformity and short stature are features; complications may arise from mal-union, bowing of long bones, contractures, scoliosis, protrusio acetabuli and (more seriously) basilar invagination.

Hyperplastic callus occurs in some and (very rarely) malignant change in this. Many patients have a normal life span, but in type (ii) above death usually occurs before 20 years of age.

REFERENCES

Smith R, Francis M J O, Houghton G R 1983 The brittle bone syndrome. London, Butterworth
Sillence D O, Senn A, Danks D M 1979 Genetic heterogeneity in osteogenesis imperfecta. Journal of Medical Genetics 16:101–116
Wynne-Davies R, Gormley J 1981 Clinical and genetic patterns in osteogenesis imperfecta. Journal of Medical Genetics 18:222

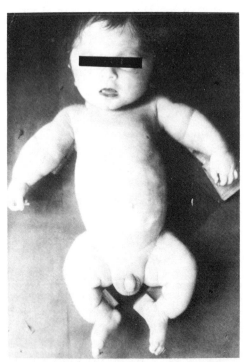

Fig. 54.1 This child died in the perinatal period. The appearance of disproportionately short limbs is produced by the telescoping of fractures.

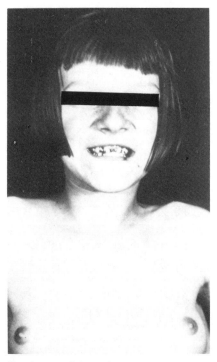

Fig. 54.2 Dentinogenesis imperfecta.

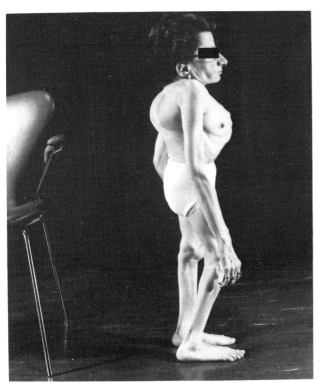

Fig. 54.3 Severe deformity, short stature, scoliosis and pectus carinatum in an adult patient.

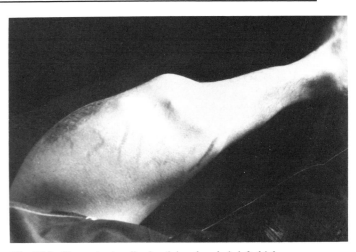

Fig. 54.4 Hyperplastic callus involving the whole left thigh.

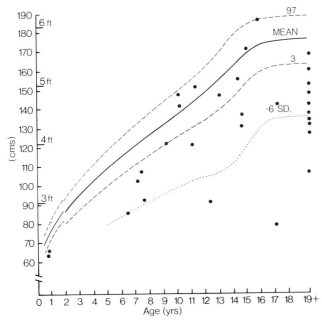

Fig. 54.5 Osteogenesis imperfecta—percentile height (males).

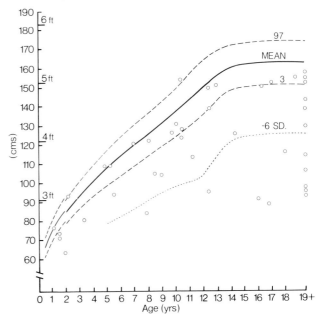

Fig. 54.6 Osteogenesis imperfecta—percentile height (females).

Whole body: Birth

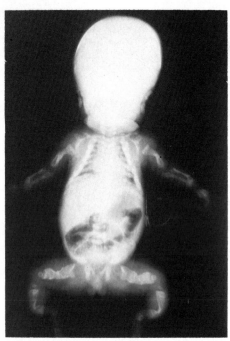

Fig. 54.7   L.M.—Neonate—Whole body
There are multiple healed or healing fractures with deformity present in all bones including the ribs, which have a beaded appearance (due to healing fractures). There is poor mineralisation of the skull vault and Wormian bones are present. The thoracic cage is narrow and bell shaped; the spine is relatively spared. This is probably the lethal (autosomal recessive) form of osteogenesis imperfecta.

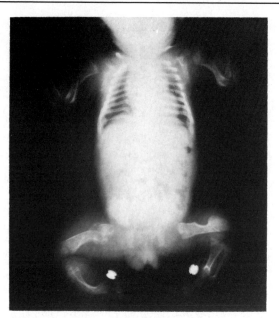

Fig. 54.8   T.K.—Neonate—Whole body
There is generalised demineralisation, and there are multiple fractures, one (recent) through the right humeral shaft, and one (old) through the left; both femora and leg bones are involved, with reduction in length and deformity. The pelvis, spine and thoracic cage are nearly normal, and this is not a lethal form of osteogenesis imperfecta (compare Fig. 54.7).

Skull: 3 months–5 years

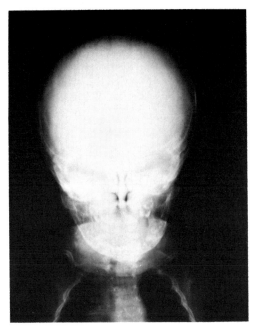

Fig. 54.9 D.S.—3 months—AP skull
Poorly mineralised vault bones with wide sutures and multiple
Wormian bones, seen well in the superior part of the sagittal suture.

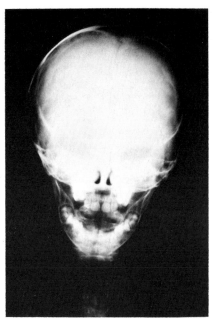

Fig. 54.10 C.L.—1 year 6 months—AP skull
There is a right parietal fracture. The anterior fontanelle is still
widely open.

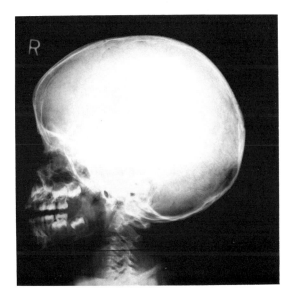

Fig. 54.11 A.M.—2 years 8 months—Lateral skull
The Wormian bones in the lambdoid suture line are well shown.
There is mild frontal bossing and general undermineralisation of the
vault.

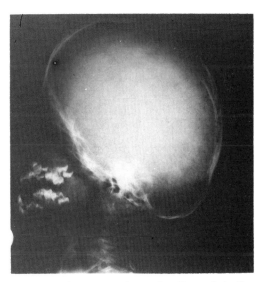

Fig. 54.12 R.McA.—5 years 6 months—Lateral skull
Gross undermineralisation accompanied by 'squashing' of the vault.
Numerous Wormian bones are present.

Skull: 5 years–Adult

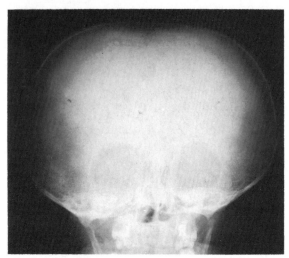

Fig. 54.13   J.W.—5 years—AP skull
Marked lateral bulging of the parietal regions ('tam-o-shanter' appearance), and basilar invagination (seen on the lateral view), associated with the severe undermineralisation. Wormian bones are seen in the suture lines.
Same patient: Fig. 54.14

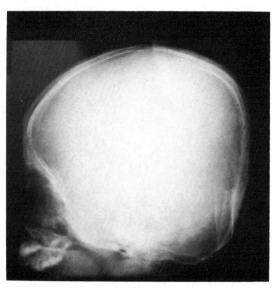

Fig. 54.14   J.W.—5 years—Lateral skull (same patient as in Fig. 54.13)
See Figure 54.13.

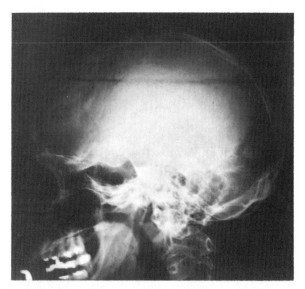

Fig. 54.15   J.G.—28 years—Lateral skull
Wormian bones are seen in the region of the lambdoid suture. Marked basilar invagination is present, causing neurological problems. (The odontoid peg is well above a line drawn between the hard palate and occiput.)
Same patient: Fig. 54.48

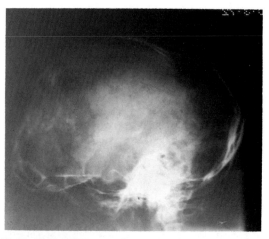

Fig. 54.16   R.M.—50 years—Lateral skull
Patchy undermineralisation of the skull vault, which is deformed, but there is no significant basilar invagination. The patient wears a hearing aid—indicating he is one of the many osteogenesis imperfecta patients developing deafness in early adult life.

Spine : 2 months–5 years

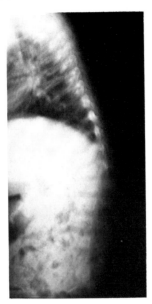

Fig. 54.17   M.V.—2 months—Lateral thoracic and lumbar spine
Mild upper lumbar kyphos. Flattening of several vertebral bodies particularly in the lower dorsal and upper lumbar regions with anterior wedging, associated with osteoporosis.

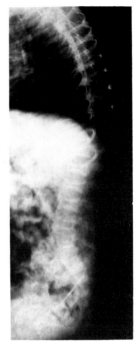

Fig. 54.18   T.M.—16 months—Lateral thoracic and lumbar spine
Marked osteoporosis and generalised biconcave flattening of the vertebral bodies. The height of the vertebral bodies is clearly less than that of the intervening disc spaces.

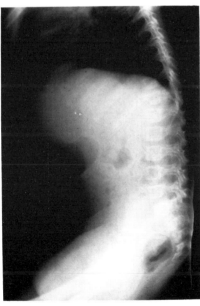

Fig. 54.19   G.D.—17 months—Lateral spine
Generalised osteoporosis, but not as serious as Figure 54.18. There is anterior wedging of T12.

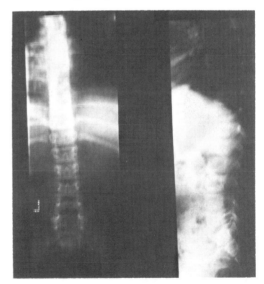

Fig. 54.20   Z.C.—5 years 6 months—AP and lateral dorsal and lumbar spine
There is slight scoliosis, but no apparent osteoporosis.
Same patient : Fig. 54.58

Spine: 6 years–Adult

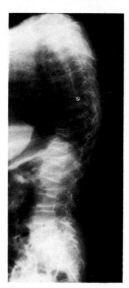

Fig. 54.21   S.B.—6 years—Lateral thoracic and lumbar spine
Marked osteoporosis with biconcave vertebrae and anterior wedging
most obvious at L1.
Same patient: Figs 54.25, 54.59

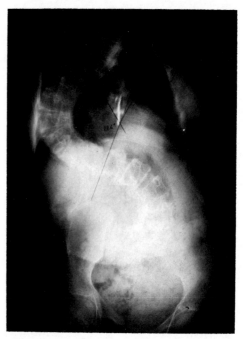

**Fig. 54.22   W.S.—11 years 7 months—AP thoracic and
lumbar spine**
Severe scoliosis and generalised osteoporosis, although there is only
slight loss of height of the vertebral bodies.
Same patient: Figs 54.37, 54.47

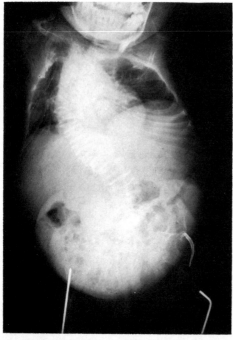

**Fig. 54.23   K.B.—13 years 6 months—AP thoracic and
lumbar spine and pelvis**
Severe osteoporosis and scoliosis. There are multiple rib deformities
associated with the scoliosis, and secondary to the fractures. The
pelvis is severely deformed, with protrusio acetabuli.
Same patient: Figs 54.44, 54.63

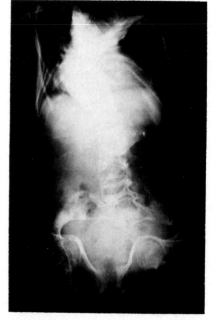

Fig. 54.24   V.H.—29 years—AP spine and pelvis
All bones are porotic and there are gross deformities of the rib cage.
The pelvis is triradiate, with pronounced protrusio acetabuli. No
recent fractures are seen.

Chest: 11 months–Adult

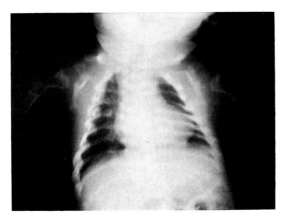

Fig. 54.25 S.B.—11 months—AP chest
Ribs slender and deformed at sites of multiple fractures, and there is expansion of the anterior ends. The thoracic cage is small in relation to the size of the heart. Both humeri show healing fractures of their mid-shafts.
Same patient: Figs 54.21, 54.59

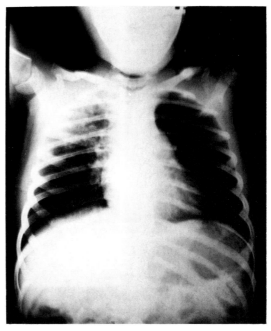

Fig. 54.26 P.S.—7 years 5 months—AP chest
Mild generalised osteoporosis, with slender ribs, an old fracture of the left clavicle and a more recent one of the right. (There is patchy consolidation in the upper zone of the right lung.)

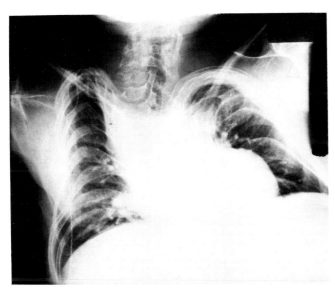

Fig. 54.27 K.M.—24 years—PA chest
Gross deformity of the thoracic cage, with bent, slender and porotic ribs. There is a kyphoscoliotic deformity but no evidence of recent fracture. (A hearing aid cord overlies the right lung field, illustrating the early onset of deafness in some osteogenesis imperfecta patients.)

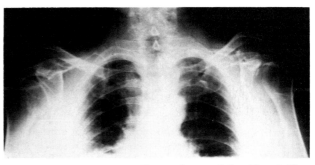

Fig. 54.28 G.L.—65 years—Upper chest and shoulder girdle
There is upward bowing of both clavicles, which appear rather long and irregularity of several ribs, indicating previous fractures. Osteoporosis present.

Pelvis and lower limb : Birth–1 year

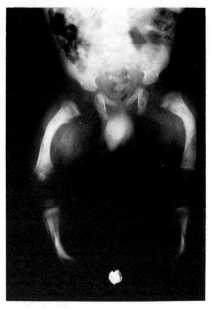

Fig. 54.29   M.L.—Neonate—Pelvis and lower limbs
There is lateral bowing of the mid-shafts of femora, tibiae and
fibulae, and medially along the concavities, there is cortical
thickening. Ossification is poor in the region of the metaphyses. Bone
maturation at knees is normal. The long bones in general have a
rather squat thick appearance.

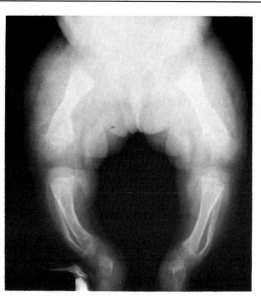

Fig. 54.30   P.O'F.—Neonate—Pelvis and lower limbs
There is generalised osteoporosis, and the long bones are broad with
thinned cortices (irregular at sites of previous fractures). There is loss
of normal modelling, particularly of the femoral necks.

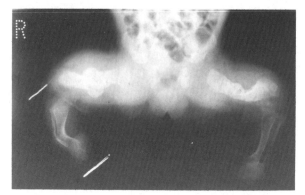

Fig. 54.31   P.M.—2 weeks—Pelvis and lower limbs
There are gross deformities of all long bones with healing fractures.

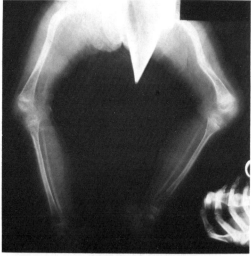

Fig. 54.32   V.J.—1 year 6 months—Pelvis and lower limbs
The long bones are slender and porotic with deformities due to
previous fractures.

Pelvis : 1–9 years

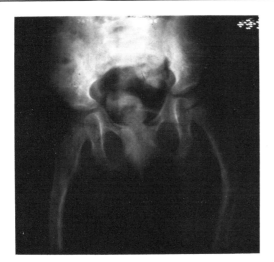

Fig. 54.33   M.S.—1 year 9 months—Pelvis
There is mild bone demineralisation but no deformity of the pelvis.

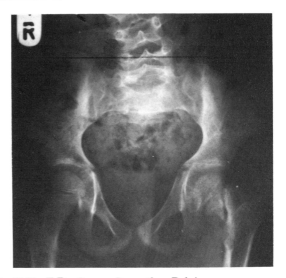

Fig. 54.34   T.B.—3 years 2 months—Pelvis
There is only slight demineralisation. The pelvic ring is asymmetrical, with protrusio acetabuli more obvious on the left side. *Same patient: Fig. 54.45*

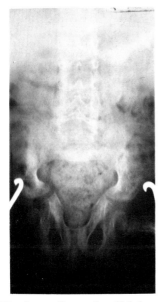

Fig. 54.35   R.M.—4 years 8 months—Pelvis
The bones are demineralised and the vertebral bodies flattened. The trabecular pattern in the pelvis is coarse and the pelvis is triradiate, with bilateral protrusio acetabuli.

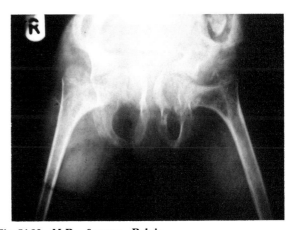

Fig. 54.36   M.R.—9 years—Pelvis
There is severe osteoporosis with severe triradiate deformity of the pelvis and bilateral protrusio acetabuli. Marked coxa vara is also present.

Pelvis : 14 years–Adult

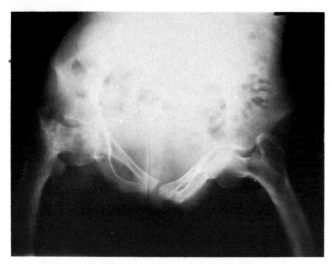

Fig. 54.37   W.S.—14 years 6 months—Pelvis
There is pelvic tilt with pronounced lumbar lordosis and bilateral coxa vara.
Same patient : Figs 54.22, 54.47

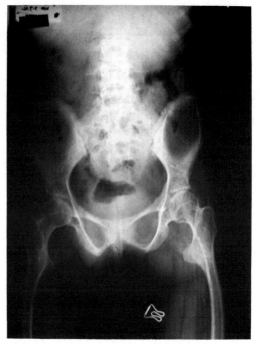

Fig. 54.38   M.T.—16 years 4 months—Pelvis
Generalised osteoporosis, with slender bones and thin cortices. No recent fractures, and no protrusio acetabuli.

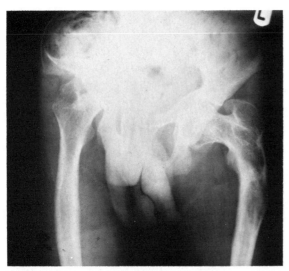

Fig. 54.39   N.T.—17 years—Pelvis
The bones are porotic and there is marked protrusio acetabuli particularly on the right, with coxa vara on the left.

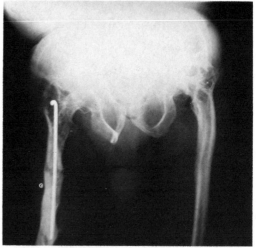

Fig. 54.40   J.S.—38 years—Pelvis
Severe pelvic deformity with compression and protrusio acetabuli.

Lower limb: 1–11 years

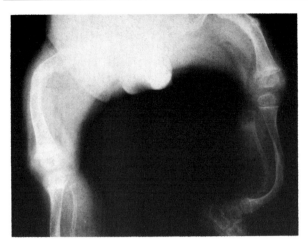

Fig. 54.41   M.H.—1–2 years—AP both femora
The bones are demineralised with multiple bowing deformities and healing fractures in both femoral shafts.

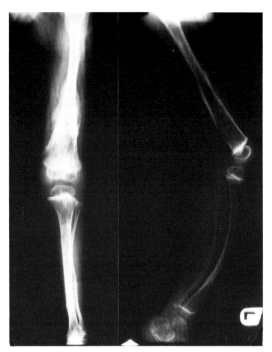

Fig. 54.42   S.R.—3 years 10 months—AP and lateral left lower limb
There is little deformity, but callus formation in the femur extends more than half way up the shaft, although no fracture line can be seen.
Same patient: Figs 54.51, 54.60, 54.65

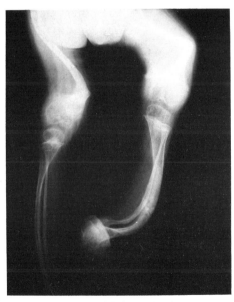

Fig. 54.43   R.B.—11 years—AP lower limbs
There are marked deformities in the left tibia and fibula, and both femora. The bones are slender and porotic, with metaphyseal flaring and there are recent fractures at the sites of maximum angulation.

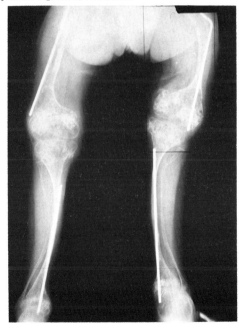

Fig. 54.44   K.B.—11 years 9 months—Both lower limbs
There is marked demineralisation, and the diaphyses are extremely slender with bowing. There is irregular ossification adjacent to the growth plates suggesting 'cystic' areas with coarse trabeculae between ('popcorn' appearance). This radiological sign was not present at an earlier age.
Same patient: Figs 54.23, 54.63

Leg : Birth–15 years

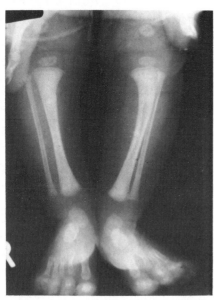

Fig. 54.45   T.B.—Neonate—AP both tibiae and fibulae
The long bones appear reasonably well-mineralised but there is an
undisplaced fracture of the right fibula.
Same patient : Fig. 54.34

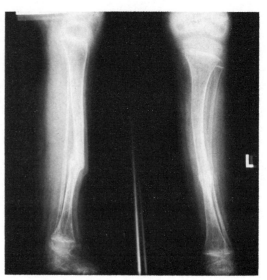

Fig. 54.46   C.C.—4 years—AP and lateral left tibia and fibula
The bones are demineralised. There is mild medial bowing of the
proximal tibial shaft and a healing fracture lower down.

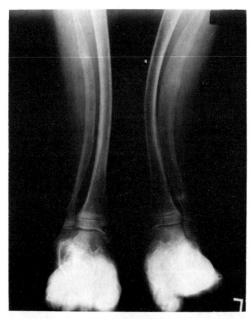

**Fig. 54.47   W.S.—9 years 6 months—AP both tibiae and
fibulae**
There is generalised osteoporosis with medial bowing of both tibial
and fibular shafts and healing fractures in the middle and distal thirds
of the left fibula. Periosteal new bone is present along the concavity of
the bowed shaft.
Same patient : Figs 54.22, 54.37

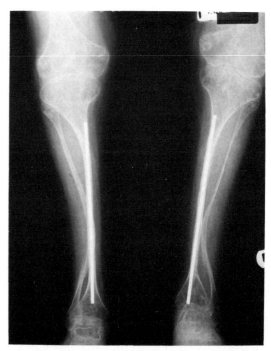

**Fig. 54.48   J.G.—15 years 10 months—AP both tibiae and
fibulae**
The metaphyses appear broad in relation to the extremely slender
diaphyses. Intramedullary nailing has prevented any more serious
deformity.
Same patient : Fig. 54.15

Some lower limb complications (largely preventable except for Fig. 54.52)

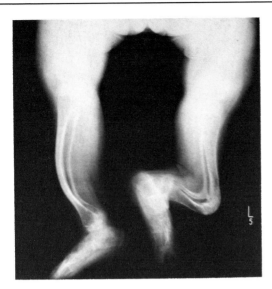

Fig. 54.49   G.R.—8 years 10 months—Both tibiae and fibulae
There is gross anterior bowing of the left tibia and fibula, which has a fracture line at the site of maximum angulation.

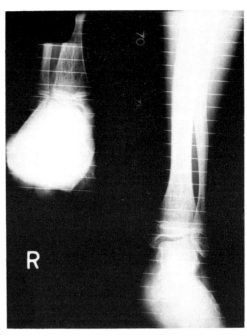

Fig. 54.50   T.D.—15 years—Ankle and foot scanogram
There is generalised bone demineralisation and severe shortening on the right, due to recurrent fractures this side.

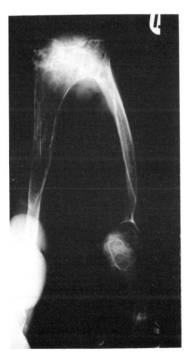

Fig. 54.51   S.R.—15 years 1 month—Left lower limb
This child with severe osteogenesis imperfecta had virtually no treatment all his life and died at the age of 18 years. The radiograph shows gross contracture and ankylosis of the knee joint.
Same patient: Figs 54.42, 54.60, 54.65

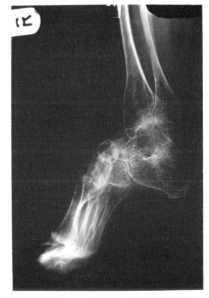

Fig. 54.52   D.W.—19 years—Lateral right ankle and foot
There is generalised osteoporosis but the pes cavus is a neurological deformity associated with basilar impression and partial paraplegia.

Upper limb: Birth–9 months

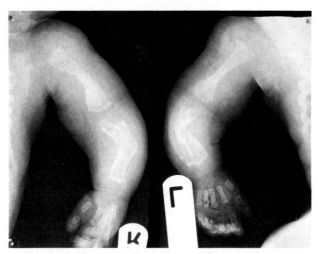

Fig. 54.53   D.B.—Neonate—Upper limbs
The long bones are short and broad with recent fractures of the right humeral shaft and distal metaphysis. Angulation of radius and ulna suggests an intra-uterine fracture.

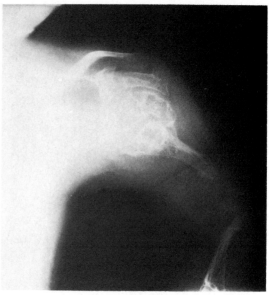

Fig. 54.54   J.R.-B.—Neonate—Left upper limb
There is total disorganisation in the region of the elbow. The humerus is shortened and the radius and ulna can barely be identified as separate bones. There is marked expansion of the proximal humeral shaft and head with a cystic (or 'popcorn') appearance of lytic areas. The shoulder joint cannot be identified.
Same patient: Fig. 54.62

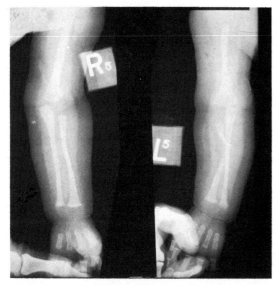

Fig. 54.55   A.B.—6 months—Upper limbs
The bones are slender and there are fractures of both radii.

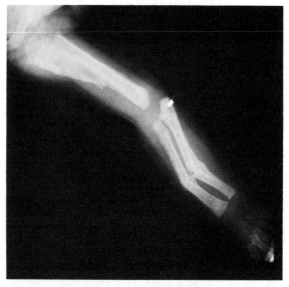

Fig. 54.56   A.D.—9 months—Upper limb
There is a healing fracture of the upper humeral shaft with excess callus and a periosteal reaction extending to the elbow. Mid-shaft fractures of the radius and ulna are present but the bones at this stage do not appear unduly porotic.

Forearm and hand : 4–10 years

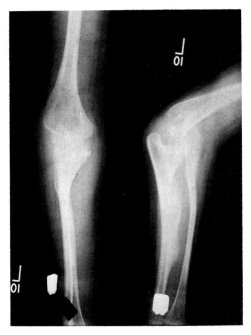

Fig. 54.57   O.C.—4 years 2 months—Left forearm
There is anterior dislocation of the elbow. Ossification is present within the interosseous ligament.

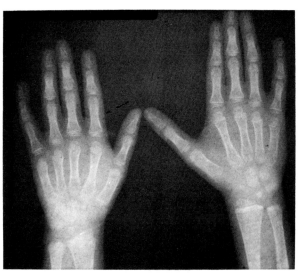

Fig. 54.58—Z.C.—5 years 6 months—Both hands
The bones are demineralised but no gross deformity is present and there is no evidence of past or present fractures.
Same patient : Fig. 54.20

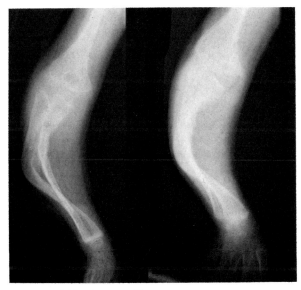

Fig. 54.59   S.B.—6 years—Forearm
There are bowing deformities of radius and ulna with cortical irregularities of the proximal ulna, which appear to represent recent greenstick fractures, and deformity of the radial head.
Same patient : Figs 54.21, 54.25

Fig. 54.60   S.R.—10 years 3 months—Left forearm
There is cross union between the radius and ulna with pronounced bowing. There appears to be ankylosis at the elbow and wrist, and the carpal ossific centres cannot be separately identified. The overall size of the carpus is reduced, and there is general osteoporosis.
Same patient : Figs 54.42, 54.51, 54.65

Upper limb: 12–19 years

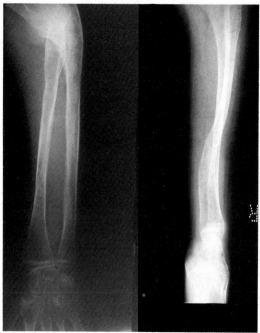

Fig. 54.61   H.S.—12 years 1 month—AP and lateral right forearm
The bones are osteoporotic and the trabecular pattern coarse. There are cortical irregularities of the proximal radius due to healing fractures, and anterior bowing of the bone.

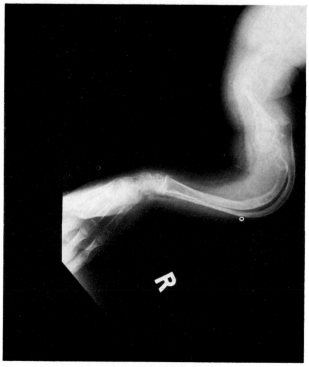

Fig. 54.62   J.R.—B.—13 years—Right arm
Gross bowing deformity is present and the humerus is short and angulated with a mid-shaft pseudarthrosis. The elbow joint is totally disorganised, and all bones are osteoporotic.
Same patient: Fig. 54.54

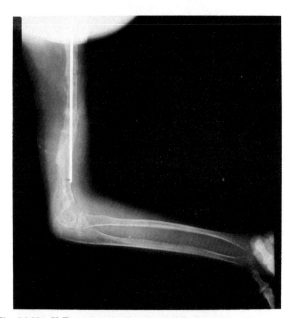

Fig. 54.63   K.B.—15 years 5 months—Left arm
All bones are porotic and thin. The humerus shows multiple osteotomies and an intramedullary nail ('Shish-kebab' operation). The forearm shows no evidence of fractures, but there is marked soft tissue wasting.
Same patient: Figs 54.23, 54.44

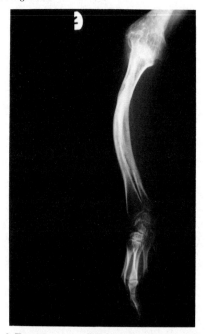

Fig. 54.64   A.B.—19 years—Right forearm
There is posterior bowing of the radius and ulna and the radial head is dislocated. Soft tissue wasting and generalised osteoporosis are present and the distal end of the ulna is rather short.
Same patient: Fig. 54.68

Hyperplastic callus

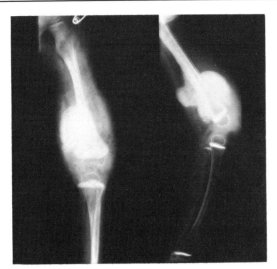

Fig. 54.65   S.R.—2 years—AP and lateral femur
There is generalised osteoporosis. The tibia and fibula are slender and
the cortices thinned. The fracture through the distal shaft of the
femur is healing with hyperplastic callus.
Same patient: Figs 54.42, 54.51, 54.60

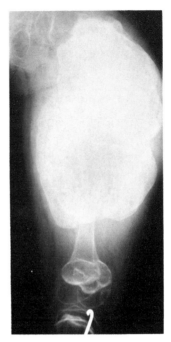

Fig. 54.66   J.M.—9 years 2 months—AP left femur
In this older child there is a gross amount of hyperplastic callus
around the shaft.
Same patient: Fig. 54.67

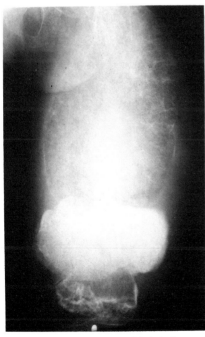

Fig. 54.67   J.M.—9 years 4 months—AP left femur (same
patient as in Fig. 54.66)
2 months later—the previously dense callus around the shaft is now
radiolucent, indicating fatty marrow infiltration. Meanwhile, there
has been a further metaphyseal fracture, with dense callus formation.

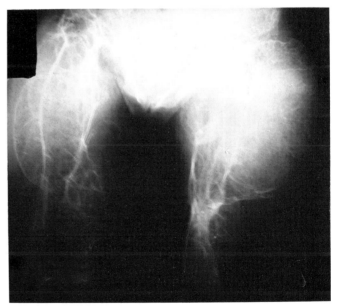

Fig. 54.68   A.B.—26 years—AP both femora
There are grotesque bony outgrowths of the femoral shafts,
representing hyperplastic callus formation around previous fractures.
The pelvis is triangular with protrusio acetabuli.
Same patient: Fig. 54.64

Malignant change in hyperplastic callus

Figs 54.69–54.72   A.E.—7–8 years

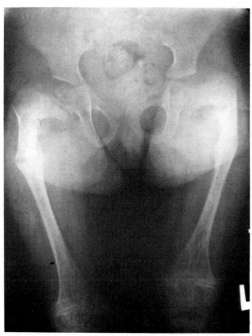

Fig. 54.69 The fractured upper femoral shaft is unremarkable for osteogenesis imperfecta.

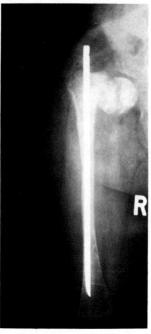

Fig. 54.70 It was treated with a Küntscher nail, but there is now an additional fracture through the femoral neck.

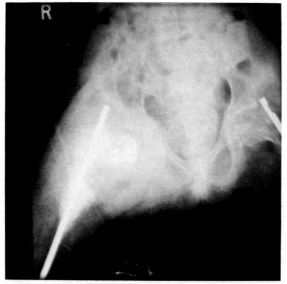

Fig. 54.71 4 months later, there is extensive patchy sclerosis involving the upper femoral shaft, neck and head, with destruction of the cortex and bony spicules extending into the surrounding soft tissue mass. The changes are now clearly those of an osteosarcoma.

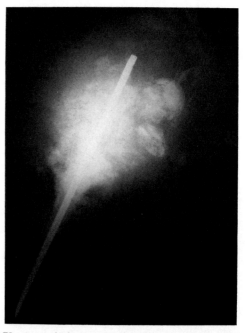

Fig. 54.72 4 months later again, there are now lytic as well as sclerotic areas involving the proximal shaft, head and neck. There is expansion of the shaft with destruction of the cortex, and new bone extending into the surrounding soft tissues. The patient died shortly afterwards.

Chapter Fifty-five
IDIOPATHIC JUVENILE OSTEOPOROSIS

The acute onset of osteoporosis in childhood, with spontaneous remission in 2–5 years.

Inheritance

Non-genetic.

Frequency

Extremely rare, possible prevalence of about 1 per million.

Clinical features

Facial appearance—normal.
Intelligence—normal.
Stature—normal initially, but osteoporosis of the vertebrae leads to platyspondyly and trunk shortening.
Presenting feature/age—pathological fractures of limbs or vertebrae, usually a year or two before puberty.
Deformities—of fractures, or bending due to softening of bone.
Associated anomalies—none.

Radiographic features

There is generalised osteoporosis with flattened or biconcave vertebrae and thin cortices. Fractures are typically in the metaphyseal area of long bones.

Bone maturation

Probably normal, apart from osteoporosis.

Biochemistry

No specific findings.

Differential diagnosis

From other osteoporoses of childhood, principally *osteogenesis imperfecta of late onset*, but this is usually associated with blue sclerae, Wormian bones and often a family history of the condition.

Non-accidental injury usually occurs at an earlier age, and osteoporosis is not a feature.

REFERENCES

Dent C E, Friedman M 1965 Idiopathic juvenile osteoporosis. Quarterly Journal of Medicine 34:177–210
Jowsey J, Johnson K A 1972 Juvenile osteoporosis. Bone findings in seven patients. Journal of Pediatrics 81:511–517
Smith R 1979 Biochemical disorders of the skeleton. Butterworth, London

Spine and lower limb : 12–14 years

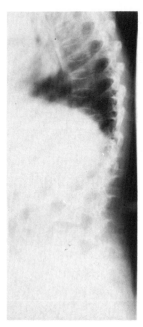

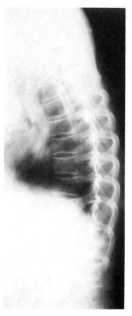

Fig. 55.1   S.R.—12 years—Lateral spine
Biconcave flattening of the vertebrae, typical of osteoporosis.
Same patient : Figs 55.2, 55.5, 55.7

Fig. 55.2   S.R.—13 years—Lateral spine (same patient as in Fig. 55.1)
One year later the vertebral bodies have regained height, but there is still diminished bone density.

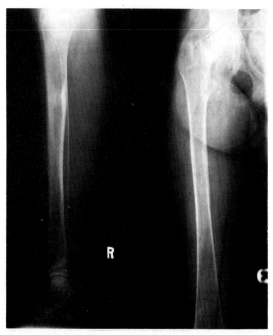

Fig. 55.3   A.M.—14 years—AP femur. Lateral tibia
The bones are osteoporotic and there are fractures of the femoral neck and tibial shaft (healing)
Same patient : Figs 55.6, 55.8

Knee and ankle: 13–14 years

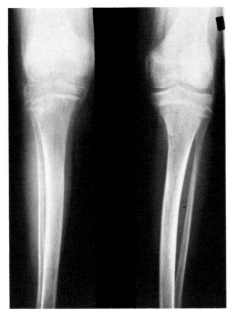

Fig. 55.4   S.R.—13 years—AP knees
There are metaphyseal fractures of the right medial femur and right lateral tibia. The left tibia is bowed medially. The cortices are thin.
Same patient: Figs 55.1, 55.2, 55.6

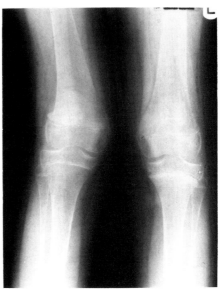

Fig. 55.5   A.M.—14 years—AP knees
There is generalised osteoporosis and a metaphyseal fracture of the right lower femur.
Same patient: Figs 55.3, 55.8

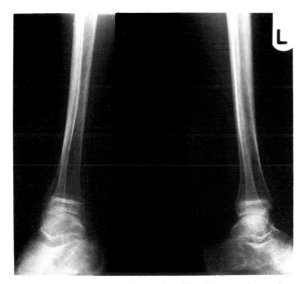

Fig. 55.6   S.R.—13 years—Lateral ankles (same patient as in Fig. 55.4)
Marked osteoporosis, but no fractures are seen.

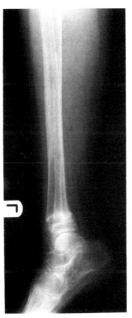

Fig. 55.7   A.M.—14 years—Lateral left ankle (same patient as in Fig. 55.5)
There is a fracture through the distal tibial metaphysis.

Chapter Fifty-six
IDIOPATHIC OSTEOLYSIS
(Hajdu–Cheney type)

Acro-osteolysis is associated with Wormian bones in the skull, multiple areas of osteoporosis and pathological fractures.

Inheritance

Autosomal dominant.

Frequency

Extremely rare.

Clinical features

Facial appearance—normal.
Intelligence—normal or retarded.
Stature—short.
Presenting feature/age—pathological fractures in childhood and pain associated with the lytic areas in the hands and feet.
Deformities—related to the fractures and softening of bone.

Radiological features

Skull—Wormian bones and delayed closure of fontanelles. Air cells are non-aerated.
Thorax—small, with slender ribs.
Long bones—thin, osteoporotic and subject to pathological fractures.
Hands/feet—acro-osteolysis of the terminal phalanges.

Bone maturation

Not known.

Differential diagnosis

From other disorders with acro-osteolysis, but the generalised osteoporosis and Wormian bones should distinguish the Hajdu–Cheney syndrome.

Osteogenesis imperfecta and *idiopathic juvenile osteoporosis* are not associated with acro-osteolysis.

Cleido-cranial dysostosis is not associated with osteoporosis.

Progress/complications

The rate of progression is variable. Little is known about the natural history, but complications are related to the multiple fractures and softening of bone.

REFERENCES

Elias A N, Pinals R S, Anderson H C 1978 Hereditary osteodysplasia with acro-osteolysis (the Hajdu–Cheney syndrome). American Journal of Medicine 65/4:627–636
Weleber R G, Beals R K 1976 The Hajdu Cheney syndrome. Report of two cases and review of the literature. Journal of Pediatrics 88/2:243–249

Skull and trunk: Birth–5 years

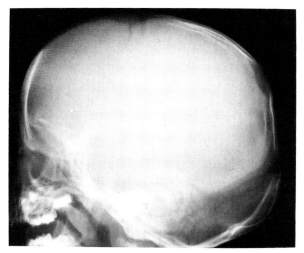

Fig. 56.1 C.L.—2 years 4 months—Lateral skull
The sutures are wide with Wormian bones in the lambdoid suture and delayed closure of the anterior fontanelle. The pituitary fossa is enlarged. The sinuses and mastoid air cells are underpneumatised.
Same patient: Figs 56.2–56.7

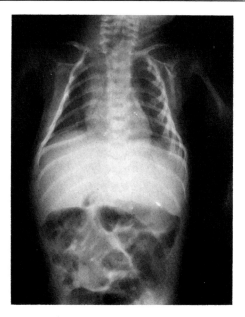

Fig. 56.2 C.L.—Neonate—AP trunk (same patient as in Fig. 56.1)
The thoracic cage is small. Varus deformities of the shoulder are present, due to fractures and there is generalised osteoporosis.

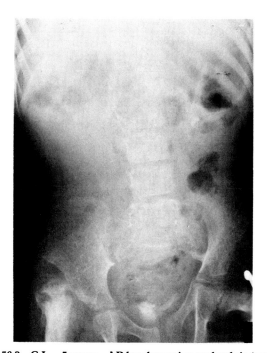

Fig. 56.3 C.L.—5 years—AP lumbar spine and pelvis (same patient as in Fig. 56.1)
There is widening of the interpedicular distances with flattening of the pedicles. There is a fracture-dislocation of the right hip with new bone formation.

Limbs : Birth–6 years

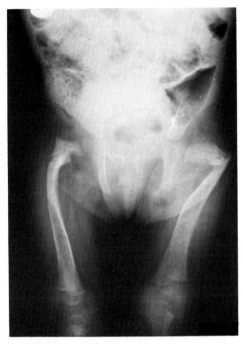

Fig. 56.4   C.L.—2 years—Pelvis and femora
There is generalised osteoporosis and thinning of both the pelvis and femora with multiple fractures and deformity.
Same patient : Figs 56.1–56.3, 56.5–56.7

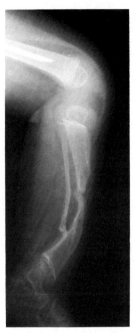

Fig. 56.5   C.L.—4 years—Lateral tibia and fibula (same patient as in Fig. 56.4)
There is marked osteoporosis and the tibia and fibula are attenuated with several fractures.

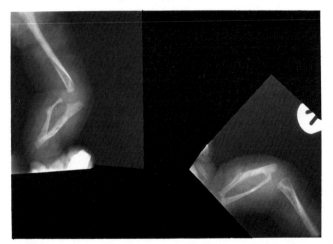

Fig. 56.6   C.L.—Neonate—Upper limbs (same patient as in Fig. 56.4)
All bones are thin and osteoporotic : the radius and ulna particularly so, with deformity and dislocation at the elbow.

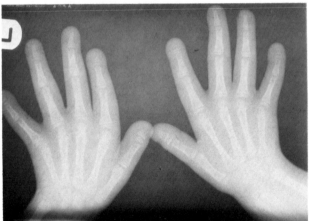

Fig. 56.7   C.L.—6 years—PA hands (same patient as in Fig. 56.4)
The left wrist is disorganised and there is acro-osteolysis of several terminal phalanges. All the hand bones are slender and osteoporotic.

Section Thirteen
SCLEROSING BONE DYSPLASIAS

Chapter Fifty-seven
OSTEOPETROSIS

Generalised osteosclerosis, bone fragility and failure of modelling; in severe cases, encroachment on marrow cavities with anaemia and later pancytopenia. There may be exacerbations and remissions of the osteosclerosis.

Inheritance

Severe congenital form—autosomal recessive.
Milder tarda form—autosomal dominant.

Frequency

Possible prevalence of 3 per million, but probably more, since milder cases will be unidentified.

Clinical features

Facial appearance—normal, although the severe form may be associated with hydrocephalus.
Intelligence—normal.
Stature—reduced in the severe form. Normal in the milder type.

Presenting feature/age—Some severe cases are stillborn.

Congenital type—failure to thrive, short stature, pathological fractures, progressive anaemia, compensatory enlargement of the liver and spleen, delayed dentition with small teeth.

Tarda type—may only be discovered incidentally in adult life on radiography. Others have pathological fractures.

Deformities—none, other than problems from fractures.
Associated anomalies—none.

Radiographic features

Skull—thickened vault and base, obstructed foramina; sometimes hydrocephalus, perhaps associated with reduction in size of the foramen magnum.

Spine—usually 'sandwich' appearance, with areas of increased density along superior and inferior margins of vertebrae on the lateral view. They may be of uniform density.
Thorax—ribs and clavicles are thick, undermodelled and sclerotic.
Long bones—expanded metaphyses ('club-shaped'), osteosclerosis, without distinction between the cortical and cancellous bone. May be a bone-within-a-bone appearance of the small bones of hands and feet, resulting from fluctuating activity in the disease process.
Pelvis—may have arcuate bands.

Bone maturation

Probably normal.

Biochemistry

Not known. Plasma calcium and phosphate are normal or inconsistently altered.

Differential diagnosis

From other dysplasias with increased bone density. In *craniometaphyseal* and *craniodiaphysial* disorders the vertebrae are not affected. In *dysosteosclerosis* there is platyspondyly, also the long bones show a clear area between the metaphysis and diaphysis. In *frontometaphyseal dysplasia* there is massive thickening of the frontal bones. *Pycnodysostosis* is differentiated by the appearance of the skull, with its open fontanelles, and the hands, with their dysplastic or absent terminal phalanges, and there is often hypoplasia of the clavicles.

Other disorders of increased bone density are diagnosed on general clinical and laboratory investigations (renal osteodystrophy, heavy metal poisoning, myelofibrosis, neoplasms, syphilis).

Progress/complications

In the severe form there is death in childhood from pancytopenia, recurrent infections, including osteomyelitis, and cranial nerve palsies. However, modern treatment with transplanted marrow cells in infancy can arrest the progress of the disease. An illustration of this is shown at the end of this chapter.

Mild cases may have no disability.

REFERENCES

Beighton P, Cremin B J 1980 Sclerosing bone disease. Berlin, Springer

Graham C B, Rudhe U, Eklof O 1973 Osteopetrosis. In: Kaufmann H J (ed) Intrinsic diseases of bones. Progress in Pediatric Radiology 4. S Karger, Basle, p 375

Loria-Cortes R, Quesada-Calva E, Cordeno-Chaverri C 1977 Osteopetrosis in children: a report of 26 cases. Journal of Pediatrics 91:43

Sief C A, Chessells J M, Levinski R J, Pritchard J, Rodgers D W, Casey A, Muller K, Hall C M 1983 Allogenic bone marrow transplantation in infantile malignant osteopetrosis. Lancet i: 437–441

Skull: Birth–2 years

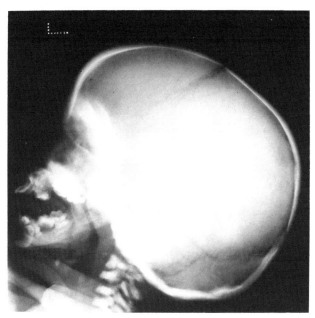

Fig. 57.1 M.W.—Neonate—Lateral skull
The vault is large with a minor degree of sclerosis. There is increased bone density of the base and in the region of the paranasal sinuses.
Same patient: Figs 57.2, 57.3, 57.8, 57.11, 57.14, 57.17, 57.23, 57.30

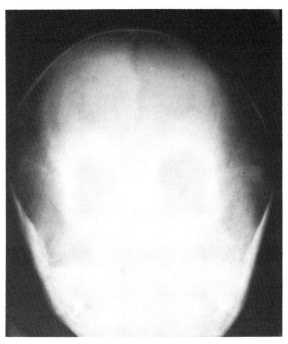

Fig. 57.2 M.W.—Neonate—AP skull (same patient as in Fig. 57.1)
There is marked sclerosis around the orbits and of the paranasal sinuses.

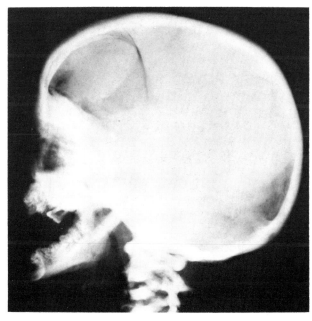

Fig. 57.3 M.W.—1 year—Lateral skull (same patient as in Fig. 57.1)
Frontal craniotomy present. There has been an increase in the thickness and sclerosis of the vault bones and base.

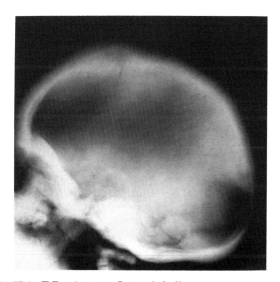

Fig. 57.4 T.D.—2 years—Lateral skull
There is generalised sclerosis with thickening of the vault over the frontal and parietal regions. There is a 'hair on end' appearance over the parietal bones and loss of differentiation into inner and outer tables, presumably a site of extramedullary haemopoeisis. The skull base is sclerotic, as are the paranasal sinuses. There is some dolichocephaly.
Same patient: Figs 57.13, 57.16, 57.22, 57.29

Skull: 2 years–Adult

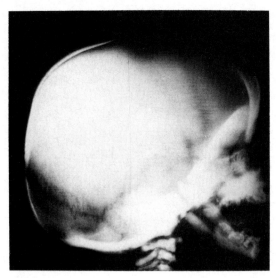

Fig. 57.5 C.B.—2 years—Lateral skull
The vault is large with wide sutures. The frontal bone is thickened and there is sclerosis of the base and in the paranasal region.
Same patient: Figs 57.6, 57.20, 57.21

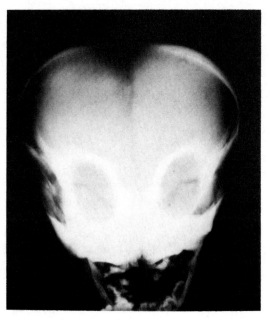

Fig. 57.6 C.B.—2 years—AP skull (same patient as in Fig. 57.5)
The anterior fontanelle is large and there are biparietal bulges. The paranasal regions and the supraorbital margins are dense.

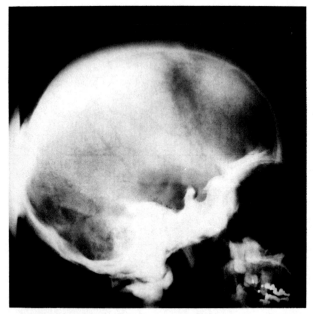

Fig. 57.7 R.P.—33 years—Lateral skull
Increased bone density of the base is present and the parietal bones are thickened. The sinuses are well aerated. (The tarda form of osteopetrosis.)
Same patient: Figs 57.12, 57.15, 57.18, 57.25, 57.31, 57.32

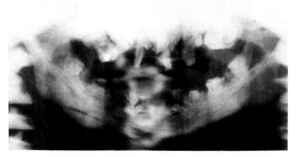

Fig. 57.8 M.W.—8 years—Orthopantomogram (a whole mandible view taken with a moving X-ray beam)
Gross abnormalities of the teeth are present.
Same patient: Figs 57.1–57.3, 57.11, 57.14, 57.17, 57.23, 57.30

Spine: 3 months–Adult

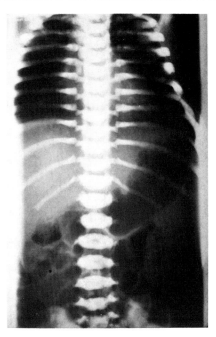

Fig. 57.9 K.M.—3 months—AP spine
Bone density is increased throughout.
Same patient: Figs 57.10, 57.28

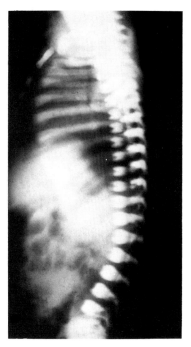

Fig. 57.10   K.M.—3 months—Lateral spine (same patient as in Fig. 57.9)
Bone density is increased. Pronounced notches are present on the anterior margins of the vertebral bodies.

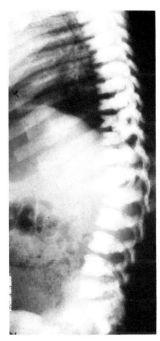

Fig. 57.11   M.W.—8 years—Lateral dorso-lumbar spine
There is increased bony sclerosis with the appearance of a 'bone-in-a-bone'; pronounced anterior notches are present in the dorsal vertebral bodies.
Same patient: Figs 57.1–57.3, 57.8, 57.14, 57.17, 57.23, 57.30

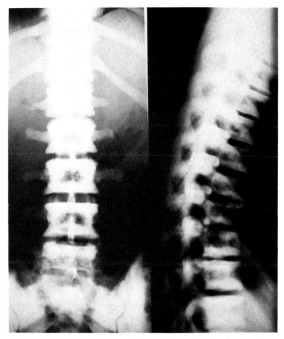

Fig. 57.12   R.P.—33 years—AP and lateral lumbar spine
There are bands of increased bone density adjacent to the vertebral end-plates. In the mid-dorsal region there is disc space narrowing with end-plate irregularity and some marginal osteophyte formation, unusual at this age.
Same patient: Figs 57.7, 57.15, 57.18, 57.25, 57.31, 57.32

Chest : 2 years–Adult

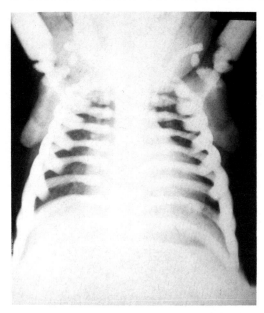

Fig. 57.13   T.D.—2 years 6 months—AP chest
There is sclerosis of all the bones with widening and abnormal
modelling of the proximal humeri and posterior ends of ribs. The
bone density in the angle of the scapulae is relatively normal.
Same patient : Figs 57.4, 57.16, 57.22, 57.29

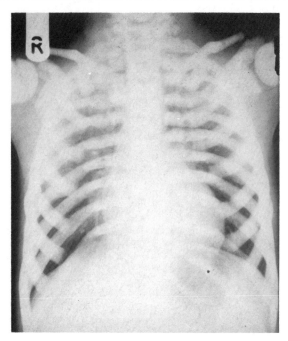

Fig. 57.14   M.W.—8 years—AP chest
Increased bone density affects the ribs and clavicles and abnormal
modelling with thickening is also present.
Same patient : Figs 57.1–57.3, 57.8, 57.11, 57.17, 57.23, 57.30

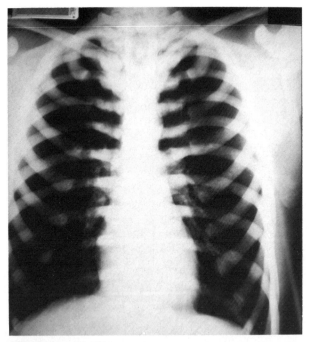

Fig. 57.15   R.P.—37 years—PA chest
There is generalised increased bone density. Bone modelling is
relatively normal.
Same patient : Figs 57.7, 57.12, 57.18, 57.25, 57.31, 57.32

Pelvis and hip: 2 years–Adult

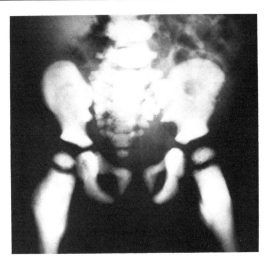

Fig. 57.16   T.D.—2 years 6 months—AP pelvis
There are crescentic bands of alternating normal and increased density in the iliac wings. The bases of the ilia are constricted and the proximal femoral metaphyses widened.
Same patient: Figs 57.4, 57.13, 57.22, 57.29

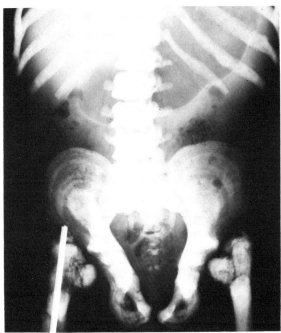

Fig. 57.17   M.W.—8 years—AP pelvis
There are concentric rings of alternating normal and increased bone density in the wings of the ilia. The pelvic ring is deformed and there are many old and recent fractures with gross coxa vara and an intra-medullary nail on the left. (It is possible some of these deformities are iatrogenic, associated with the patient's low calcium diet.)
Same patient: Figs 57.1–57.3, 57.8, 57.11, 57.14, 57.23, 57.30

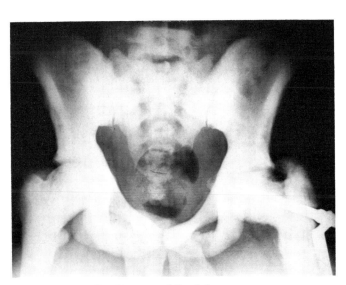

Fig. 57.18   R.P.—32 years—AP pelvis
Here there is a uniform increase in bone density with relatively normal modelling, usual in the tarda form. A recent femoral neck fracture has been treated with a nail plate.
Same patient: Figs 57.7, 57.12, 57.15, 57.25, 57.31, 57.32

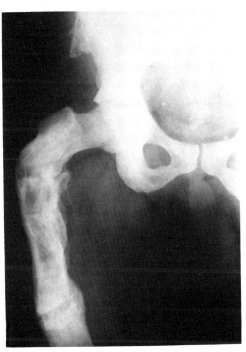

Fig. 57.19   G.G.—54 years 9 months—AP right hip and upper femur
There is considerable varus deformity due to multiple fractures of the femoral shaft. (The tarda form of osteopetrosis.)
Same patient: Figs 57.26, 57.35

Lower limb: 2–8 years

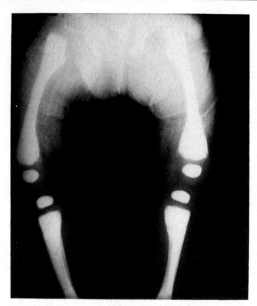

Fig. 57.20   C.B.—2 years—AP lower limbs
There is a marked uniform increase in bone density. Metaphyses are widened with loss of normal modelling. There is mild fraying of the distal femoral metaphyses, perhaps due to the known occasional association with rickets.
Same patient : Figs 57.5, 57.6, 57.21

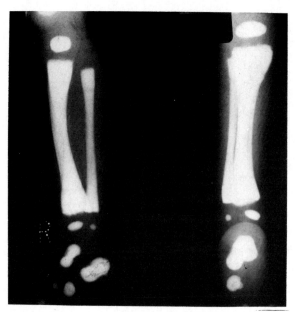

Fig. 57.21   C.B.—2 years—Right tibia and fibula (same patient as in Fig. 57.20)
Increased bone density with 'bone-in-a-bone' appearance. Loss of modelling and irregularity of metaphyses, similar to the femora in Figure 57.20.

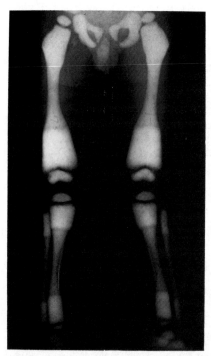

Fig. 57.22   T.D.—2 years 6 months—AP lower limbs
Here there are striking broad, dense bands at the ends of the long bones with relatively normal diaphyses. The long bones show a faint 'bone-in-a-bone' appearance.
Same patient : Figs 57.4, 57.13, 57.16, 57.29

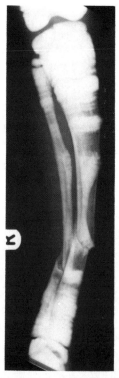

Fig. 57.23   M.W.—8 years—Lateral right tibia and fibula
There is a marked 'bone-in-a-bone' appearance. This child was treated for many years with a low calcium diet which appeared to be partially successful in giving rise to areas of normal bone density.
Same patient : Figs 57.1–57.3, 57.8, 57.11, 57.14, 57.17, 57.30

Lower limb: 11 years–Adult

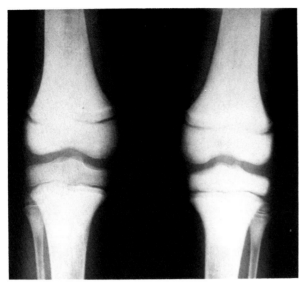

Fig. 57.24 A.J.—11 years—AP knees
There is increased bone density but modelling appears to be normal.
There are metaphyseal bands of normal bone density which could be
due to a remission with a phase of normal bone growth, or possibly to
healing rickets.

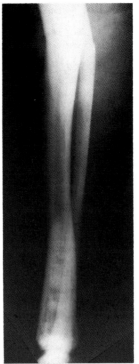

Fig. 57.25 R.P.—31 years—Right tibia and fibula
There are alternating bands of increased density with abnormal
modelling although the mid-diaphyseal shaft appears normal. There
is a fracture through the proximal diaphysis.
Same patient: Figs 57.7, 57.12, 57.15, 57.18, 57.31, 57.32

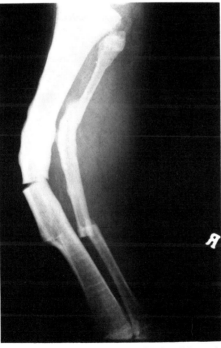

Fig. 57.26 G.G.—54 years 2 months—Right tibia and fibula
Increased bone density with a recent mid-shaft transverse fracture
and evidence of old fractures, with considerable deformity.
Same patient: Figs 57.19, 57.35

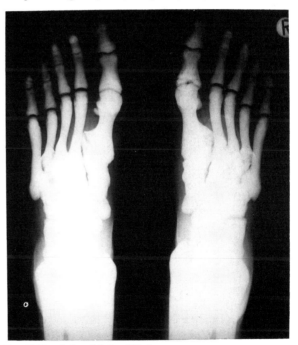

Fig. 57.27 A.S.—Adult—PA feet
There is generalised increased bone density and failure of modelling.
Healing stress fractures are present through the proximal shafts of
both 4th metatarsals. (The tarda form of osteopetrosis.)
Same patient: Figs 57.33, 57.34

Upper limb : 3 months–Adult

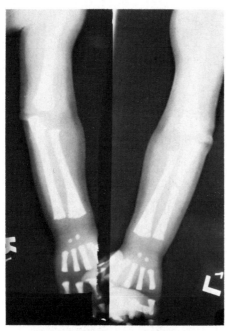

Fig. 57.28   K.M.—3 months—AP upper limbs
There is a generalised increase in bone density but with normal modelling.
Same patient : Figs 57.9, 57.10

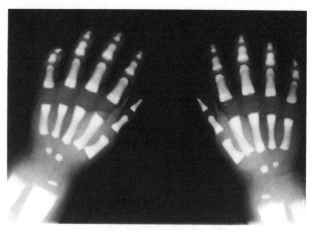

Fig. 57.29   T.D.—2 years 6 months—PA hands
There is increased bone density and the appearance of a 'bone-within-a-bone' in the phalanges and metacarpals. The distal radial and ulnar metaphyses are frayed (? associated rickets) and there is loss of normal modelling.
Same patient : Figs 57.4, 57.13, 57.16, 57.22

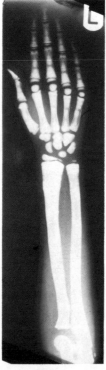

Fig. 57.30   M.W.—8 years—PA left forearm and hand
Increased bone density is present. Alternating bands are present at the distal radial and ulnar metaphyses and here there is loss of normal modelling. In the metacarpals and phalanges there is the 'bone-within-a-bone' appearance.
Same patient : Figs 57.1–57.3, 57.8, 57.11, 57.14, 57.17, 57.23

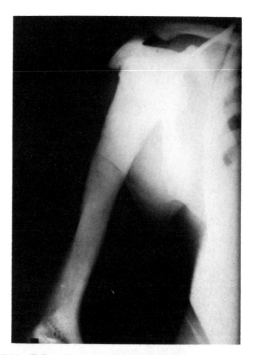

Fig. 57.31   R.P.—32 years—Right humerus
There is uniform osteosclerosis and the deltoid tuberosity is rather prominent.
Same patient : Figs 57.7, 57.12, 57.15, 57.18, 57.25, 57.32

Upper limb: Adult

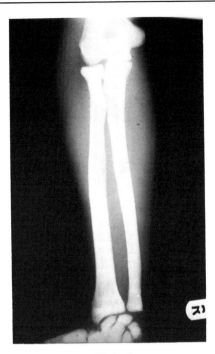

Fig. 57.32   R.P.—37 years—Right forearm
Increased bone density with narrow and irregular medullary spaces.
Same patient: Figs 57.7, 57.12, 57.15, 57.18, 57.25, 57.31

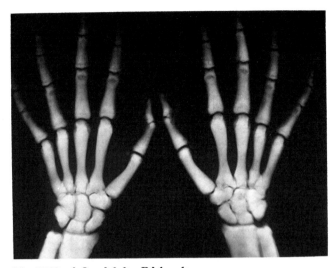

Fig. 57.33   A.S.—Adult—PA hands
There is generalised increased bone density with small areas of
normal density at the distal phalangeal metaphyses.
Same patient: Figs 57.27, 57.34

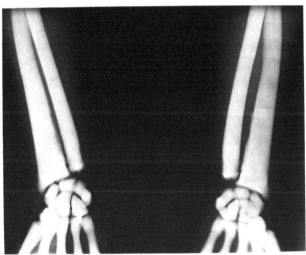

Fig. 57.34   A.S.—Adult—Forearms and wrists (same patient as
in Fig. 57.33)
There is loss of the normal diaphyseal constriction of the radius and
ulna and bone density is increased.

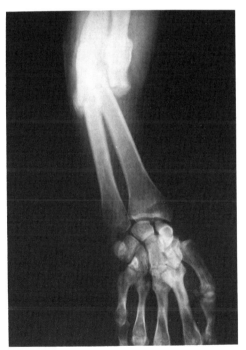

Fig. 57.35   G.G.—51 years 7 months—Forearm and wrist
There is minimal increased bone density but fractures are present in
the mid-radial and ulnar shafts and old ones in the mid-shafts of the
second, third and fourth metacarpals.
Same patient: Figs 57.19, 57.26

Case history: E.G.—Skull

Patient E.G. presented at the age of eight months with hydrocephalus, bilateral optic atrophy secondary to optic nerve compression, a left facial nerve palsy and some (questionable) deafness. He had the characteristic radiological changes of osteopetrosis.

The progressive improvement of the bone changes following a bone-marrow transplant from a compatible donor is demonstrated in Figures 57.36–57.41.

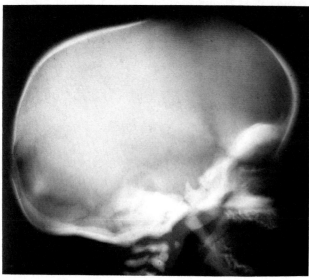

Fig. 57.36   E.G.—8 months—Lateral skull
The vault is large, with wide sutures and a large anterior fontanelle. There is a generalised increase in bone density, especially involving the skull base.

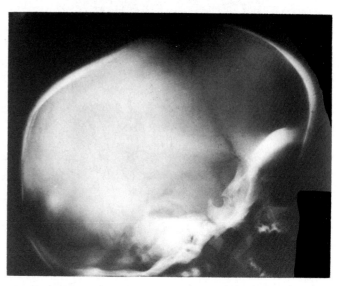

Fig. 57.37   E.G.—1 year 1 month—Lateral skull
2 months following bone-marrow transplant. The vault remains large but the markings and trabecular pattern of the vault bones is more normal. The skull base is also less sclerotic (best seen around the pituitary fossa).

Case history: E.G.—Limbs

Fig. 57.38   E.G.—8 months—AP knee and tibia and fibula
There are alternating bands of increased bone density at the poorly
modelled metaphyses. There is also a radiolucent subperiosteal line
down the diaphyses medially. The alternating sclerosis and normal
bone density is a usual feature of osteopetrosis.

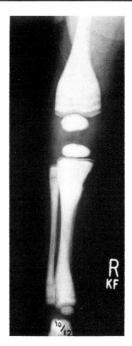

Fig. 57.39   E.G.—10 months—AP knee and tibia and fibula
1 month after bone-marrow transplant. The new bone at the
metaphyses is of normal density. There is also a wider strip of
subperiosteal bone (normal density) along the diaphyses. Bone
resorption is occurring at the medial proximal tibial metaphysis,
giving rise to more normal metaphyseal modelling.

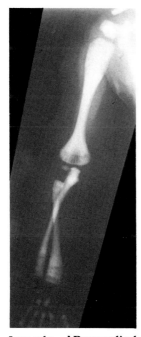

Fig. 57.40   E.G.—8 months—AP upper limb
There are alternating bands of normal and increased-density bone at
the abnormally broad metaphyses.

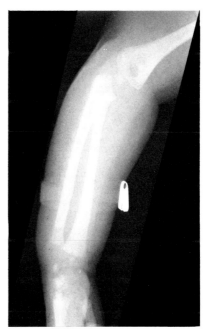

Fig. 57.41   E.G.—1 year 1 month—Forearm
4½ months after the bone-marrow transplant, the previously sclerotic
bone is more normal in density and the metaphyses are normally
modelled. A medullary cavity is now clearly visible in the diaphyses.

Case history: H.E.—Skull and trunk

Patient H.E. presented with severe infantile osteopetrosis, hepatosplenomegaly and bilateral optic nerve atrophy.

1 year 8 months later, following a bone marrow transplant, he shows bones of normal density and modelling.

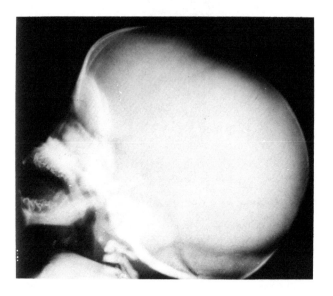

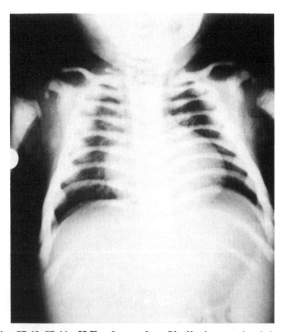

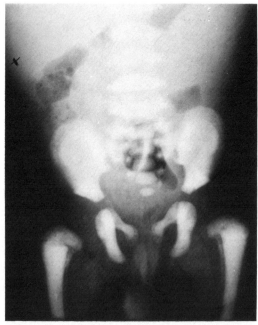

Figs 57.42–57.44   H.E.—6 months—Skull, chest and pelvis
There are characteristic changes of the severe infantile form of osteopetrosis. Healing fractures are present at the proximal femoral metaphyses. The spleen and liver are large.

Case history : H.E.—Upper limb

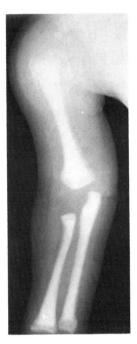

Fig. 57.45   H.E.—6 months—Upper limb
Pre-bone-marrow transplant. Osteosclerosis and a proximal humeral metaphyseal fracture with callus are present.

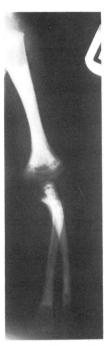

Fig. 57.46   H.E.—13 months—Upper limb
5 months after the bone-marrow transplant. Bone of normal density is present at the metaphyses. Medullary cavities are now visible.

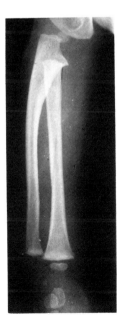

Fig. 57.47   H.E.—2 years 4 months—Upper limb
1 year 8 months post-transplant. There is now normal bone density and modelling, with good cortical and medullary differentiation.

Chapter Fifty-eight
PYCNODYSOSTOSIS

Short-limbed short stature, persistent open fontanelles, osteosclerosis with increased bone fragility, dysplastic facial bones and mandible, hypoplastic clavicles and dysplastic terminal phalanges.

Inheritance

Autosomal recessive.

Frequency

Possible prevalence of under 1 per million.

Clinical features

Facial appearance—bulging frontal and occipital regions, small face, micrognathos with premature or delayed eruption of teeth.

Intelligence—normal.

Stature—reduced, with some disproportionate limb shortening.

Presenting feature/age—infancy—pathological fractures, short stature. Mild cases may only be discovered incidentally in adult life.

Deformities—related to the pathological fractures.

Associated anomalies—none.

Radiographic features

Skull—persistent open fontanelles, widened suture lines, Wormian bones. Small maxillae and the angle of the mandible more obtuse than normal.

Spine—increased density of vertebrae (rarely 'sandwich' appearance of osteopetrosis). May be failure of segmentation at C1/C2 or L5/S1. Spondylolisthesis possible.

Thorax—hypoplastic clavicles, sometimes absent. Osteosclerotic and undermodelled ribs.

Long bones—osteosclerosis and some under-modelling throughout.

Hands—terminal phalanges ragged or absent. This may occur as an active process in adult life.

Bone maturation

Probably normal.

Biochemistry

Not known.

Differential diagnosis

From *osteopetrosis* and other disorders of increased bone density. The appearance of the skull, clavicle and hands typifies pycnodysostosis. Also, the osteosclerosis is uniform or patchy in pycnodysostosis, not striped, as it may be in osteopetrosis.

Cranio-cleido-dysostosis also has Wormian bones and absent clavicles, but in this disorder there is no increased bone density or fragility.

In *renal osteodystrophy* there may be osteosclerosis with a 'rugger jersey' appearance of the spine, but the other features of osteomalacia and hyperparathyroidism differentiate it.

Progress/complications

Pathological fractures may or may not occur. There is no sclerosis of the base of the skull, hence cranial nerve palsies do not develop. Osteomyelitis, particularly of the mandible, may occur with problems at the temporo-mandibular joint.

REFERENCES

Beighton P, Cremin B J 1980 Sclerosing bone dysplasias. Berlin, Springer

Elmore S M 1967 Pycnodysostosis: review. Journal of Bone and Joint Surgery 49A:153

Maroteaux P, Fauré C 1973 Pycnodysostosis. In: Kaufmann H J (ed) Intrinsic diseases of bones. Progress in Pediatric Radiology 4: Basel, S Karger, pp 403–413

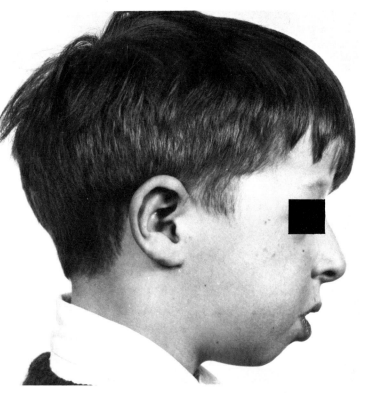

Fig. 58.1 Normal appearance apart from micrognathos.

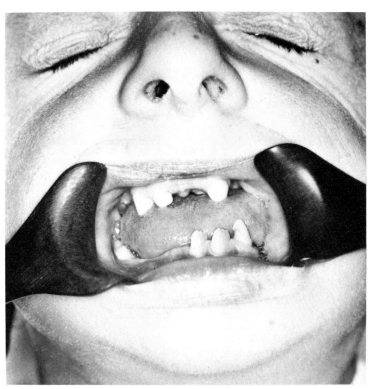

Fig. 58.2 Disordered eruption of teeth.

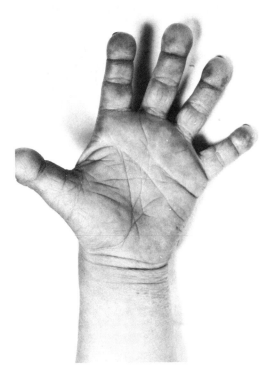

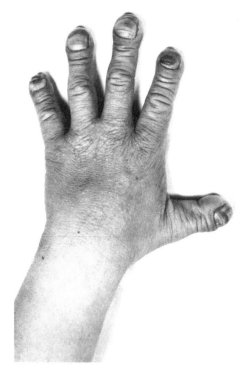

Figs 58.3 and 58.4 Expanded terminal phalanges accompanying the osteolytic process here.

Skull: 2 months–Adult

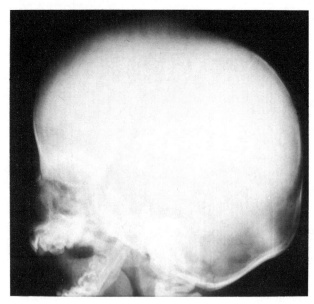

Fig. 58.5   O.R.—2 months—Lateral skull
The anterior fontanelle is large. Some increased bone density is present. There is loss of the normal angle of the mandible.
Same patient: Figs 58.12, 58.22, 58.28

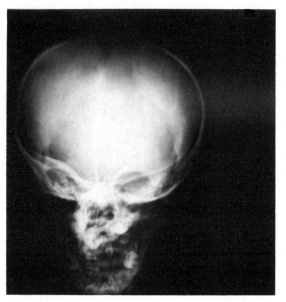

Fig. 58.6   I.D.—7 years—AP skull
The vault is large. The sutures are wide with some sutural (Wormian) bones. The parietal bones are rather dense and bulge laterally. The teeth are missing or dysplastic.
Same patient: Figs 58.7, 58.19, 58.29

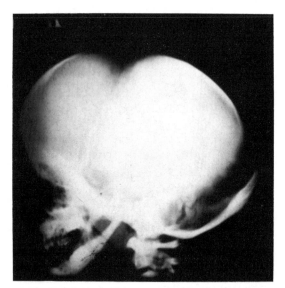

Fig. 58.7   I.D.—7 years—Lateral skull (same patient as in Fig. 58.6)
The sutures are wide and the vault large with generalised increase in bone density. There is loss of the normal mandibular angle, the body and ramus forming a straight line.

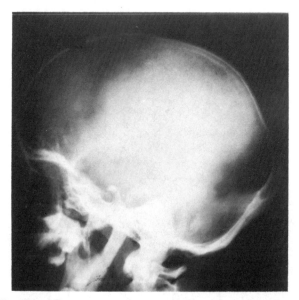

Fig. 58.8   J.D.—43 years—Lateral skull (uncle of child in Fig. 58.6)
The fontanelle is closed in this patient (it can remain open throughout adult life). The parietal bones show increased density, but not the base. The body and ramus of the mandible are in one straight line.
Same patient: Fig. 58.30

Spine: 9 years–Adult

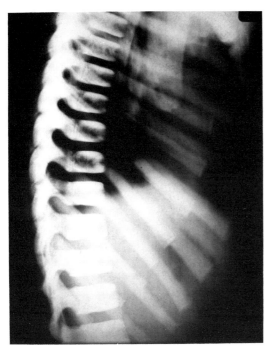

Fig. 58.9 **T.P.—9½ years—Lateral dorsal spine** (brother of patients in Figs 58.11, 58.13 and 58.31)
There is increased bone density in the region of the vertebral end-plates giving the sandwich or 'rugger jersey' appearance, similar to osteopetrosis.
Same patient: Figs 58.10, 58.15, 58.24, 58.33

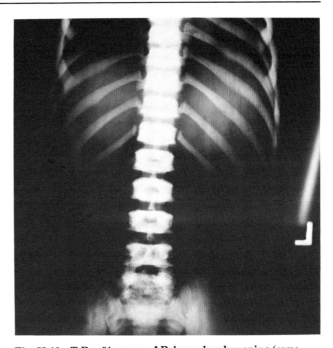

Fig. 58.10 **T.P.—9½ years—AP dorso-lumbar spine** (same patient as in Fig. 58.9)
There is increased bone density with normal modelling of the ribs and spine.

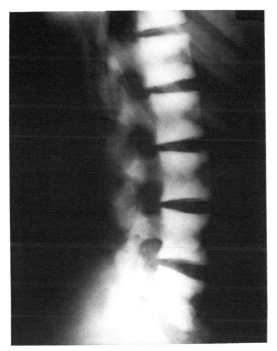

Fig. 58.11 **J.P.—29 years—Lateral lumbar spine**
The sandwich appearance persists adjacent to the disc spaces. Normal bone is present in the centre of the anterior vertebral margins.
Same patient: Figs 58.14, 58.16, 58.17, 58.20, 58.21, 58.23, 58.25, 58.32

Chest: 2 months–Adult

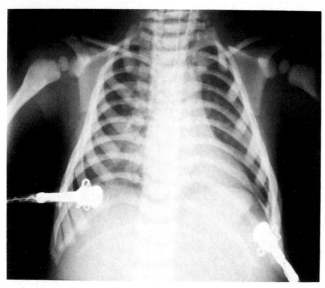

Fig. 58.12   O.R.—2 months—AP chest
There is mild hypoplasia of the distal ends of the clavicles, otherwise
bone modelling and density appear normal.
Same patient: Figs 58.5, 58.22, 58.28

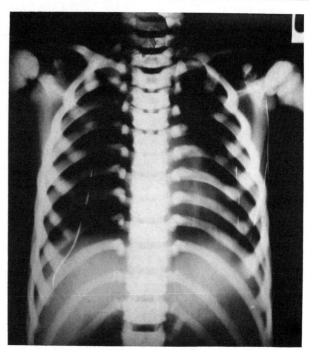

Fig. 58.13   G.P.—7 years—PA chest (brother of patients in Figs
58.14, 58.15 and 58.31)
There is generalised and uniform increased bone density. The distal
ends of the clavicles are hypoplastic.
Same patient: Figs 58.18, 58.26, 58.27, 58.34

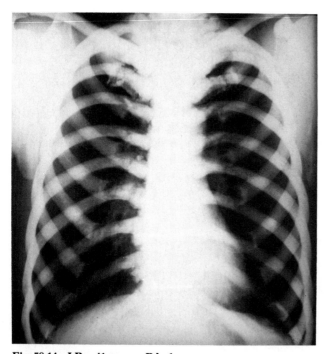

Fig. 58.14   J.P.—41 years—PA chest
Generalised and uniform increase in bone density. Modelling appears
normal.
Same patient: Figs 58.11, 58.16, 58.17, 58.20, 58.21, 58.23, 58.25, 58.32

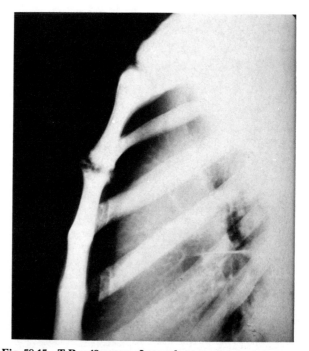

Fig. 58.15   T.P.—43 years—Lateral sternum
There is some fluffy ossification around one of the sternal joints,
possibly a uniting fracture.
Same patient: Figs 58.9, 58.10, 58.24, 58.33

Hips: 8 years–Adult

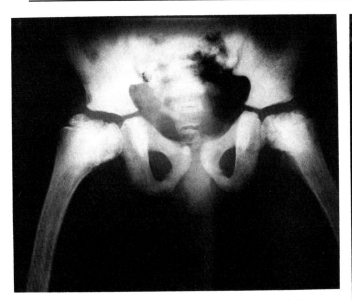

Fig. 58.16   J.P.—8 years—AP hips (brother of patients in Figs 58.9, 58.18 and 58.31)
There is generalised osteosclerosis with coxa vara.
Same patient: Figs 58.11, 58.14, 58.17, 58.20, 58.21, 58.23, 58.25, 58.32

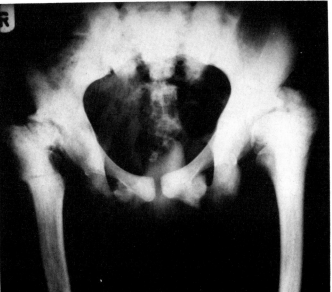

Fig. 58.17   J.P.—18 years—AP hips (same patient as in Fig. 58.16)
There has been a right upper femoral osteotomy for the coxa vara deformity. On the left the greater trochanter impinges on the ilium. Premature osteoarthritis is present. There are lateral osteophytes at the acetabular margins, the femoral heads are flattened and necks short.

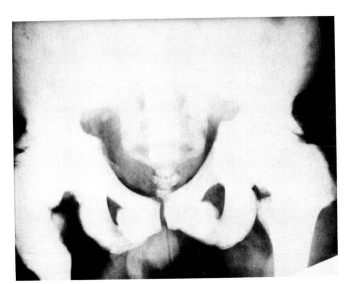

Fig. 58.18   G.P.—Adult—AP hips (brother of patients in Figs 58.9, 58.16 and 58.31)
There is a marked uniform increase in bone density and joint spaces are virtually obliterated. The femoral heads are flattened and the necks short and deformed. There is marked acetabular lipping.
Same patient: Figs 58.13, 58.26, 58.27, 58.34

Spine and lower limb: 7 years–Adult

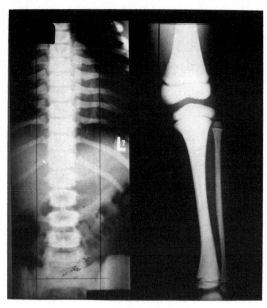

Fig. 58.19A   I.D.—7 years—AP spine
Increased bone density but normal modelling.

Fig. 58.19B   I.D.—7 years—AP tibia and fibula
Generalised increased bone density, modelling is normal and there is no delay in maturation.
Same patient: Figs 58.6, 58.7, 58.29

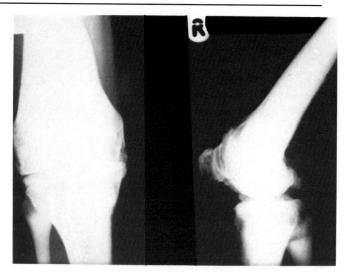

Fig. 58.20   J.P.—36 years—AP and lateral right knee (brother of patients in Figs 58.9, 58.13 and 58.31)
Increased bone density is present, and osteoarthritis with joint space narrowing, marginal osteophytes and loose bodies in the joint. There is abnormal modelling of the distal femoral metaphysis perhaps due to a previous fracture.
Same patient: Figs 58.11, 58.14, 58.16, 58.17, 58.21, 58.23, 58.25, 58.32

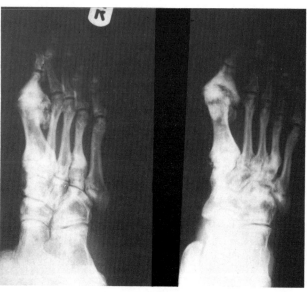

Fig. 58.21   J.P.—36 years—PA right foot (same patient as in Fig. 58.20)
There is here a patchy increase in bone density. The first metatarsophalangeal joint is virtually obliterated. There is marked splaying of the base of the proximal phalanx of the first toe.

Limbs : 2 months–Adult

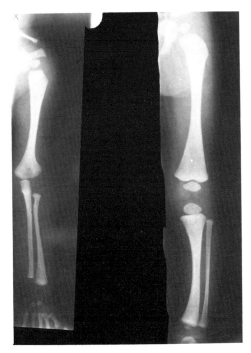

Fig. 58.22A O.R.—2 months—AP upper limb
Increased bone density is present and mild hypoplasia of the distal end of the clavicle. There is some increase in the metaphyseal flaring.

Fig. 58.22B O.R.—2 months—AP lower limb
There is a diffuse increase in bone density with mild lateral bowing of the tibia.
Same patient : Figs 58.5, 58.12, 58.28

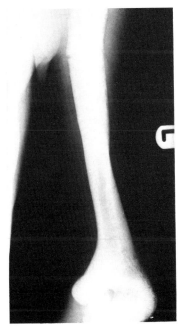

Fig. 58.24 T.P.—35 years—AP humerus (brother of patients in Figs 58.13, 58.23 and 58.31)
There is increased bone density and a break in the medial cortex of the proximal humeral shaft.
Same patient : Figs 58.9, 58.10, 58.15, 58.33

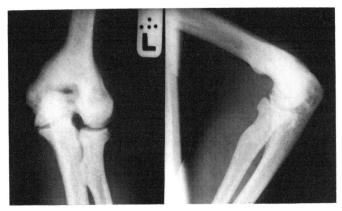

Fig. 58.23 J.P.—30 years—AP and lateral left elbow (brother of patients in Figs 58.13, 58.24 and 58.31)
There is increased bone density with deformity of the distal humeral metaphysis due to a previous fracture here.
Same patient : Figs 58.11, 58.14, 58.16, 58.17, 58.20, 58.21, 58.25, 58.32

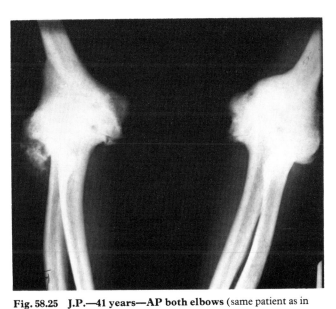

Fig. 58.25 J.P.—41 years—AP both elbows (same patient as in Fig. 58.23)
There is increased bone density and lateral bowing of the radii and ulnae. There is loss of normal modelling of the distal humeral metaphyses and on the right at least there has been a supracondylar fracture. There is marked osteoarthritis with narrowing of the joint spaces, marginal osteophytes and loose bodies in the joints.

Upper limb: Adult

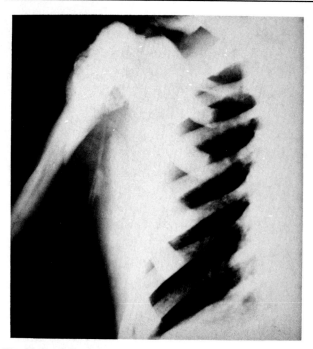

Fig. 58.26   G.P.—45 years—AP shoulder (brother of patients in Figs 58.9, 58.11 and 58.31)
There is increased bone density and a fluffy cortical margin present around the humeral head. The joint space is virtually obliterated, and there have probably been fractures in this region.
Same patient: Figs 58.13, 58.18, 58.27, 58.34

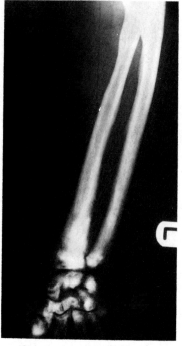

Fig. 58.27   G.P.—45 years—AP left forearm (same patient as in Fig. 58.23)
There is patchy increased bone density, especially involving the distal radial shaft and carpal bones.

Hand: 2 months–Adult

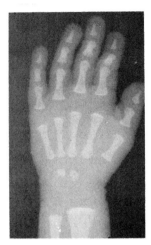

Fig. 58.28 O.R.—2 months—PA left hand
There is a uniform increase of bone density and splaying of the
metaphyses. The terminal phalanges are pointed and hypoplastic.
Two carpal ossification centres are already present.
Same patient: Figs 58.5, 58.12, 58.22

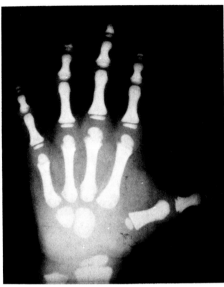

Fig. 58.29 I.D.—5 years—PA left hand
There is increased bone density with splayed metaphyses. The
terminal phalanges show osteolysis with some soft tissue swelling.
Carpal ossification is retarded.
Same patient: Figs 58.6, 58.7, 58.19

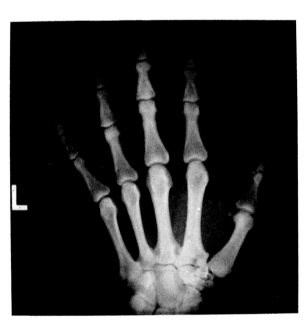

Fig. 58.30 J.D.—43 years—PA left hand
Normal bone modelling, but acro-osteolysis is present especially of
the terminal phalanx of the index finger.
Same patient: Fig. 58.8

Hand: Adult

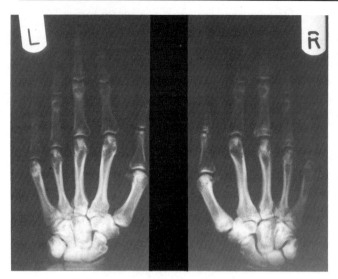

Fig. 58.31   R.P.—43 years—PA both hands
There is patchy increased bone density (terminal phalanges are not visible).

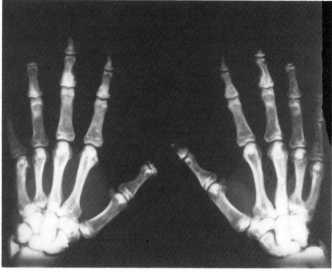

Fig. 58.32   J.P.—36 years—PA both hands
There is acro-osteolysis of all terminal phalanges. Joint space narrowing is present at the distal interphalangeal joint of the right thumb. The carpal bones are sclerotic, but the metacarpals and digits only patchily so.
Same patient : Figs 58.11, 58.14, 58.16, 58.17, 58.20, 58.21, 58.23, 58.25

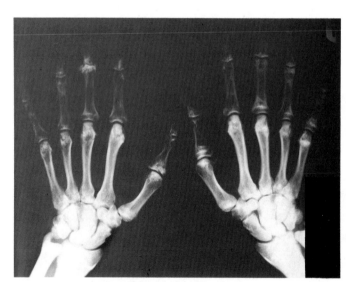

Fig. 58.33   T.P.—45 years—PA both hands
There is a patchy increase in bone density, and subluxation and joint space narrowing at the proximal interphalangeal joint of the left middle finger.
Same patient : Figs 58.9, 58.10, 58.15, 58.24

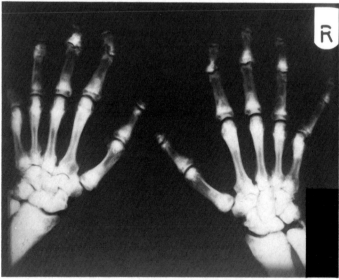

Fig. 58.34   G.P.—48 years—PA both hands
Patchy osteosclerosis is present. There is osteolysis involving all the terminal and some middle phalanges. Erosions are present around the bases of the middle phalanges.
Same patient : Figs 58.13, 58.18, 58.26, 58.27

Chapter Fifty-nine
METAPHYSEAL DYSPLASIA
(Pyle's disease)

One of the more benign of the sclerosing bone dysplasias; 'metaphyseal' dysplasia is a bad name, since the skull is affected as well as the metaphyses. Possibly some malalignment of long bones, with or without fragility.

Inheritance

Autosomal recessive.

Frequency

Rare. Probable prevalence of under 1 per million population.

Clinical features

Facial appearance—normal, except that teeth may be dysplastic or misplaced.

Intelligence—normal.

Stature—there are reports of tall stature with abnormally long limbs, but no figures are available.

Presenting feature/age—probably an incidental finding on radiography.

Deformities—probably none. Long bone malalignment sometimes, e.g. genu valgum.

Associated anomalies—none.

Radiographic features

Skull—some sclerosis of both vault and base.

Spine—normal.

Shoulder girdle—clavicles thickened at their medial end.

Pelvis—thickened pubis and ischium.

Long bones—poor modelling of all long bones, including hand. The metaphyses are expanded, the smooth enlargement extending well into the diaphyses.

Bone maturation

Probably normal.

Biochemistry

Not known.

Differential diagnosis

Principally from *cranio-metaphyseal dysplasia*, a more severe disease with massive thickening of the skull and complications associated with this.

Other sclerosing bone dysplasias also with metaphyseal expansion, in particular the milder (dominant) form of osteopetrosis.

Progress/complications

Few if any complications develop, with the exception of genu valgum or other minor limb deformity. Cranial nerve palsies are not a feature.

REFERENCES

Beighton P, Cremin B J 1980 Sclerosing bone dysplasia. Berlin, Springer

Gorlin R J, Koszalka M F, Spranger J 1970 Pyle's disease (familial metaphyseal dysplasias). Journal of Bone and Joint Surgery 52A:347

Skull and chest: 5 years–Adult

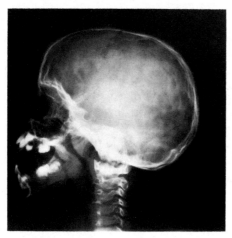

Fig. 59.1   C.B.—5 years—Lateral skull
The vault is not thickened but there is mild sclerosis of the base.
There is prognathos and some enlargement of the mandible.
Same patient: Figs 59.3, 59.5, 59.6

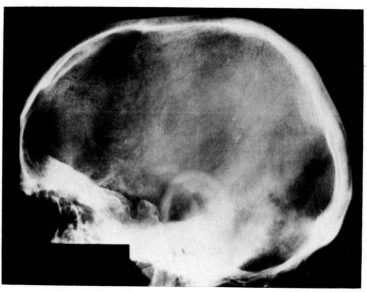

Fig. 59.2   G.—Adult—Lateral skull
There is mild sclerosis of the base and vault.
Same patient: Figs 59.4, 59.7–59.9

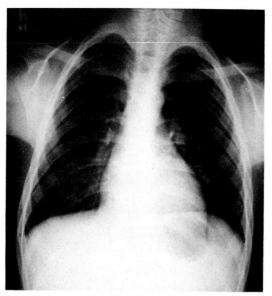

Fig. 59.3   C.B.—5 years—AP chest (same patient as in Fig. 59.1)
Slight widening of the ribs but otherwise no real abnormality is
present.

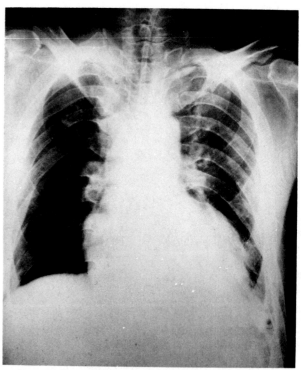

Fig. 59.4   G.—adult—PA chest (same patient as in Fig. 59.2)
There is marked widening of the ribs and medial halves of the
clavicles.

Lower limb: 5 years–Adult

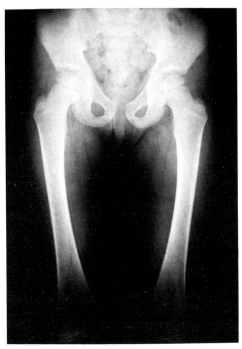

Fig. 59.5   C.B.—5 years—AP pelvis and femora
There is mild loss of modelling only.
Same patient: Figs 59.1, 59.3, 59.6

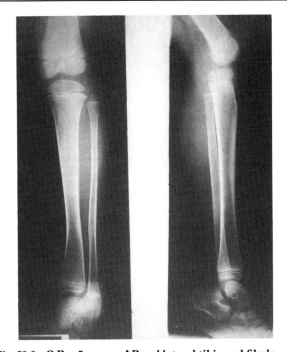

Fig. 59.6   C.B.—5 years—AP and lateral tibia and fibula
(same patient as in Fig. 59.1)
There is mild expansion at the metaphyses with some loss of
alignment, but no sclerosis.

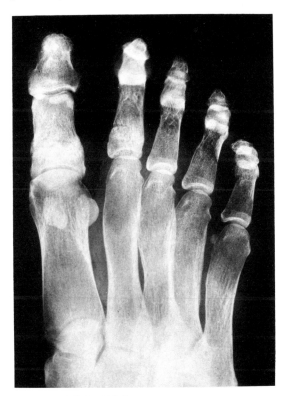

Fig. 59.7   G.—Adult—PA foot
The widening of the metaphyses extends well into the diaphyses, but
there is no sclerosis.
Same patient: Figs 59.2, 59.4, 59.8, 59.9

Upper limb: Adult

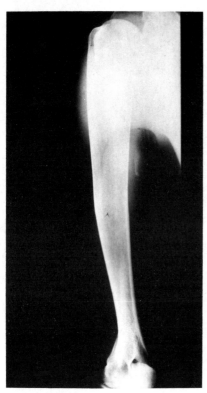

Fig. 59.8   G.—Adult—AP humerus
The metaphyseal expansion is only at the upper end of the humerus,
extending to the diaphysis. The lower end is normal.
Same patient: Figs 59.2, 59.4, 59.7, 59.9

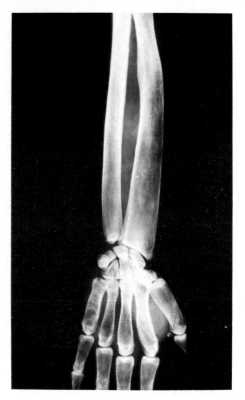

Fig. 59.9   G.—Adult—AP forearm (same patient as in Fig. 59.2)
There is almost complete lack of modelling of the ulna, but the
abnormality of modelling of the radius affects only its lower two-
thirds. There is also some loss of modelling of the hand bones.

Chapter Sixty
CRANIOMETAPHYSEAL DYSPLASIA

Similar to Pyle's disease, but the cranial changes are more marked than the metaphyseal.

Inheritance

There is an autosomal dominant form, which is less severe than the much rarer autosomal recessive.

Frequency

Probably under 0.1 per million prevalence.

Clinical features

Facial appearance—overgrowth of frontal region, nasion and mandible in severe cases, with irregularity of teeth.
Intelligence—normal.
Stature—probably normal, but there are no figures.
Presenting feature/age—skull and facial deformity in infancy or childhood; cranial nerve palsies (facial or auditory, usually).
Deformities—of skull only.
Associated anomalies—none.

Radiographic features

Skull—mild cases (autosomal dominant) may only show some frontal enlargement or thickening of the nasion area. Severe cases are grossly abnormal with massive thickening of the vault, facial bones and mandible (one of the disorders formerly called 'leontiasis ossea'). Paranasal sinuses may be obliterated, and teeth irregular.
Spine—normal.
Clavicles—widened.
Long bones—some metaphyseal expansion with cortical thinning. In early childhood it is the *diaphysis* which is abnormal, showing thickening and sclerosis.
Pelvis—probably normal.

Bone maturation

Probably normal, but no data.

Biochemistry

Not known.

Differential diagnosis

From the other sclerosing bone dysplasias. Mild cases are similar to *Pyle's disease*, but here the pelvis tends to be thickened and the changes in the long bones more marked, also cranial nerve palsies are not a feature.

In early childhood, craniometaphyseal dysplasia is difficult to distinguish from *craniodiaphyseal dysplasia*, since they both have diaphyseal thickening and sclerosis of long bones.

The *severe autosomal recessive form of craniometaphyseal dysplasia* has marked overgrowth and sclerosis of the base of the skull, paranasal sinuses and sometimes the mandible, similar to craniodiaphyseal dysplasia.

Progress/complications

There is no complication from the metaphyseal disorders; the skull overgrowth may cause cosmetic problems only, but, at its worst there may be VIIth and VIIIth nerve paralysis, blocked sinuses and severely disordered dentition.

REFERENCES

Beighton P, Cremin B J 1980 Sclerosing bone dysplasias. Springer, Berlin
Millard D R, Maisels D O, Batstone J H F, Yates B W 1967 Craniofacial surgery in craniometaphyseal dysplasia. American Journal of Surgery 113:615
Penchaszadeh V B, Gutierrez E R, Figueroa E P 1980 Autosomal recessive craniometaphyseal dysplasia. American Journal of Medical Genetics 5:43–55
Rimoin D L, Woodruff S L, Holman B L 1969 Craniometaphyseal dysplasia (Pyle's disease): autosomal dominant inheritance in a large kindred. Birth Defects: Original Article Series 5:96

Skull : 4 months

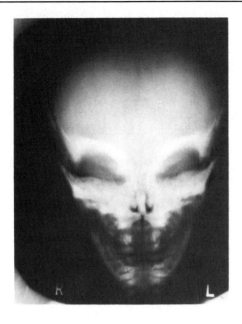

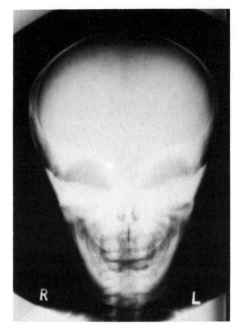

Fig. 60.1   **Fig. 60.2**

Fig. 60.1   R.C.—4 months—AP skull (overpenetrated to show base)
Fig. 60.2   R.C.—4 months—AP skull (normal penetration to show vault)
Figures 60.1 and 60.2 show thickening and sclerosis of the base and paranasal region with only mild sclerosis of the vault.
This is probably the severe autosomal recessive form.
Same patient : Figs 60.3–60.5, 60.8–60.13, 60.15, 60.16

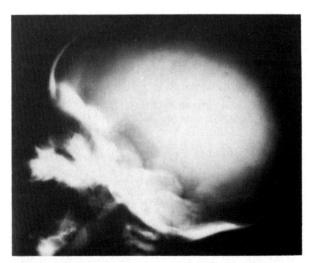

Fig. 60.3   R.C.—4 months—Lateral skull (same patient as in
Fig. 60.1)
There is marked sclerosis of the base and maxilla but this is less
apparent in the vault.

Skull : 1–2 years

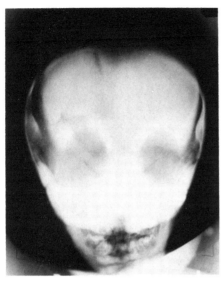

Fig. 60.4   R.C.—1 year—AP skull
The sclerosis has progressed and there has now been a right frontal craniotomy to decompress the involved cranial nerves.
Same patient : Figs 60.1–60.3, 60.5, 60.8–60.13, 60.15, 60.16

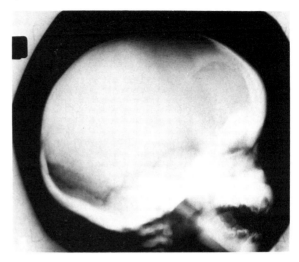

Fig. 60.5   R.C.—1 year—Lateral skull (same patient as in Fig. 60.4)
See Figure 60.4.

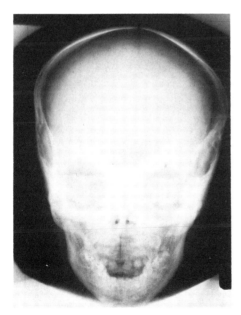

Fig. 60.6   R.S.—2 years—AP skull
Sclerosis of base and vault are seen but this is less severe than in the preceding case. This is probably the milder autosomal dominant form.
Same patient : Figs 60.7, 60.14, 60.17, 60.18

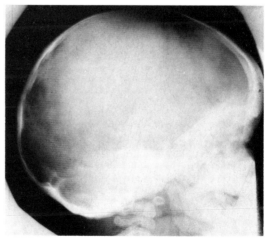

Fig. 60.7   R.S.—2 years—lateral skull (same patient as in Fig. 60.6)
See Figure 60.6.

Spine and chest : 4 months–2 years

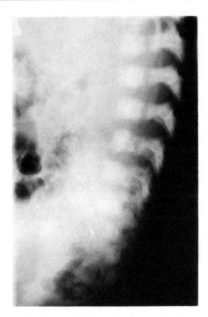

Fig. 60.8   R.C.—4 months—Lateral lumbar spine
Patchy sclerosis throughout but otherwise normal.
Same patient : Figs 60.1–60.5, 60.9–60.13, 60.15, 60.16

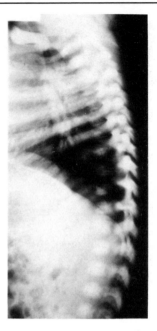

Fig. 60.9   R.C.—4 months—Lateral dorso-lumbar spine
(same patient as in Fig. 60.8)
See Figure 60.8.

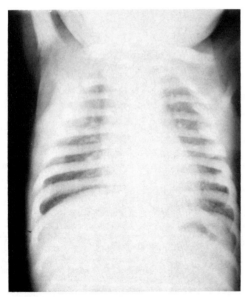

Fig. 60.10   R.C.—4 months—AP chest (same patient as in Fig. 60.8)
There is increased bone density but only a mild modelling deformity.

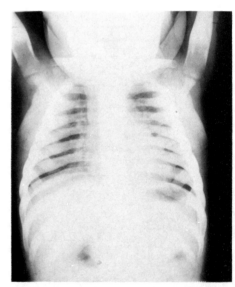

Fig. 60.11   R.C.—2 years—AP chest (same patient as in Fig. 60.8)
By this age the sclerosis and enlargement of clavicles and ribs is very marked. The upper humeral shaft also shows sclerosis and a curious expansion.

Lower limb : 4 months–2 years

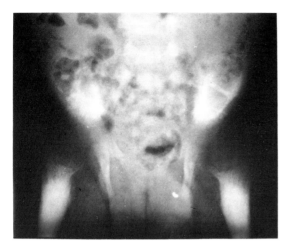

Fig. 60.12 R.C.—4 months—Pelvis
The pelvis is narrow with poorly formed acetabula and patchy sclerosis. The capital femoral epiphyses have not yet appeared and the trochanteric region is dense.
Same patient : Figs 60.1–60.5, 60.8–60.11, 60.13, 60.15, 60.16

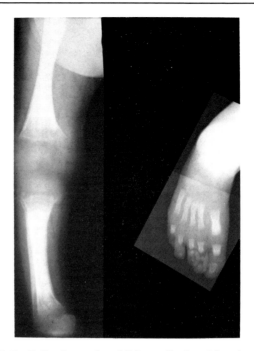

Fig. 60.13 R.C.—4 months—AP lower limb and foot (same patient as in Fig. 60.12)
The sclerosis is confined to the *diaphyseal* areas of the long bones at this age. The foot is normal.

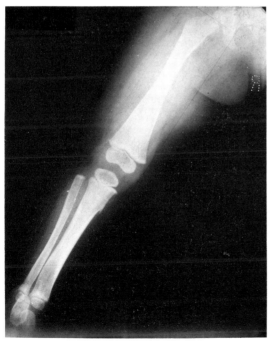

Fig. 60.14 R.S.—2 years—AP right lower limb
There is some irregularity of modelling and the sclerosis is more extensive by this age.
Same patient : Figs 60.6, 60.7, 60.17, 60.18

Upper limb: 4 months–2 years

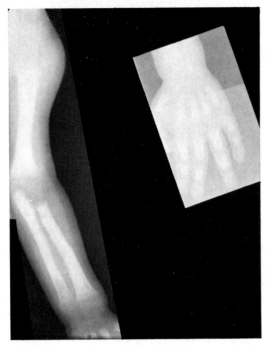

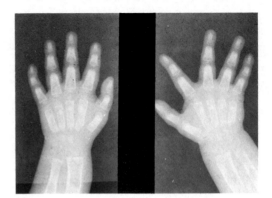

Fig. 60.16   R.C.—2 years—PA hands (same patient as in Fig. 60.15)
There is some irregularity of modelling and unusual islands of sclerosis in the metacarpals and phalanges.

Fig. 60.15   R.C.—4 months—AP upper limb and hand
As with the lower limb the sclerosis is principally diaphyseal at this age. The hand is normal.
Same patient: Figs 60.1–60.5, 60.8–60.13, 60.16

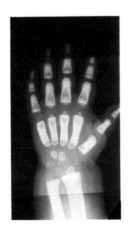

Fig. 60.17   R.S.—2 years—PA hand
The lower radius and ulna show sclerosis of the metaphyseal areas, but in the metacarpals and phalanges the sclerosis is still diaphyseal.
Same patient: Figs 60.6, 60.7, 60.14, 60.18

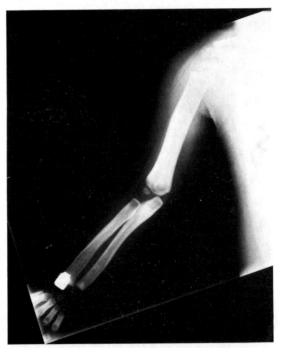

Fig. 60.18   R.S.—2 years—AP upper limb (same patient as in Fig. 60.17)
Some failure of modelling is present but the sclerosis involves the whole humeral shaft.

Chapter Sixty-one
FRONTOMETAPHYSEAL DYSPLASIA

The principal feature is massive overgrowth of the frontal region of the skull, but there are mild dysplastic changes in the rest of the skeleton.

Inheritance

Uncertain, but some cases may be autosomal dominant, or perhaps X-linked recessive.

Frequency

Extremely rare; under 0.1 per million prevalence.

Clinical features

Facial appearance—bizarre appearance, with enlarged and overhanging supraorbital ridges.

Intelligence—normal.

Presenting feature/age—prominent frontal region in early childhood, perhaps micrognathia, and bowing of long bones.

Deformities—apart from the skull deformity there may be bowing of long bones, or a 'wavy' appearance of alternating expansion and constriction, and associated genu valgum, varum or recurvatum. Finger contractures have been reported, also scoliosis.

Associated anomalies—none reported, except later developing deafness, which is presumably related to sclerosis of the skull.

Radiographic features

Skull—there is probably patchy sclerosis of the whole vault and base, but the main bony overgrowth is of the frontal region, particularly around the supraorbital ridges. There is obliteration of paranasal sinuses, and there may be a small mandible with dental problems.

Spine—There may be some irregular platyspondyly, but congenital vertebral anomalies also occur (fusion, absence, etc).

Thorax—slender ribbon-like appearance of ribs.

Pelvis—this may show a constriction of the base of the ilia, above the acetabula, which are themselves normal.

Long bones—failure of metaphyseal modelling is the main defect, the bones appearing wavy and bowed.

Carpus and tarsus—fusion and 'erosion' have been reported.

Bone maturation

No data.

Biochemistry

Not known.

Differential diagnosis

From *other sclerosing bone dysplasias* but the localised nature of the frontal sclerosis, together with only mild generalised modelling defects should differentiate fronto-metaphyseal dysplasia.

Osteodysplasty (Melnick–Needles syndrome) is more difficult to differentiate, indeed there may be overlap of the two conditions, merely illustrating differing expressions of the same gene.

Progress/complications

The main problem is a cosmetic one; deafness may develop in childhood or middle age.

REFERENCES

Beighton P, Cremin B J 1980 Sclerosing bone dysplasias. Berlin, Springer
Danks D M, Mayne V 1974 Frontometaphyseal dysplasia: a progressive disease of bone and connective tissue. Birth Defects: Original Article Series X (12):57

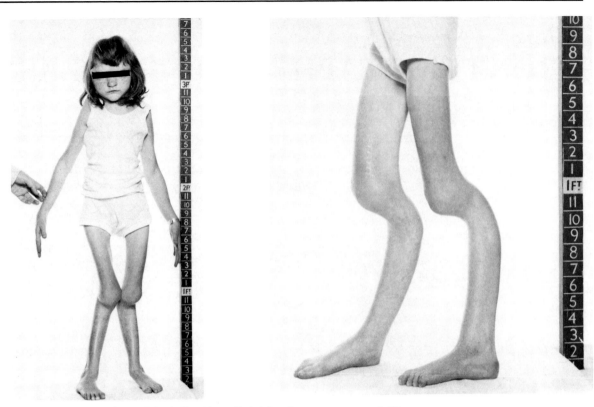

Figs 61.1 and 61.2 Enlarged frontal area of the skull and lower limb deformity—severe genu and tibia recurvatum.

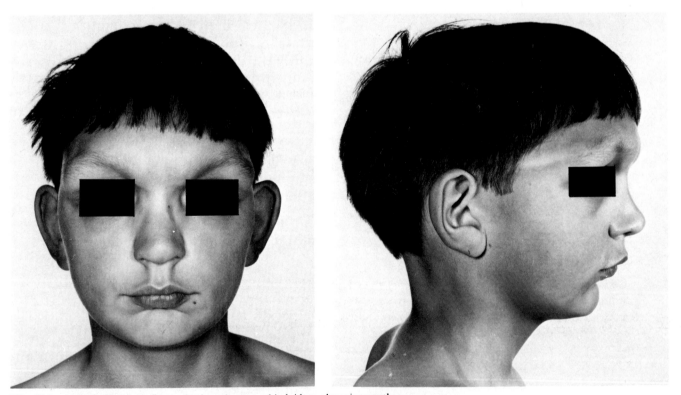

Figs 61.3 and 61.4 Hypertelorism and enlarged supra-orbital ridges, also micrognathos.

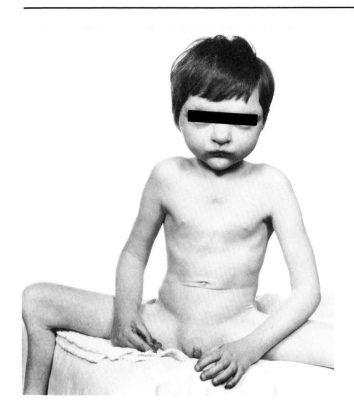
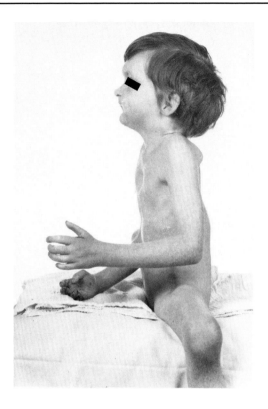
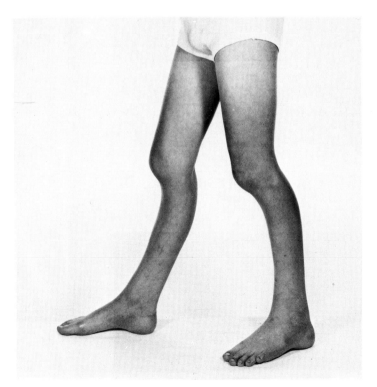

Figs 61.5, 61.6 and 61.7 Similar to Figures 61.1 and 61.2, but the lower-limb deformity is not so severe.

Skull: 3 years

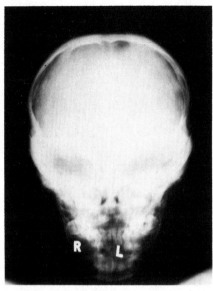

Fig. 61.8   S.B.—3 years—AP skull
All views show sclerosis and thickening of the vault and base.
Same patient: Figs 61.9, 61.10, 61.15, 61.19, 61.23, 61.26–61.28

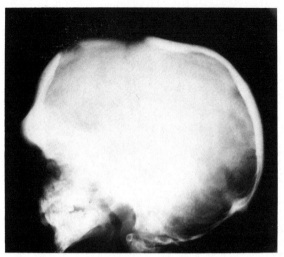

Fig. 61.9   S.B.—3 years—Lateral skull (same patient as in Fig. 61.8)
The pronounced frontal overhang is clearly shown and the anterior fontanelle remains widely open. There is disproportion between the size of the vault and the facial bones.

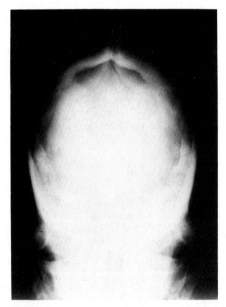

Fig. 61.10   S.B.—3 years—Towne's view (same patient as in Fig. 61.8)
See Figures 61.8 and 61.9.

Skull: 9–10 years

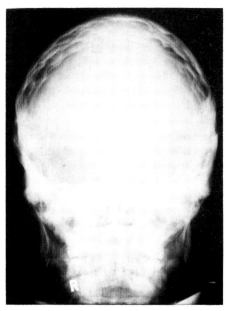

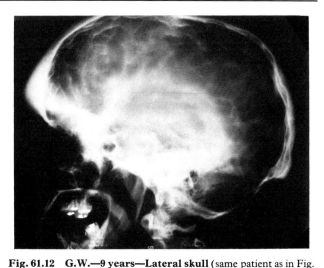

Fig. 61.12   G.W.—9 years—Lateral skull (same patient as in Fig. 61.11)
The main sites of sclerosis and thickening are the frontal and orbital regions. There is some micrognathia with malocclusion.

Fig. 61.11   G.W.—9 years—AP skull
There is no great sclerosis or thickening of the vault, but pronounced convolutional markings are present. (This is not a sign of raised intracranial pressure in childhood.)
Same patient: Figs 61.12, 61.16, 61.20, 61.21, 61.24, 61.29

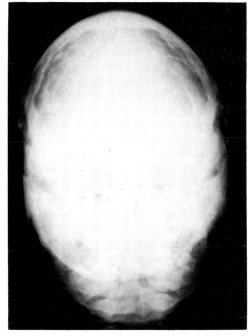

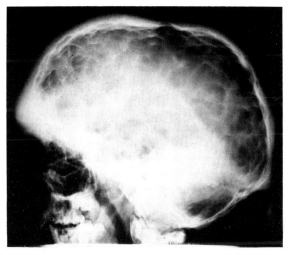

Fig. 61.14   M.L.—10 years 6 months—Lateral skull (same patient as in Fig. 61.13)
There is marked disproportion between the size of the vault and facial bones, with sclerosis chiefly of the frontal region and base.

Fig. 61.13   M.L.—10 years 6 months—AP skull
Similar to Figure 61.11.
Same patient: Figs 61.14, 61.17, 61.18, 61.22, 61.25, 61.30

Spine: 3–10 years

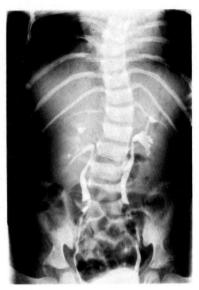

Fig. 61.15   S.B.—3 years—AP spine and pelvis
The whole length of the ribs is irregularly narrowed and 'thread-like'. The interpedicular distances are unusually wide and there is scoliosis. The bases of the ilia show marked constriction above the acetabula, which are normal.
Same patient: Figs 61.8–61.10, 61.19, 61.23, 61.26–61.28

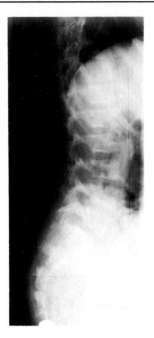

Fig. 61.16   G.W.—9 years—Lateral thoracic and lumbar spine
Little abnormality apart from scalloping of the posterior vertebral border.
Same patient: Figs 61.11, 61.12, 61.20, 61.21, 61.24, 61.29

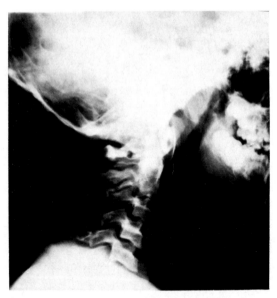

Fig. 61.17   M.L.—10 years 6 months—Lateral cervical spine
The cervical vertebrae are unequal in size with some platyspondyly.
Same patient: Figs 61.13, 61.14, 61.22, 61.25, 61.30

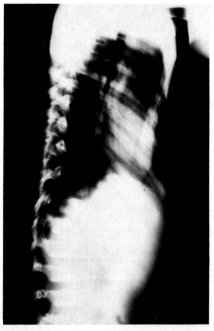

Fig. 61.18   M.L.—10 years 6 months—Lateral spine (same patient as in Fig. 16.17)
There is platyspondyly and some posterior scalloping.

Chest and pelvis : 3–10 years

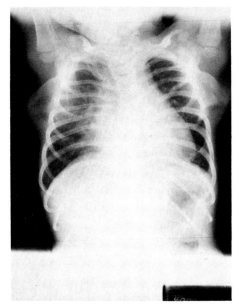

Fig. 61.19 S.B.—3 years—AP chest
There are thin and poorly modelled ribs but no apparent sclerosis.
Same patient : Figs 61.8–61.10, 61.15, 61.23, 61.26–61.28

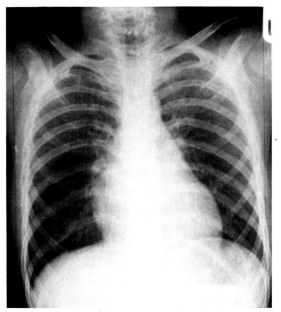

Fig. 61.20 G.W.—9 years—AP chest
No abnormality is present apart from some irregular modelling of the
lower borders of most ribs.
Same patient : Figs 61.11, 61.12, 61.16, 61.21, 61.24, 61.29

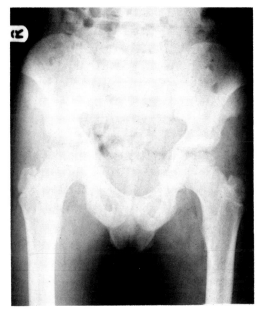

Fig. 61.21 G.W.—9 years—Pelvis (same patient as in Fig. 61.20)
There appears to be a bony ridge extending from the wing of the ilia
towards the base. The hips are normal.

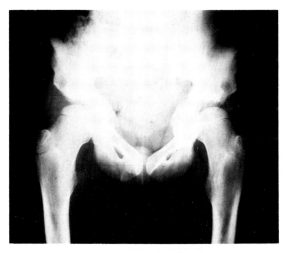

Fig. 61.22 M.L.—10 years 6 months—Pelvis
There is unusual modelling of the ilia and the pubis and ischium are
thickened. The hip joints are normal apart from some thickening of
the femoral necks.
Same patient : Figs 61.13, 61.14, 61.17, 61.18, 61.25, 61.30

Lower limb: 4–10 years

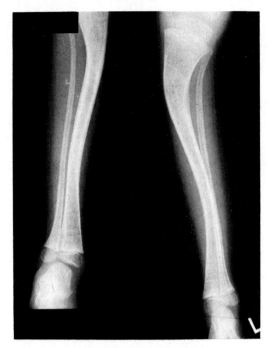

Fig. 61.23   S.B.—4 years—AP tibiae and fibulae
Both bones have a 'wavy' appearance although modelling and density
are normal.
Same patient: Figs 61.8–61.10, 61.15, 61.19, 61.26–61.28

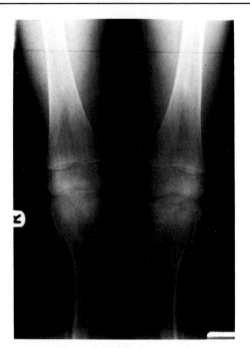

Fig. 61.24   G.W.—9 years—AP knees
There is mild medial bowing of both upper tibiae.
Same patient: Figs 61.11, 61.12, 61.16, 61.20, 61.21, 61.24

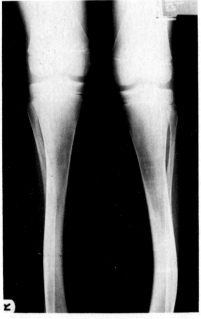

**Fig. 61.25   M.L.—10 years 6 months—AP knees and upper
tibiae**
There is some lack of modelling of the tibiae with lateral bowing.
Same patient: Figs 61.13, 61.14, 61.17, 61.18, 61.22, 61.30

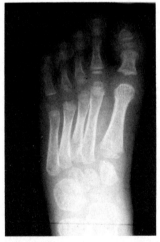

Fig. 61.26   S.B.—4 years—PA foot (same patient as in Fig. 61.23)
The foot appears to be normal apart from the short pointed terminal
phalanges.

Upper limb: 3–10 years

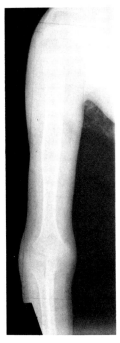

Fig. 61.27   S.B.—3 years—AP humerus and elbow
There is no abnormality apart from expansion of the mid-humeral shaft.
Same patient: Figs 61.8–61.10, 61.15, 61.19, 61.23, 61.26, 61.28

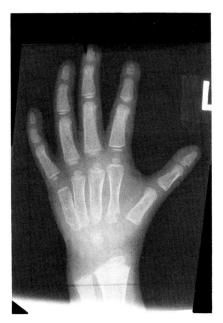

Fig. 61.28   S.B.—3 years—PA left hand (same patient as in Fig. 61.27)
All the hand bones are abnormally modelled and the terminal phalanges are unusually short. Bone maturation is delayed at about 18 months.

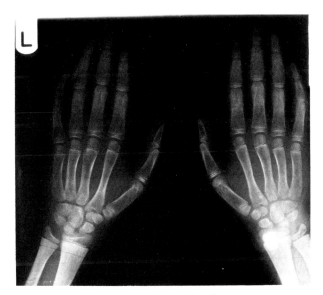

Fig. 61.29   G.W.—9 years—PA both hands
Failure of modelling is principally phalangeal. Bone maturation is not delayed.
Same patient: Figs 61.11, 61.12, 61.16, 61.20, 61.21, 61.24

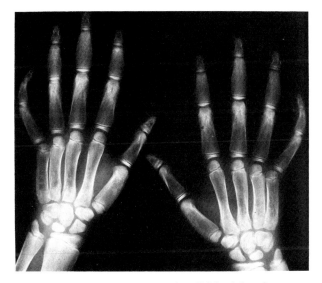

Fig. 61.30   M.L.—10 years 6 months—PA both hands
There are bizarre modelling deformities; note particularly the expansion of the middle phalanges. All terminal phalanges are disproportionately short.
Same patient: Figs 61.13, 61.14, 61.17, 61.18, 61.22, 61.25

Chapter Sixty-two
OSTEODYSPLASTY
(Melnick–Needles syndrome)

Cranio-facial abnormalities accompanied by 'wavy' shafts of long bones and irregular brachydactyly or lengthening of hand bones.

Inheritance

Usually autosomal dominant, but there is a rare autosomal recessive form which is lethal in the perinatal period.

Frequency

Extremely rare.

Clinical features

Facial appearance—micrognathos and malalignment of teeth, sometimes exophthalmos.

Intelligence—normal.

Stature—may be normal, or somewhat reduced.

Presenting feature/age—in infancy, or only incidentally in adult life, on account of facial appearance or abnormal bone modelling.

Deformities—scoliosis, protrusio acetabuli, long bone irregularity.

Associated anomalies—deafness and speech problems.

Radiographic features

Skull—patchy sclerosis of vault and base, thickened supraorbital ridges, lack of air cells, micrognathos, irregularities of dentition—or none of these signs. In the rare recessive form there may be virtually no mineralisation of the vault.

Spine—there may be abnormal modelling of vertebral bodies, but this is not a major feature.

Thorax—asymmetrical and irregular constrictions of the ribs, particularly at their posterior ends.

Pelvis—bizarre modelling, with constricted base of the ilia, protrusio acetabuli, thick ischia and a grooved appearance of the iliac wings—or one of them. The hips may be subluxed or dislocated, and the femoral necks thickened.

Long bones—the main characteristic is a wavy 'S-shaped' outline with multiple constrictions as well as bowing.

Hands—the terminal phalanges may be short and thick, and the other hand bones either shortened or lengthened.

Bone maturation

Not known, but probably normal.

Biochemistry

Not known.

Differential diagnosis

The major diagnostic difficulty is with *frontometaphyseal dysplasia*. Apart from this there is no other condition with the 'S-shaped' appearance of the long bones, but the skull changes need to be differentiated from other sclerosing bone dysplasias. The rare recessive (lethal) form is similar to *hypophosphatasia* with its absence of cranial mineralisation but the biochemical changes in the latter will distinguish it.

Progress/complications

There may be none in mildly affected individuals. Possible complications are related to dentition, scoliosis, pelvic disproportion, limb malalignment and deafness due to cranial nerve compression and recurrent ear infections.

REFERENCES

Beighton P, Cremin B J 1980 Sclerosing bone dysplasias. Berlin, Springer

Kozlowski K, Mayne V, Danks D M 1973 Precocious type of osteodysplasia. A new autosomal recessive form. Acta Radiologica 14:171–176

Skull and spine : 11 years

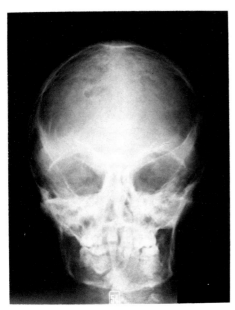

Fig. 62.1 C.N.—11 years—AP skull
There are patchy areas of sclerosis in the vault.
Same patient : Figs 62.2–62.8

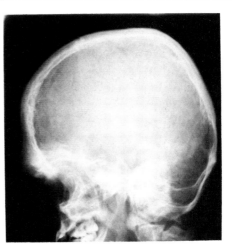

Fig. 62.2 C.N.—11 years—Lateral skull (same patient as in Fig. 62.1)
There is brachycephaly with thickening of the vault, supra-orbital ridge and base. The sphenoid sinus and mastoid air cells are not aerated.

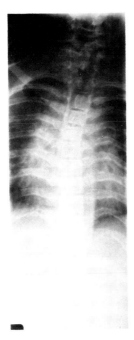

Fig. 62.3 C.N.—11 years—AP spine (same patient as in Fig. 62.1)
The vertebrae look normal in this projection, apart from slight scoliosis. The ribs have asymmetrical irregular constrictions at their posterior ends.

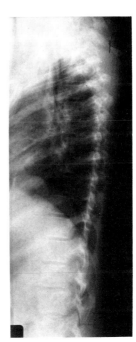

Fig. 62.4 C.N.—11 years—Lateral thoracic and lumbar spine (same patient as in Fig. 62.1)
The vertebrae are rather tall, and the disc spaces narrowed.

Spine and limbs : 11 years

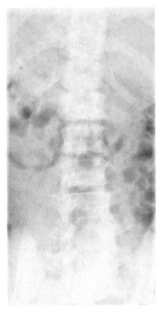

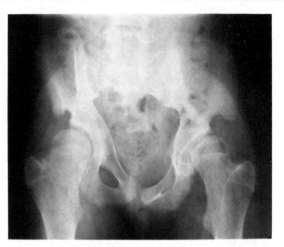

Fig. 62.6   C.N.—11 years—Pelvis (same patient as in Fig. 62.5)
The modelling of the whole pelvis is abnormal and asymmetrical—
the bases of the ilia are constricted; the anterior superior spines too
low; the ischia 'heavy'; there is protrusio acetabuli and bilateral
subluxation of the hips. The femoral necks are also thickened.

Fig. 62.5   C.N.—11 years—AP spine
There is abnormal modelling of the lumbar vertebrae and posterior
ends of the ribs, also spina bifida occulta of L5 and the sacrum.
Same patient : Figs 62.1–62.4, 62.6–62.8

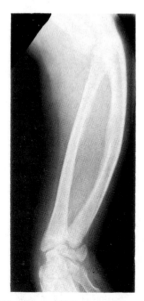

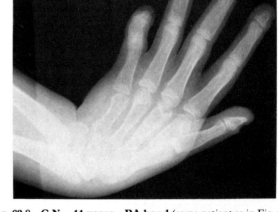

Fig. 62.8   C.N.—11 years—PA hand (same patient as in Fig. 62.5)
There is ulnar deviation of the wrist and the hand bones are irregular
in length—the terminal phalanges are short and broad, the middle
phalanges unusually long and the 5th metacarpal is short.

Fig. 62.7   C.N.—11 years—AP forearm (same patient as in Fig.
62.5)
The radial head is dislocated, and the ulna bowed medially, there is
also some irregularity of the shafts.

Severe autosomal recessive form

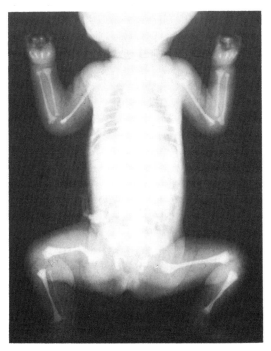

Fig. 62.9   G.H.—Neonate—AP whole body
All ribs and long bones are very thin, some of them with a 'wavy' appearance, and there is a right femoral fracture. The pelvis shows constriction above the acetabulum, but no other abnormality.
Same patient: Figs 62.10–62.12

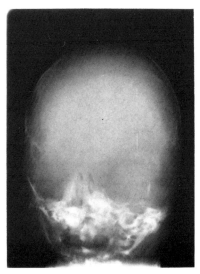

Fig. 62.10   G.H.—Neonate—AP skull (same patient as in Fig. 62.9)
There is virtually no ossification of the vault.

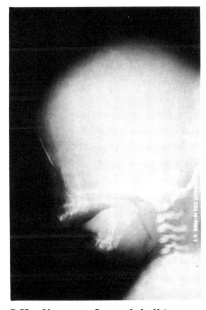

Fig. 62.11   G.H.—Neonate—Lateral skull (same patient as in Fig. 62.9)
Similar to Figure 62.10. The teeth are poorly formed and the mandible short, with absence of the angle between the body and ramus.

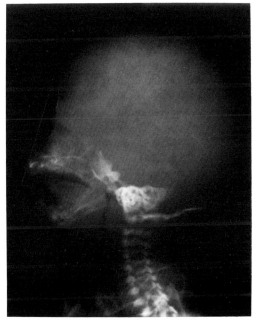

Fig. 62.12   G.H.—Neonate—Lateral skull (same patient as in Fig. 62.9)
Similar to Figure 62.11

Chapter Sixty-three
ENGELMANN'S DISEASE
(progressive diaphyseal dysplasia)

Limb pain, weakness and hypotonia associated with diaphyseal thickening of the long bones and sometimes sclerosis of the skull.

Inheritance

Autosomal dominant.

Frequency

Prevalence under 1 per million.

Clinical features

Facial appearance—normal.
Intelligence—normal.
Stature—normal or increased.
Presenting feature/age—in infancy, failure to thrive, weakness, awkward gait, reduced muscle mass, limb pain.
Deformities—none.
Range of severity—wide. Some discovered only incidentally in adult life.
Associated anomalies—none.

Radiographic features

Skull—thickened vault and base, with narrowed foramina.
Spine—not usually involved, but upper cervical vertebrae may be sclerotic. Striations are sometimes present.
Thorax—ribs only occasionally involved.

Long bones—the middle two quarters of the long bones are thickened on both periosteal and medullary sides, but metaphyses and epiphyses are normal. Principally lower limb involvement. Hands and feet not affected.

Bone maturation

Probably normal.

Biochemistry

Not known.

Differential diagnosis

Metaphyseal and *craniometaphyseal disorders* are apparent by reason of their metaphyseal involvement from the middle years of childhood, but craniometaphyseal dysplasia shows diaphyseal thickening in infancy.
Infantile cortical hyperostosis occurs at a younger age, and regresses.
Hyperphosphatasia, chronic osteomyelitis and *syphilis* are excluded by laboratory tests.
Paget's disease is asymmetrical.

Progress/complications

At its most severe there is increasing muscle weakness and debility. Sometimes pressure on optic and auditory nerves.

REFERENCES

Beighton P, Cremin B J 1980 Sclerosing bone dysplasias, Berlin, Springer
Smith R, Walton R J, Corner B D, Gordon I R S 1977 Clinical and biochemical studies in Engelmann's disease (progressive diaphyseal dysplasia). Quarterly Journal of Medicine 46:273–294

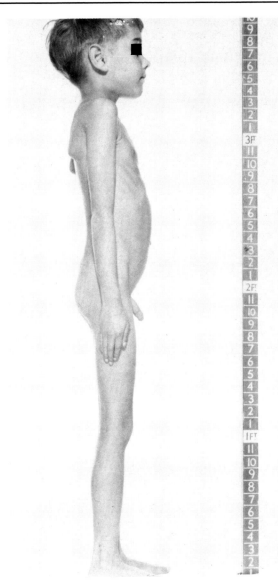

Fig. 63.1 Poor musculature, but normal body proportions.

Skull: 5 years–Adult

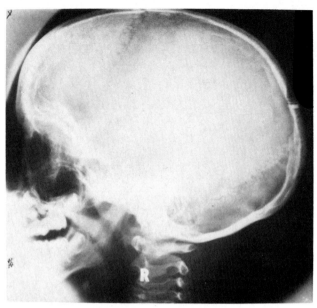

Fig. 63.2   R.H.—5½ years—Lateral skull
There is mild increased bone density and thickening of the frontal
bone and base. There is no pneumatisation of the frontal sinuses.
Same patient: Figs 63.3, 63.10, 63.14, 63.18–63.20

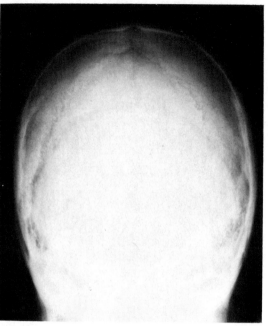

Fig. 63.3   R.H.—5½ years—Skull (Townes view) (same patient
as in Fig. 63.2)
There is very mild increase in the thickness of the vault.

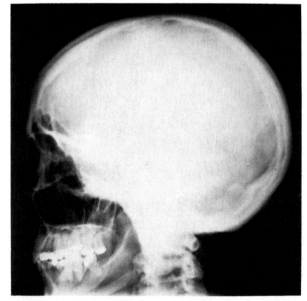

Fig. 63.4   W.P.—Adult—Lateral skull
There is thickening of the vault and sclerosis of the base of the skull,
more obvious than in the previous patient.
Same patient: Figs 63.6–63.8, 63.11, 63.12, 63.15–63.17, 63.21

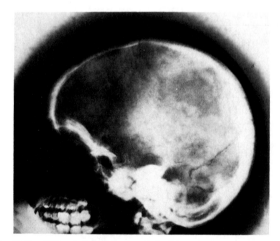

Fig. 63.5   C.S.—Adult—Lateral skull
There is patchy sclerosis and thickening of the vault and base.
Same patient: Figs 63.9, 63.13

Spine and chest : Adult

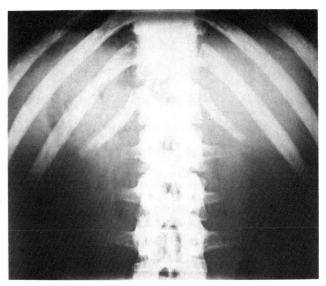

Fig. 63.6 W.P.—Adult—Lumbar spine
There is patchy sclerosis of the vertebral bodies and ribs, with loss of normal modelling.
Same patient : Figs 63.4, 63.7, 63.8, 63.11, 63.12, 63.15–63.17, 63.21

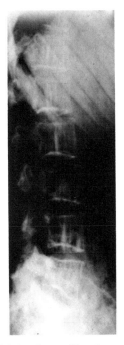

Fig. 63.7 W.P.—Adult—Lateral lumbar spine (same patient as in Fig. 63.6)
There is loss of the normal lumbar lordosis and vertical striations are present in the vertebral bodies.

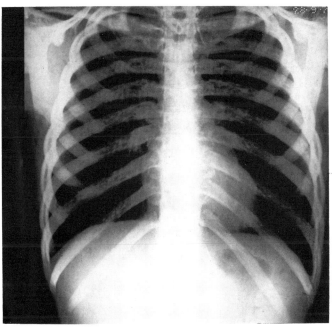

Fig. 63.8 W.P.—Adult—AP thorax (same patient as in Fig. 63.6)
There is patchy sclerosis, thickening and under-modelling of the ribs particularly at their posterior ends.

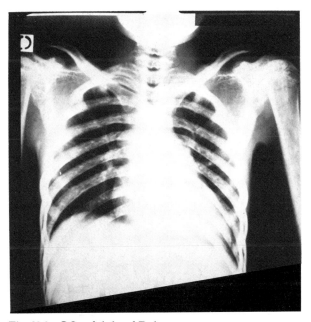

Fig. 63.9 C.S.—Adult—AP chest
Patchy sclerosis involves the ribs, scapulae and upper humeri.
Same patient : Figs 63.5, 63.13

Pelvis: 5 years–Adult

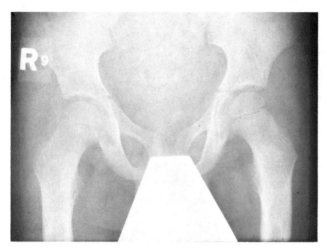

Fig. 63.10   R.H.—5½ years—AP pelvis
The pelvis and head and neck of femur show normal modelling and
bone density, that is, the epiphyses and metaphyses are normal, but
thickening of the femoral shaft can just be seen.
Same patient: Figs 63.2, 63.3, 63.14, 63.18–63.20

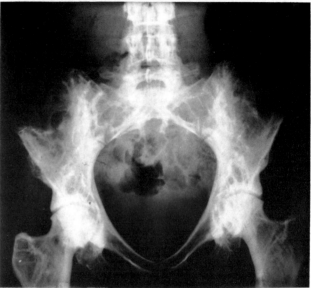

Fig. 63.11   W.P.—Adult—AP pelvis
There are striations of increased bone density, which appear to follow
the lines of stress in the pelvis.
Same patient: Figs 63.4, 63.6–63.8, 63.12, 63.15–63.17, 63.21

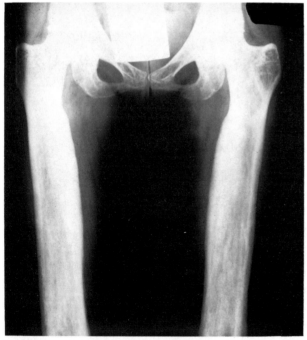

Fig. 63.12   W.P.—Adult—Lower limbs (same patient as in Fig.
63.11)
The femoral diaphyses are widened and under-modelled, mainly due
to increased cortical thickness on the medial side. The femoral necks
and heads have normal modelling.

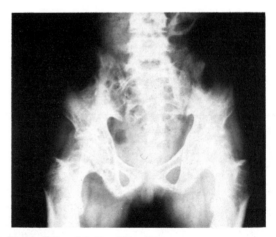

Fig. 63.13   C.S.—Adult—AP pelvis
There are striations of increased bone density throughout the pelvis.
The proximal femoral shafts show pronounced cortical thickening.
Same patient: Figs 63.5, 63.9

Lower limb : 5½ years–Adult

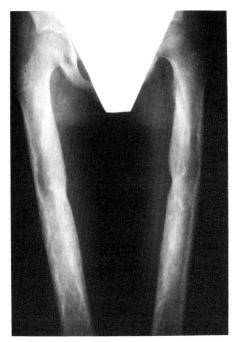

Fig. 63.14   R.H.—5½ years—AP femora
There is patchy increased bone density with diaphyseal widening and irregular cortical thickening of both femora. Femoral heads and necks are normal.
Same patient : Figs 63.2, 63.3, 63.10, 63.18–63.20

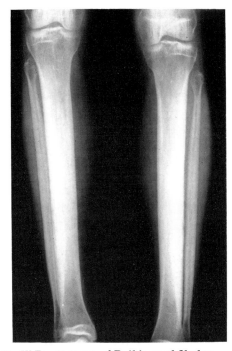

Fig. 63.15   W.P.—14 years—AP tibiae and fibulae
Diaphyses are wide and sclerotic with cortical thickening. The metaphyses are normal. The fibulae are short.
Same patient : Figs 63.4, 63.6–63.8, 63.11, 63.12, 63.16, 63.17, 63.21

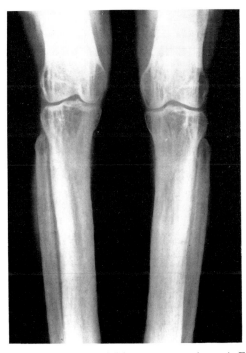

Fig. 63.16   W.P.—Adult—AP knees (same patient as in Fig. 63.15)
Epiphyses have now fused but vertical striations of increased bone density are present around the knee joint, although the modelling remains normal in this region. Diaphyseal modelling is abnormal with thickened cortices and increased bone density.

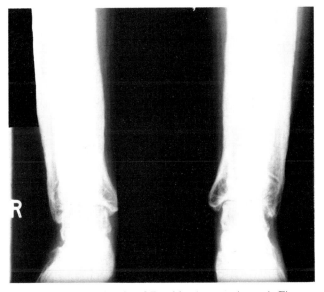

Fig. 63.17   W.P.—Adult—AP ankles (same patient as in Fig. 63.15)
There are radiating striations of increased bone density extending from the ankle joint. The diaphyses show greater thickening and sclerosis compared with Figure 63.15, at a younger age.

Upper limb : 5½ years–14

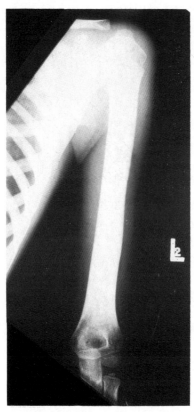

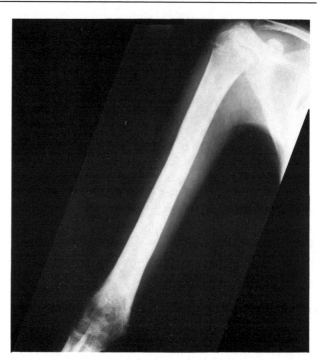

Fig. 63.19   R.H.—5½ years—AP right humerus (same patient as in Fig. 63.18)
Same as the left humerus (Fig. 63.18).

Fig. 63.18   R.H.—5½ years—Left humerus
The diaphyseal cortex is thickened with increased bone density.
Normal epiphyses and metaphyses.
Same patient : Figs 63.2, 63.3, 63.10, 63.14, 63.19, 63.20

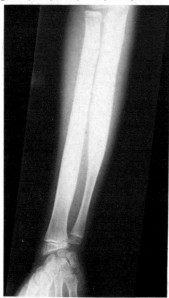

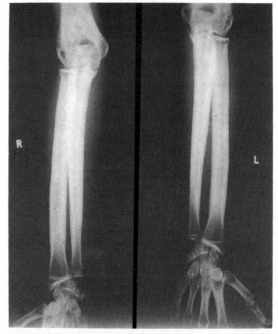

Fig. 63.20   R.H.—5½ years—AP right forearm (same patient as in Fig. 63.18)
There is pronounced widening of the radial and ulnar diaphyses with thickened, irregular and sclerotic cortices. The metaphyses and epiphyses are normal. It is interesting that the distal shaft, particularly of the ulna, has been spared so far.

Fig. 63.21   W.P.—14 years—AP arms
The diaphyses are wide due to thickening of the cortex. There is loss of normal modelling, but the metaphyses and epiphyses are normal.
Same patient : Figs 63.4, 63.6–63.8, 63.11, 63.12, 63.15–63.17

Chapter Sixty-four
CRANIODIAPHYSEAL DYSPLASIA

This is the most severe of the group, with progressive overgrowth of all areas of the skull, facial bones and mandible leading to mental retardation and cranial nerve palsies. Accompanied by diaphyseal thickening of the long bones—of little significance in comparison with the skull changes.

Inheritance

Probably autosomal recessive.

Frequency

Very rare indeed—under 0.1 per million prevalence.

Clinical features

Facial appearance—progressive, irregular distortion, with particularly severe changes in the nasal region: another disease once labelled 'leontiasis ossea'.

Intelligence—initially normal, but mental retardation and epilepsy develop.

Stature—initially normal, but later said to be stunted.

Presenting feature/age—facial deformity in early childhood, chiefly around the nose.

Associated anomalies—none.

Radiographic features

Skull—progressive thickening of all areas of the skull, obliteration of all sinuses and nasal passages.

Spine—may be normal, but the neural arch is sometimes sclerotic.

Limb girdles—some irregularity and thickening but this is of no great significance.

Long bones—diaphyseal thickening, variable in degree and sometimes regressing with age.

Bone maturation

Probably normal, but no data.

Biochemistry

Not known.

Differential diagnosis

From other rare sclerosing dysplasias of bone, but the great severity of the skull changes, with progressive worsening, together with *diaphyseal* rather than metaphyseal expansion should differentiate it. However, in *craniometaphyseal dysplasia* the young child may show some diaphyseal cortical thickening.

Progress/complications

All complications are related to the massive overgrowth of the skull; in particular nerve palsies of the IInd, VIIth and VIIIth cranial nerves are to be expected. Later mental retardation and epilepsy develop.

REFERENCES

Beighton P, Cremin B J 1980 Sclerosing bone dysplasias, Berlin, Springer
Tucker A S, Klein L, Antony G J 1976 Craniodiaphyseal dysplasia: evolution over a 5 year period. Skeletal Radiology 1/1:47–53

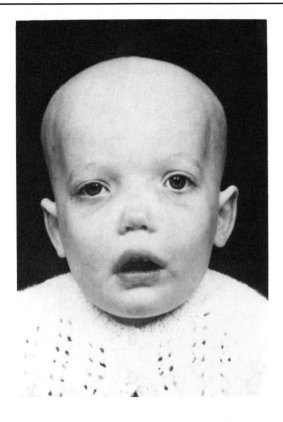
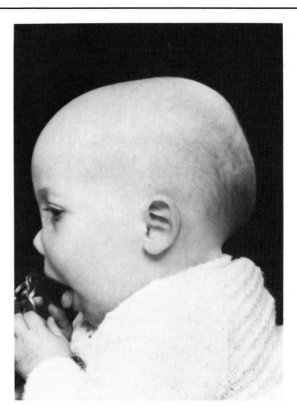
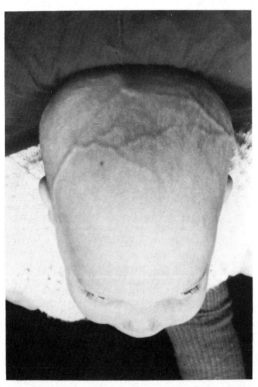

Figs 64.1, 64.2 and 64.3   The thickening of the skull with distortion of the nasal region is already seen at this young age (1 year).

Skull: 1–7 years

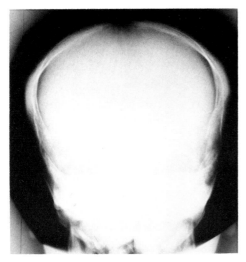

Fig. 64.4   K.B.—1 year—AP skull
The vault is thickened.
Same patient: Figs 64.5, 64.9, 64.12, 64.15

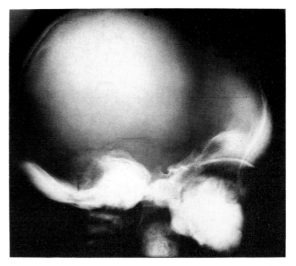

Fig. 64.5   K.B.—1 year—Lateral skull (same patient as in Fig. 64.4)
In addition to the pronounced thickening of the vault there is sclerosis with expansion of the base and paranasal sinuses.

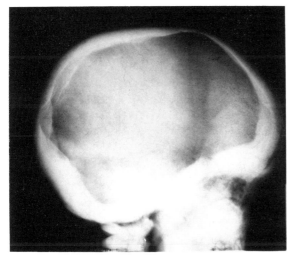

Fig. 64.6   J.W.—5 years—Lateral skull
Marked sclerosis and bony overgrowth involves the vault, base and paranasal sinuses.
Same patient: Figs 64.8, 64.10, 64.11, 64.13, 64.16, 64.17

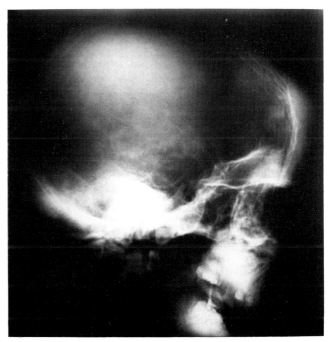

Fig. 64.7   J.H.—7 years—Lateral skull
Sclerosis of the maxilla and mandible are clearly seen, as well as of the skull. There is also malalignment of teeth.
Same patient: Fig. 64.14

Spine and chest: 1–5 years

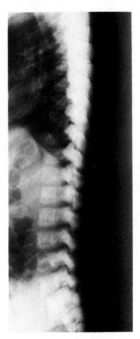

Fig. 64.8   J.W.—5 years—Lateral spine
There is patchy sclerosis of the vertebral bodies with a 'bone-in-a-bone' appearance. The sclerosis of the neural arches is even more marked.
Same patient: Figs 64.6, 64.10, 64.11, 64.13, 64.16, 64.17

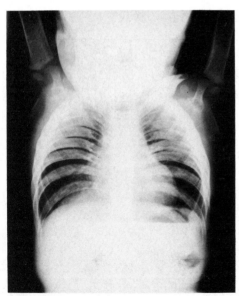

Fig. 64.9   K.B.—1 year—AP chest
The ribs show uniform widening and the middle thirds of the clavicles show cortical thickening and sclerosis. There is diaphyseal widening of the upper humeral shafts.
Same patient: Figs 64.4, 64.5, 64.12, 64.15

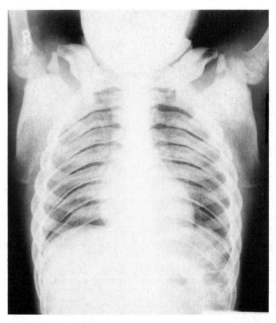

Fig. 64.10   J.W.—5 years—AP chest (same patient as in Fig. 64.8)
The ribs are widened and the clavicles show expansion with cortical sclerosis.

Pelvis and lower limb: 1–7 years

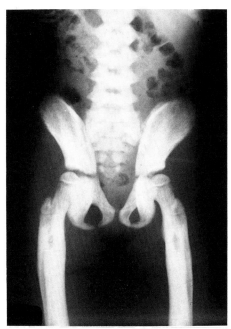

Fig. 64.11   J.W.—5 years—AP pelvis
The iliac wings appear greatly narrowed and there is sclerosis of the spine, bases of the ilia and ischia. The femoral shafts are widened and sclerotic and medullary cavities narrowed, but the capital femoral epiphyses and femoral necks are normal.
Same patient: Figs 64.6, 64.8, 64.10, 64.13, 64.16, 64.17

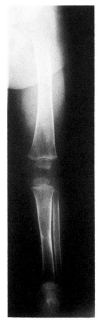

Fig. 64.12   K.B.—1 year—AP left lower limb
The diaphyses are wide but there is no sclerosis or cortical thickening at this age.
Same patient: Figs 64.4, 64.5, 64.9, 64.15

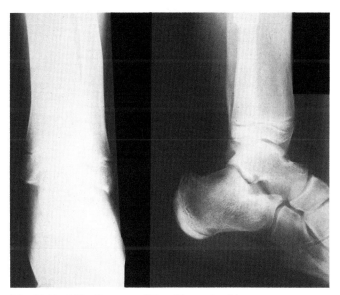

Fig. 64.14   J.H.—7 years—AP and lateral ankle
The diaphyses are wide and slightly bowed. Cortices are thickened but sclerosis is not marked.
Same patient: Fig. 64.7

Fig. 64.13   J.W.—5 years—AP left lower limb (same patient as in Fig. 64.11)
The diaphyses are wide with sclerosis, cortical thickening and narrowing of the intramedullary cavities.

Upper limb: 1–5 years

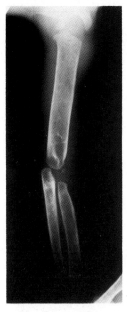

Fig. 64.15   R.B.—1 year—AP upper limb
Marked expansion of the diaphyses is present but there is no sclerosis
at this age.
Same patient: Figs 64.4, 64.5, 64.9, 64.12

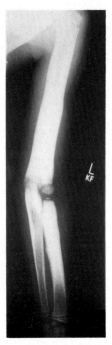

Fig. 64.16   J.W.—5 years—AP upper limb
The cortex of the diaphyses is irregularly thickened and sclerotic.
Same patient: Figs 64.6, 64.8, 64.10, 64.11, 64.13, 64.17

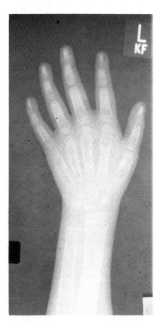

Fig. 64.17   J.W.—5 years—PA left hand (same patient as in Fig.
64.16)
There is patchy sclerosis and the diaphyses are wide.

Chapter Sixty-five
OSTEOPATHIA STRIATA
(with cranial sclerosis)

It is only the condition of linear bone striations associated with cranial sclerosis which is described here. Striations alone may be a feature of many 'increased bone density' syndromes, or occur alone, and *per se*, are of no importance.

Inheritance

Autosomal dominant.

Frequency

Extremely rare. Probably under 0.1 per million prevalence.

Clinical features

Facial appearance—normal, or enlarged frontal region.
Intelligence—normal.
Stature—normal.
Presenting feature/age—variable. The condition may be discovered incidentally, or deformity of the skull with or without deafness may present first.
Deformities—none apart from the skull.
Associated anomalies—none.

Radiographic features

Skull—increased density of vault, either generalised, or limited to anterior or posterior regions. Obliteration of sinuses.

Spine—probably normal, or there may be sclerosis off end-plates.
Long bones—striations, particularly around the knee and femoral neck. Patchy sclerosis of diaphyses.
Carpus/tarsus—patchy sclerosis.

Bone maturation

Probably normal.

Biochemistry

Not known.

Differential diagnosis

From many other disorders of increased bone density which may show linear striations occasionally. *Focal dermal hypoplasia* shows bone striations in early infancy, but skin changes (pigmentation, atrophy and fat herniation) should differentiate it.

Progress/complications

Apart from deafness, there appear to be no complications and the radiographic appearance of the long bones does not alter with age.

REFERENCES

Beighton P, Cremin B J 1980 Sclerosing bone dysplasias. Berlin, Springer
Franklyn P P, Wilkinson D 1978 Two cases of osteopathic striata, deafness and cranial osteopetrosis. Annals of Radiology 21:91–93

Skull: 10 years

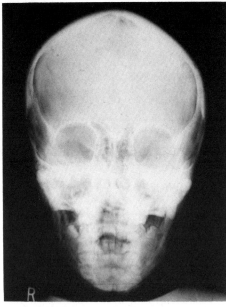

Fig. 65.1　N.C.—10 years—AP skull
There is uniform thickening and sclerosis of the vault and base.
Same patient: Figs 65.2, 65.9, 65.10, 65.12

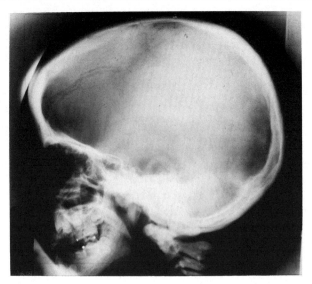

Fig. 65.2　N.C.—10 years—Lateral skull (same patient as in Fig. 65.1)
See Figure 65.1.

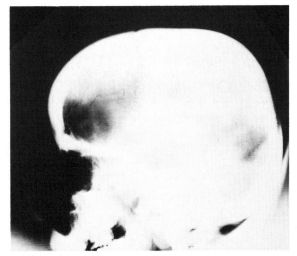

Fig. 65.3　B.L.—Lateral skull
This is an older child than that in Figure 65.2, with more extensive sclerosis of the vault and base.

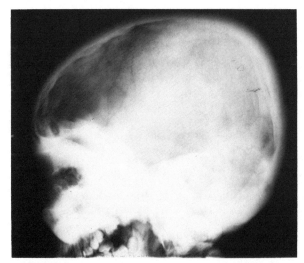

Fig. 65.4　Unknown male—10 years—Lateral skull
Most of the sclerosis involves the base, orbital regions and maxillae.
Same patient: Figs 65.5–65.8, 65.11, 65.13

Spine and chest: 10 years

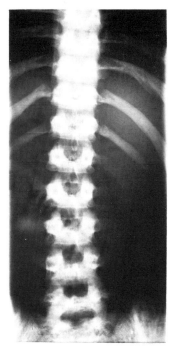

**Fig. 65.5 Unknown male—10 years—AP lower thoracic/
lumbar spine**
There is patchy increase in bone density throughout.
Same patient: Figs 65.4, 65.6–65.8, 65.11, 65.13

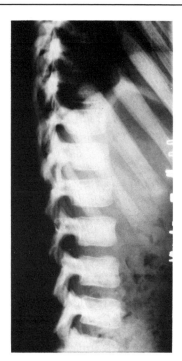

Fig. 65.6 Unknown male—10 years—Lateral lumbar spine
(same patient as in Fig. 65.5)
The sclerosis, unusually, does not involve the end-plates, but the
central part of the vertebrae.

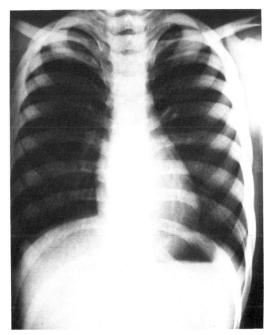

Fig. 65.7 Unknown male—10 years—PA chest (same patient as
in Fig. 65.5)
The ribs and clavicles are broad and sclerotic.

Pelvis and lower limb: 10 years

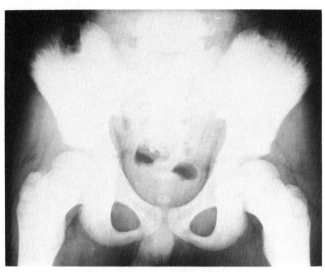

Fig. 65.8   Unknown male—10 years—Pelvis
There are radiating striations in the iliac wings together with
generalised sclerosis.
Same patient: Figs 65.4–65.7, 65.11, 65.13

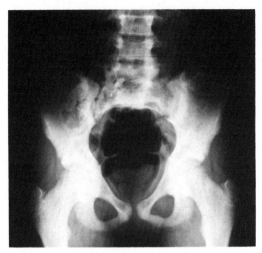

Fig. 65.9   N.C.—10 years—Pelvis
The pelvis is normal apart from mild sclerosis but there are striations
of the femoral necks.
Same patient: Figs. 65.1, 65.2, 65.10, 65.12

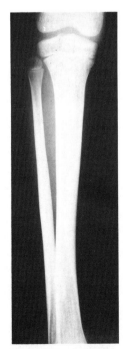

Fig. 65.10   N.C.—10 years—AP right lower limb (same patient
as in Fig. 65.9)
There are striations in the metaphyseal regions.

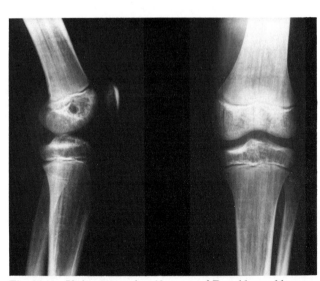

Fig. 65.11   Unknown male—10 years—AP and lateral knee
(same patient as in Fig. 65.8)
The striations extend through all the metaphyses and can also be seen
(less marked) in the epiphyses.

Upper limb: 10 years

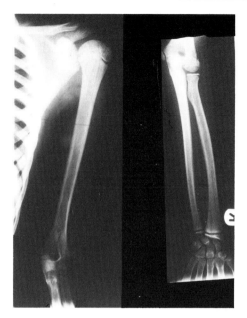

Fig. 65.12 N.C.—10 years—AP right upper limb
Striations are present in the metaphyses.
Same patient: Figs 65.1, 65.2, 65.9, 65.10

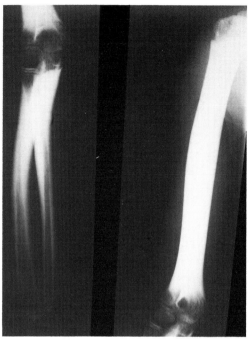

Fig. 65.13 Unknown male—10 years—AP right upper limb
Striations cannot be seen but there is failure of modelling and
increased density throughout.
Same patient: Figs 65.4–65.8, 65.11

Chapter Sixty-six
OSTEOPOIKILOSIS

A spotted appearance of bones which is associated with skin papules in about 10% of cases. It is of little clinical importance.

Inheritance

Autosomal dominant.

Frequency

Extremely rare, under 0.1 per million prevalence.

Clinical features

Facial appearance—normal.
Intelligence—normal.
Stature—normal.
Presenting feature/age—skin lesions, if present, otherwise the spotted appearance of the bones is discovered incidentally.
Deformities—none.
Associated anomalies—none.

Radiographic features

Skull—normal.
Spine—normal.
Thorax—normal.
Long bones—multiple sclerotic areas affect the ends of the bones rather than the diaphyses.

Limb girdles—not usually involved, with the exception of the pelvis.
Carpus/tarsus—may be affected.

Bone maturation

Probably normal.

Biochemistry

Not known.

Differential diagnosis

No real difficulty, with the possible exception of *melorheostosis*, in which skin lesions may also be a feature. However, its linear streaks of increased bone density should differentiate it.

Progress/complications

Both enlargement and regression has been reported, but the condition appears to be essentially benign.

REFERENCES

Beighton P, Cremin B J 1980 Sclerosing bone dysplasias. Berlin, Springer

Osteopoikilosis

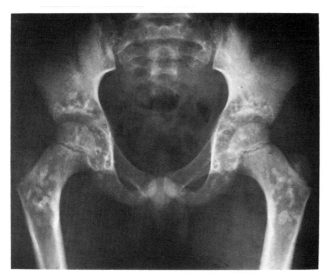

Fig. 66.1 A.G.—13 years—Pelvis
The spotted appearance is present in both metaphyses and epiphyses.

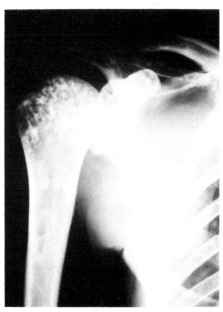

Fig. 66.2 G.A.—Adult—Right shoulder
There is 'stippling' around the shoulder joint, and of a lesser density, down the shaft of the humerus.

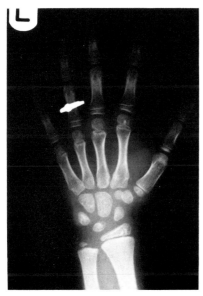

Fig. 66.3 J.F.—8 years—PA left hand
There are small islands of increased density in the lower radial epiphysis, hamate, triquetral, lunate and second metacarpal.

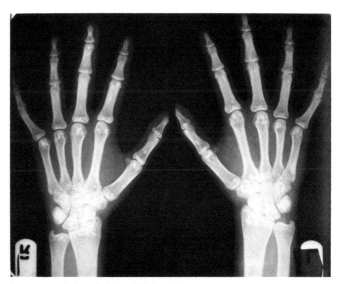

Fig. 66.4 A.A.—Adult—PA hand
The stippling is seen adjacent to all joints.

Chapter Sixty-seven
MELORHEOSTOSIS

Usually unilateral, but always asymmetrical bony lesions, with linear streaks of hyperostosis developing along the main axis of the long bones, associated with pain, soft tissue contractures, fibrosis and skin abnormalities.

Inheritance

Non-genetic.

Frequency

Possibly prevalence of 1 per million.

Clinical features

Facial appearance—normal.
Intelligence—normal.
Stature—normal, although the growth of individual bones may be affected.
Presenting feature/age—indurated skin lesions and contractures may be present at birth, but more usually appear during childhood. Pain is an early feature, with bony swelling around affected joints.
Deformities—contractures, particularly of palmar and plantar fascia. Bony swellings, with ectopic deposits of bone around joints limiting their movement.
Range of severity—variable. Some are discovered only incidentally in adult life.
Associated anomalies—none.

Radiographic features

Skull, spine and thorax—rarely affected.
Long bones—sclerotic linear streaks along the long axis of the bone, often likened to dripping candle-grease. Frequently only one bone, or one limb is involved. Streaks and spots may affect the epiphyses and bone is deposited in soft tissues around the joints.

Bone maturation

Normal.

Biochemistry

Not known.

Differential diagnosis

Not usually in doubt, but before bony lesions develop it can be confused with other disorders associated with contractures.

Progress/complications

Lesions progress during childhood and more slowly during adult life. Pain and limitation of joint movement are the main disability. Malignant change has not been reported.

REFERENCES

Campbell C J, Papdemitriou T, Bonfiglio M 1968 Melorheostosis: a report of the clinical, roentgenographic and pathological findings in fourteen cases. Journal of Bone and Joint Surgery 50A:1281

Younge D, Drummond D, Herring J, Cruess R L 1979 Melorheostosis in children. Clinical features and natural history. Journal of Bone and Joint Surgery Series B 61/4:415–418

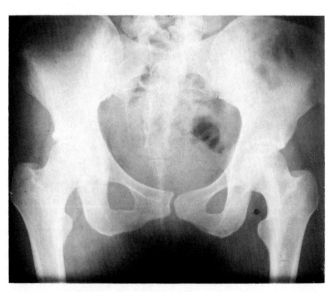

Fig. 67.1   W.D.—Adult—AP pelvis
A wide linear sclerotic streak is present in the head and extending down the neck of the left femur.

Lower limb: Adult

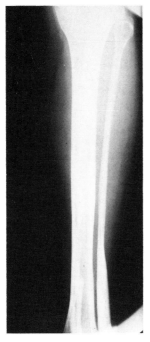

Fig. 67.2   W.—Adult—AP left tibia and fibula
There is a linear sclerotic streak extending down the lower quarter of the tibia and thickening of the lower fibula.

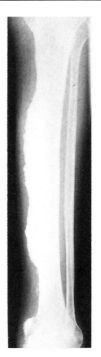

Fig. 67.3   Unknown—Adult—AP left tibia and fibula
The irregularly expanded and sclerotic area of the tibial diaphysis does not appear to affect the lateral aspect of the bone.

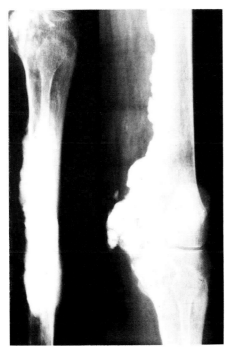

Fig. 67.4   W.H.—Adult—Laterals of knee, tibia and fibula
Irregular masses of sclerotic bone appear to be 'stuck' to the surface of the shafts and there will clearly be interference with joint movement.
Same patient: Figs 67.7, 67.8

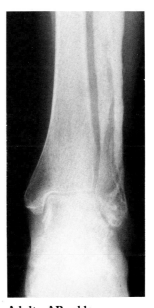

Fig. 67.5   A.S.—Adult—AP ankle
The tibia appears normal, whereas the fibula has irregular, thickened areas overlying the cortex.

Foot : 4 years–Adult

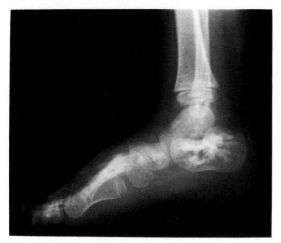

Fig. 67.6   P.W.—4 years—Lateral foot and ankle
There are sclerotic bands along the axis of the calcaneum and in one
or two of the other tarsal bones.

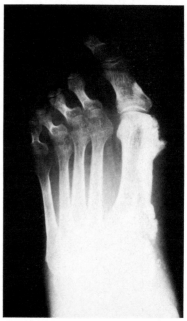

Fig. 67.7   W.H.—Adult—PA foot
Only the hallux is involved and the sclerotic areas are confined to the
medial border.
Same patient : Figs 67.4, 67.8

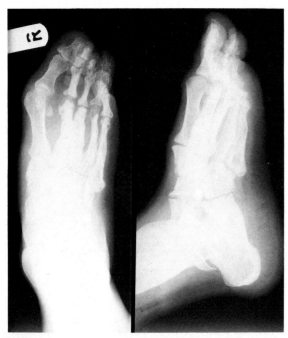

Fig. 67.8   W.H.—Adult—PA and lateral right foot (same
patient as in Fig. 67.7)
The whole tarsus and foot are involved in the sclerotic process with
the exception of the hallux. The third ray is abnormally short,
probably because of premature fusion of those epiphyses affected by
the disease.

Upper limb : 6 months–Adult

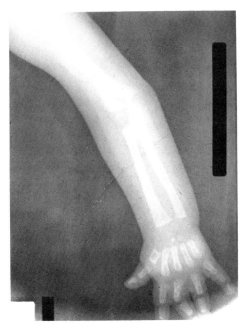

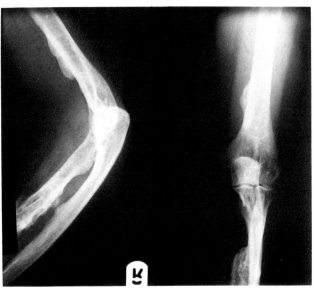

Fig. 67.10   M.McC.—63 years—AP and lateral right elbow
The elbow joint is normal, but lumps of sclerotic bone are present on
the shafts of the humerus, radius and ulna.
Same patient : Fig. 67.11

Fig. 67.9   Unknown—6 months—AP upper limb
Finger contractures are present and there are rather ill-defined linear
bands of sclerotic bone seen most obviously in the lower humerus, but
also present in the forearm and hand. The first metacarpal is short.

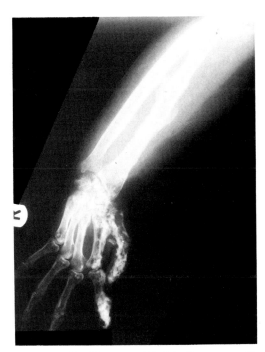

Fig. 67.11   M.McC.—63 years—AP right forearm and hand
(same patient as in Fig. 67.10)
The radius is seen again (compare Fig. 67.10) but the hand shows not
only streaks of sclerosis with irregular expansion, but osteopoikilosis
in the third, fourth and fifth rays and where modelling is normal.

Chapter Sixty-eight
FAMILIAL HYPERTROPHIC OSTEOARTHROPATHY (pachydermoperiostitis)

The commoner adult form is associated with skin lesions, sclerosis of the skull and long bones and clubbing of the fingers and toes. In the infant, there is clubbing of the fingers/toes and periosteal new bone formation.

Inheritance

Autosomal dominant, but autosomal recessive cases have been described also.

Frequency

Not known, but rare.

Clinical features

Facial appearance—in adults there is seborrhoeic hyperplasia of the skin. In the infant (rare form) there are eczematous skin eruptions.

Intelligence—normal.

Stature—normal.

Presenting feature/age—the common form presents in adolescence, with slowly progressive sclerosis of long bones. In the infant there is periosteal new bone formation.

Deformities—none.

Associated anomalies—none apart from the clubbing.

Radiographic features

Skull—in the infant there is defective cranial ossification and Wormian bones. In the adult, sclerosis of the vault and base.

Spine—sclerosis of the end-plates; at a later age intervertebral foramina may be narrowed.

Thorax—thickened ribs in the adult.

Long bones—in the infant there is periosteal new bone formation, which is organised by the age of 5 years. In the adult the thickening starts at the distal ends of the forearms and legs later extending to the whole diaphysis.

Hands/feet—in the infant the terminal phalanges are hypoplastic, but with thickened soft tissues, and this feature remains. In the adult, there is no bone hypoplasia but soft tissue hyperplasia only (clubbing).

Bone maturation

Not known, probably normal.

Biochemistry

Not known.

Differential diagnosis

From other conditions in which clubbing of the fingers/toes is a feature—principally *secondary hypertrophic osteoarthropathy* but this is associated with pulmonary disease, and is non-familial.

In the infant, conditions with defective cranial ossification and Wormian bones need to be differentiated, particularly *pycnodysostosis, osteogenesis imperfecta* and *Menkes syndrome*. The normal bone density and clubbing of the digits should distinguish pachydermoperiostitis.

REFERENCES

Chamberlain D S, Whitaker J, Silverman F N 1965 Idiopathic osteoarthropathy and cranial defects in children (familial idiopathic osteoarthropathy). Am J Radiol 93:408

Cremin B J 1970 Familial idiopathic osteoarthropathy of children: A case report and progress. Br J Radiol 43:568

Currarino G, Tierney R C, Giesel R G, Weihl C 1961 Familial idiopathic osteoarthropathy. Am J Radiol 85:633

Skull: Birth–18 months

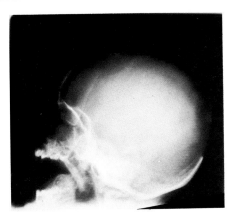

Fig. 68.1 W.A.—Neonate—Lateral skull
The sutures are wide and multiple Wormian (sutural) bones are present in the lambdoid and sagittal sutures.
Same patient: Figs 68.2–68.8

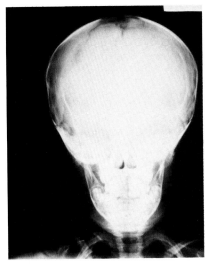

Fig. 68.2 W.A.–9 months—AP skull (same patient as in Fig. 68.1)
Large Wormian bones are present in the sagittal suture.

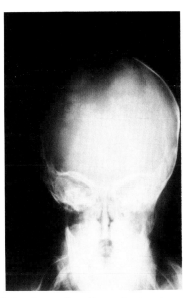

Fig. 68.3 W.A.—9 months—Towne's view (same patient as in Fig. 68.1)
The sutures and anterior fontanelle are wide open and multiple Wormian bones are present.

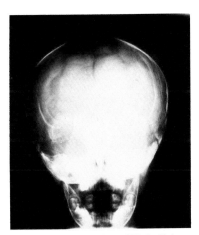

Fig. 68.4 W.A.—1 year 6 months—AP skull (same patient as in Fig. 68.1)
The anterior fontanelle is still open and Wormian bones are present.

Limbs : 9 months–2 years

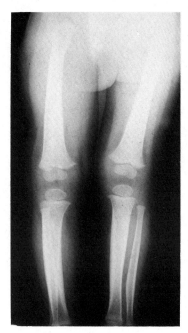

Fig. 68.5   W.A.—1 year 6 months—AP lower limbs
There are multiple layers of periosteal new bone formation along
the shafts of all long bones.
Same patient : Figs 68.1–68.4, 68.6–68.8

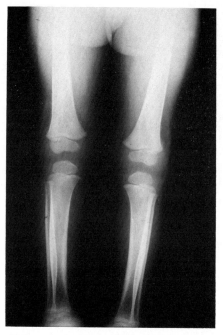

Fig. 68.6   W.A.—2 years 6 months—AP lower limbs (same
patient as in Fig. 68.5)
The periosteal new bone has organised with residual modelling
deformities and widening of the shafts.

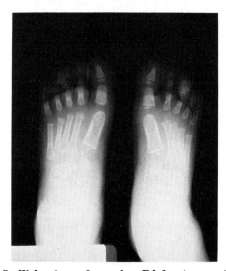

Fig. 68.7   W.A.—1 year 6 months—PA feet (same patient as
in Fig. 68.5)
The terminal phalanges are hypoplastic and pointed, but the soft
tissues over the tufts are thickened (clubbing)

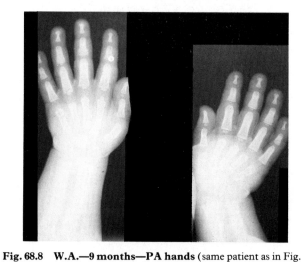

Fig. 68.8   W.A.—9 months—PA hands (same patient as in Fig.
68.5)
Modelling, particularly of the terminal phalanges, is abnormal with
especially pronounced tufts.

Adult

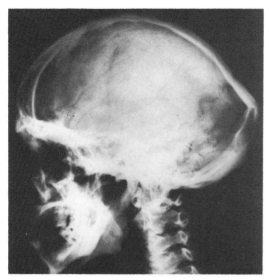

Fig. 68.9   H.F.—50 years—Lateral skull
There is patchy sclerosis of the vault, and marked thickening of the base.
Same patient: Figs 68.10, 68.11

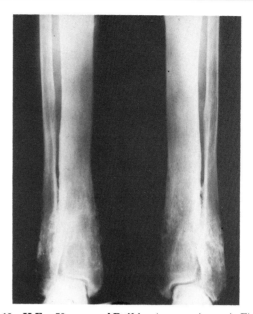

Fig. 68.10   H.F.—50 years—AP tibiae (same patient as in Fig. 68.9)
There is patchy sclerosis and thickening throughout the shaft with cortical irregularity.

Fig. 68.11   H.F.—50 years—AP forearm (same patient as in Fig. 68.9)
Similar to Figure 68.10.

Section Fourteen
TUMOUR-LIKE BONE DYSPLASIAS

Chapter Sixty-nine
DIAPHYSEAL ACLASIS
(multiple hereditary exostoses)

Multiple exostoses, most commonly at the ends of the long bones, but also involving the pectoral and pelvic girdles, ribs, and (rarely) vertebrae.

Inheritance

Autosomal dominant, probably full penetrance, but mildly affected individuals are difficult to identify. Mild and severe cases occur within the same family.

Frequency

One of the commonest of the skeletal dysplasias with a probable prevalence of around 9 per million index patients, or 23 per million including affected relatives.

Clinical features

Facial appearance—normal.
Intelligence—normal.
Stature—unaffected in mild cases, but some three-quarters of patients are below average height.
Presenting feature/age—occasionally bony lumps are noted at birth, but more usually in early childhood. Nerve compression (eg lateral popliteal) or fracture of an exostosis are sometimes presenting features.
Deformities—related to the bony outgrowths, and also to disproportionate growth between radius and ulna, tibia and fibula, with deformity at the elbow, wrist, knee and ankle (valgus, varus, bowing, shortening of individual bones).
Associated anomalies—none.

Radiographic features

Skull—normal.
Spine—usually normal, but occasional exostoses.
Long bones—cartilage-capped exostoses at the ends, arising from the juxta-epiphyseal region, pointing away from the epiphyses. As the child grows they appear to move down the diaphysis.
Ribs, pectoral and pelvic girdles—frequently the site of exostoses, from the minute to very large bony masses which obstruct joint movement.

Bone maturation

Probably normal.

Biochemistry

Not known.

Differential diagnosis

The only other disorder with multiple exostoses exactly similar to diaphyseal aclasis is the *tricho-rhino-phalangeal syndrome*, and they are not an invariable feature. This syndrome is characterised by an unusual facies (long philtrum, wide mouth), sparse, slow growing hair, cone epiphyses in the hands and sometimes Perthes-like changes in the capital femoral epiphyses.

In *Ollier's disease* there may be a similar forearm deformity (short ulna, curved radius), but there are no exostoses; cartilage 'islands' are present in the shaft of long bones, and the disease is sometimes confined to one bone and is typically asymmetrical.

Small exostoses at the ends of long bones can occur in *fibrodysplasia ossificans progressiva*; within the epiphysis in *dysplasia epiphysealis hemimelica*; and rarely in other generalised skeletal dysplasias.

Progress/complications

The exostoses cease growth with the growth of the child, and it is unusual for new ones to appear in adult life. Deformities are noted above. Chondrosarcoma is the most serious complication, but probably occurs only in under 2% of affected individuals over the age of 20 years. Malignant change is more likely in the exostoses of the pectoral and pelvic girdles than in those affecting the long bones.

REFERENCES

Shapiro F, Simon S, Climcher M J 1979 Hereditary multiple exostoses: anthropometric, roengenographic and clinical aspects. Journal of Bone and Joint Surgery 61A:815–824
Solomon L 1963 Hereditary multiple exostoses. Journal of Bone and Joint Surgery 45B:292
Voutsinas S, Wynne-Davies R 1983 The infrequency of malignant disease in diaphyseal aclasis and neurofibromatosis. Journal of Medical Genetics 20:345–349

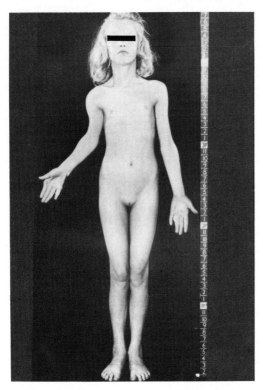

Fig. 69.1 Normal height and body proportions, but with cubitus valgus and deformity of the upper arm. Small exostoses can be seen on the fingers of both hands.

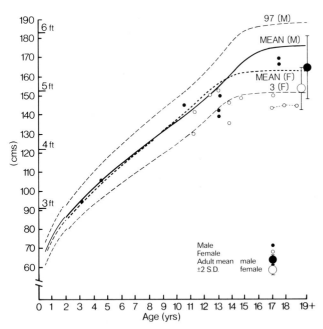

Fig. 69.2 Diaphyseal aclasis—percentile height.

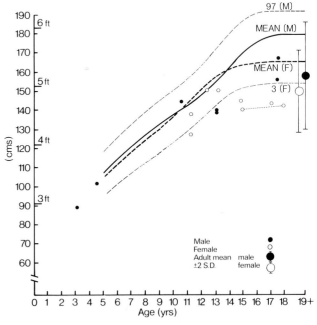

Fig. 69.3 Diaphyseal aclasis—percentile span.

Spine: 6–9 years

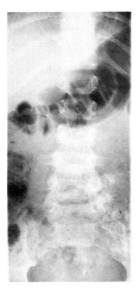

Fig. 69.4 S.C.—6 years—AP lower dorsal and lumbar spine
There is a large exostosis overlying the left sacro-iliac joint, extending
up to L3.
Same patient: Figs 69.5–69.7, 69.11, 69.14, 69.20

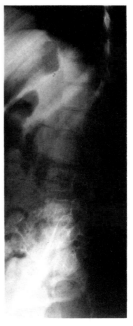

**Fig. 69.5 S.C.—6 years—Lateral lower dorsal and lumbar
spine** (same patient as in Fig. 69.4)
The bone mass extends forwards into the abdomen (causing ureteral
obstruction). There is probably also an exostosis on the antero-
inferior border of L3.

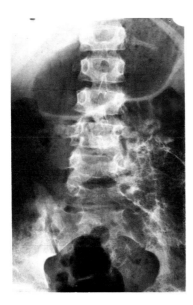

Fig. 69.6 S.C.—9 years 3 months—AP lumbar spine (same
patient as in Fig. 69.4)
The exostosis overlying the sacro-iliac region has enlarged.

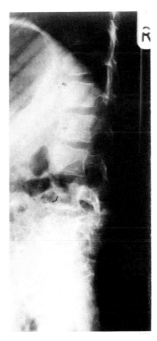

Fig. 69.7 S.C.—9 years 3 months—Lateral lumbar spine
(same patient as in Fig. 69.4)
In addition to the increase in size of the pelvic exostosis, the one on
L3 has also enlarged.

Thorax : 12 years–Adult

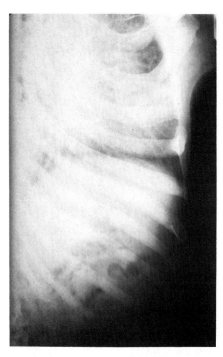

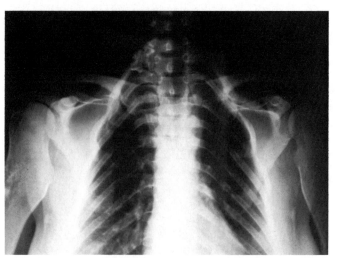

Fig. 69.9   C.S.—28 years—AP shoulders
There are exostoses arising from both proximal humeral diaphyses.
Another is present at the posterior end of the right first rib. They
show patchy calcification, indicating the cartilaginous nature of part
of them.

Fig. 69.8   C.C.—12 years 4 months—Oblique view of ribs
There is a bony exostosis, probably arising from the scapula.
Same patient : Fig. 69.25

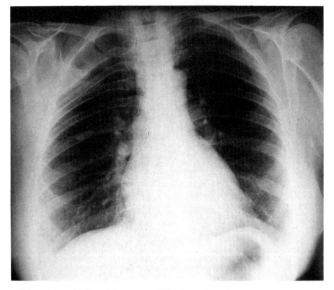

Fig. 69.10   D.B.—61 years—PA chest
There is widening of the right second to third intercostal space caused
by an exostosis, probably arising from the 2nd rib. There is an area of
increased density in the posterior right seventh rib, presumably
caused by an exostosis overlying this area.

Pelvis: 4–9 years

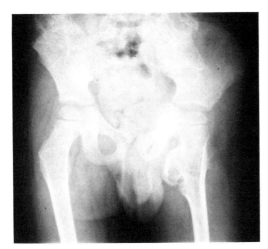

Fig. 69.11 S.C.—4 years—Pelvis
There are large exostoses on both iliac wings, and also in the left intertrochanteric region.
Same patient: Figs 69.4–69.7, 69.14, 69.20

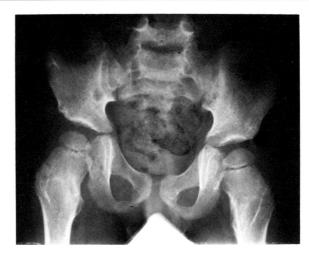

Fig. 69.12 J.K.—4 years 10 months—Pelvis
The femoral necks are broad with abnormal modelling ('sessile' exostoses). Cortical irregularity of the right iliac wing demonstrates early exostosis development here.

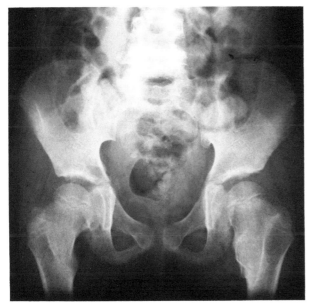

Fig. 69.13 S.K.—6 years 8 months—Pelvis
The femoral necks are short and broad and there is a small exostosis in the region of the left lesser trochanter.
Same patient: Figs 69.22, 69.46

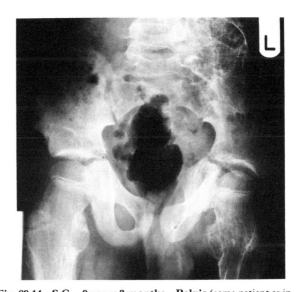

Fig. 69.14 S.C.—9 years 3 months—Pelvis (same patient as in Fig. 69.11)
The right iliac exostosis has been excised, the left is now larger. The left trochanteric exostosis was also excised but has recurred, and there is now a sessile bony growth on the right.

Pelvis: 14 years–Adult

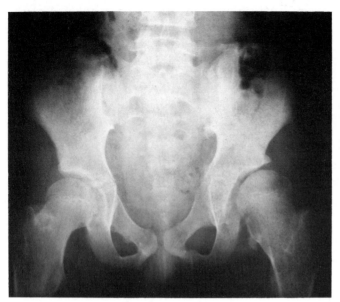

Fig. 69.15   A.Y.—14 years 10 months—Pelvis
Multiple exostoses arise medially from the left femoral neck and also around the right greater trochanter. The left femoral head is incompletely covered.
Same patient: Figs 69.26, 69.48

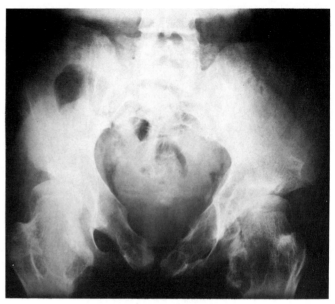

Fig. 69.16   E.P.—15 years 3 months—Pelvis
The femoral heads are flattened and incompletely covered, and the necks broad. Exostoses arise from the trochanters and the symphysis pubis. There are unusual radiating striations in the iliac wings, presumably representing the bases of small exostoses, and the whole pelvic configuration is asymmetrical.
Same patient: Fig. 69.32

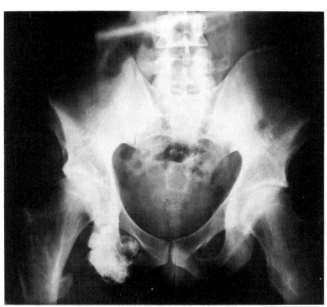

Fig. 69.17   R.T.—41 years 9 months—Pelvis
A large exostosis arises from the right ischium, and smaller ones from the trochanters. The increased density of the large exostosis is presumably calcification in its cartilage cap.

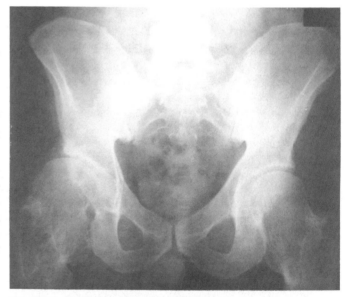

Fig. 69.18   R.F.—50 years—Pelvis
The femoral necks are broad and the femoral heads flat and poorly covered. Multiple exostoses are present around the trochanters.

Lower limb : 2–6 years

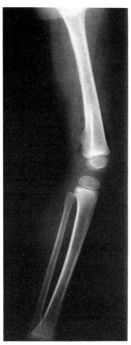

Fig. 69.19   J.K.—2 years 10 months—AP and lateral left lower limb
There is a small cartilage capped exostosis arising from the distal femoral metaphysis, pointing away from the knee joint.

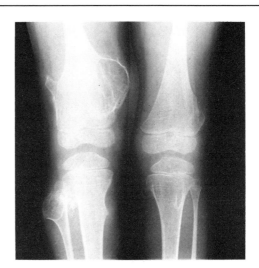

Fig. 69.20   S.C.—5 years 6 months—AP knees
There are large sessile and pedunculated exostoses around the right knee, and smaller growths on the left.
Same patient : Figs 69.4–69.7, 69.11, 69.14

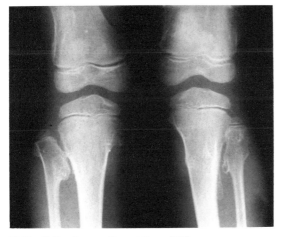

Fig. 69.21   P.S.—5 years 9 months—AP knees
Exostoses arise from the metaphyses around the joints. The epiphyses show some flattening and abnormal modelling.
Same patient : Fig. 69.31

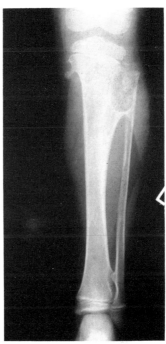

Fig. 69.22   S.K.—6 years 8 months—AP left tibia and fibula
Exostoses are present at the proximal and distal ends of the leg bones.
Same patient : Figs 69.13, 69.46

Lower limb: 9–14 years

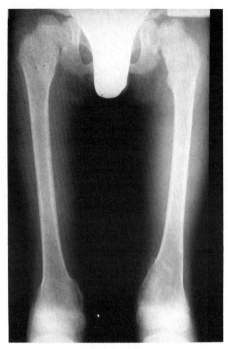

Fig. 69.23   C.S.—9 years 8 months—AP femora
The femoral necks and intertrochanteric regions are broad and the
capital femoral epiphyses rather flat. There are exostoses at the distal
femoral metaphyses associated with abnormal modelling here.
Same patient: Figs 69.24, 69.39

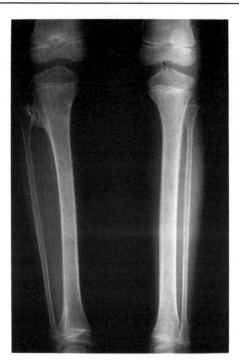

Fig. 69.24   C.S.—9 years 8 months—AP tibiae and fibulae
(same patient as in Fig. 69.23)
There is a large exostosis arising from the right upper tibia causing
lateral displacement of the fibula. The small lucent streaks in the
distal tibial metaphyses represent the bases of small exostoses.

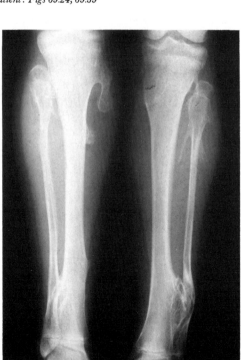

Fig. 69.25   C.C.—12 years—AP tibiae and fibulae
There are multiple exostoses at both ends of the bones. The one
arising from the lower end of the left tibia is causing compression
and deformity of the adjacent fibula.
Same patient: Fig. 69.8

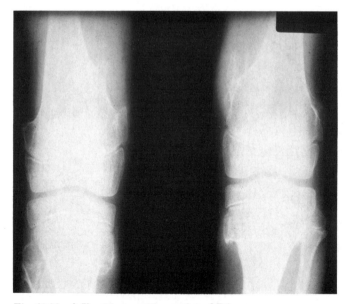

Fig. 69.26   A.Y.—14 years 10 months—AP knees
Exostoses are arising from all bones. The proximal end of the right
fibula is disproportionately short.
Same patient: Figs 69.15, 69.48

Knee: 19 years–Adult

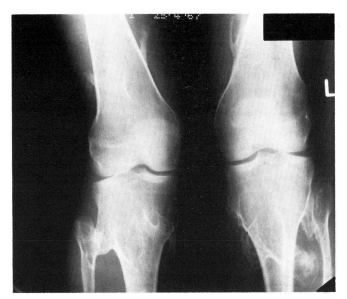

Fig. 69.27   X.X.—19 years—AP knees
Multiple exostoses.

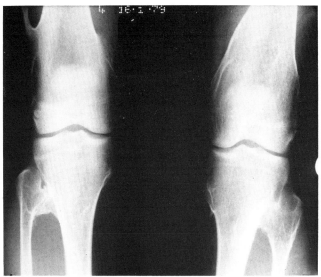

Fig. 69.28   A.D.—40 years—AP knees
There are multiple exostoses and the metaphyses are wide and
abnormally modelled.

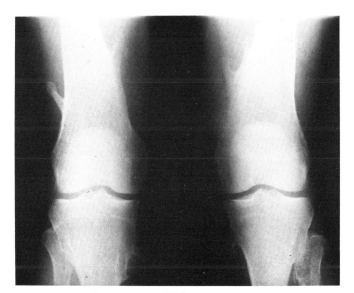

Fig. 69.29   J.C.—52 years—AP knees
Multiple exostoses.

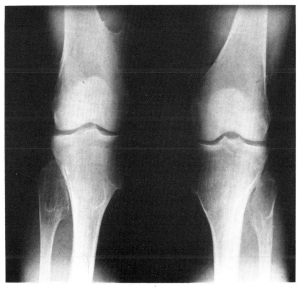

Fig. 69.30   D.B.—61 years—AP knees
Multiple exostoses.
Same patient: Fig. 69.10

Ankle: 5 years–Adult

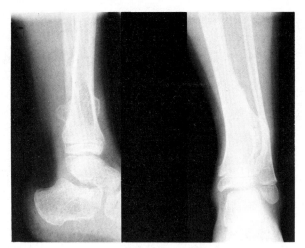

Fig. 69.31   P.S.—5 years 9 months—AP and lateral left ankle
There are several exostoses arising from the distal tibia and fibula, causing some bowing of the tibia.
Same patient: Fig. 69.32

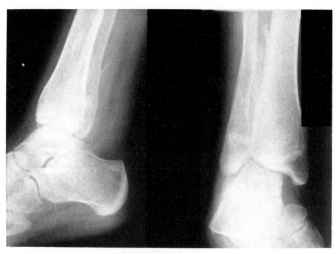

Fig. 69.32   E.P.—15 years 3 months—AP and lateral right ankle
There are multiple small exostoses arising from the distal tibia and fibula. The lateral side of the tibial epiphysis is flattened and the talus tilted.
Same patient: Fig. 69.16

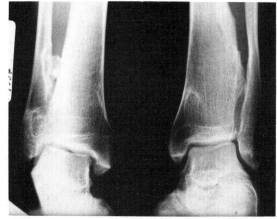

Fig. 69.33   T.R.—Adult—AP ankles
Exostoses arise from the distal tibiae and there is some secondary deformity of the fibulae and tilting of the tali. The ankle mortice is poorly formed, especially on the lateral side.

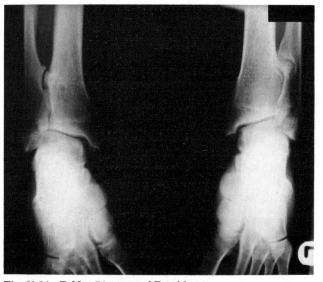

Fig. 69.34   R.M.—54 years—AP ankles
Exostoses arise from both distal tibiae and there are deformities of the adjacent fibulae. The mortice is poorly formed.

Scapula: 8–13 years

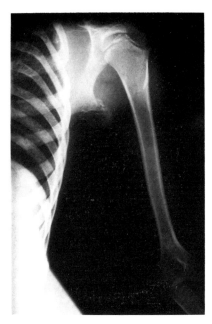

Fig. 69.35 A.E.—8 years—AP left scapula
A large cartilage-capped exostosis arises laterally from the inferior angle of the scapula.

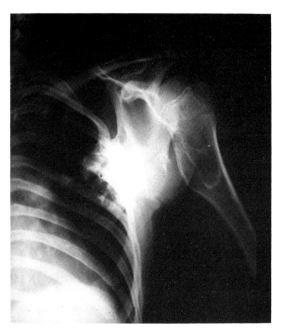

Fig. 69.36 C.S.—13 years 5 months—AP scapula
A large irregular exostosis arises from the antero-medial edge of the wing of the scapula, and another is present in the proximal humerus.
Same patient: Figs 69.37, 69.47

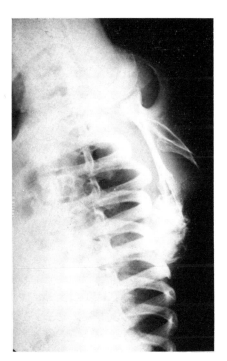

Fig. 69.37 C.S.—13 years 5 months—Lateral scapula (same patient as in Fig. 69.36)
See caption to Figure 69.36

Humerus : 6 years–Adult

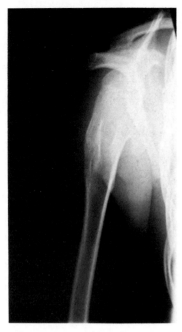

Fig. 69.38 J.J.—6 years—AP shoulder
There is widening of the proximal humerus with cortical thickening
and irregularity, and a large exostosis arises from the medial border.
Same patient : Fig. 69.40

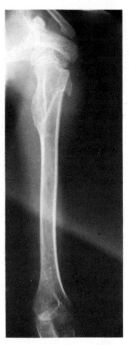

Fig. 69.39   C.S.—9 years 8 months—AP humerus
Exostoses arise from the proximal humerus.
Same patient : Figs 69.23, 69.24

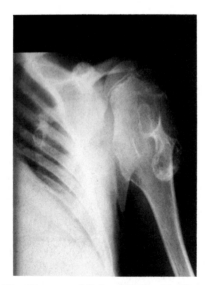

Fig. 69.40   J.J.—10 years—AP shoulder (same patient as in Fig.
69.38)
There is widening of the proximal humerus and a large cartilage-
capped exostosis is present. There is a second exostosis on the medial
border of the scapula.

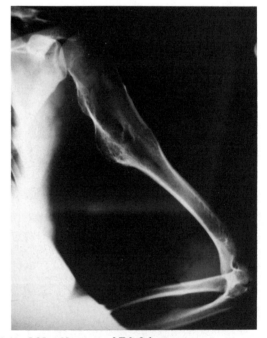

Fig. 69.41   J.M.—28 years—AP left humerus
In this adult patient exostoses arise from the proximal and mid-
humerus and are associated with loss of normal bone modelling.

Forearm : 3 years–Adult

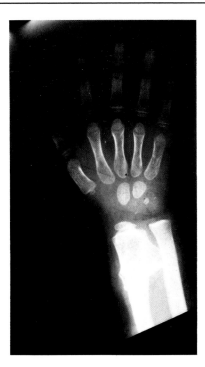

Fig. 69.42 Unknown—3 years 6 months—PA wrist and hand
Large exostosis of lower radius abutting against the ulna.

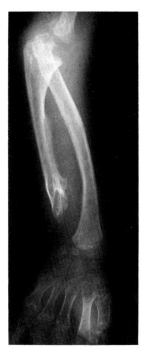

Fig. 69.43 C.R.—9 years—AP forearm
There is lateral bowing of the radius and the distal end of the ulna is short and pointed (reverse Madelung deformity), with an exostosis at its distal end. The interosseous ridge is unusually prominent.

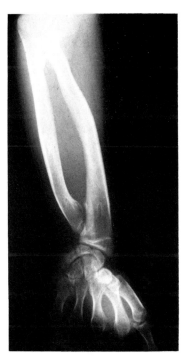

Fig. 69.44 D.C.—17 years 6 months—AP forearm
There is lateral bowing of the radius with some thickening of the area of attachment of the interosseous membrane. The distal ulna is short and bowed, with a very small epiphysis.

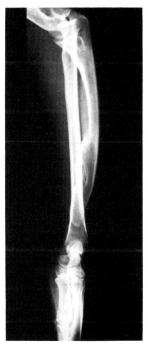

Fig. 69.45 G.L.—Adult—AP forearm
Exostoses are present arising from the mid-shaft of the ulna and the distal ulna is short and pointed.

Hand: 9–19 years

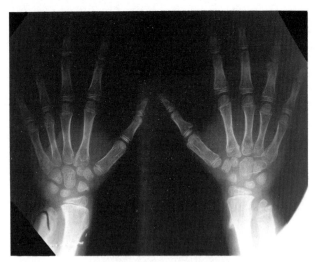

Fig. 69.46   S.K.—9 years—PA both hands
There are exostoses at the lower ends of the radius and ulna, and
smaller ones on the right first metacarpal and proximal phalanx.
Same patient: Figs 69.13, 69.22

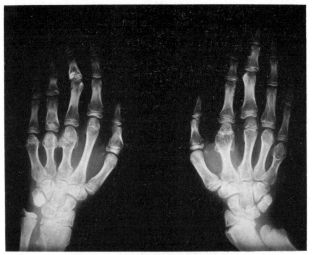

Fig. 69.47   C.S.—13 years 5 months—PA both hands
There is asymmetrical shortening of metacarpals 3, 4 and 5 on the
left, and 2, 4 and 5 on the right, not all associated with obvious
exostoses. There is also disordered growth (nearly cone-shaped) of
the middle phalanx of the left middle finger.
Same patient: Figs 69.36, 69.37

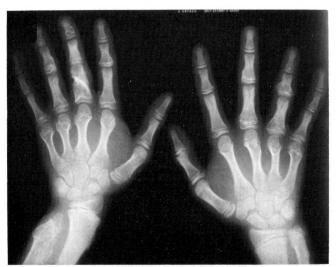

Fig. 69.48   A.Y.—14 years 10 months—PA hands
Multiple phalangeal exostoses are present. Some metacarpals show
shortening due to premature fusion of the epiphyses, and a chevron
deformity of the epiphysis is present in the middle phalanx of the
right ring finger. The distal right ulna is short and pointed.
Same patient: Figs 69.15, 69.26

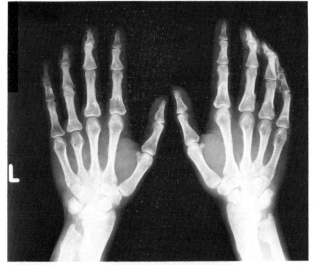

Fig. 69.49   D.P.—19 years—PA both hands
The fourth metacarpal on the left is short and there are several small
exostoses on the digits (mainly proximal phalanges, but also middle).

Chapter Seventy
OLLIER'S DISEASE
(enchondromatosis)

A characteristically asymmetrical disorder sometimes unilateral, or even confined to the radial or ulnar (tibial or fibular) side of the limb. Unossified cartilage remains in the metaphysis and diaphysis, often forming huge tumour masses.

Inheritance

Non-genetic. (Rare milder form is autosomal dominant.)

Frequency

Possible prevalence of 1 per million.

Clinical features

Facial appearance—normal.
Intelligence—normal.
Stature—normal, although affected bones are short and there may be asymmetrical limb shortening.
Presenting feature/age—bony lumps, usually in childhood.
Deformity—may be severe—large tumour masses can develop and may require resection. A forearm deformity may be caused by distal shortening of the ulna accompanied by lateral bowing of the radius.
Range of severity—variable. There is a rare, milder form of the disease, of autosomal dominant inheritance, principally affecting the hands and feet.
Associated anomalies—none.

Radiographic features

Skull and spine—very rarely affected.
Appendicular skeleton—affected bones show irregular translucent areas in the metaphyses, with streaks of cartilage alternating with areas of increased density. The diaphysis is not usually extensively involved. The epiphyses are normal in childhood, but later, areas of increased density and mottling become apparent.

Bone maturation

Normal.

Biochemistry

Not known.

Differential diagnosis

Solitary enchondromata need differentiating. The forearm deformity of Ollier's disease is similar to that in severe cases of *diaphyseal aclasis*, but exostoses elsewhere differentiate the latter. In *polyostotic fibrous dysplasia* the individual lytic areas show a thickened (perhaps 5 mm) sclerotic lobulated margin, unlike the pencil-thin sclerotic outline in the lesions of Ollier's disease.

Progress/complications

Lesions develop during growth, but new ones do not usually occur after puberty. Fractures may occur through abnormal bone. Asymmetry of limb length and massive swellings are the main disability. Malignant change is not usually a feature but has been reported (cf. Maffucci's disease, Chapter 71).

REFERENCES

Mainzer F, Minagi H, Steinbach H L 1971 The variable manifestations of multiple enchondromatosis. Radiology 99:377
Murray A M, Cruickshank B 1960 Dyschondroplasia. Journal of Bone and Joint Surgery 42B:344

Pelvis: 5–12 years

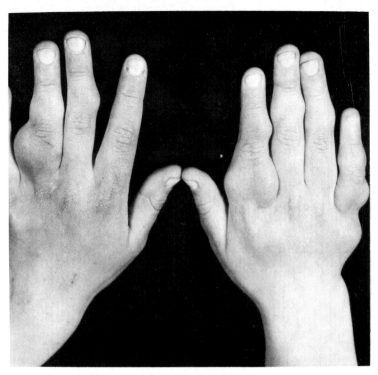

Fig. 70.1   Enchondromata distorting most fingers.

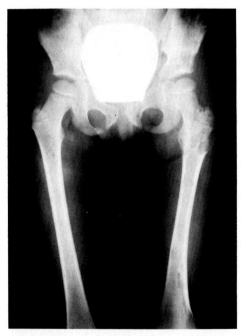

Fig. 70.2   C.V.—5 years—AP pelvis and femora
There are two small radiolucent lesions in the upper right femoral diaphysis, but the left side is more severely affected in the ilium, femoral neck and trochanteric region, as well as in the distal metaphysis. The capital femoral epiphyses are normal.
Same patient: Figs 70.4, 70.8, 70.9

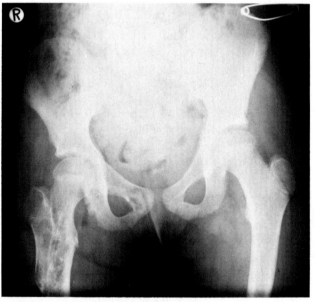

Fig. 70.3   E.W.—12 years—AP pelvis
There is pelvic tilt to the right and probably limb shortening this side. On the right there are well defined radiolucent areas in the ischium, pubic ramus and proximal femoral diaphysis. These are associated with abnormal modelling and bone expansion. The left side is normal.
Same patient: Figs 70.5, 70.6

Lower limb: 5–14 years

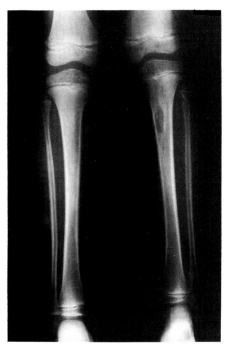

Fig. 70.4   C.V.—5 years—AP tibiae and fibulae
The right side is normal. On the left there is one large oval shaped
clearly defined lytic area in the upper tibial shaft. Smaller lytic areas
are present in the distal metaphyses.
Same patient: Figs 70.2, 70.8, 70.9

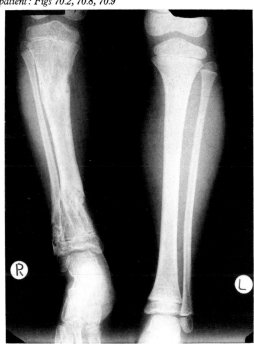

Fig. 70.6   E.W.—12 years—AP tibiae and fibulae (same patient
as in Fig. 70.5)
The left tibia and fibula appear normal. On the right both bones are
shortened with expansion and abnormal modelling associated with
well defined lytic lesions in the meta-diaphyses. This patient has
unilateral distribution of the disease involving the right side of the
pelvis and the whole right lower limb.

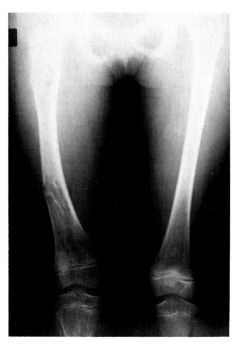

Fig. 70.5   E.W.—12 years—AP femora
There is lateral bowing of the right femur and in the proximal and
distal meta-diaphyses there are well defined radiolucent areas causing
some bone expansion and abnormal modelling. Some of these have a
longitudinal orientation, suggesting that there has been extension of
the lesion during the growing period. The right femur is shorter than
the normal left one.
Same patient: Figs 70.3, 70.6

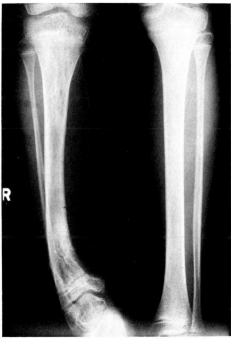

Fig. 70.7   T.F.—14 years—AP tibiae and fibulae
Normal left leg. On the right there is shortening and lateral bowing of
the distal tibia and fibula. In the upper and lower meta-diaphyses
there are irregular areas of patchy lysis and sclerosis.

Hand and foot: 5–11 years

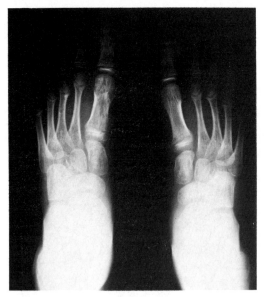

Fig. 70.8 C.V.—5 years—PA feet
Normal right foot. Small lytic areas are present throughout the first and third rays of the left foot.
Same patient: Figs 70.2, 70.4, 70.9

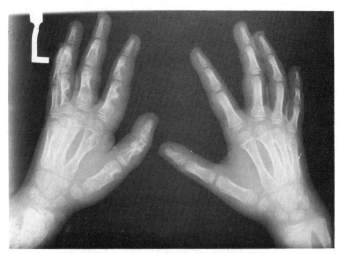

Fig. 70.9 C.V.—5 years—Oblique views of both hands
Normal right hand. There are multiple well-defined lytic areas in the left hand and most of the metacarpals and phalanges and the distal ulna is shortened.
Same patient: Figs 70.2, 70.4, 70.8

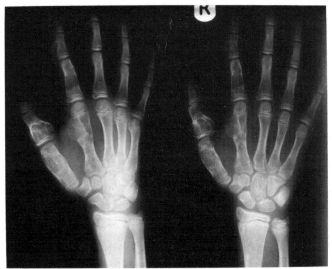

Fig. 70.10 T.L.—8 years—PA and oblique right hand
There are expanding osteolytic areas in the metaphyses and diaphyses of the metacarpals and phalanges of the thumb and index finger. The middle, ring and little fingers are normal. This demonstrates a radial ray distribution.

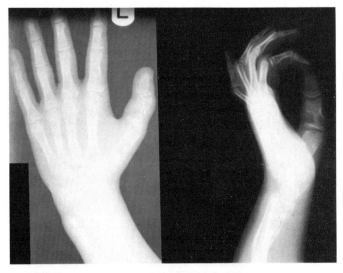

Fig. 70.11 N.B.—11 years—PA and lateral left hand
There are expanding, well defined osteolytic areas in the metacarpals and phalanges of the thumb and index finger. This demonstrates a radial ray distribution. The distal radial metaphysis shows similar osteolytic areas and there is overall shortening of the radius. There has been an ulnar osteotomy with shortening and screw fixation to prevent dislocation at the wrist.

Chapter Seventy-one
MAFFUCCI'S DISEASE

Bone lesions are similar to those of Ollier's disease, but are associated with multiple haemangiomata.

Inheritance

Non-genetic.

Frequency

Extremely rare (more so than Ollier's).

Clinical features

Similar to Ollier's disease; with the addition of haemangiomata.

Radiographic features

As in Ollier's disease, there are irregular translucent areas in the metaphyses, alternating with streaks of increased density, occurring in asymmetrical or isolated areas of the skeleton. In addition, soft tissues around the bony lesions are enlarged, and contain multiple phleboliths (circular calcification within the venous channels).

Progress/complications

There may be progression of the lesions after puberty. Malignant change is reported in either bony or vascular tissue in 15–18% of cases.

REFERENCES

Lewis R J, Ketchan A S 1973 Maffucci's syndrome: functional and neoplastic significance. Case report and a review of the literature. Journal of Bone and Joint Surgery 55A: 1465

Maffucci's disease

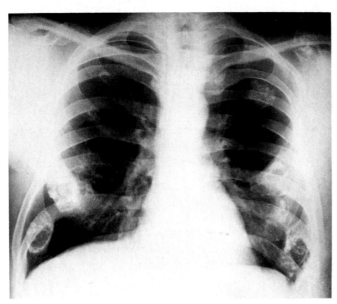

Fig. 71.1   T.C.—29 years 8 months—PA chest
There is expansion of the anterior ends of several ribs (4, 5 and 6 on
the right and 2, 4, 5 and 6 on the left), with radiolucent areas and
spotty calcification.
Same patient: Figs 71.2, 71.3

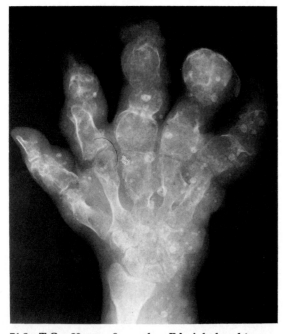

Fig. 71.2   T.C.—29 years 8 months—PA right hand (same
patient as in Fig. 71.1)
There is marked deformity with expansion of the metacarpals and
phalanges and distortion of joint spaces. Multiple rounded calcific
opacities are present which are phleboliths within a soft tissue venous
malformation. The distal end of the ulna is short.

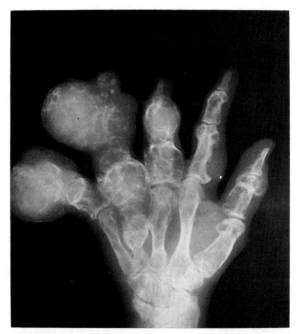

Fig. 71.3   T.C.—29 years 8 months—PA left hand (same patient
as in Fig. 71.1)
There is again marked expansion with radiolucent areas within the
metacarpals and phalanges but with gross distortion of the middle,
ring and little fingers (suggesting an ulnar ray distribution). There is
no obvious soft tissue calcification on this side.

Chapter Seventy-two
DYSPLASIA EPIPHYSEALIS HEMIMELICA

Osteocartilaginous outgrowths in the region of a joint, usually in the lower limb and sometimes confined to the lateral or medial half.

Inheritance

Non-genetic.

Frequency

Very rare, perhaps a prevalence of 1 per million. Males are affected three times more frequently than females.

Clinical features

Facial appearance—normal.
Intelligence—normal.
Stature—normal, although the length of an affected bone may be altered.
Presenting feature/age—deformity and a bony swelling in the region of a joint, sometimes with pain and locking. Usually in early childhood, but may be in infancy.
Deformities—only those related to the bony outgrowths.
Associated anomalies—none.

Radiographic features

Skull—normal.
Spine—normal.
Thorax—normal.
Upper limb—usually normal, but any epiphyses can be involved, as can the carpal bones.
Lower limb—the commonest sites are the tarsus, lower tibia and lower femur.

The exostosis projects from the epiphyseal area, and this may protrude into the joint. The medial side is affected more often than the lateral side and in two-thirds of cases the lesions are multiple. The metaphysis may be streaked.

Bone maturation

Probably normal.

Biochemistry

Not known.

Differential diagnosis

Ollier's disease—but here there is marked involvement of the metaphysis, and the epiphysis only rarely in severe cases.
Diaphyseal aclasis—this is distinguished by its normal epiphyses, invariably multiple involvement and often a dominant family history.

Progress/complications

Growth of the lesion usually ceases with growth of the child. Fracture and loose bodies in the joint may occur. Secondary osteoarthritis is inevitable if there is joint deformation. Malignant change has not been reported.

REFERENCES

Carlson D H, Wilkinson R H 1979 Variability of unilateral epiphyseal dysplasia (dysplasia epiphysealis hemimelica). Radiology 133/2: 369–373
Fasting O J, Bjerkreim I 1976 Dysplasia epiphysealis hemimelica. Acta Orthopaedica Scandinavica 47/2:217–225
Kettlekamp D B, Campbell C J, Bonfiglio M 1966 Dysplasia epiphysealis hemimelica. Journal of Bone and Joint Surgery 48A:746

Pelvis and hip: 9 months–13 years

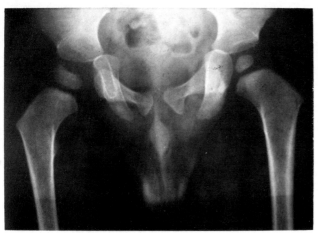

Fig. 72.1   L.R.—9 months—Pelvis
The right capital femoral epiphysis is enlarged, and deformed superiorly by a partially ossified mass. The ischium appears to be displaced medially, also.
Same patient: Figs 72.2, 72.3

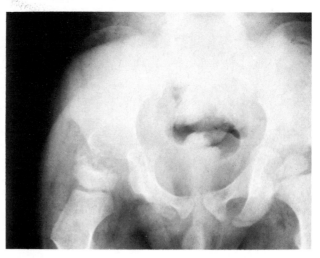

Fig. 72.2   L.R.—2 years—Pelvis (same patient as in Fig. 72.1)
15 months later, there has been an increase in the size of the mass, and it is no longer contained by the acetabulum.

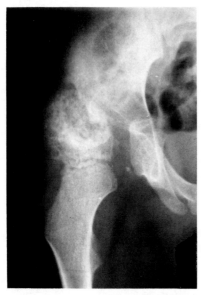

Fig. 72.3   L.R.—5 years—Right hip (same patient as in Fig. 72.1)
The cartilaginous mass is now very large, with spotty ossification, and appears to be almost entirely outside the acetabulum.

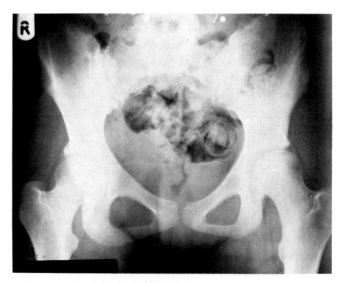

Fig. 72.4   N.D.—13 years—Pelvis
There is patchy sclerosis above the left acetabulum, extending upwards towards the iliac crest.
Same patient: Figs 72.7, 72.12, 72.13

Knee: 8–13 years

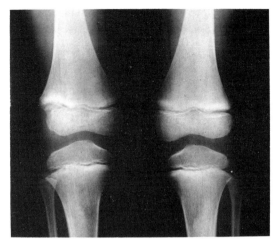

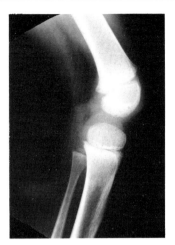

Fig. 72.5   C.O.—8 years—AP knees
There is the beginning of an osteocartilaginous mass on the lateral side of the right femoral epiphysis.
Same patient: Fig. 72.6

Fig. 72.6   C.O.—8 years—Lateral right knee (same patient as in Fig. 72.5)
Only minor irregularity of the femoral epiphysis can be seen.

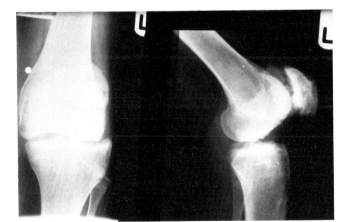

Fig. 72.7   N.D.—13 years—AP and lateral left knee
   AP—there is a large mass overlying the lateral side of the lower femur and epiphysis and associated sclerosis and joint narrowing of the adjacent tibial epiphysis.
   Lateral—the femoral and tibial epiphyses and the patella are all sclerotic and irregular.
Same patient: Figs 72.4, 72.12, 72.13

Ankle: 3–7 years

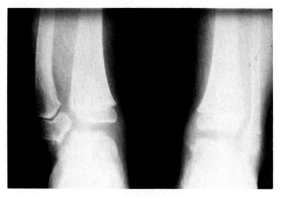

Fig. 72.8   J.W.—3 years—AP ankles
There is lateral bowing of the right fibula with diastasis of the distal tibiofibular joint. Both the shaft and epiphysis show bony overgrowth. The articular margin of the talus is irregular.
Same patient: Figs 72.9–72.11

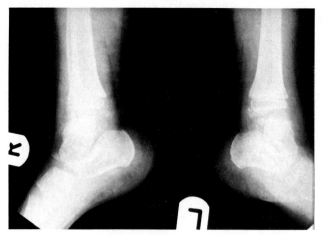

Fig. 72.9   J.W.—3 years—Lateral ankles (same patient as in Fig. 72.8)
There are bony outgrowths around the talus, calcaneum and lower tibial epiphysis on both sides. The fibula is bowed forwards.

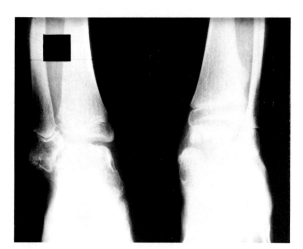

Fig. 72.10   J.W.—7 years—AP ankles (same patient as in Fig. 72.8)
4 years later the lesion on the right has changed little with growth. On the left the ankle is now more clearly involved, with premature fusion of the lateral epiphyseal plate, joint space narrowing, sclerosis and articular irregularity.

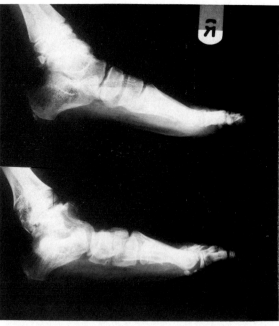

Fig. 72.11   J.W.—7 years—Lateral ankles and feet (same patient as in Fig. 72.8)
There is severe joint space narrowing of the ankles with articular irregularity.

Ankle: 13 years

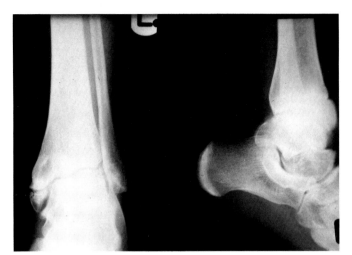

Fig. 72.12   N.D.—13 years—AP and lateral left ankle
The bony overgrowth is affecting mainly the region of the medial malleolus, and there is adjacent sclerosis.
Same patient: Figs 72.4, 72.7, 72.13

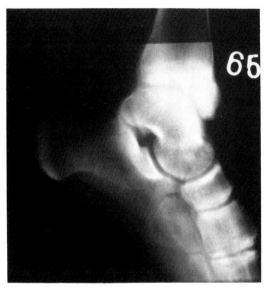

Fig. 72.13   N.D.—13 years—Lateral tomogram left ankle
(same patient as in Fig. 72.12)
There is marked sclerosis on both sides of the ankle joint.

Section Fifteen
FIBROUS DISORDERS

Chapter Seventy-three
FIBRODYSPLASIA OSSIFICANS PROGRESSIVA (myositis ossificans progressiva)

Characterised by calcification and then ossification of fasciae, aponeuroses and tendons, secondarily affecting muscle, and a short hallux (more rarely thumb) caused by absence of one or fusion of two phalanges; sometimes also anomalies of cervical vertebrae.

Inheritance

Autosomal dominant, but most cases are sporadic. There is a predominance of males.

Frequency

Extremely rare, probably under 0.1 per million prevalence in the population.

Clinical features

Facial appearance—normal.
Intelligence—normal.
Stature—normal.
Presenting feature/age—the deformity of the hallux or other bony abnormality is present from birth. Usually in late childhood hard swellings develop, often in the region of the shoulder girdle. They are not painful, and may subside, to be replaced later by others. On occasion they are cyst-like and can break down and discharge, recurring even if excised.
Deformities—these are secondary to the developing masses of hard tissue covering (usually) the muscles of the back—torticollis and 'scoliosis'.
Associated anomalies—hallux and cervical anomalies have been noted above.

Radiographic features

Cervical vertebrae—small, with thick pedicles and massive spinous processes; the bodies later become irregular. Fusion of spinous processes appears to be progressive.

Hip joints—acetabular dysplasia is associated with apparently enlarged femoral heads and necks.
Hallux—shortening with fusion of phalanges, or absence of one, to give a variable deformity.
Hands—the first metacarpals are usually short, and sometimes also the middle phalanges of other digits. There may be extra epiphyseal centres, and possibly carpal fusions.

This disorder is a disease principally of soft tissues, and calcified masses, preceded by soft tissue swellings, develop usually over the trunk and neck, proximal limbs and sometimes over the masseter: not usually over distal limb muscles, nor does it affect cardiac or involuntary muscle.

Bone maturation

Some patients show delayed and others irregular maturation.

Biochemistry

Not known.

Differential diagnosis

There is no similar disease.

Progress/complications

There is progressive rigidity of the chest with respiratory complications. If the disease has started early, then death occurs probably before the age of 20. Milder cases have exacerbations and remissions over a longer period.

REFERENCES

Cremin B, Connor J M, Beighton P 1980 The radiological spectrum of fibrodysplasia ossificans progressiva. Clinical Radiology 33:499–508
Hall C M, Sutcliffe J 1979 Fibrodysplasia ossificans progressiva. Annales de Radiologie 22:119–123
Rogers J G, Geho W B 1979 Fibrodysplasia ossificans progressiva: a survey of forty two cases. Journal of Bone and Joint Surgery 61:909–914

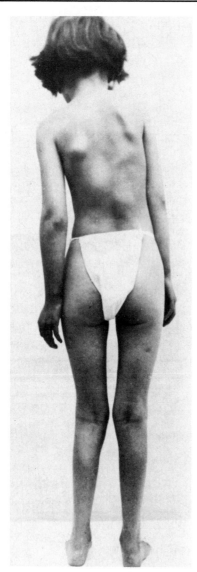

Fig. 73.1 Obvious masses in the soft tissues of the back.

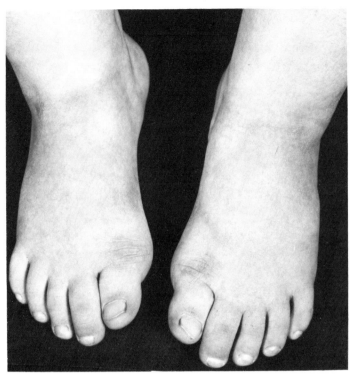
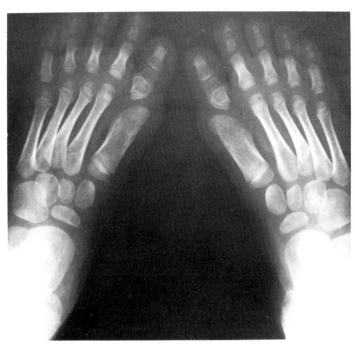

Figs 73.2 and 73.3   Clinical photograph and radiograph to show the short hallux with bony deformity of the first metatarsal and proximal phalanx.

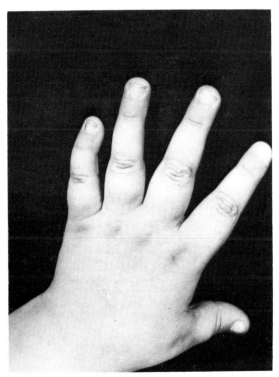
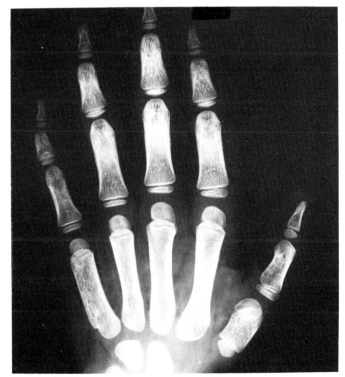

Figs. 73.4 and 73.5   A similar deformity of the thumb, not invariably present in this disease.

Cervical spine : Childhood

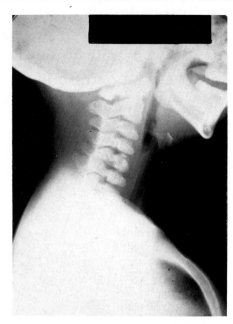

Fig. 73.6   B.H.—Childhood—Lateral cervical spine
The vertebral bodies are small, the pedicles widened and the spinous processes massive. There is no soft tissue calcification yet.
Same patient : Fig. 73.15

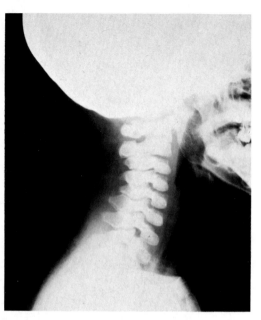

Fig. 73.7   F.D.—5 years—Lateral cervical spine
Similar to Figure 73.6, but the vertebral bodies are more irregular. There is a soft tissue swelling posteriorly, not yet calcified.
Same patient : Figs 73.10, 73.20, 73.30, 73.31

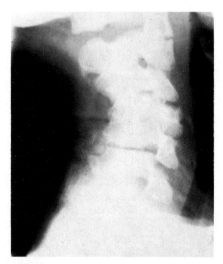

Fig. 73.8   U.N.—Childhood—Lateral cervical spine
Similar to Figure 73.7.
Same patient : Figs 73.17, 73.24

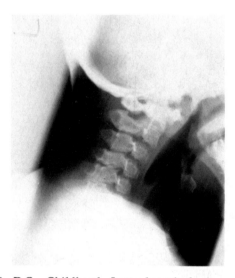

Fig. 73.9   D.G.—Childhood—Lateral cervical spine
There are fusions of the massive spinous processes from C2–C5. The bodies have reduced antero-posterior diameters and are markedly irregular in outline and alignment.

Cervical spine : 15 years–Adult

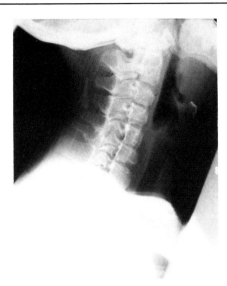

Fig. 73.10   F.D.—15 years—Lateral cervical spine
The soft tissue lump seen previously has now ossified. The massive pedicles and spinous processes are progressively fusing.
Same patient : Figs 73.7, 73.20, 73.30, 73.31

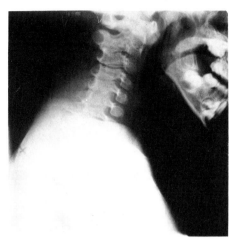

Fig. 73.11   A.S.—Adolescent—Lateral cervical spine
Some of the vertebrae are fused and the bodies are unusually small.
No soft tissue swelling.
Same patient : Fig. 73.14

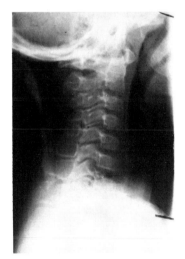

Fig. 73.12   A.M.—Adolescent—Lateral cervical spine
Ossification of the fascia overlying the posterior neck muscles is clearly seen.
Same patient : Figs 73.16, 73.19, 73.29

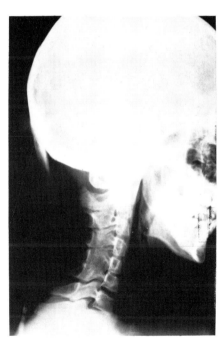

Fig. 73.13   M.B.—Adult—Lateral cervical spine
Similar to Figure 73.12, but also showing fused spinous processes and small vertebral bodies.
Same patient : Figs 73.23, 73.28

Chest: Childhood and adolescence

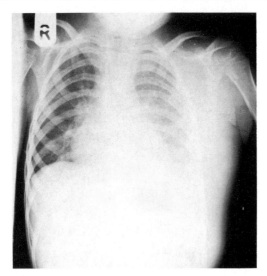

Fig. 73.14 A.S.—Childhood—AP chest
There is a large soft tissue swelling in the region of the left shoulder
girdle and chest (no calcification yet).
Same patient: Fig. 73.11

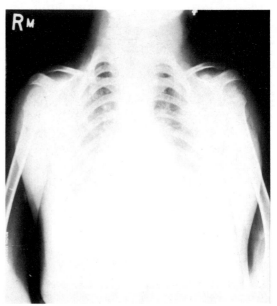

Fig. 73.15 B.H.—Childhood—AP chest
Rather similar to Figure 73.14, but there is here some ossification in
the swelling.
Same patient: Fig. 73.16

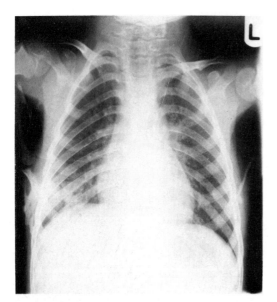

Fig. 73.16 A.M.—Adolescent—AP chest
Soft tissue ossification overlies much of the thorax and both axillae.
(Same date of X-ray as Fig. 73.12.)
Same patient: Figs 73.12, 73.19, 73.29

Pelvis : Childhood–15 years

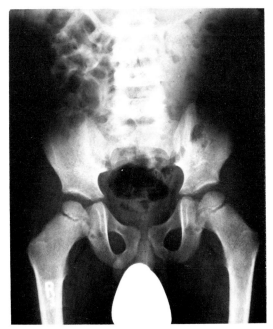

Fig. 73.17   U.N.—Childhood—Pelvis
Normal.
Same patient : Figs 73.8, 73.24

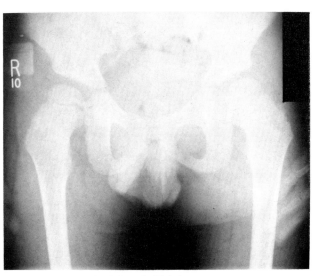

Fig. 73.18   P.S.—Childhood—Pelvis
The capital femoral epiphyses are slightly flattened, and the necks broad. There is also a minor degree of acetabular dysplasia.
Same patient : Fig. 73.25

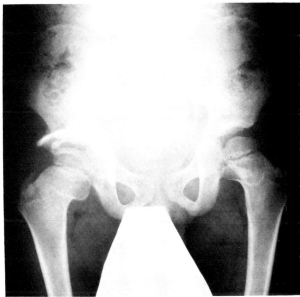

Fig. 73.19   A.M.—Childhood—Pelvis
There is bizarre ossification along the upper border of the right acetabulum, extending laterally (similar to a 'shelf' operation for congenital dislocation of the hip).
The femoral necks are irregularly thickened.
Same patient : Figs 73.12, 73.16, 73.29

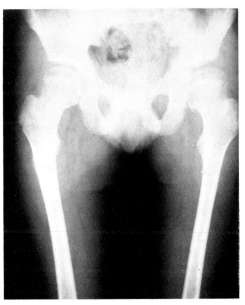

Fig. 73.20   F.D.—15 years—Pelvis
There is acetabular dysplasia, but the femoral heads and necks are disproportionately large both in relation to the pelvis and the femoral shafts. There are a few small areas of ossification, and probably a tiny exostosis on the upper right femoral shaft.
Same patient : Figs 73.7, 73.10, 73.30, 73.31

Pelvis and lower limb

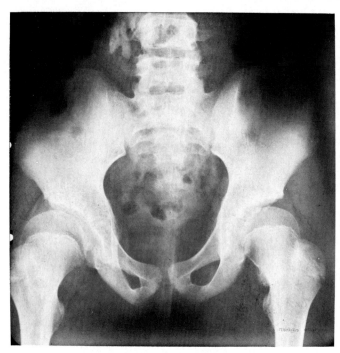

Fig. 73.21   Unknown—Adolescent—Pelvis
The acetabula are shallow and the large femoral heads are very poorly
covered. There is some soft tissue ossification overlying the fourth
lumbar vertebra and the left femoral neck.

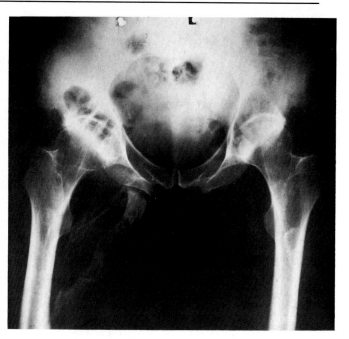

Fig. 73.22   S.L.—Adult—Pelvis
The main abnormality here is of massive soft tissue ossification in the
right adductor region, as well as narrowed joint spaces and
osteoarthritis.

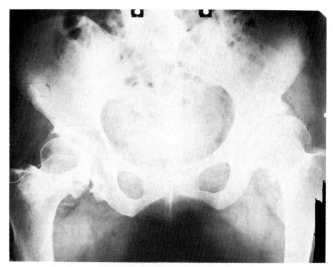

Fig. 73.23   M.B.—Adult—Pelvis
There are large femoral heads associated with acetabular dysplasia,
and also soft tissue ossification around both hip joints, more severe on
the right.
Same patient : Figs 73.13, 73.28

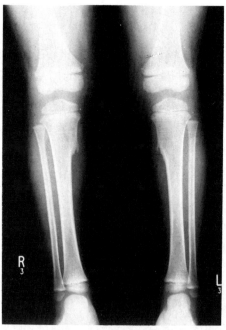

Fig. 73.24   U.N.—Childhood—AP knees, tibiae and fibulae
The soft tissues and alignment of the long bones are normal. There
are small 'exostoses' on the medial aspect of both upper tibiae,
associated with abnormal metaphyseal modelling here.
Same patient : Figs 73.8, 73.17

Feet

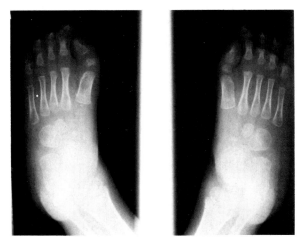

Fig. 73.25 P.S.—Childhood—PA feet
The bones are normal apart from a short first metatarsal with a pseudoepiphysis at its distal end, and only one phalanx—giving hallux valgus.
Same patient : Fig. 73.18

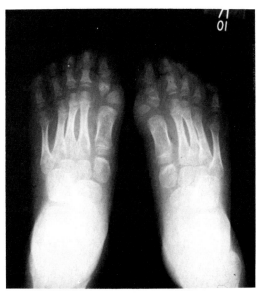

Fig. 73.26 Unknown—Childhood—PA feet
The first metatarsals are short, with distal pseudoepiphyses, but here there are two hallux phalanges, somewhat out of alignment with the metatarsal.
Same patient : Fig. 73.26

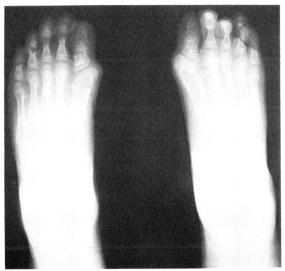

Fig. 73.27 Unknown—Adolescent—PA feet
Same patient as in Figure 73.26 but older—the metatarsal pseudoepiphysis has fused on the left.

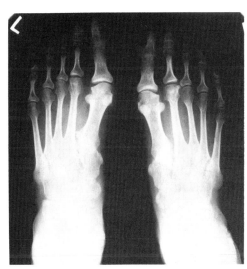

Fig. 73.28 M.B.—Adult—PA feet
Here the metatarsals are normal, but there is fusion of the proximal and distal phalanges.
Same patient : Figs 73.13, 73.23

Upper limb: 5 years–Adolescence

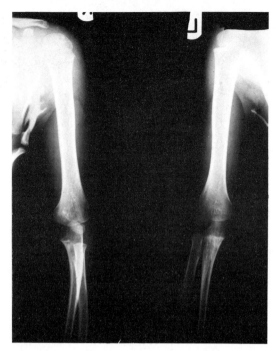

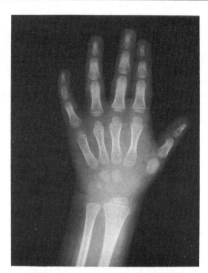

Fig. 73.30 F.D.—5 years—PA left hand
The first metacarpal is short and oval in shape, and the index middle
phalanx is also rather short. Bone maturation is retarded by 1–2 years.
Same patient: Figs 73.7, 73.10, 73.20, 73.31

Fig. 73.29 A.M.—Childhood—AP humeri and elbows
Soft tissue ossification is present in the scapular region, but the limb
bones are normal.
Same patient: Figs 73.12, 73.16, 73.19

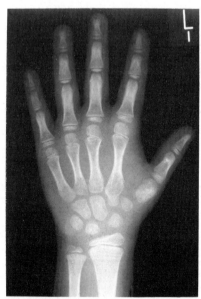

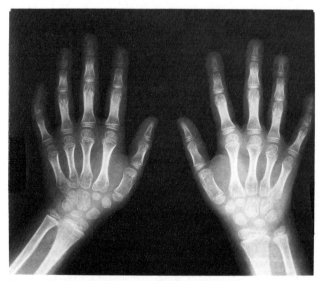

Fig. 73.32 W.—Adolescent—PA hands
The first metacarpals are only a little short, and have distal as well as
proximal epiphyses. The only other abnormality is clinodactyly
associated with a short middle phalanx of the little finger.

Fig. 73.31 F.D.—11 years—PA left hand (same patient as in Fig.
73.30)
The first metacarpal is short and thick; the only other abnormality is
fusion of the trapezoid to the base of the second metacarpal. Bone
maturation is retarded by about 3 years.

Chapter Seventy-four
POLYOSTOTIC FIBROUS DYSPLASIA
(including Albright's syndrome)

Patchy, irregular replacement of bone by fibrous tissue, associated with sometimes gross deformities and pathological fractures.

Albright's syndrome includes skin pigmentation and (in girls) endocrine problems with precocious puberty.

Inheritance

Nearly always non-genetic, but there is one family report of autosomal dominant inheritance.

Frequency

Possible prevalence of 2–3 per million population. Albright's syndrome occurs in only about 2% of all cases.

Clinical features

Facial appearance—if the skull and facial bones are affected then there will be asymmetrical deformity. Pigmentation may occur in the mucous membranes of the mouth.

Intelligence—normal.

Stature—may be normal but growth of individual affected bones can be increased or epiphyses fuse early leading to shortening.

Presenting feature/age—skin pigmentation (if a feature) will be present at birth. More usually, the presenting feature is pain and deformity from pathological fracture of affected bone. In Albright's syndrome the first sign may be vaginal bleeding in early childhood.

Deformities—bending of affected bones may become very severe and bizarre in appearance.

Associated anomalies—none.

Radiographic features

Any bone may be involved, the fibrous tissue spreading throughout and replacing bone.

There are patches of rarefaction and sclerosis but unaffected bone appears quite normal, i.e. this is a condition of multiple lesions rather than a generalised disorder of bone growth.

Epiphyses are not usually involved.

Bone maturation

Normal.

Biochemistry

Not known.

Differential diagnosis

Primary hyperparathyroidism is a little similar, but has a generalised, not patchy effect and is uncommon in childhood, also it can be identified by biochemical tests.

Progress/complications

Lesions enlarge and progress until epiphyses close, only very occasionally do they resolve. If the skull is affected cranial nerve foramina may become obliterated. Fractures and deformity make this a particularly crippling disease, at its worst.

Other endocrine problems have been reported (pituitary and thyroid and also diabetes mellitus and rickets).

REFERENCES

Evans G A, Park W M 1978 Familial multiple non-osteogenic fibromata. Journal of Bone and Joint Surgery 60B:416–419

Harris W H, Dudley R, Barry R J 1962 The natural history of fibrous dysplasia. Journal of Bone and Joint Surgery 44A;207

McArthur R G, Hayles A B, Lambert P W 1979 Albright's syndrome with rickets. Mayo Clinic Proceedings 54/5:313–320

Shires R, Whyte M P, Avioli L V 1979 Idiopathic hypothalamic hypogonadotropic hypogonadism with polyostotic fibrous dysplasia. Archives of Internal Medicine 139/10:1187–1189

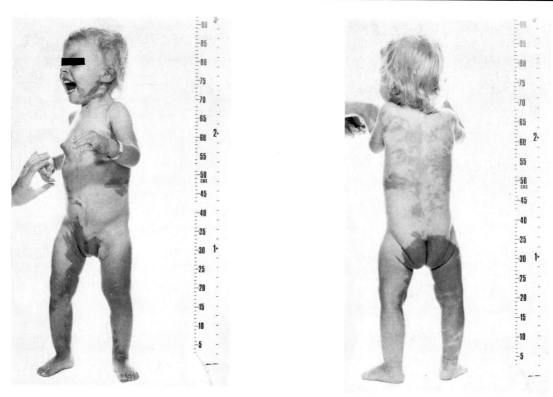

Figs 74.1 and 74.2   A 3-year-old girl with Albright's syndrome, showing precocious puberty and skin pigmentation.

Skull: 1–5 years

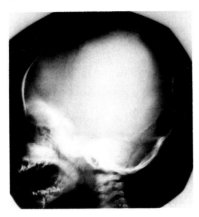

Fig. 74.3   U.F.—1 year 3 months—Lateral skull (Albright's syndrome)
There is increased density of the base and maxilla.
Same patient: Figs 74.5, 74.6, 74.8, 74.9, 74.13, 74.19–74.22, 74.30

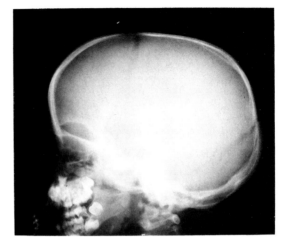

Fig. 74.4   V.G.—3 years 8 months—Lateral skull
There is uniform increased density of the vault. This patient had endocrine disorders: Cushing's and hyperparathyroidism.
Same patient: Figs 74.14, 74.29

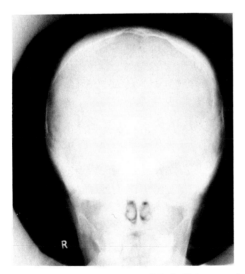

Fig. 74.5   U.F.—5 years 2 months—AP skull (same patient as in Fig. 74.3)
There is now thickening and sclerosis of the vault in addition to the base.

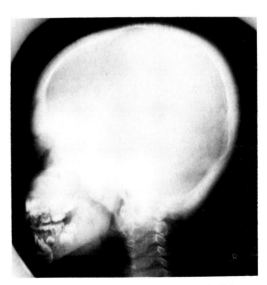

Fig. 74.6   U.F.—5 years 2 months—Lateral skull (same patient as in Fig. 74.3)
The base and maxilla show a further increase in sclerosis, with 'floating' teeth.

Skull: 8–14 years

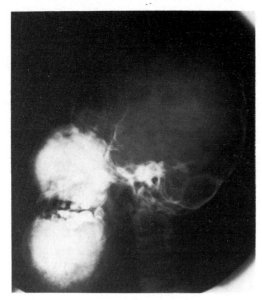

Fig. 74.7   C.L.—8 years—Lateral skull (cherubism)
Marked bony overgrowth of the maxilla and mandible, with
distortion and sclerosis, also 'floating' teeth.

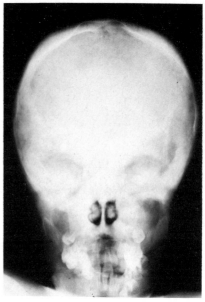

Fig. 74.8   U.F.—8 years 6 months—AP skull
There has been further overgrowth and sclerosis of the vault, base
and facial bones.
Same patient: Figs 74.3, 74.5, 74.6, 74.9, 74.13, 74.19–74.22, 74.30

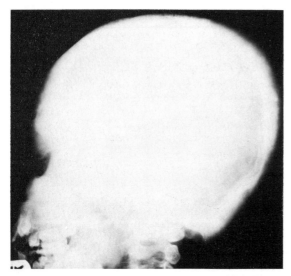

**Fig. 74.9   U.F.—8 years 6 months—Lateral skull (same patient
as in Fig. 74.8)**
Lateral view of Figure 74.8

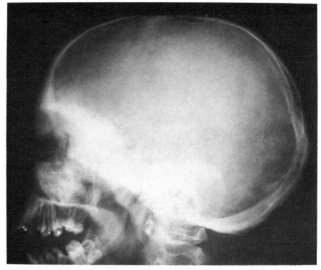

Fig. 74.10   K.B.—14 years—Lateral skull
There is thickening and sclerosis of the base of the skull and the
maxillary antra are opaque and dense.
Same patient: Figs 74.17, 74.18, 74.25, 74.26, 74.28

Pelvis: 1–3 years

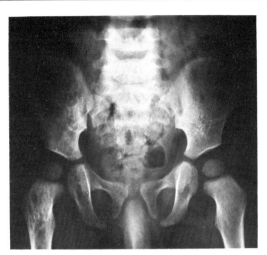

Fig. 74.11 S.M.—1 year 7 months—Pelvis
Ill-defined lytic areas are present in the right femoral neck.
Same patient: Figs 74.12, 74.15, 74.16, 74.23, 74.27

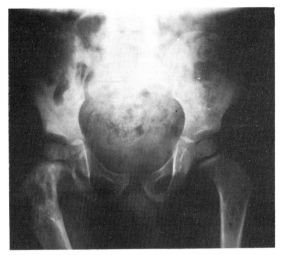

Fig. 74.12 S.M.—2 years—Pelvis (same patient as in Fig. 74.11)
Lytic areas can now be seen in both upper femora. These have a sclerotic margin extending over several millimetres, unlike other bony cystic lesions.

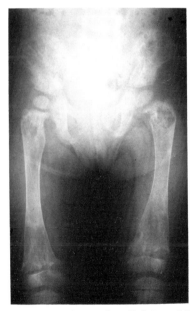

Fig. 74.13 U.F.—2 years 6 months—Pelvis and femora
There are multiple lytic areas in the ilia and upper and lower femora.
Figs 74.3, 74.5, 74.6, 74.8, 74.9, 74.19–74.22, 74.30

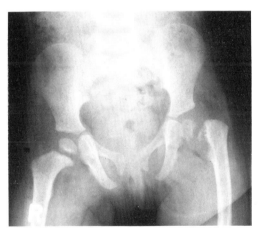

Fig. 74.14 V.G.—3 years 6 months—Pelvis
This patient has a pathological fracture through a lytic area in the left femoral neck.
Same patient: Figs 74.4, 74.29

Pelvis and hip: 4–12 years

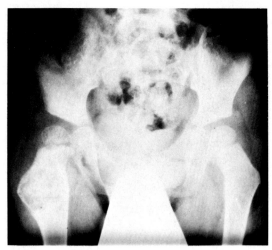

Fig. 74.15   S.M.—4 years—Pelvis
There are lytic areas in both upper femora, the right trochanteric region being expanded.
Same patient: Figs 74.11, 74.12, 74.16, 74.23, 74.27

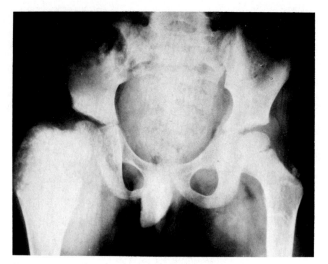

Fig. 74.16   S.M.—5 years 3 months—Pelvis
Same patient as in Figure 74.15 with progression of the lesions. The epiphyses are characteristically spared.

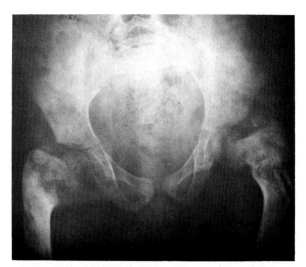

Fig. 74.17   K.B.—7 years—Pelvis
This child shows patchy areas of lysis and sclerosis, and secondary coxa vara.
Same patient: Figs 74.10, 74.18, 74.25, 74.26, 74.28

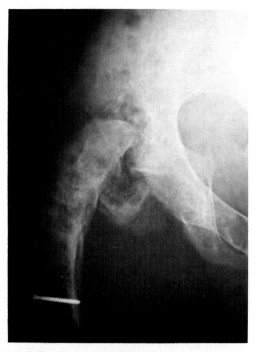

Fig. 74.18   K.B.—12 years—Right hip
Same patient as in Figure 74.17 with progression of the lesion, pathological fracture of the neck and upward displacement of the femur.

Lower limb: 4–8 years

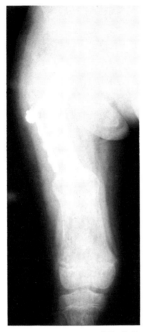

Fig. 74.19 U.F.—4 years 6 months—AP right femur
There is fibrous dysplasia throughout the femur, with internal fixation of a fracture.
Same patient: Figs 74.3, 74.5, 74.6, 74.8, 74.9, 74.13, 74.20–74.22, 74.30

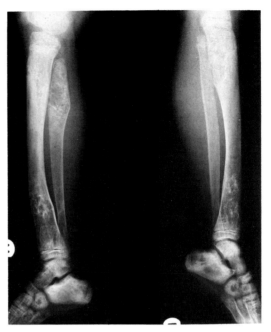

Fig. 74.20 U.F.—5 years—Lateral tibiae and fibulae (same patient as in Fig. 74.19)
This shows all the characteristic features of lysis, marginal sclerosis and expansion of bone.

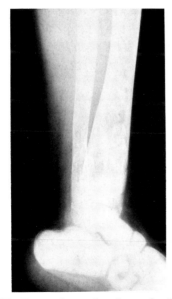

Fig. 74.21 U.F.—7 years 6 months—Lateral ankle (same patient as in Fig. 74.19)
There are lesions in the tibia, fibula and all the tarsal bones.

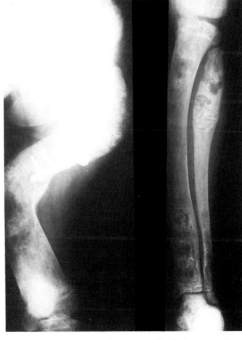

Fig. 74.22 U.F.—8 years—AP lower limb (same patient as in Fig. 74.19)
The softening and deformity of bone is becoming more severe.

Lower limb: 1–21 years

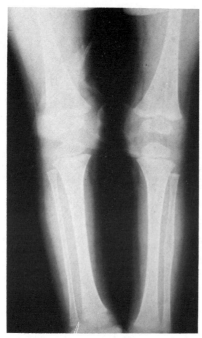

Fig. 74.23   S.M.—1 year 7 months—AP tibiae
These are almost normal, but there is expansion of the mid-shaft of the left fibula.
Same patient: Figs 74.11, 74.12, 74.15, 74.16, 74.27

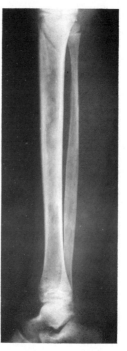

Fig. 74.24   J.A.—7 years 3 months—Lateral tibia and fibula
Areas of lysis and sclerosis are scattered throughout both diaphyses.

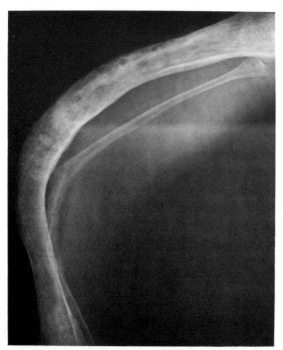

Fig. 74.25   K.B.—21 years—Lateral tibia and fibula
There is a gross bowing deformity with fibrous dysplasia throughout.
Same patient: Figs 74.10, 74.17, 74.18, 74.26, 74.28

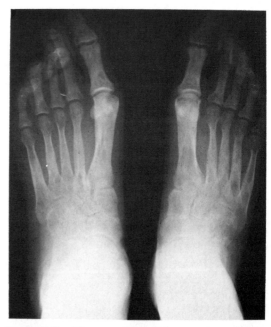

Fig. 74.26   K.B.—21 years—PA feet (same patient as in Fig. 74.25)
There is patchy sclerosis and lysis with some abnormal modelling of metatarsals.

Upper limb: 1–19 years

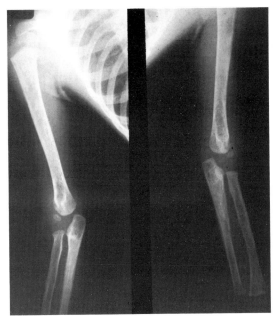

Fig. 74.27 S.M.—1 year 7 months—AP upper limb
There are patchy areas of lysis in the diaphyses of all bones, but
normal modelling at this early age.
Same patient: Figs 74.11, 74.12, 74.15, 74.16, 74.23

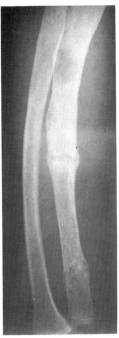

Fig. 74.28 K.B.—19 years—AP radius and ulna
The ulna is the more severely affected, and shows a transverse healing
fracture of the mid-shaft.
Same patient: Figs 74.10, 74.17, 74.18, 74.25, 74.26

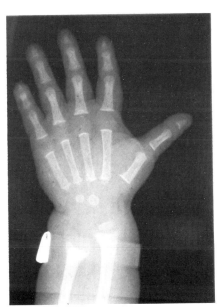

Fig. 74.29 V.G.—2 years—PA left hand
There is acro-osteolysis with pointing of the terminal phalanges,
(probably hyperparathyroidism) and also a rachitic lesion at the lower
end of the ulna, and to a lesser extent of the radius.
Same patient: Figs 74.4, 74.14

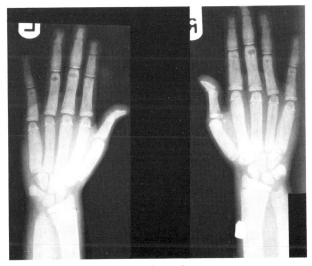

Fig. 74.30 U.F.—8 years—PA hands
There is generalised lack of bone modelling with sclerosis, but there
are also a few lytic areas. Bone maturation is greatly advanced (about
12 years) in this patient with Albright's syndrome.
Same patient: Figs 74.3, 74.5, 74.6, 74.8, 74.9, 74.13, 74.19–74.22

Chapter Seventy-five
NEUROFIBROMATOSIS
(von Recklinghausen's disease)

The principal features of von Recklinghausen's disease are *café-au-lait* skin patches and neurofibromata, but a proportion of patients have skeletal disorders: scoliosis, congenital pseudarthrosis of the tibia and local gigantism.

Inheritance

Autosomal dominant.

Frequency

Difficult to estimate, since so many individuals are only mildly affected and are not diagnosed. Estimates as high as 1 in 2–3000 births have been made, but individuals with skeletal involvement will be very much less than this—a possible prevalence of 22 per million population index patients or 28 per million including affected relatives.

Clinical features

Facial appearance—normal.

Intelligence—there is some evidence that these patients are often in the lower normal range if not mentally retarded.

Stature—probably unaffected, unless there is severe scoliosis.

Presenting feature/age—rarely there are signs at birth—pseudarthrosis of tibia, rarely of other bones, fibrous replacement of the ulna and occasionally *café-au-lait* patches. More usually the latter develop during the first decade, as do the nerve tumours. Plexiform neurofibromata and local gigantism of both soft tissue and bone may be the presenting signs.

Deformities—relating to overgrowth of tissues, scoliosis and the tibial pseudarthrosis.

Radiographic features

Spine—Enlarged exit foramina. Posterior scalloping of vertebral bodies, not necessarily associated with neurofibromata.

Thorax—'ribbon' ribs.

Limb—overgrowth

Tibia—a pseudarthrosis may already be present at birth or the tibia may be sclerotic and bowed (anteriorly), only breaking subsequently.

Soft tissue changes—overgrowth and, on arteriography, large vessel stenosis.

Bone maturation

Normal.

Biochemistry

Not known.

Differential diagnosis

There is no condition similar to fully developed neurofibromatosis, but if bony lesions are present they need to be differentiated from fibrous dysplasia of bone (*per se*); the scoliosis may be confused with simple idiopathic scoliosis. In each case, the additional features of neurofibromatosis (*café-au-lait* patches and neurofibromata) differentiate it.

Progress/complications

These relate to the bony deformities, to nerve entrapment from the neurofibromata and, more seriously, to malignant disease, possibly in about 4% of patients over the age of 21 years. A variety of malignant neoplasms occur in neurofibromatosis, for example: glioma, astrocytoma, liposarcoma, non-lymphatic leukaemia; malignant change in a subcutaneous nodule has not been reported.

REFERENCES

Carey J C, Laub J M, Hall B D 1979 Penetrance and variability in neurofibromatosis: a genetic study of 60 families. Birth Defects: Original Article Series XV (5B):271–281

Holt J F, Wright E M 1948 The radiographic features of neurofibromatosis. Radiology 51:647

Voutsinas S, Wynne-Davies R 1983 The infrequency of malignant disease in diaphyseal aclasis and neurofibromatosis. Journal of Medical Genetics 20:345–349

Winter R B, Moe J H, Bradford D S, Lonstein J E, Pedras C V, Weber A H 1979 Spine deformity in neurofibromatosis. Journal of Bone and Joint Surgery 61A:677–694

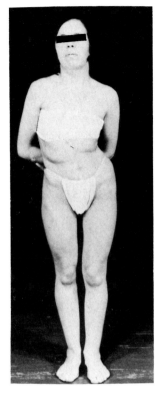
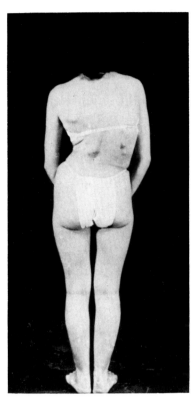

Figs 75.1 and 75.2   Right thoracic scoliosis in an adolescent girl, showing also several *café-au-lait* patches.

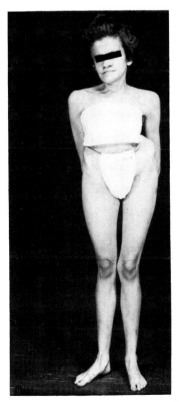

Figs 75.3 and 75.4   A more severe spinal curvature (untreated) in an older woman.

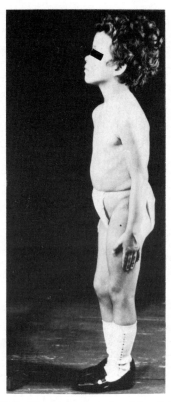
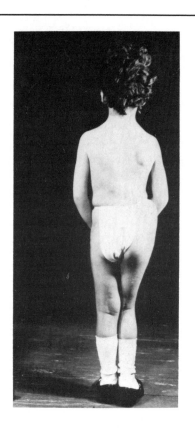

Figs 75.5 and 75.6   There is hypertrophy of the whole right lower limb.

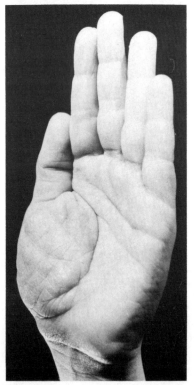
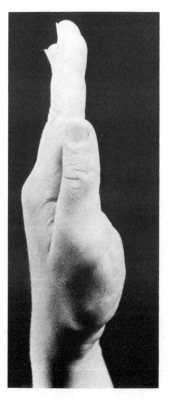

Figs 75.7 and 75.8   Thickened neurofibromatous tissue involving the thenar eminence.

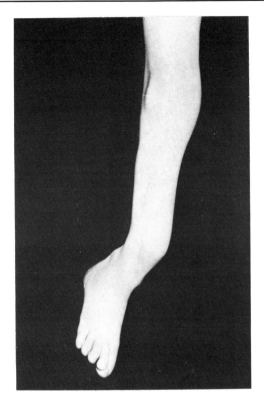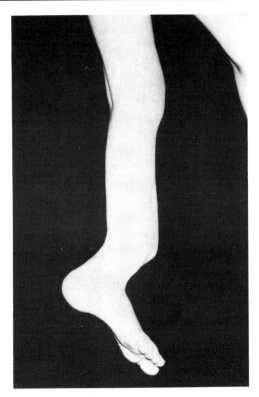

Figs 75.9 and 75.10 Marked anterior angulation of the tibia, the site of a congenital pseudarthrosis.

Cervical spine : 7 years

Figs 75.11–75.14   J.S.—7 years—Cervical spine to show large exit foramina

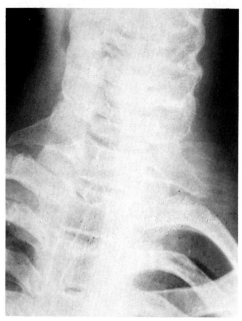

Fig. 75.11   AP
Not well shown in the AP view.

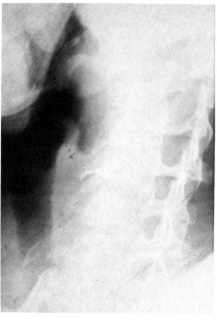

Fig. 75.12   Oblique
Large exit foramina are well shown, particularly from C2–C4.

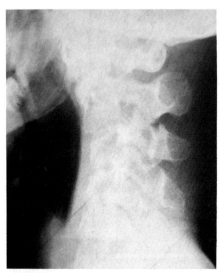

Fig. 75.13   Lateral
Exit foramina cannot be seen on this projection. There is a reversal of the normal cervical curve and irregular modelling of the bodies.

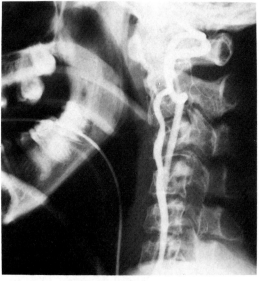

Fig. 75.14   Left vertebral angiogram
There is anterior displacement of the left vertebral artery at the level of C2 and C3. At C2 there has been reflux down the normally sited *right* vertebral artery. There are two or three localised areas of vertebral artery stenosis.

Spine: 3–9 years

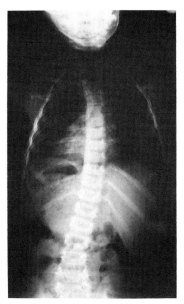

Fig. 75.15   A.N.—3 years 9 months—AP spine
Left lumbar scoliosis.

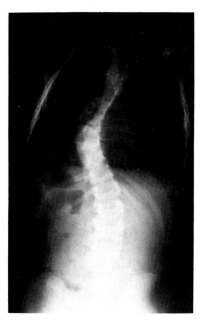

Fig. 75.16   J.R.—5 years—AP spine
Left thoracic scoliosis.

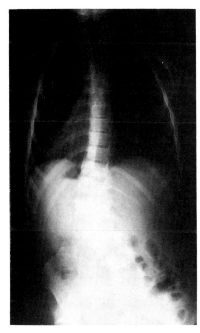

Fig. 75.17   S.B.—5 years—AP spine
Left thoraco-lumbar scoliosis, a typically short curve.

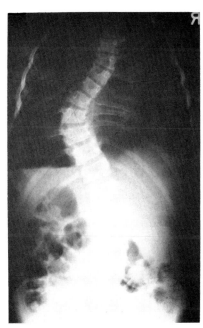

Fig. 75.18   A.D.—9 years—AP spine
Left thoracic scoliosis.

Spine and thorax: 3 years–Adult

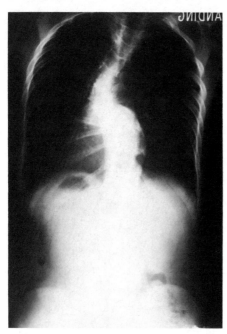

Fig. 75.19   H.B.—15 years—AP spine
Thoracic kyphoscoliosis.

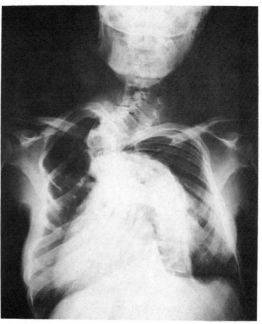

Fig. 75.20   E.F.—30 years—AP spine
Very severe, untreated scoliosis. There are also rib anomalies at the upper left thorax, probably associated with intercostal neurofibromata rather than scoliosis.
Same patient: Fig. 75.21

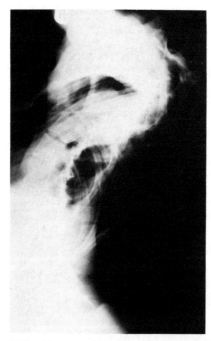

Fig. 75.21   E.F.—33 years—Lateral spine (same patient as in Fig. 75.20)
Lateral view showing gross kyphosis (see clinical photographs).

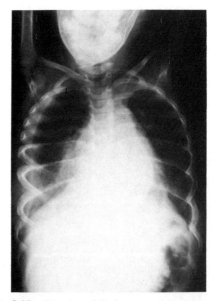

Fig. 75.22   S.M.—3 years—AP thorax
There is irregular narrowing of the posterior ends of most ribs ('ribbon' appearance), presumably associated with intercostal neurofibromata. There is also marked pleural thickening.

Lower limb: 2–10 years

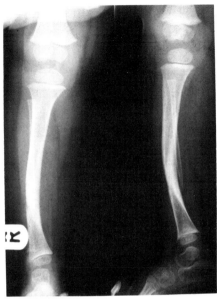

Fig. 75.23   J.D.—2 years—AP and lateral right tibia and fibula
There is sclerosis of the tibia at the junction of the middle and lower thirds, with anterior bowing and an incomplete fracture line can just be seen.
Same patient: Figs 75.24, 75.25

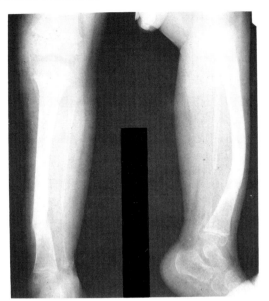

Fig. 75.24   J.D.—4 years—AP and lateral right tibia and fibula (same patient as in Fig. 75.23)
The bowing and sclerosis have increased, and the fracture line is clearly visible. On this view it can now be seen that there is absence of part of the fibular shaft, with tapered bone ends—indicating a pseudarthrosis in this area.

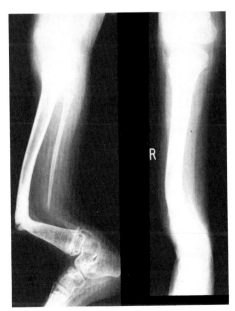

Fig. 75.25   J.D.—10 years—Lateral and AP right tibia and fibula (same patient as in Fig. 75.23)
Pseudarthrosis with gross anterior angulation.

Pseudarthrosis : Birth–2 years

Figs 75.26–75.32   A.V.—Series of seven radiographs to show progression and treatment of tibial pseudarthrosis

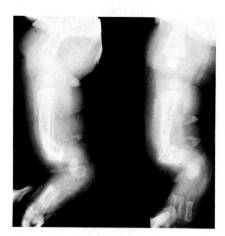

Fig. 75.26   Neonate
There is already an established pseudarthrosis.

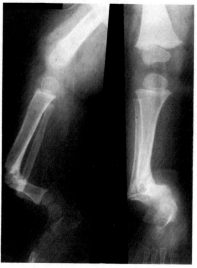

Fig. 75.27   1 year 6 months
The deformity and sclerosis have increased.

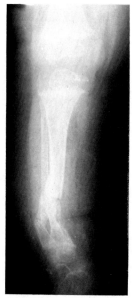

Fig. 75.28   2 years—AP
There has been correction of the deformity, and bone grafting.

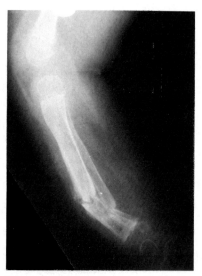

Fig. 75.29   2 years
Lateral view of Figure 75.28.

Pseudarthrosis : 3–4 years

Figs 75.26–75.32   A.V.—Series of seven radiographs to show progression and treatment of tibial pseudarthrosis

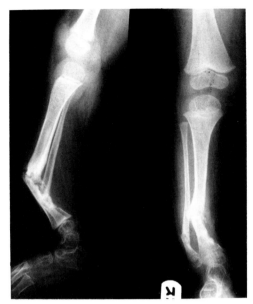

Fig. 75.30   3 years
The deformity is recurring, with reabsorption of bone around the fracture site.

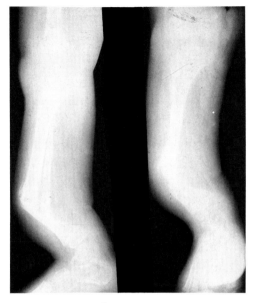

Fig. 75.31   3 years 6 months
Worsening of the deformity.

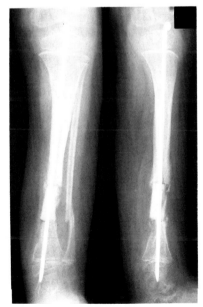

Fig. 75.32   4 years
Correction osteotomy and an intramedullary nail.

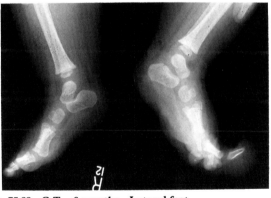

Fig. 75.33   G.T.—6 months—Lateral feet
There is a macro-hallux on the left.

Section Sixteen
MALFORMATION SYNDROMES

Chapter Seventy-six
ACROCEPHALO-SYNDACTYLY
(and similar)

There are many craniostenosis syndromes, distinguished by their congenital limb anomalies and differing inheritance. In all of them there is premature fusion of (variable) cranial sutures, with or without mental retardation.

Inheritance

The two most frequent are *Apert's syndrome*, of autosomal dominant inheritance, and *Carpenter's syndrome*, which is autosomal recessive.

Frequency

About 0.5 per million prevalence.

Clinical features

Facial appearance—bizarre, with a steep forehead, hypoplastic middle third of face, sometimes asymmetry, hypertelorism, malocclusion, strabismus. Since variable cranial sutures are involved, the facial appearance varies.

Intelligence—in all the syndromes there is a risk of mental retardation; this most usually develops in Apert's, less often in Carpenter's.

Stature—this may be normal, but marked limb shortening may be present, associated with congenital limb anomalies.

Presenting feature/age—congenital, with craniosynostosis present at birth, or developing during the first year or so. Limb anomalies are apparent at birth.

Deformities—related to the limb anomalies.

Associated anomalies—not infrequently visceral and genito-urinary anomalies co-exist.

Radiographic features

Skull—in *Apert's syndrome:* often the coronal suture is prematurely fused, but others may be involved. Turricephaly develops, with flattening of the occiput. The middle third of the face is flattened, with relative prognathos. In *Carpenter's syndrome:* the sagittal and lambdoid sutures are usually the first to close, sometimes

with asymmetry and a bizarre skull deformity. Increased digital markings indicate raised intracranial pressure (in the presence of fusion of suture lines).

Spine—fusion or other congenital anomalies of vertebrae may occur.

Limb anomalies—in *Apert's syndrome* there is usually syndactyly of the second to fourth digits, leaving the first and fifth free. Characteristically the distal parts of the digits are joined, and the proximal is free. The thumb metacarpal is probably normal, but there may be only one phalanx, trapezoid in shape. Symphalangism and fusion of carpal bones is not unusual. The feet show comparable deformities. In *Carpenter's syndrome* the syndactyly may be less evident (third and fourth digits only) and polydactyly is usual, post-axial in the hand, sometimes pre-axial in the foot.

Bone maturation

Normal.

Biochemistry

Not known.

Differential diagnosis

The craniostenosis syndromes form a characteristic group: for diagnosis of individual disorders reference should be made to a specialist text.

Progress/complications

The most serious are related to raised intracranial pressure, mental retardation and to complications from visceral anomalies. Progressive synostosis of hand and foot bones has been reported.

REFERENCES

Blank C E 1960 Apert's syndrome (a type of acrocephalo-syndactyly): observation on a British series of thirty-nine cases. American Journal of Human Genetics 24:152

Cohen M M 1975 An etiologic and nosologic overview of craniosynostosis syndromes. Birth Defects: Original Article Series (2) XII:137–189

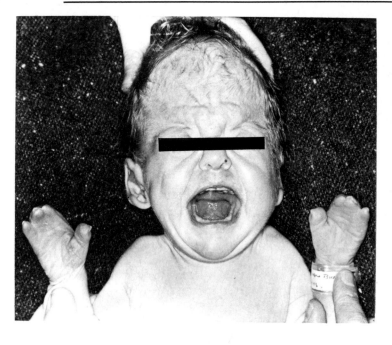

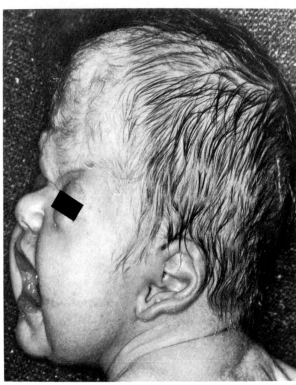

Figs 76.1 and 76.2   A newborn child with Apert's syndrome, showing the tower skull and massive syndactyly.

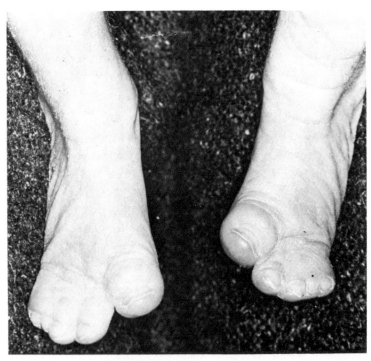

Fig. 76.3   Similar massive syndactyly of the toes.

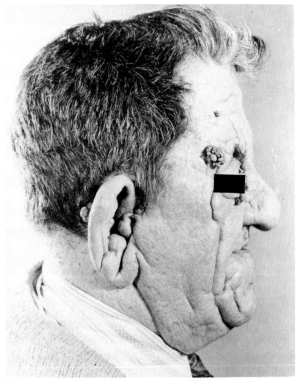

Fig. 76.4   Carpenter's syndrome, with less marked skull deformity.

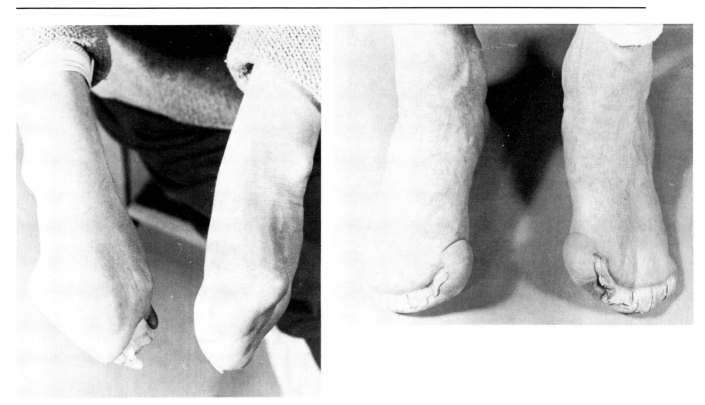

Figs 76.5 and 76.6   The same patient, showing massive syndactyly of the hands and toes, with polydactyly of the latter.

Skull: Birth–14 years

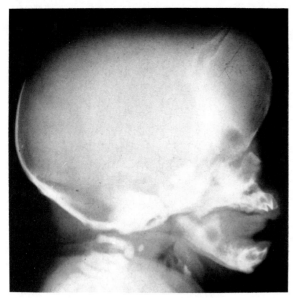

Fig. 76.7   Unknown—Neonate—Lateral skull
The coronal suture is just open, but the straight line appearance and surrounding sclerosis indicates imminent fusion.
Same patient: Fig. 76.8

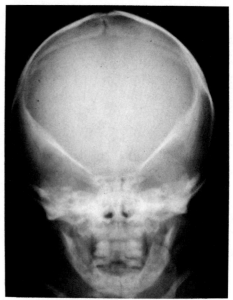

Fig. 76.8   Unknown—Neonate—Lateral skull (same patient as in Fig. 76.7)
There is the same appearance of coronal and sagittal sutures.

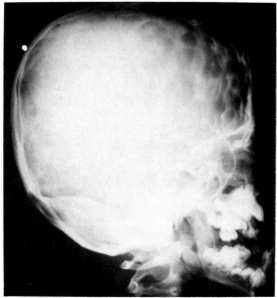

Fig. 76.9   D.H.—6 years—Lateral skull
Pronounced convolutional markings and tower-shaped skull are present, resulting from premature fusion of the coronal suture. There is likely to be raised intracranial pressure in this patient.

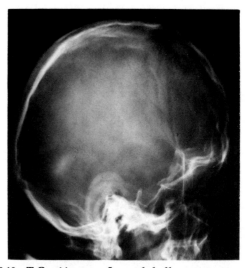

Fig. 76.10   T.C.—14 years—Lateral skull
This shows the typical tower-shaped skull (turricephaly) due to premature fusion of the coronal suture, but not the sagittal. There is no increase in the convolutional markings.
Same patient: Fig. 76.12

Hand and foot: 2 years–Adult

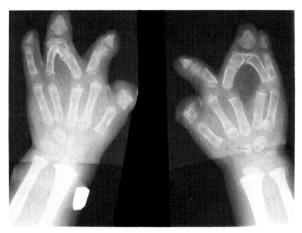

Fig. 76.11   M.C.—2 years—PA hands
Bone modelling is normal, but there is syndactyly with bony fusion
principally involving the distal ends of the fingers. The thumb is short
with only one short deformed phalanx.
Same patient: Fig. 76.14

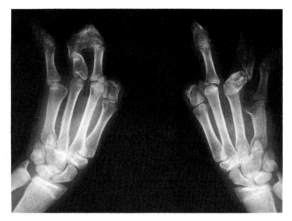

Fig. 76.12   T.C.—10 years—PA hands
There are only four digits; both thumbs have only one distorted
phalanx and on the right there is distal fusion of the two middle digits.
It is difficult to be certain but there appears to be fusion of some
carpal bones.
Same patient: Fig. 76.10

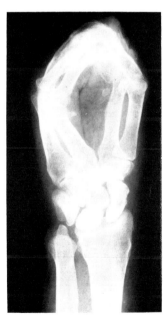

Fig. 76.13   M.C.—50 years—PA hand (Carpenter's syndrome)
This man is untreated and shows well the distal fusion of digits; the
'mitten' deformity.

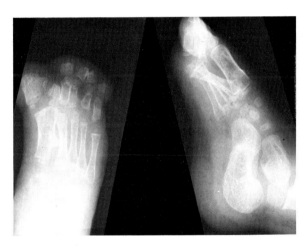

Fig. 76.14   M.C.—2 years—AP and lateral foot (same patient as
in Fig. 76.11)
There is pre-axial polydactyly and bony fusion of the third and fourth
digits distally. The deformed foot with calcaneo-cuboid fusion is well
shown.

Chapter Seventy-seven
CLEIDOCRANIAL DYSPLASIA
(craniocleido dysostosis)

Usually stated to be a disorder affecting the membrane bones, principally the clavicles, vault of skull, symphysis pubis and also terminal phalanges.

Inheritance

Autosomal dominant.

Frequency

Very rare. Less than 1 per million prevalence.

Clinical features

Facial appearance—bulging frontal and parietal regions, low nasal bridge.
Intelligence—normal.
Stature—some shortness, but not marked.
Presenting feature/age—disordered eruption of teeth; absent clavicles allow shoulders to be approximated anteriorly.
Deformities—drooping shoulders; pectus excavatum; coxa vara.
Associated anomalies—usually none. Absent or hypoplastic radius and fibula have been reported.

Radiographic features

Skull—Wormian bones, open fontanelles in childhood, but these close later. Late and incomplete development of accessory nasal sinuses and mastoid air cells. Failure of fusion of mandibular symphysis. Dental anomalies.
Spine—retarded ossification of vertebrae, biconvexity remaining for longer than usual.
Shoulder girdle—clavicles partially or completely absent, together with the muscles normally attached; although, rarely, these may be normal. If the lateral part is present, it may overlap the medial, as in pseudarthrosis of the clavicle. Scapulae may be small and deformed.
Pelvis/hips—wide symphysis pubis sometimes with coxa vara. The sacroiliac joint and triradiate cartilages are also wide.

Femur—rarely, pseudarthrosis of the shaft occurs.
Hands—short terminal phalanges and short middle phalanges of the index and little fingers. There may be coned epiphyses and pseudoepiphyses.

Bone maturation

Skull and vertebrae, shoulder girdle and pelvis are all delayed. No other information.

Biochemistry

Not known.

Differential diagnosis

In *pycnodysostosis* there are also short or absent clavicles, but here the bone is dense and abnormally fragile.

In *osteogenesis imperfecta* there are Wormian bones, but the clavicles and pelvis are not deficient. Features of osteoporosis are present.

Congenital pseudarthrosis of the clavicle is almost invariably right sided and is an isolated deformity (also of autosomal dominant inheritance). Delineation from cleidocranial dysplasia can only be made if the latter has signs elsewhere in the skeleton.

An example of familial congenital pseudarthrosis of the clavicle is shown at the end of the chapter.

Progress/complications

No serious complications. Some instability of the shoulder joints may develop.

REFERENCES

Eventov I, Reider-Grosswasser I, Weiss S 1979 Cleidocranial dysplasia. A family study. Clinical Radiology 30/3:323–328
Fauré C, Maroteaux P 1973 Cleidocranial dysplasia. In: Kaufmann H J (ed) Intrinsic diseases of bones. Progress in Pediatric Radiology 4, S Karger, Basel, pp 211–237

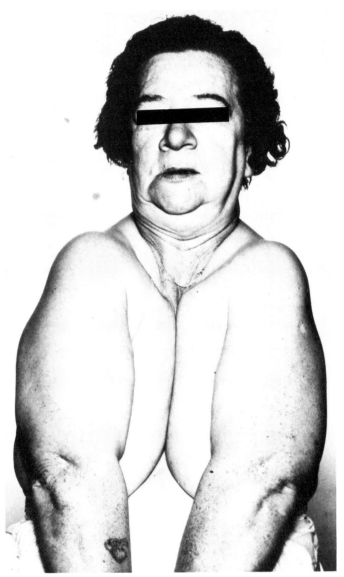

Fig. 77.1 Absence of the clavicles allows approximation of the shoulders.

Skull : Birth

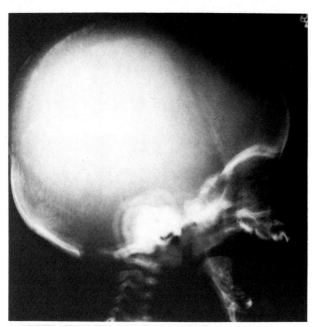

Fig. 77.2   Fe.A.—Neonate—Lateral skull (sib of children in Figs 77.3, 77.5 and 77.6)
Multiple Wormian bones are present and the anterior fontanelle is larger than usual for a neonate.
Same patient: Figs 77.13, 77.22

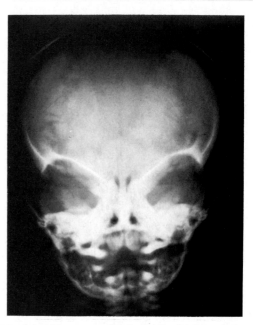

Fig. 77.3   J.A.—Neonate—AP skull (sib of children in Figs 77.2, 77.5 and 77.6)
The sutures are wide and there are Wormian (sutural) bones.
Same patient: Figs 77.4, 77.14, 77.29

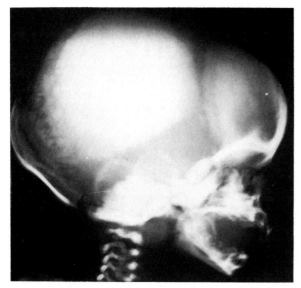

Fig. 77.4   J.A.—Neonate—Lateral skull (same patient as in Fig. 77.3)
The sutures are wide with a large anterior fontanelle. Multiple Wormian bones are very well shown.

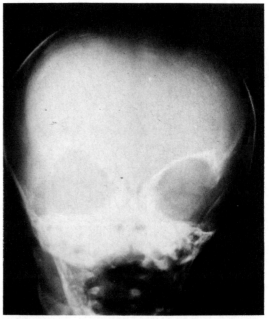

Fig. 77.5   M.A.—Neonate—AP skull (sib of children in Figs 77.2, 77.3 and 77.6)
Wide sutures and Wormian bones are present, and there is some lateral bulging of the parietal bones.
Same patient: Figs 77.8, 77.20, 77.23, 77.24, 77.28, 77.33, 77.35

Skull: 6 months–1 year

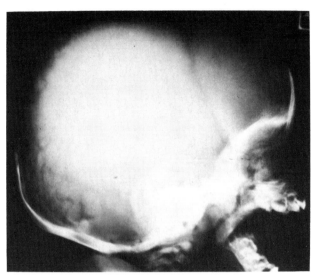

Fig. 77.6 F.A.—6 months—Lateral skull (sib of children in Figs 77.2, 77.3 and 77.8)
The sutures are wide with multiple Wormian bones. The anterior fontanelle is large.
Same patient: Figs 77.7, 77.21

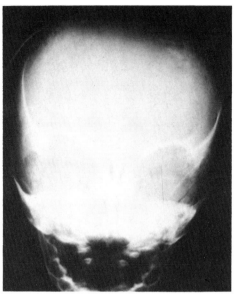

Fig. 77.7 F.A.—6 months—AP skull (same patient as in Fig. 77.6)
Wide sutures—especially the sagittal—and Wormian bones. There is lateral bulging of the parietal bones.

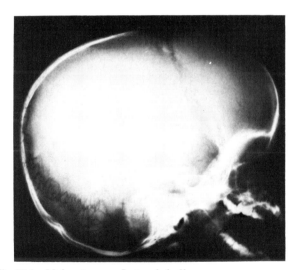

Fig. 77.8 M.A.—1 year—Lateral skull
The anterior fontanelle remains large. Wormian bones are present in the lamboid suture.
Same patient: Figs 77.5, 77.20, 77.23, 77.24, 77.28, 77.33, 77.35

Skull: 4–8 years

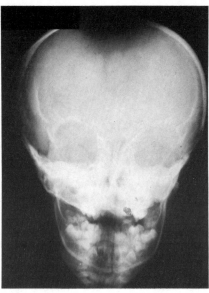

Fig. 77.9   S.W.—4 years 4 months—AP skull
The anterior fontanelle remains large (normally closed by the age of 2 years).
Same patient: Figs 77.10, 77.15, 77.19, 77.27, 77.32, 77.34

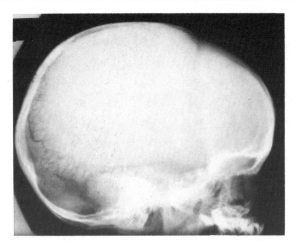

Fig. 77.10   S.W.—4 years 4 months—Lateral skull (same patient as in Fig. 77.9)
Large anterior fontanelle. Wormian bones are present in the lambdoid suture.

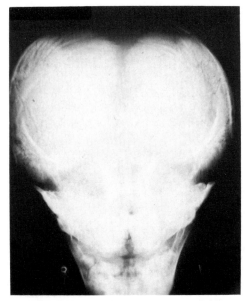

Fig. 77.11   N.W.—8 years—AP skull
Wide sutures and anterior and posterior fontanelles are still present with multiple Wormian bones. Lateral bulging of parietal bones is noted.
Same patient: Figs 77.12, 77.17, 77.18, 77.25, 77.26

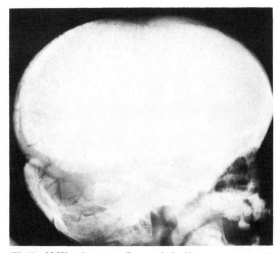

Fig. 77.12   N.W.—8 years—Lateral skull (same patient as in Fig. 77.11)
Large fontanelles, wide sutures and multiple Wormian bones.

Chest: Birth–Adult

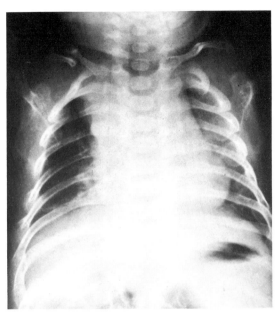

Fig. 77.13 Fe.A.—Neonate—AP chest
There is a pseudarthrosis of the right clavicle with overlap of the fragments.
Same patient: Figs 77.2, 77.22

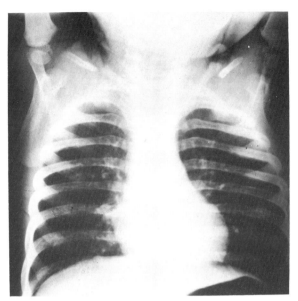

Fig. 77.14 J.A.—6 months—AP chest
There are pseudarthroses of both clavicles and hypoplasia of their distal ends.
Same patient: Figs 77.3, 77.4, 77.29

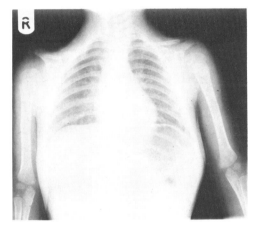

Fig. 77.15 S.W.—4 years 4 months—AP chest
There is a 'pseudarthrosis' of the right clavicle, with absence of the medial half and hypoplasia of the lateral part. There is mild hypoplasia of the lateral end of the left clavicle. Both humeri are undermodelled with lack of the normal metaphyseal flaring.
Same patient: Figs 77.9, 77.10, 77.19, 77.27, 77.32, 77.34

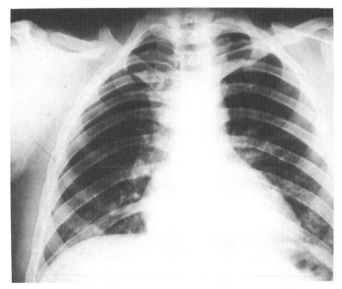

Fig. 77.16 F.A.—Adult—PA chest (father of children in Figs 77.5, 77.6, 77.13 and 77.14)
There is a pseudarthrosis of the right clavicle with some hypoplasia of its lateral part. As a result the right scapula is low with lateral rotation of its inferior angle.

Spine and shoulders : 3–4 years

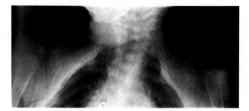

Fig. 77.17   N.W.—3 years—AP shoulder girdle
Both clavicles are absent. This occurs in only 10% of patients with
cleido-cranial dysostosis.
Same patient : Figs 77.11, 77.12, 77.18, 77.25, 77.26

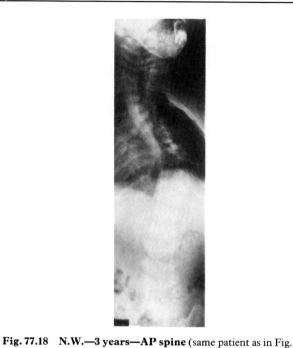

Fig. 77.18   N.W.—3 years—AP spine (same patient as in Fig.
77.17)
There is a lower dorsal scoliosis convex to the right. There is widening
of the interpedicular distances, abnormal modelling of the vertebral
bodies, and wide separation between the pedicles and body.

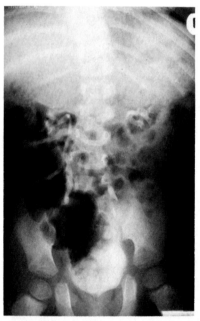

**Fig. 77.19   S.W.—4 years 4 months—Intravenous urogram
including AP spine and pelvis**
There is a mild scoliosis convex to the left centred at L1. The
symphysis pubis is wide with hypoplastic pubic rami. The iliac wings
are narrow. The capital femoral epiphyses are large and rounded.
Bilateral duplex kidneys are present.
Same patient : Figs 77.9, 77.10, 77.15, 77.27, 77.32, 77.34

Pelvis: Birth–6 months

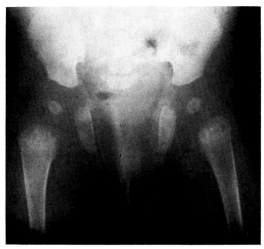

Fig. 77.20 M.A.—Neonate—AP pelvis
Advanced maturation of the capital femoral epiphyses, with wide separation from the short femoral necks. The metaphyses show mild irregularity. The symphysis pubis is wide with hypoplastic pubic rami.
Same patient: Figs 77.5, 77.8, 77.23, 77.24, 77.28, 77.33, 77.35

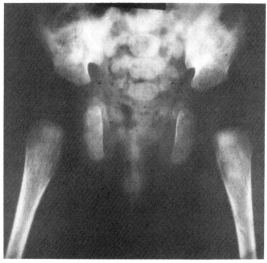

Fig. 77.21 F.A.—Neonate—AP pelvis
Short femoral necks and hypoplastic pubic rami with a wide symphysis pubis.
Same patient: Figs 77.6, 77.7

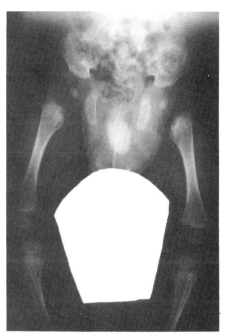

Fig. 77.22 Fe.A.—Neonate—AP pelvis
Advanced maturation of capital femoral epiphyses with wide separation from the short femoral necks, and metaphyseal irregularity. There are hypoplastic pubic rami and a wide symphysis pubis.
Same patient: Figs 77.2, 77.13

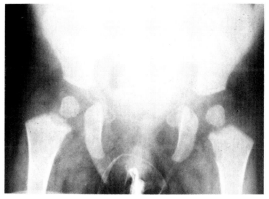

Fig. 77.23 M.A.—6 months—AP pelvis (same patient as in Fig. 77.20)
Large rounded capital femoral epiphyses and short femoral necks; hypoplastic pubic rami and wide symphysis pubis.

Pelvis: 9 months–4 years

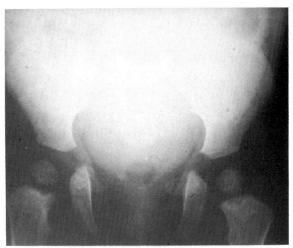

Fig. 77.24   M.A.—9 months—AP pelvis
Rounded capital femoral epiphyses and short femoral necks;
hypoplastic pubic rami.
Same patient: Figs 77.5, 77.8, 77.20, 77.23, 77.28, 77.33, 77.35

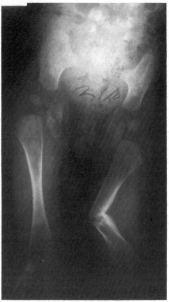

Fig. 77.25   N.W.—2 years 6 months—AP pelvis and femora
There is a pseudarthrosis of the femoral shaft on the left. The capital
femoral epiphyses are rounded with short femoral necks and poor
modelling.
Same patient: Figs 77.11, 77.12, 77.17, 77.18, 77.26

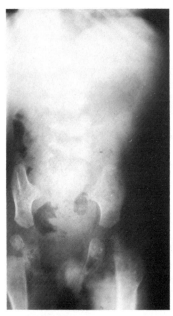

Fig. 77.26   N.W.—2 years 6 months—AP spine, pelvis (same
patient as in Fig. 77.25)
The iliac wings are narrow and there is virtually no ossification of the
pubic rami. The spine shows abnormal modelling with unusual
separation of the pedicles from the rounded vertebral bodies.

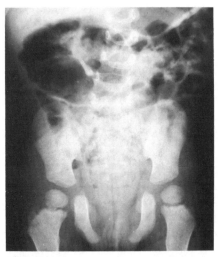

Fig. 77.27   S.W.—4 years 4 months—AP pelvis
The iliac wings are narrow. The symphysis pubis is wide and the
pubic rami hypoplastic. The sacroiliac joint space is widened, as is the
triradiate cartilage. The capital femoral epiphyses appear large and
rounded.
Same patient: Figs 77.9, 77.10, 77.15, 77.19, 77.32, 77.34

Pelvis: 6–18 years

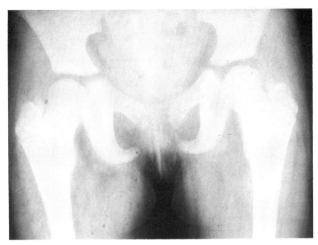

Fig. 77.28   M.A.—6 years—AP pelvis
Wide symphysis pubis. Large capital femoral epiphyses and greater
trochanters, short femoral necks in valgus position.
Same patient: Figs 77.5, 77.8, 77.20, 77.23, 77.24, 77.33, 77.35

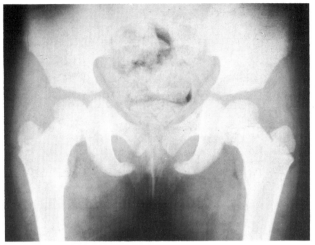

Fig. 77.29   J.A.—7 years—AP pelvis
Wide symphysis pubis, large capital femoral epiphyses and greater
trochanters.
Same patient: Figs 76.3, 76.4, 76.14

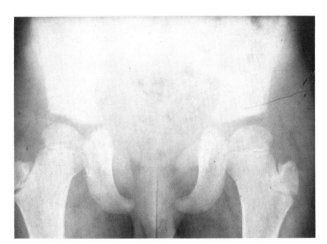

Fig. 77.30   Unknown—About 11 years—AP pelvis
Wide symphysis pubis and hypoplastic pubic rami, short femoral
necks in valgus position. Narrow iliac wings.

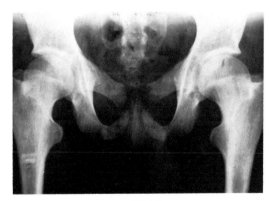

Fig. 77.31   A.T.—18 years—AP hips
Large capital femoral epiphyses with prominent trochanters, short
femoral necks.

Limbs: 2–8 years

Fig. 77.32   S.W.—4 years 4 months—AP tibiae and fibulae
Rounded epiphyses and abnormal metaphyseal modelling; also minor
irregularities of the fibular diaphyses.
Same patient: Figs 77.9, 77.10, 77.15, 77.19, 77.27, 77.34

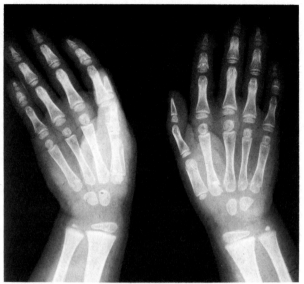

Fig. 77.33   M.A.—2 years—PA hands
There are cone epiphyses of the middle phalanges of the index and
little fingers. The tufts are pointed and the epiphyses of the terminal
phalanges are large and rounded. There are pseudo-epiphyses at the
base of the second and fifth metacarpals.
Same patient: Figs 77.5, 77.8, 77.20, 77.23, 77.24, 77.28, 77.35

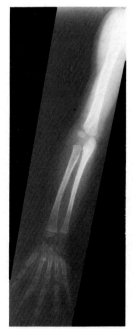

Fig. 77.34   S.W.—4 years 4 months—AP upper limb
Slight undermodelling of the long bones. There is a pseudo-epiphysis
at the base of second metacarpal.
Same patient: Figs 77.9, 77.10, 77.15, 77.19, 77.27, 77.32

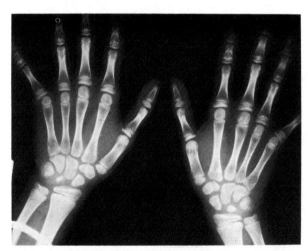

Fig. 77.35   M.A.—8 years—PA hands (same patient as in Fig.
77.33)
There are cone epiphyses of the middle phalanges and mild
clinodactyly on the left.

Familial congenital pseudarthrosis of clavicle

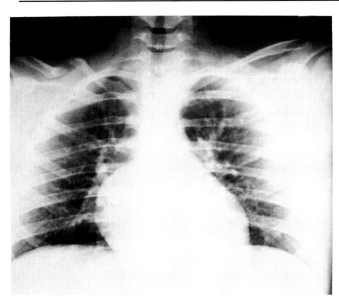

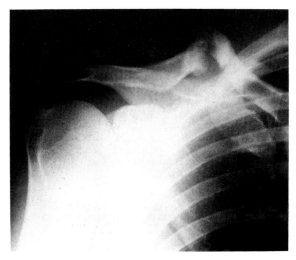

Fig. 77.37   W.W.—41 years—AP right shoulder (father of child in Fig. 77.36)
There is a pseudarthrosis of the clavicle.
Same patient : Fig. 77.38

Fig. 77.36   D.W.—12 years 3 months—AP chest (daughter of patient in Fig. 77.37)
There is a pseudarthrosis of the right clavicle.

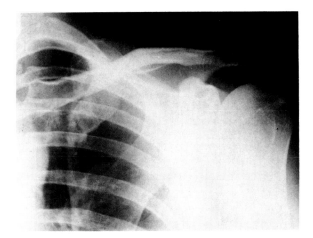

Fig. 77.38   W.W.—41 years—AP left shoulder (same patient as in Fig. 77.37)
Normal clavicle.

Chapter Seventy-eight
CRANIO-CARPO-TARSAL DYSPLASIA
(Freeman-Sheldon syndrome—
'whistling face' syndrome)

The association of cranial and facial anomalies (microstomia, long philtrum, small nose) with finger contractures and talipes equinovarus.

Inheritance

Autosomal dominant, but autosomal recessive cases have also been reported.

Frequency

Probably a good deal commoner than the currently reported figures of about 1 per million prevalence, since many cases will be diagnosed only as localised foot or hand deformities.

Clinical features

Facial appearance—characteristic, with small pursed mouth ('whistling'), long philtrum and small nose. Sometimes enophthalmus and hypertelorism.
Intelligence—normal.
Stature—may be short, but not markedly so.
Presenting feature/age—hand and foot deformities are present at birth, which should alert the clinician to the less obvious facial anomalies.
Deformities—those associated with the hand and foot deformities, including ulnar deviation of the hand.
Associated anomalies—sometimes congenital dislocation of the hips and scoliosis.

Radiographic features

Skull—steeply sloping anterior cranial fossa, some with the 'helmet' sign (downward prolongation of the frontal bones).
Spine—scoliosis, and vertebrae may be unusually tall.
Hands—palm-clutched thumbs and contractures.
Feet—talipes equino-varus, sometimes a vertical talus and an anteriorly-displaced second metatarsal associated with this.

Bone maturation

Probably normal.

Biochemistry

Not known.

Differential diagnosis

Arthrogryposis multiplex congenita, but this is usually non-genetic and involves the more proximal parts of the limbs as well as hands and feet. The face is normal.

Progress/complications

Only those associated with the hand and foot deformities and scoliosis.

REFERENCES

O'Connell D J, Hall C M 1977 Cranio-carpo-tarsal dysplasia. A report of seven cases. Radiology 123: 719–722
Rinsky L A, Bleck E E 1978 Freeman-Sheldon ('whistling face') syndrome. Journal of Bone and Joint Surgery 58A: 148

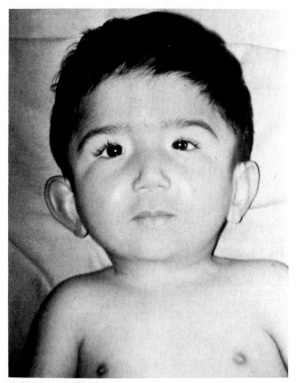

Fig. 78.1 Characteristic facies, with hypertelorism, a small nose, long philtrum and rather pursed ('whistling') mouth.

Skull: Infancy–Childhood

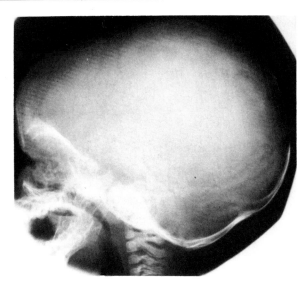

Fig. 78.2 S.G.—Infant—Lateral skull
The skull is normal apart from the 'helmet' sign, downward prolongation of the frontal bone.
Same patient: Figs 78.6, 78.7, 78.12

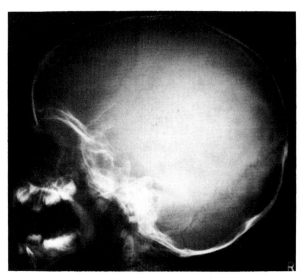

Fig. 78.3 E.H.—Infant—Lateral skull
Similar to Figure 78.2.
Same patient: Figs 78.9, 78.10

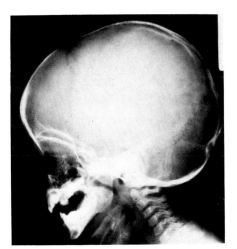

Fig. 78.4 B.M.—1 year—Lateral skull
The 'helmet' sign is present—also micrognathos.

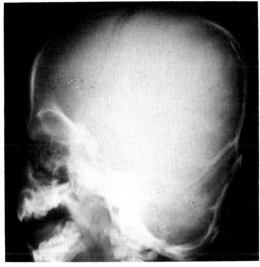

Fig. 78.5 M.C.—Childhood—Lateral skull
The 'helmet' sign is well shown. In addition, brachycephaly and occipital flattening are present. This combination is probably due to premature fusion of sutures and a post-natal positional deformity associated with hypotonia.
Same patient: Figs 78.8, 78.13

Spine and pelvis : Childhood

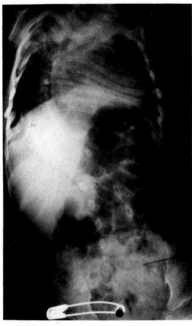

Fig. 78.6   S.G.—Childhood—AP spine
It is difficult to see individual vertebrae in the presence of left thoracic scoliosis.
Same patient: Figs 78.2, 78.7, 78.12

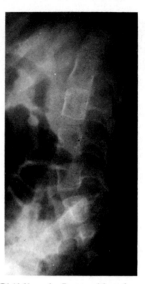

Fig. 78.7   S.G.—Childhood—Lateral lumbar spine (same patient as in Fig. 78.6)
The vertebrae have a decreased antero-posterior diameter and are unusually tall.

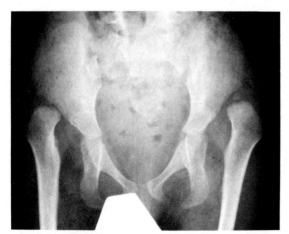

Fig. 78.8   M.C.—2 years—Pelvis
There is bilateral dislocation of the hips but otherwise normal bony modelling.
Same patient: Figs 78.5, 78.13

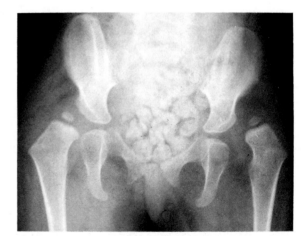

Fig. 78.9   E.H.—2 years 6 months—Pelvis
Similar to Figure 78.8.
Same patient: Figs 78.3, 78.10

Foot and hand : Childhood

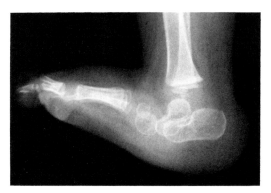

Fig. 78.10   E.H.—2 years—Lateral foot
There is a vertical talus with 'rocker-bottom' foot.
Same patient : Figs 78.3, 78.9

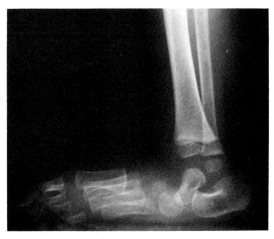

Fig. 78.11   J-M. C.—Childhood—Standing lateral foot
Similar to Figure 78.10

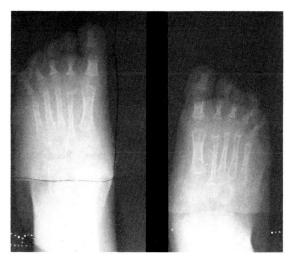

Fig. 78.12   S.G.—PA feet
There is proximal displacement of the second metatarsals associated with the vertical tali.
Same patient : Figs 78.2, 78.6, 78.7

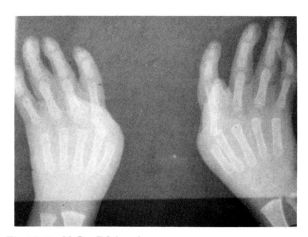

Fig. 78.13   M.C.—PA hands
There is ulnar deviation with 'palm-clutched' thumbs. Finger contractures will probably develop later.
Same patient : Figs 78.5, 78.8

Chapter Seventy-nine
CORNELIA DE LANGE SYNDROME

Short stature, a characteristic facies with eyebrows meeting in the middle, mental retardation and congenital limb absence, often involving the ulnar side.

Inheritance

Uncertain; nearly all cases are sporadic, but there is some evidence for small chromosome defects.

Frequency

Very rare, with an estimated prevalence of about 2 per million.

Clinical features

Facial appearance—low hairline, eyebrows meeting in the middle and a small nose with anteverted nostrils and an in-turned upper lip.
Intelligence—mental retardation.
Stature—short.
Presenting feature/age—congenital limb defects and facial appearance are noted at birth.
Deformities—mostly limb 'absence' defects, short first metacarpals, sometimes syndactyly and dislocated radial heads.
Associated anomalies—congenital heart and other visceral anomalies.

Radiographic features

Skull—microcephaly.
Spine—normal.
Limbs—characteristically there are ulnar 'absence' defects, but phocomelia, variable brachydactyly or short metacarpals and other minor anomalies are frequent.

Bone maturation

Not known.

Biochemistry

Not known.

Differential diagnosis

There is no other condition with similar features.

Progress/complications

Most children fail to thrive, are mentally retarded and do not survive infancy.

REFERENCES

McArthur R G, Edwards J H 1967 De Lange syndrome. Report of 20 cases. Canadian Medical Association Journal 96:1185
Pashayan H, Whelan D, Guttman S, Fraser F C 1969 Variability of the de Lange syndrome: report of 3 cases and genetic analysis of 54 families. Journal of Paediatrics 75:853

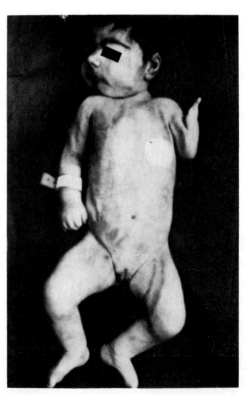

Fig. 79.1 Eyebrows meeting in the middle, anteverted nostrils, inturned upper lip and absence defects of the left upper limb—characteristically these are on the ulnar side.

Skull and chest: Birth–9 years

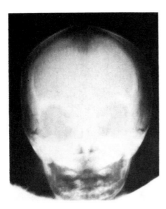

Fig. 79.2   K.E.—Neonate—AP skull
The vault is small and sutures wide.
Same patient: Figs 79.3–79.9, 79.13

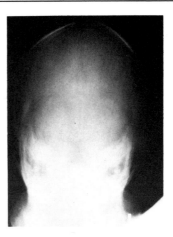

Fig. 79.3   K.E.—Neonate—Towne's view (same patient as in Fig. 79.2)
Similar to Figure 79.2.

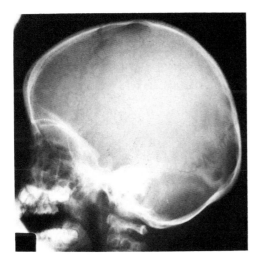

Fig. 79.4   K.E.—4 years—Lateral skull (same patient as in Fig. 79.2)
Normal apart from microcephaly.

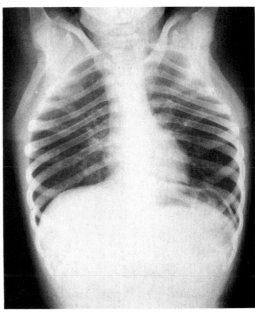

Fig. 79.5   K.E.—9 years—AP chest (same patient as in Fig. 79.2)
The lateral diameter is wider than normal, particularly in its upper part.

Barium swallow

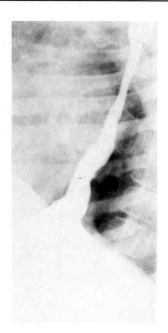
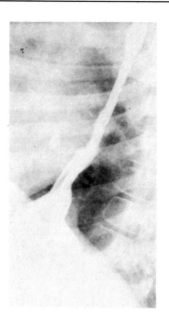

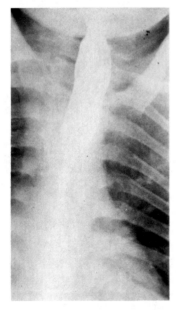
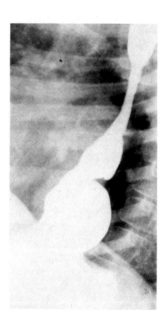

Figs 79.6–79.9   K.E.—9 years—Barium swallow
The series shows a long stricture in the middle of the oesophagus. This is
associated with a fixed sliding hiatus hernia with gastro-oesophageal reflux.
Same patient: Figs 79.2–79.5, 79.13

Limbs: Birth–6 years

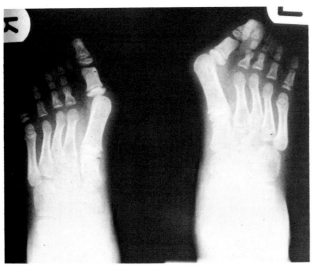

Fig. 79.10 J.A.—6 years—PA feet
There is bilateral hallux valgus, and some shortening of the second, third and fourth metatarsals with premature fusion of their epiphyses.
Same patient: Fig. 79.12

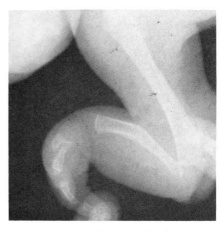

Fig. 79.11 B.S.—Neonate—AP upper limb
The humerus appears normal, the main defect being complete absence of the ulna, and ulnar three digits. The radius is short, bowed and dislocated at the elbow. The first metacarpal is shortened.

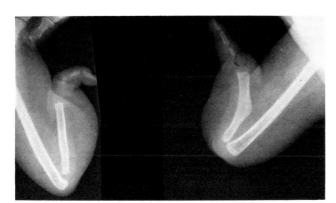

Fig. 79.12 J.A.—Neonate—AP both upper limbs (same patient as in Fig. 79.10)
Similar ulnar-side absence, but here only the thumb (with a short metacarpal) is present.

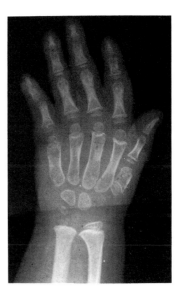

Fig. 79.13 K.E.—3 years 10 months—PA left hand
The only abnormality here is the short first metacarpal.
Same patient: Figs 79.2–79.9

Chapter Eighty
LARSEN'S SYNDROME

Characterised by multiple congenital dislocations, usually including knees and hips, together with a depressed nasal bridge and bulging forehead. Variable other structural defects both of bone and soft tissue.

Inheritance

Both autosomal dominant and recessive forms have been described.

Frequency

Probably rare, with a prevalence of less than 1 per million: it may be somewhat over-diagnosed, not all cases of congenital knee dislocation being Larsen's syndrome.

Clinical features

Facial appearance—characteristic with hypertelorism, prominent forehead and depressed nasal bridge.

Intelligence—most are normal.

Stature—some delay in development, but height is not markedly reduced.

Presenting feature/age—multiple dislocations present at birth; less commonly, talipes equino-varus, cervical vertebral anomalies, hydrocephalus, finger deformities and respiratory difficulties related to 'flabby' cartilage of the trachea.

Deformities—related to the congenital dislocations, foot and hand anomalies and cervical vertebral defects.

Associated anomalies—noted above.

Radiographic features

Spine—may be congenital segmentation anomalies.
Hips and knees—congenital dislocations.

Feet—some have double ossification centres for the calcaneum; if present this is diagnostic.

Wrists/hands—sometimes additional ossification centres for the carpal bones, 'delta' phalanx of thumb and short metacarpals.

Bone maturation

Probably normal.

Biochemistry

Not known.

Differential diagnosis

From other conditions of disordered connective tissue presenting with congenital dislocation of joints but the characteristic facies of Larsen's syndrome should differentiate it.

Progress/complications

If there are no respiratory difficulties and the problems of congenital dislocations can be overcome then these patients have little disability after childhood.

REFERENCES

Fauré C, Lascaux J P, Montagne J Ph 1976 Le syndrome de Larsen: à propos de trois observations nouvelles. Annales de Radiologie 19/6:629–636
Steel H H, Kohl E J 1972 Multiple congenital dislocations associated with other skeletal anomalies (Larsen's syndrome) in three siblings. Journal of Bone and Joint Surgery 54A:75

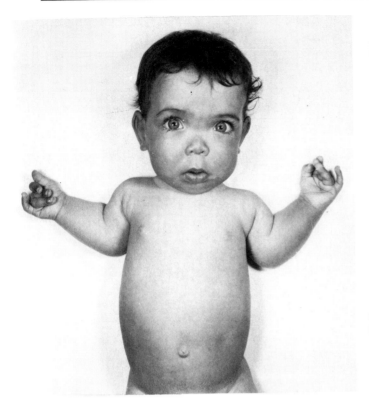
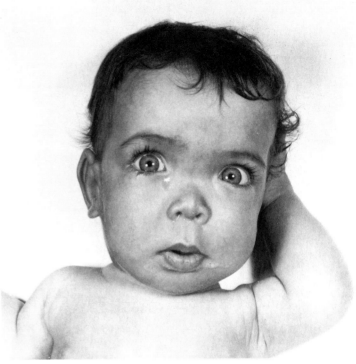
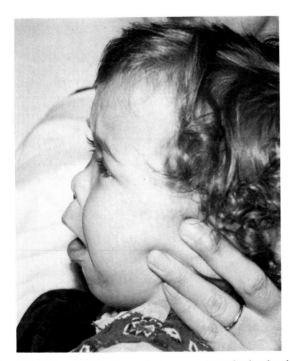

Figs 80.1, 80.2 and 80.3 Hypertelorism, prominent forehead and depressed nasal bridge.

Spine, pelvis and femora : Birth–1 year

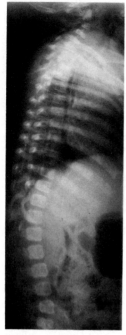

Fig. 80.4   H.McI.—Neonate—Lateral spine
Normal apart from a mild thoraco-lumbar kyphos and coronal clefts
at the level of L5 and S1.
Same patient: Figs 80.7–80.12, 80.14, 80.15

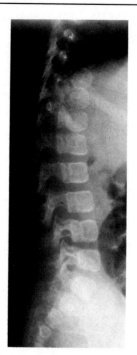

Fig. 80.5   E.C.—1 year—Lateral spine
Coronal cleft vertebrae are present from T10 to L4.

Fig. 80.6   Unknown—Neonate—AP pelvis and femora
Both hips and knees are dislocated.

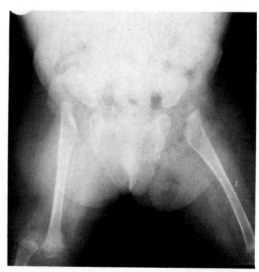

Fig. 80.7   H.McI.—1 year—AP pelvis and femora (same
patient as in Fig. 80.4)
Both hips and knees are dislocated.

Lower limb: Birth–2 years

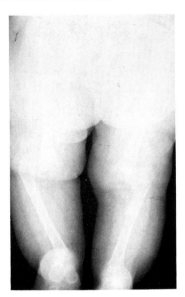

Fig. 80.8   H.McI.—Neonate—AP lower limb
In addition to bilateral knee dislocation both fibulae are absent.
Same patient: Figs 80.4, 80.7, 80.9–80.12, 80.14, 80.15

Fig. 80.9   H.McI—Neonate—Lateral ankles (same patient as in Fig. 80.8)
There are double ossification centres of the calcaneum.

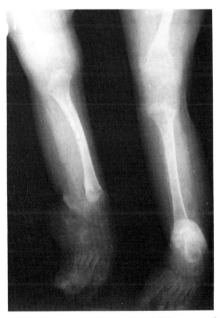

Fig. 80.10   H.McI.—2 years 1 month—AP knees and ankles
Same patient as in Figure 80.8, following treatment.

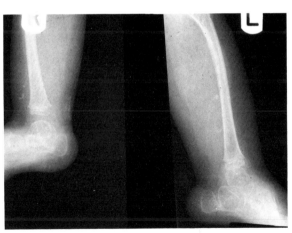

Fig. 80.11   H.McI.—2 years 7 months—Lateral ankles (same patient as in Fig. 80.8)
The double ossification centres of the calcaneum are beginning to fuse—there is mild anterior bowing of the tibia.

Upper limb and feet : Birth–4 years

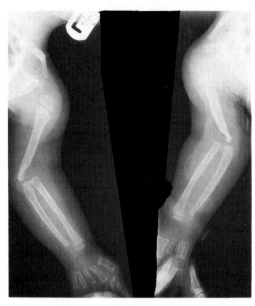

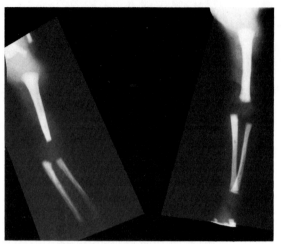

Fig. 80.13   Unknown—Neonate—AP upper limbs
Similar to Figure 80.12.

Fig. 80.12   H.McI.—Neonate—AP upper limbs
The humeri are short and tapered at the lower ends, with elbow
dislocation. There is also mild shortening and tapering of the ulna.
Same patient : Figs 80.4, 80.7–80.11, 80.14, 80.15

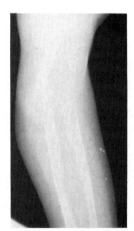

Fig. 80.14   H.McI.—2 years 8 months—AP elbow
Same patient as in Figure 80.12, still showing elbow dislocation.

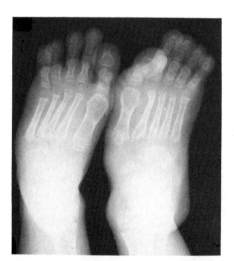

Fig. 80.15   H.McI.—4 years 3 months—PA feet (same patient as
in Fig. 80.12)
No abnormality apart from shortening of both phalanges of the
halluces.

Chapter Eighty-one
MENKES' SYNDROME
(kinky hair disease)

A disorder of copper metabolism, with severe mental retardation, peculiar hair and focal cerebral and cerebellar degeneration, also with widespread effects in the skeleton.

Inheritance

X-linked recessive.

Frequency

A frequency of 2–3 per million live births has been reported.

Clinical features

Facial appearance—micrognathia; stubby or 'kinky', whitish hair.

Intelligence—mental retardation, progressive.

Stature—retarded growth.

Presenting feature/age—within a month or two of birth there is severe neurological deterioration, with seizures, spasticity, opisthotonos.

Deformities—none, except associated with neurological deficiency.

Associated anomalies—patchy stenosis or obliteration of arteries.

Radiographic features

Skull—microcephaly, Wormian bones.
Spine—normal.

Long bones—metaphyseal spurring and periosteal new bone growth: as the neurological signs progress, the long bones become slender and porotic.

Bone maturation

No data.

Biochemistry

A defect in the intestinal absorption of copper.

Differential diagnosis

From other cerebral disorders of infancy, but the characteristic hair, with biochemical tests should distinguish Menkes' syndrome.

Progress/complications

The prognosis is poor, with rapid decerebration.

REFERENCES

Danks D M, Campbell P E, Stevens B J, Mayne V, Cartwright E 1972 Menkes' kinky hair syndrome. An inherited defect in copper absorption with widespread effects. Pediatrics 50:188–201

Menkes J H, Alter M, Steigleder G K, Weakley D R, Sung J H 1962 A sex-linked recessive disorder with retardation of growth, peculiar hair and focal cerebral and cerebellar degeneration. Pediatrics 29:764–779

Stanley P, Gwinn J L, Sutcliffe J 1976 The osseous abnormalities in Menkes' syndrome. Annals of Radiology 19:167–172

Williams D M, Atkin C L, Frens D B, Bray P F 1977 Menkes kinky hair syndrome—studies of copper metabolism and long term copper therapy. Pediatric Research 11:823–826

Skull : Birth–2 years

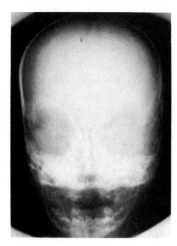

Fig. 81.1   K.B.—Neonate—AP skull
No abnormality seen on this projection.
Same patient : Figs 81.2, 81.9, 81.13

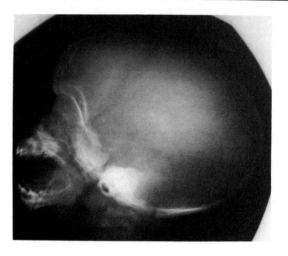

Fig. 81.2   K.B.—Neonate—Lateral skull (same patient as in Fig. 81.1)
Wormian bones are present in the lambdoid suture.

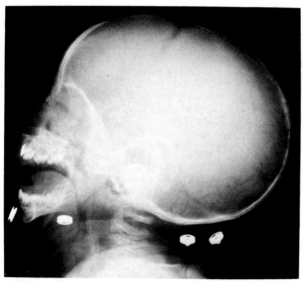

Fig. 81.3   G.N.—1 year 6 months—Lateral skull
The Wormian bones in the lambdoid suture are well shown.
Same patient : Figs 81.8, 81.10, 81.14

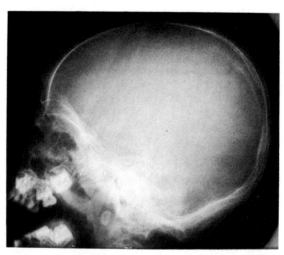

Fig. 81.4   C.S.—2 years 5 months—Lateral skull
There is microcephaly—the vault being small in relation to the facial bones, and it is thickened over the frontal and occipital regions.
Same patient : Figs 81.5, 81.6, 81.7, 81.11, 81.12, 81.15

Skull and trunk: 2–5 years

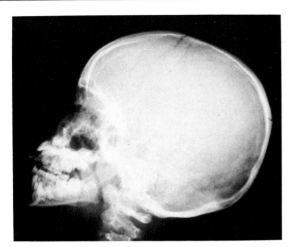

Fig. 81.5   C.S.—5 years 2 months—Lateral skull
There is microcephaly, the vault being small in relation to the facial bones.
Same patient: Figs 81.4, 81.6, 81.7, 81.11, 81.12, 81.15

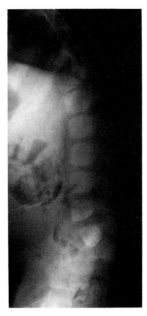

Fig. 81.6   C.S.—2 years 5 months—Lateral thoracic and lumbar spine (same patient as in Fig. 81.5)
The vertebral bodies are oval and taller than normal (probably associated with hypotonia).

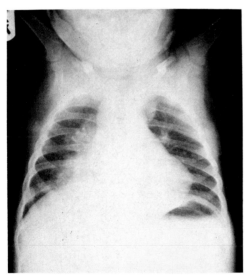

Fig. 81.7   C.S.—2 years 5 months—AP chest (same patient as in Fig. 81.5)
Apart from winging of the scapula there is no skeletal abnormality. The right lung shows patchy consolidation.

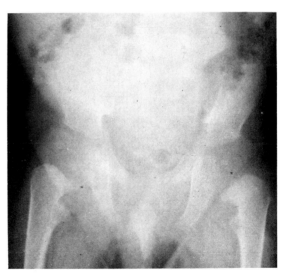

Fig. 81.8   G.N.—4 years—Pelvis
No abnormality, although the capital femoral epiphyses are only just appearing.
Same patient: Figs 81.3, 81.10, 81.14

Lower limb: Birth–5 years

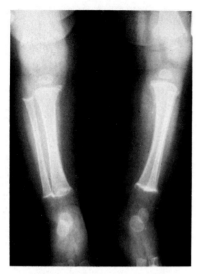

Fig. 81.9   K.B.—Neonate—AP tibiae and fibulae
Tiny spurs are seen at the lateral sides of each metaphysis.
Same patient: Figs 81.1, 81.2, 81.13

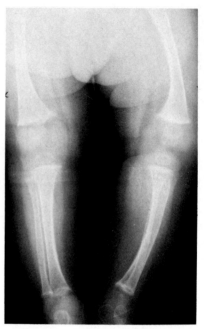

Fig. 81.10   G.N.—4 years—AP lower limbs
Similar to Figure 81.9.
Same patient: Figs 81.3, 81.8, 81.14

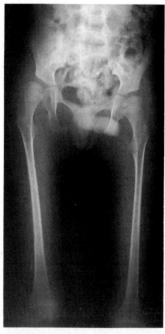

Fig. 81.11   C.S.—5 years 2 months—AP pelvis and femora
The disease has progressed, and the slender, osteoporotic bones with
subluxation of the right hip, together with muscle wasting indicate
neurological deterioration.
Same patient: Figs 81.4–81.7, 81.12, 81.15

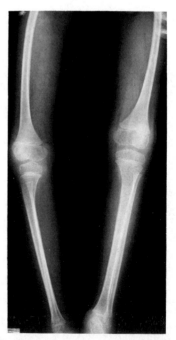

Fig. 81.12   C.S.—5 years 2 months—AP lower limb (same
patient as in Fig. 81.11)
Similar to Figure 81.11.

Upper limb : Birth–5 years

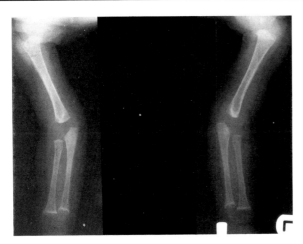

Fig. 81.13   K.B.—Neonate—AP upper limbs
The metaphyseal spurring is well shown at the shoulder and wrist.
Same patient: Figs 81.1, 81.2, 81.9

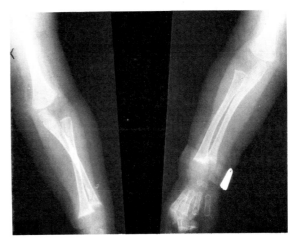

Fig. 81.14   G.N.—4 years—AP upper limbs
The metaphyseal areas at the lower end of the radius and ulna are more generally affected, not just showing lateral spurs.
Same patient: Figs 81.3, 81.8, 81.10

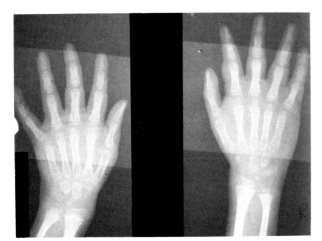

Fig. 81.15   C.S.—5 years 2 months—PA hands
The bones are slender and porotic, indicating neurological degeneration.
Same patient: Figs 81.4–81.7, 81.11, 81.12

Chapter Eighty-two
NAIL-PATELLA SYNDROME
(hereditary osteo-onychodysplasia)

A syndrome characterised by absence or hypoplasia of the nails and patellae, with other skeletal features sometimes—iliac horns, dislocation of radial heads and small scapulae.

Inheritance

Autosomal dominant, the mutant gene being linked to the locus determining the ABO blood groups.

Frequency

Birth incidence of 1 in 50 000. Probably a prevalence of about 1 per million population.

Clinical features

Facial appearance—normal.
Intelligence—normal.
Stature—normal.
Presenting feature/age—nail dysplasia is probably noticed first, at birth. If small patellae are present, then recurrent dislocation occurs during childhood.
Deformities—dislocation of radial head.
Associated anomalies—noted above.

Radiographic features

Skull—normal.
Spine—normal.
Thorax—normal thoracic cage, but scapulae may be small with a thick lateral border.
Pelvis—iliac horns arise from the middle of the iliac wings.
Knee joints—the small or absent patellae are best seen on a 'skyline' view.

Bone maturation

Normal.

Biochemistry

Not known.

Differential diagnosis

There is no similar disorder.

Progress/complications

Little disability apart from the dislocation of patellae. A later association (middle age) is renal dysfunction, proteinuria and subsequent renal failure, said to affect about one-third of patients.

REFERENCES

Duncan J G, Souter W A 1963 Hereditary onycho-osteodysplasia. Journal of Bone and Joint Surgery 45B:242
Eisenberg K S, Potter D E, Bovill E G 1972 Osteo-onychodystrophy with nephropathy and renal osteodystrophy. A case report. Journal of Bone and Joint Surgery 54A:1301
Pussell B A, Charlesworth J A, Macdonald G J, Baker W 1977 The nail patella syndrome. A report of a family. Australian and New Zealand Journal of Medicine 7/1:20–23
Valdueza A F 1973 The nail-patella syndrome. A report of 3 families. Journal of Bone and Joint Surgery 55B:145
Wright L A, Fred H L 1969 Fatal renal disease associated with hereditary osteoonychodysplasia. Southern Medical Journal 62:833

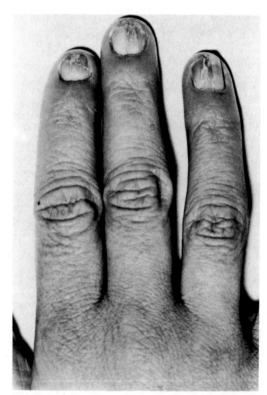

Fig. 82.1   Dysplastic nails.

Pelvis and knee: 11–15 years

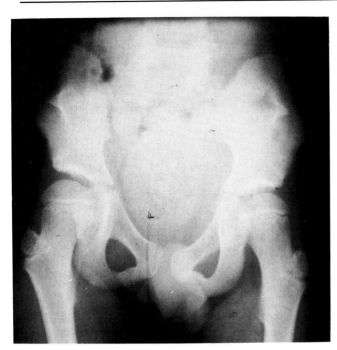

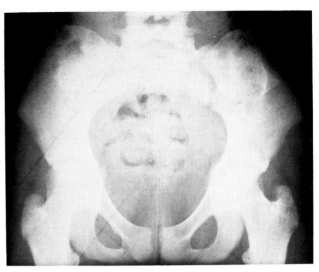

Fig. 82.3 M.H.—14 years 9 months—Pelvis
Similar to Figure 82.2.
Same patient: Fig. 82.5

Fig. 82.2 C.C.—11 years—Pelvis
Clearly defined iliac horns are present.
Same patient: Fig. 82.4

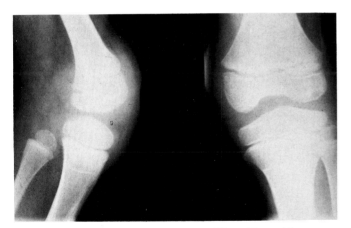

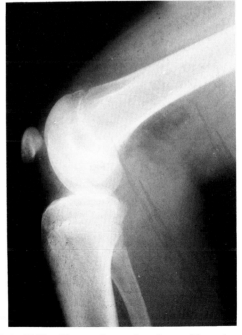

Fig. 82.4 C.C.—11 years 3 months—AP and lateral knee
(same patient as in Fig. 82.2)
Complete absence of patella.

Fig. 82.5 M.H.—15 years 6 months—Lateral knee (same
patient as in Fig. 82.3)
There is a hypoplastic patella.

Knee: 24 years

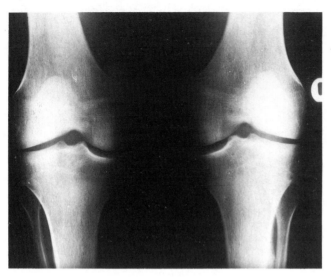

Fig. 82.6   J.S.—24 years—AP knees
The patellae are small, otherwise no abnormality.
Same patient : Figs 82.7, 82.8

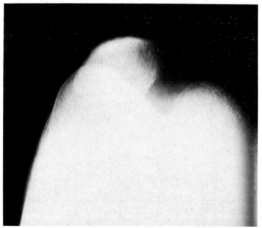

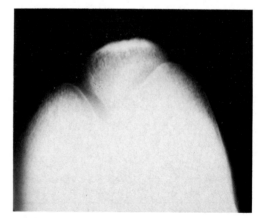

Fig. 82.7   J.S.—24 years—Skyline view left knee (same patient as in Fig. 82.6)
There are small, laterally-placed patellae.

Fig. 82.8   J.S.—24 years—Skyline view right knee (same patient as in Fig. 82.6)
Similar to Figure 82.7.

Chapter Eighty-three
ORAL-FACIAL-DIGITAL SYNDROME
(including Mohr syndrome)

Two or more syndromes characterised by clefts of the palate and lip, lobulated tongue and variable hand deformities—syndactyly, polydactyly, brachydactyly etc.

Inheritance

The oral-facial-digital syndrome occurs in females only, and is thought to be of X-linked dominant inheritance, lethal in the male. The Mohr syndrome, similar clinically, is likely to be of autosomal recessive inheritance.

Frequency

Extremely rare—prevalence probably under 0.1 per million.

Clinical features

Facial appearance—central cleft of the upper lip, tongue divided into several lobes, numerous hyperplastic frenulae, irregular and asymmetrical clefts of the palate and supernumerary or absent teeth.

Intelligence—about half the patients are mentally retarded.

Stature—normal.

Presenting feature/age—birth, with facial and hand anomalies.

Deformities—usually syndactyly, some with polydactyly, brachydactyly, clinodactyly.

Associated anomalies—in the Mohr syndrome there is conductive hearing loss and invariably pre-axial polydactyly of the feet.

Radiographic features

Skull—normal apart from cleft palate.
Spine—normal.

Long bones—there may be shortening of the tibia/fibula (in which case, talipes equino-varus is present) and radius/ulna.

Hands—post-axial polydactyly and variable brachydactyly or other signs.

Feet—in the Mohr syndrome there is pre-axial polydactyly.

Bone maturation

Probably normal.

Biochemistry

Not known.

Differential diagnosis

From other syndromes affecting the head, face and digits, but the lobulated tongue and other oral signs should distinguish the OFD syndrome.

Progress/complications

None of note.

REFERENCES

Burn J, Dezateux C, Hall C, Baraitser M 1984 Orofaciodigital syndrome with mesomelic limb shortening. Journal of Medical Genetics 21(3): 189–192

Rimoin D L, Edgerton M T 1967 Genetic and clinical heterogeneity in the oral-facial-digital syndromes. Journal of Pediatrics 71: 94–102

Smith D W 1982 Recognisable patterns of human malformation, 3rd Edn. W B Saunders, Philadelphia

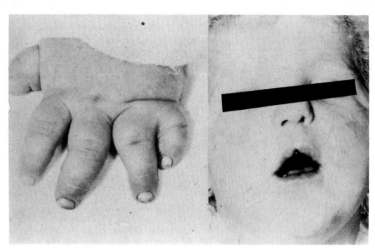

Fig. 83.1   Central cleft of the upper lip and minor syndactyly of the ring and little fingers.

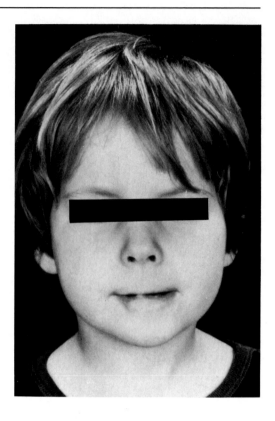

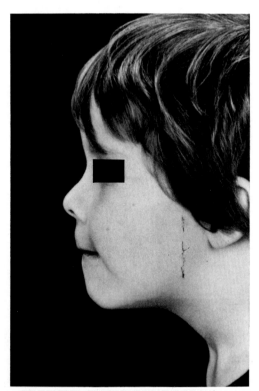

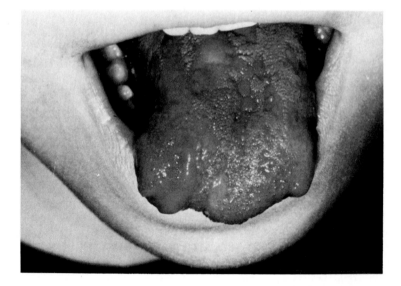

Figs 83.2, 83.3 and 83.4   Slight abnormality of the mouth and a lobulated tongue.

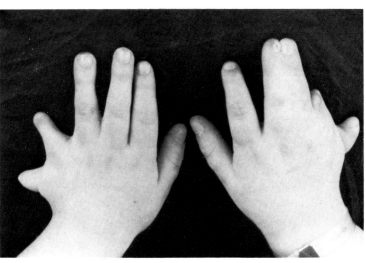

Fig. 83.5 Syndactyly of the middle and ring fingers of the right hand and bilateral post-axial polydactyly.

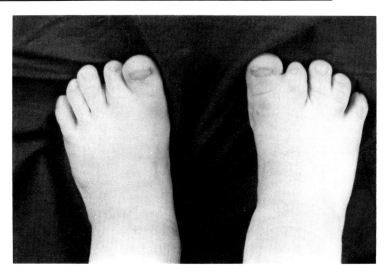

Fig. 83.6 Syndactyly of the first and second toes.

Skull and spine : Birth

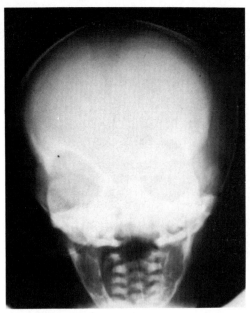

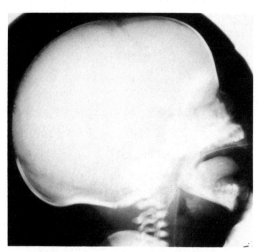

Fig. 83.8   A.B.—Neonate—Lateral skull (same patient as in Fig. 83.7)
Lateral view of Figure 83.7.

Fig. 83.7   A.B.—Neonate—AP skull (Mohr syndrome)
The fontanelles and sutures are unusually wide and there is a cleft of the hard palate
Same patient : Figs 83.8–83.10

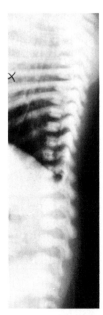

Fig. 83.9   A.B.—Neonate—Lateral spine (same patient as in Fig. 83.7)
Normal.

Lower limb : Birth–8 years

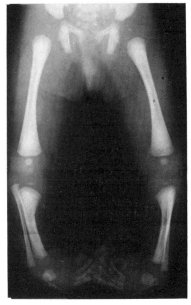

Fig. 83.10 A.B.—Neonate—AP lower limbs (Mohr syndrome)
The femora are normal. Both tibiae, but not fibulae are shortened and there is talipes equino-varus. There is also pre-axial polydactyly of both feet.
Same patient: Figs 83.7–83.9, 83.14, 83.16

Fig. 83.11 R.P.—8 years—AP right lower limb
There is shortening of the tibia and fibula—particularly the latter. There is an area of dysplastic bone growth at the junction of the upper and middle thirds of the tibia.
Same patient: Figs 83.12, 83.15, 83.17

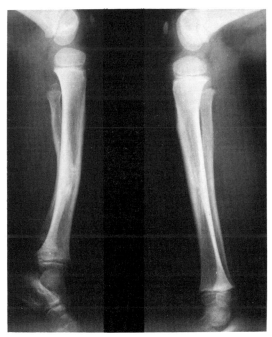

Fig. 83.12 R.P.—8 years—Lateral tibiae and fibulae (same patient as in Fig. 83.11)
Lateral view of Figure 83.11; the left lower limb is less severely affected.

Fig. 83.13 R.B.—Neonate—PA feet (sister of child in Fig. 83.10; Mohr syndrome)
There is talipes equino-varus, pre-axial polydactyly and a short, rounded first metatarsal.

Upper limb : Birth–8 years

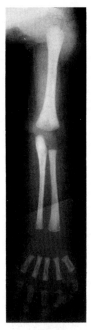

Fig. 83.14   A.B.—Neonate—AP upper limb (Mohr syndrome)
Normal arm and forearm.
Same patient : Figs 83.7–83.10, 83.16

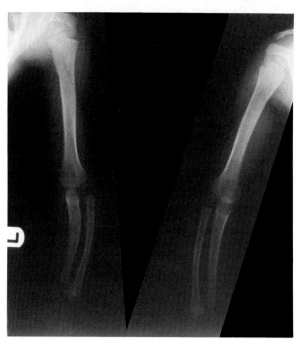

Fig. 83.15   R.P.—8 years—AP upper limbs
Shortening of both radius and ulna with subluxation of the radial
heads.
Same patient : Figs 83.11, 83.12, 83.17

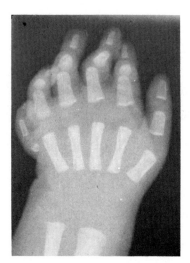

Fig. 83.16   A.B.—Neonate—PA hand (same patient as in Fig.
83.14)
There is post-axial polydactyly and several of the middle phalanges
are unusually short.

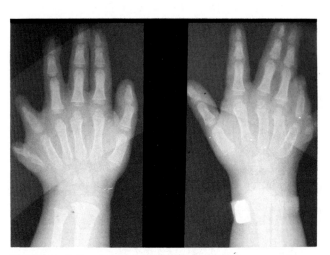

Fig. 83.17   R.P.—6 years—PA hands (same patient as in Fig.
83.15)
There is post-axial polydactyly and symphalangism of the little finger
on the left.

The carpal centres show retarded maturation.

Chapter Eighty-four
OTO-PALATO-DIGITAL SYNDROME

The main characteristics are deafness, cleft palate and variable digital anomalies.

Inheritance

Probably X-linked dominant.

Frequency

Very rare, probably under 0.1 per million prevalence.

Clinical features

Facial appearance—broad nasal root and sometimes teeth anomalies, hypertelorism.

Intelligence—mental retardation.

Stature—short.

Presenting feature/age—cleft palate is noted at birth, as are the structural hand (foot) anomalies.

Deformities—associated with the limb anomalies only.

Associated anomalies—little information, apart from the known features of the syndrome.

Radiographic features

Skull—normal apart from the cleft palate.

Spine—normal.

Hands and feet—irregular anomalies of modelling and length. There may be fusion of carpal or tarsal bones.

Bone maturation

Not known.

Biochemistry

Not known.

Differential diagnosis

From other syndromes affecting head/face and digits, but the triad of signs noted here should distinguish the oto-palato-digital syndrome.

Progress/complications

Conductive deafness is the main problem. It is unlikely the life span is diminished.

REFERENCES

Gall J C, Stern A M, Poznanski A K, Garn S M, Weinstein E D, Hayward J R 1972 Oto-palatal-digital syndrome. Comparison of clinical and radiographic manifestations in males and females. American Journal of Human Genetics 24:24

Salinas C F, Jorgenson R J, Lorenzo R L 1979 Variable expression in oto-palato-digital syndrome; cleft palate in female. Birth Defects: Original Article Series, XV (5B): 329–345

Oto-palato-digital syndrome

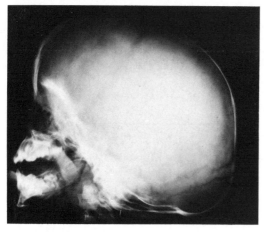

Fig. 84.1   J.P.M.—Lateral skull
Normal apart from unusual prominence of the occiput.
Same patient: Figs 84.2–84.4

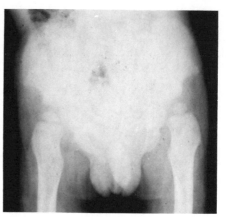

Fig. 84.2   J.P.M.—Pelvis (same patient as in Fig. 84.1)
There is unusual flaring of the iliac wings and acetabular dysplasia.

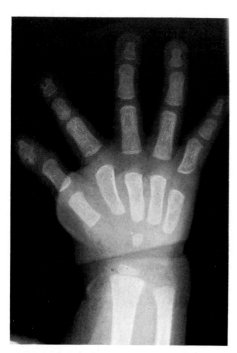

Fig. 84.3   J.P.M.—PA hand (same patient as in Fig. 84.1)
All the hand bones are abnormally modelled with irregular
shortening.

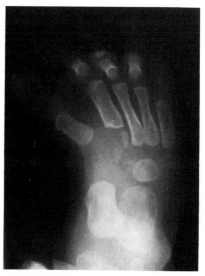

Fig. 84.4   J.P.M.—PA foot (same patient as in Fig. 84.1)
There is adduction of the hallux with absent ossification of the
proximal phalanx. The 2nd metatarsal is abnormally modelled as is its
basal epiphysis.

Chapter Eighty-five
RUBINSTEIN-TAYBI SYNDROME

Malformation of the thumbs and halluces, with mental retardation and a characteristic facies.

Inheritance

Most cases are sporadic, but there appears to be a small risk for sibs (1%).

Frequency

Possible population incidence of about 3 per million.

Clinical features

Facial appearance—a beaked nose with the philtrum extending below the alae. Palpebral fissures slope downwards.

Intelligence—severe mental retardation.

Stature—short.

Presenting feature/age—the characteristic face and thumb anomaly are seen at birth.

Deformities—none apart from those associated with the hand/foot defects.

Associated anomalies—rib fusions, cervical spondylolisthesis, and pulmonary stenosis have been reported.

Radiographic features

Skull—microcephaly, sometimes with Wormian bones.

Spine—cervical spondylolisthesis, with tetraplegia, has been reported.

Hands/feet—the chief deformity is broadening and irregularity of the phalanges.

Bone maturation

Not known.

Biochemistry

Not known.

Differential diagnosis

From other mental retardation syndromes with digital anomalies, but the facies in the Rubinstein Taybi syndrome is quite characteristic.

Progress/complications

The most serious complication appears to be related to the cervical anomaly, if present.

REFERENCES

Johnson C F 1966 Broad thumbs and broad great toes with facial abnormalities and mental retardation. Journal of Pediatrics 68:942

Simpson N E, Brissenden J E 1973 The Rubinstein-Taybi syndrome: familial and dermatoglyphic data. American Journal of Medical Genetics 25:225

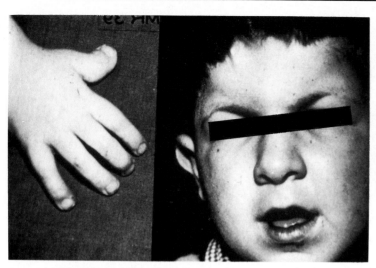

Figs 85.1 and 85.2   Short, broad and irregular thumb and characteristic facies with long philtrum and beaked nose.

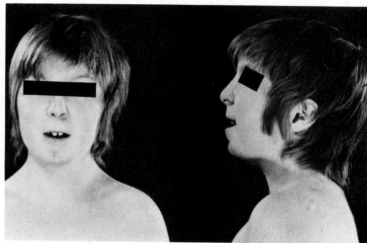

Figs 85.3 and 85.4   A long philtrum extending below the alae.

Skull and pelvis: 4 months–8 years

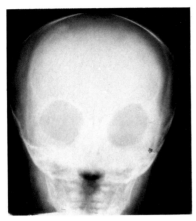

Fig. 85.5 J.T.—4 months—AP skull
Probably normal.
Same patient: Figs 85.6, 85.9

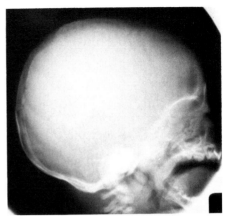

Fig. 85.6 J.T.—4 months—Lateral skull (same patient as in Fig. 85.5)
Probably normal.

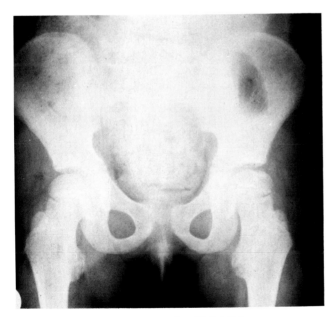

Fig. 85.7 M.C.—8 years 4 months—Pelvis/hips
The pelvis is normal, but there is irregularity of the capital femoral epiphyses as well as of the lateral part of the growth plates.
Same patient: Figs 85.8, 85.10, 85.11

Hands and feet : 4 months–8 years

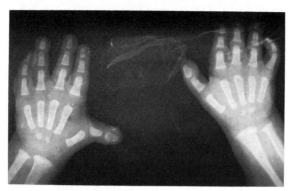

Fig. 85.8   M.C.—4 months—PA hands
There is deformity of the thumb with broad terminal phalanges.
Same patient : Figs 85.7, 85.10, 85.11

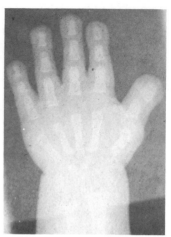

Fig. 85.9   J.T.—4 months—PA left hand
Similar to Figure 85.8.
Same patient : Fig. 85.6

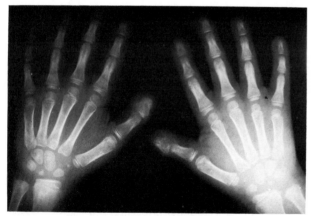

Fig. 85.10   M.C.—8 years 4 months—PA hands (same patient as
in Fig. 85.8)
The terminal phalanges of the thumbs are short and widened as are,
to a lesser extent, the other terminal phalanges.

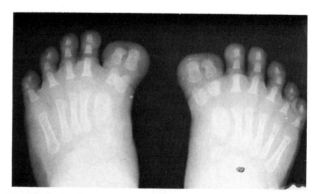

Fig. 85.11   M.C.—2 years—PA feet (same patient as in Fig. 85.8)
This patient has pre-axial polydactyly on the left and pre- and post-
axial polydactyly on the right. All hallux terminal phalanges are
broadened.

Chapter Eighty-six
TRICHO-RHINO-PHALANGEAL SYNDROME

Fine, sometimes very sparse hair, a characteristic face with a wide mouth, long philtrum and beaked nose, with brachydactyly. Some develop Perthes'-like changes in the hips, some have diaphyseal aclasis-like lesions.

Inheritance

Autosomal dominant.

Frequency

Possible prevalence of 5–10 per million, depending on the number of affected relatives.

Clinical features

Facial appearance—see clinical photographs.

Intelligence—normal, although a type with microcephaly and mental retardation has been described.

Stature—probably normal.

Presenting feature/age—any one of the clinical features may first draw attention to the disorder, in childhood or adult life.

Deformities—of the hips, in relation to Perthes'-like changes. Of the shafts of long bones if exostoses develop.

Range of severity—wide. Some may only exhibit minor facial changes and brachydactyly.

Associated anomalies—possible. Little information is available.

Radiographic features

Skull—usually normal, except in the type with microcephaly.

Spine—normal.

Pelvis/hip joints—may be normal, but the capital femoral epiphyses can show bilateral flattening and fragmentation, with secondary deformity of the head and neck of the femur.

Long bones—may be normal. Some cases show exostoses, usually of the large sessile type.

Hands/wrists—cone shaped epiphyses with premature epiphyseal fusion, giving shortening of the affected metacarpals and digits. Sometimes the phalangeal epiphyses are unusually dense ('ivory' epiphyses).

Bone maturation

Probably normal.

Biochemistry

Not known.

Differential diagnosis

From *Perthes' disease* and other epiphyseal disorders of the hip and *diaphyseal aclasis*. The characteristic facies and hands will differentiate this syndrome.

Progress/complications

None known, apart from the hip disability.

REFERENCES

Giedion A, Burdea M, Fruchter Z, Melori T, Trosc V 1973 Autosomal dominant transmission of the tricho-rhino phalangeal syndrome. Helvetica Paediatrica Acta 28:249–259

Fontaine G, Maroteaux P, Farriaux J P, Richard J, Roelens B 1970 Le syndrome tricho-rhino-phalangien. Archives Françaises de Pédiatrie 27:635–647

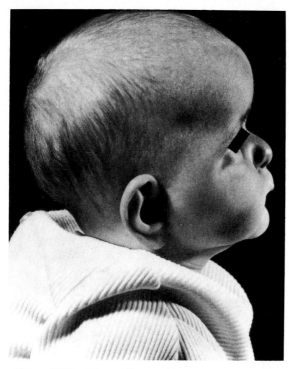
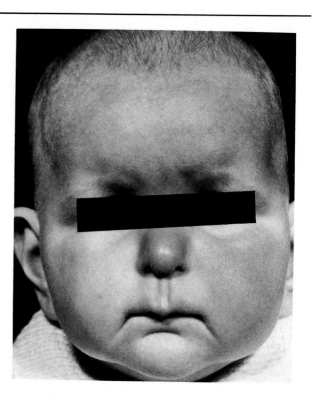

Figs 86.1 and 86.2   A long philtrum with some beaking of the nose.

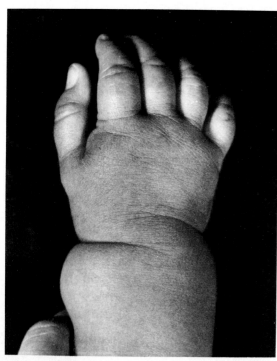

Fig. 86.3   Brachydactyly.

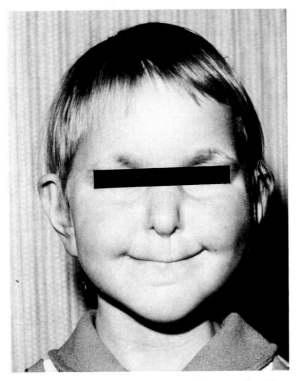
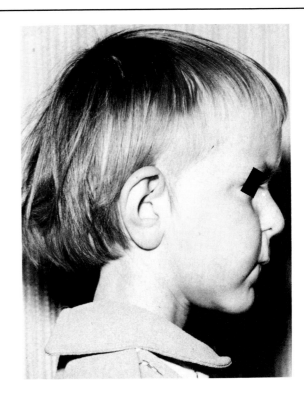

Figs 86.4 and 86.5 Fine, sparse hair, long philtrum and a wide mouth.

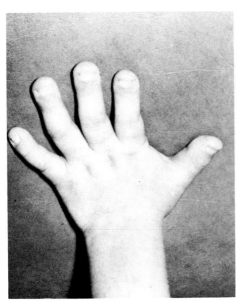
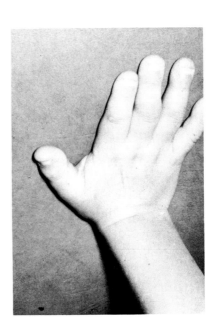

Figs 86.6 and 86.7 Premature fusion of epiphyses giving short, stubby hands.

Skull and pelvis : 7 years

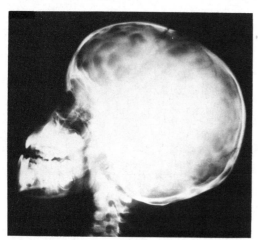

Fig. 86.8   K.H.—7 years—Lateral skull
Normal sutures. Rather prominent convolutional markings.
Prominent mandible with abnormal bite.
Same patient: Figs 86.9, 86.14–86.16

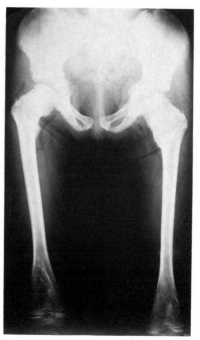

Fig. 86.9   K.H.—7 years—AP pelvis and femora (same patient
as in Fig. 86.8)
The capital femoral epiphyses are small, fragmented and flattened.
The femoral necks are broadened with multiple exostoses here and at
the distal metaphyses. These are associated with abnormal modelling.
The right femur is short.

Pelvis : 15 years–Adult

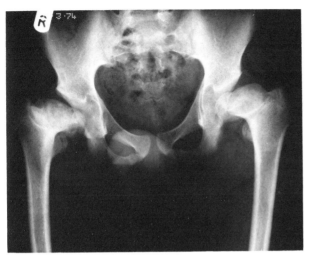

Fig. 86.10   P.G.—15 years—AP hips
The acetabular fossae are shallow and slope steeply. The capital femoral epiphyses are underdeveloped for this age and also appear to have slipped, with marked coxa vara.
Same patient : Figs 86.12, 86.22

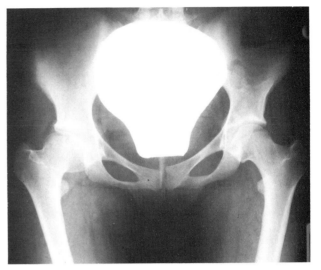

Fig. 86.11   C.B.—17 years—AP pelvis (sister of children in Figs 86.20 and 86.21)
The femoral heads are flattened and the necks are broad with coxa vara. Premature osteoarthritis is present.
Same patient : Fig. 86.23

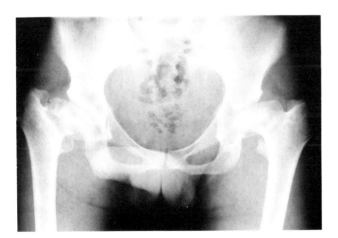

Fig. 86.12   P.G.—20 years—AP hips (same patient as in Fig. 86.10)
The femoral heads are flattened and are poorly covered by the dysplastic acetabula. The femoral necks are short and there is coxa vara.

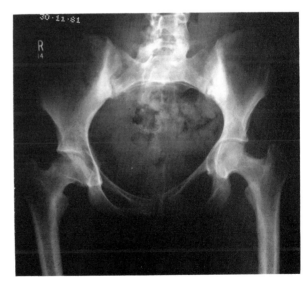

Fig. 86.13   D.M.—Adult—AP pelvis
There is asymmetry of the pelvis with the right iliac wing larger than the left. The femoral heads are flattened and there is bilateral coxa vara.
Same patient : Fig. 86.25

Lower limb and shoulder : 4 years

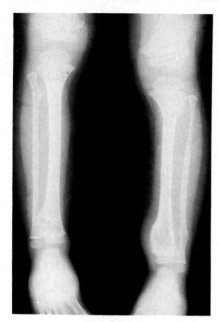

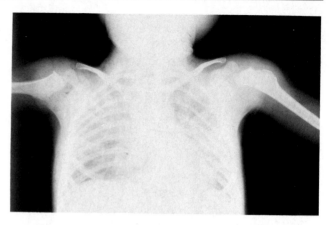

Fig. 86.15   K.H.—4 years—AP shoulders and upper chest
(same patient as in Fig. 86.14)
There are multiple exostoses at both proximal humeral metaphyses.

Fig. 86.14   K.H.—4 years—AP tibiae and fibulae
Multiple exostoses are present at the metaphyses, indistinguishable
from the sessile exostoses of diaphyseal aclasis.
Same patient : Figs 86.8, 86.9, 86.15, 86.16

Hand: 4–11 years

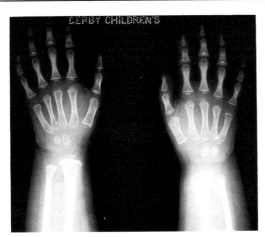

Fig. 86.16   K.H.—4 years—PA hands
There are cone epiphyses of the middle phalanges and some
shortening. Exostoses are present at the distal radial and ulnar
metaphyses on the right.
Bone maturation = 2 years
Same patient: Figs. 86.8, 86.9, 86.14, 86.15

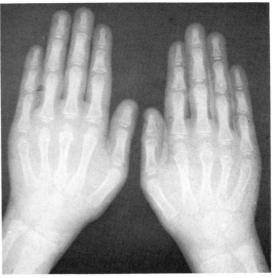

Fig. 86.17   G.H.—8 years—PA hands
There are cone epiphyses at the second and fifth middle phalanges
with some shortening. 'Ivory' epiphyses are present in several middle
and terminal phalanges.
Bone maturation = 5 years

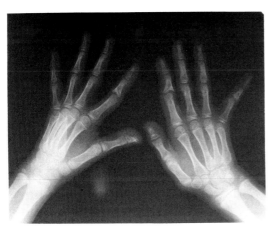

Fig. 86.18   A.P.—10 years—PA hands (brother of child in Fig.
86.19)
Cone epiphyses are present at the second and fifth middle phalanges
and at the terminal ones of the thumb. There is premature fusion at
these sites with shortening. The distal radial epiphyses are somewhat
flattened.
Bone maturation = 7 years 10 months

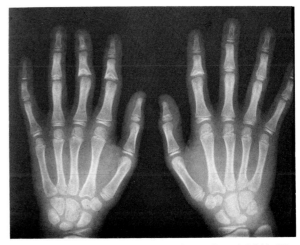

Fig. 86.19   B.P.—11 years—PA hands (brother of child in Fig.
86.18)
Cone epiphyses are present on the left at the second, third and fifth
middle phalanges and on the right at the second and fifth only. These
are associated with premature fusion and shortening.
Bone maturation = 7 years

Hand: 11–15 years

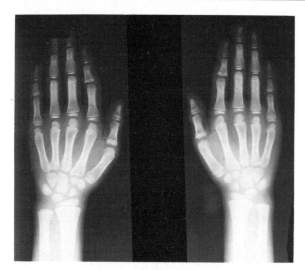

Fig. 86.20   A.B.—11 years—PA hands (brother of children in Figs 86.11 and 86.21)
There are cone epiphyses at the middle phalanges of the second and fifth digits, and at the proximal and distal phalanges of the thumbs.
Bone maturation = 6 years

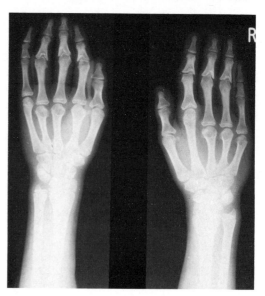

Fig. 86.21   A.B.—14½ years—PA hands (sister of children in Figs 86.11 and 86.20)
On the left there is shortening of the second, third and fourth metacarpals due to premature epiphyseal fusion. Cone epiphyses are present at the second, third and fourth middle phalanges. **On the right** there is shortening of the third, fourth and fifth metacarpals, with cone epiphyses at the second proximal phalanx, and the middle phalanges of the second, third and fourth digits.
Bone maturation = 14 years

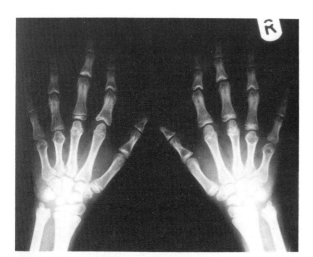

Fig. 86.22   P.G.—15 years—PA hands
There is deviation of the index and middle fingers. Premature fusion of cone epiphyses can be seen of the middle and some of the proximal phalanges.
Bone maturation = 13 years
Same patient: Figs 86.10, 86.12

Hand: 16 years–Adult

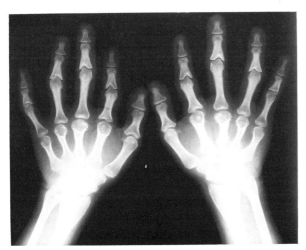

Fig. 86.23   C.B.—16 years 10 months—PA hands
There is shortening of the second to fifth metacarpals bilaterally. The middle phalanges have chevron deformities at their bases, at the site of previous cone epiphyses.
Same patient: Fig. 86.11

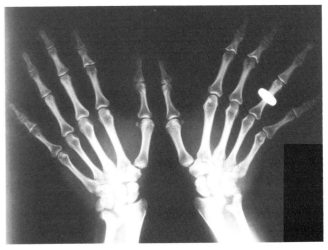

Fig. 86.24   V.P.—Adult—PA hands (mother of patients in Figs 86.18 and 86.19)
Both index fingers show shortening of the middle phalanges and on the right the middle finger is similar. The lower end of the ulna is malformed and short.

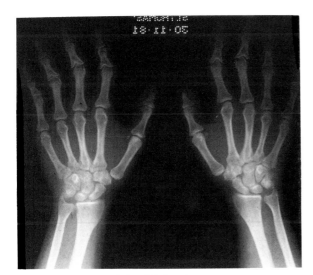

Figs. 86.25   D.M.—Adult—PA hands
The second to fifth digits are ulnar-deviated at the proximal interphalangeal joints, and the middle phalanges are short and deformed.
Same patient: Fig. 86.13

INDEX

α-acetyl-glucosaminidase, xiii, 360
α-L-iduronidase, xiii, 343, 356
α-mannosidase, xiii, 363
Absence or hypoplasia
    of clavicle, 362, 378, 454, 458, 461, 584–585, 589–590
    of fibula, xiii, 227–228, 318, 436, 584, 607
    genital, 220
    of odontoid, 61, 66–67, 86, 145, 239, 258, 262, 274, 277, 291, 345, 352, 368, 372, 375, 519, 535–537
    of patella, 614–616
    of phalanges, 301–303, 454, 463, 483, 486, 547–548, 584
    of radius, xiii, 584
    of ribs, 394, 454, 458
    of tibia, 223–225
    of ulna, xiii, 600–601, 603
    phocomelia, 600
Accessory carpal centres
    advanced ossification centres, 273, 289, 294, 308, 362, 372, 384, 565, 591
    delayed ossification centres, 19, 34–35, 61, 83–85, 115, 127, 160, 163, 176, 254, 256, 273, 300, 349, 355, 359, 371, 384, 463, 483, 556, 622, 635–636
    dysharmonious ossification centres, 258, 271, 273, 294, 362, 384
    See also Pseudoepiphyses, Polydactyly
Acetabulum
    dysplasia (uncovered femoral head), 19, 24–25, 104, 109–110, 156, 163, 229, 231, 233, 235, 239, 248–249, 299, 339, 362, 365, 379, 473, 486, 524, 547, 553–554, 624, 633
    horizontal roof, 24, 43, 61, 65, 74, 86, 92–95, 110, 181, 185, 193, 195, 265, 274, 276, 282, 284
    protrusio, 19, 26, 283, 400, 404, 412, 418, 421–422, 484, 486
    scalloped (irregular), 19, 24, 123, 153, 174, 248
Achondrogenesis, xiv, 311, 320–321
Achondroplasia, xiii, 181–199, 200, 229, 239
Acrocephalo-syndactyly, 579–583
Acrodysostosis (peripheral dysostosis), 233–236
Acromesomelic dwarfism, xiii, 229–232, 233
Acro-osteolysis, 434, 436, 463–464, 565
Albers-Schönberg disease (osteopetrosis), 439–453
Albright's syndrome (polyostotic fibrous dysplasia), 557–565
Alkaline phosphatase, 393, 400
Amino acid disorder, xiii, See also Homocystinuria
Anaemia, 156, 439
Aneurysm, aortic, 325
Ankles
    dislocated, 223, 225, 227–228

valgus, 19, 30–31
Ankylosis, 425, 427
Aortic aneurysm, 325
Apert's syndrome (acrocephalo-syndactyly), 579–583
Arachnodactyly, 325, 327, 334–335, 340
Arthrogram, 61, 76
Arthrogryposis multiplex congenita, 596
Asphyxiating thoracic dysplasia (Jeune's disease), 301, 304–308, 311
Atlanto-axial instability, 61–62, 86, 368, 372. See also Odontoid dysplasia, Skull
Autosomal dominant inheritance, 19, 36, 49, 61, 86, 104, 131, 151, 161, 164, 171, 181, 200, 213, 220, 223, 227, 233, 239, 274, 289, 325, 393, 411, 434, 439, 469, 471, 475, 484, 488, 501, 506, 512, 533, 547, 557, 566, 579, 584, 596, 604, 614, 629
Autosomal recessive inheritance, 19, 49, 104, 116, 145, 156, 181, 229, 239, 258, 274–275, 295, 301, 304, 311, 335, 343, 356, 360, 363, 366, 372, 385, 393, 400, 411, 414, 439, 454, 465, 469–470, 484, 487, 495, 512, 579, 596, 604, 617. See also X-linked inheritance

β-galactosidase, xiii, 385
Basilar invagination, 411–412, 416, 425. See also Paraplegia
'Battered baby', 411, 431
Biochemistry in
    homocystinuria, 335
    Hunter syndrome, 350
    Hurler syndrome, 343
    hyperphosphatasia, 400
    hypophosphataemic rickets, 404
    hypophosphatasia, 393
    mannosidosis, 363
    Marfan syndrome, 325
    Menkes syndrome, 609
    metaphyseal chondrodysplasia (type Schmid), 131
    metaphyseal chondrodysplasia (type McKusick), 145
    mucolipidoses, 385
    osteogenesis imperfecta, 411
    prenatal diagnosis, xiii
    pseudohypoparathyroidism, 407
    Sanfilippo syndrome, 360
    Scheie syndrome, 356
Bipartite patella, 150
Blindness, 36, 164. See also Eye
Bone
    constrictions
        diaphyses, 274, 469, 480, 484
        ilium, 445, 484, 487
    decreased density (osteoporosis), 116, 124, 126–128, 335, 338, 340, 386, 393–399, 411–436, 609, 612–613

increased density
    generalised, 439–515
    patchy, 400, 404, 501, 506, 557
    See also Epiphyses, Metaphyses, Osteoarthritis, Sclerosis
maturation
    advanced, 273, 289, 294, 308, 362, 372, 384, 565, 591
    delayed, 19, 34–35, 61, 83–86, 99, 104, 115, 127, 151, 156, 160, 161, 163, 171, 176, 239, 254, 256, 272–274, 295, 300, 311, 320, 343, 349–350, 355, 359, 366, 371–372, 384, 393, 404, 463, 483, 547, 556, 584, 622, 635–636
    dysharmonious, 258, 271, 273, 294, 311, 314–319, 362, 384, 547
marrow, 156, 439–440, 450–453
pain, 434, 484, 488, 508, 539, 557. See also Osteoarthritis
stippling, 506–507, 511
striation, 163, 488, 491–493, 501–505, 508–509, 533. See also Exostoses
Bow legs see Genu varum, Tibia varum
Bowing of
    clavicle, 261, 419, 475
    femur, 56, 131, 137–141, 145, 147, 153, 249, 265, 301, 312–313, 397, 403, 406, 412, 420, 423, 445, 475, 535, 562
    fibula, 185, 209–210, 265, 284, 313, 403, 420, 423–425, 475, 482, 535, 542, 563–564, 573
    humerus, 281, 313, 428
    radius, 211, 213, 215–216, 218, 222, 227, 261, 288, 313, 398–399, 427–428, 531, 603
    tibia, 56–57, 132, 139–141, 265, 313, 403, 420, 423–425, 433, 475, 482, 528, 535, 564, 573, 607
    ulna, 211, 222, 288, 313, 399, 427–428, 486, 531
Brachycephaly, 53, 151, 233, 352, 485, 597
Brachydactyly, 165, 167, 229, 233, 236, 241, 484, 600, 617, 622, 629–630. See also Fingers, Metacarpals, Phalanges
Bruising, 411

Café-au-lait patches, 557, 566–567
Caffey's disease see Infantile cortical hyperostosis
Calcification
    ectopic, 41, 49, 407, 547
    premature, 54, 220
    See also Ossification centres
Callus, hyperplastic, 412–413, 429–430
Camptomelic dwarfism, 412
Capital femoral epiphyses see Hip, Epiphyses
Carpal bones
    accessory, 85, 604

Carpal bones (*cont.*)
  enlarged, 340
  erosion of, 436, 475
  fusion of, 128, 272, 301–302, 475, 547, 556,
    579, 583, 623
  ossification
    advanced, 273, 289, 294, 308, 362, 372,
      384, 565, 591
    delayed, 19, 34–35, 61, 83–85, 115, 127,
      160, 163, 176, 254, 256, 273, 300,
      349, 355, 359, 371, 384, 463, 483,
      556, 622, 635–636
    dysharmonious, 258, 271, 273, 294, 362,
      384
Carpenter's syndrome (acrocephalo-
    polysyndactyly), 573–583
Cartilage hair hypoplasia (metaphyseal
    chondrodysplasia, type McKusick),
    145–150
Cataract, 49, 61, 289, 325, 407
Cervical spine
  atlanto-axial instability, 62, 67, 258, 276,
    368, 372, 375
  flexion/extension radiographs, 368
  foramina, enlarged, 570
  hypoplasia, 61, 67, 454, 547, 550–551
  lordosis, 570
  odontoid absence or hypoplasia, 61, 66–67,
    86, 91, 145, 239, 258, 262, 274, 277,
    291, 368, 372, 375
  platyspondyly, 67, 86, 91, 161, 262, 274,
    297, 480
  sclerosis, 488
  spinous processes, enlarged, 547, 551
    fusion of, 547, 550–551
  spondylolisthesis, 625
Cherubism, 560
Chest *see* Thorax
Chevron deformity *see* Cone epiphyses, Knee
Chondrodysplasia calcifans punctata
    (Conradi's disease), 49–58
Chondro-ectodermal dysplasia (Ellis–van
    Creveld syndrome), 301–303, 304, 311
Chondrosarcoma, 519
Chromosome anomalies, xiii, 600
Clavicle
  absent or hypoplastic, 362, 378, 454, 458,
    461, 584–585, 589–590
  broad, 343, 347, 354, 356, 360, 362, 365,
    369, 378, 385, 439, 465–466, 469, 472,
    498, 503
  long, 234, 261, 281, 419
  pseudarthrosis, 584, 589, 595
Cleft lip, 301, 617–618
Cleft palate, 36, 39, 61, 86, 258, 289, 617, 620
Cleido-cranial dysplasia (cranio-cleido
    dysostosis), 434, 584–595
Clinodactyly, 35, 160, 212, 273, 334, 556,
    594, 617, 637
Clubbing of fingers and toes, 512, 514
Clubfoot *see* Foot, Talipes
Coarse facies, 343–345, 351, 356, 366, 385
Collagen defect, xiii, 325, 411
Cone epiphyses, 19, 148–149, 168, 232–233,
    236, 273, 295, 300–301, 304, 308, 584,
    594, 629, 635–637
Congenital contractural arachnodactyly, 325
Conradi's disease (chondrodysplasia calcifans
    punctata), 49–58

Contractures of
  elbow, 33, 181, 200
  fingers, 116, 118, 127–128, 273, 325, 327,
    334, 349, 357, 359–360, 362, 475, 511,
    596, 599
  hand, 128, 325, 334, 360, 596
  hip, 61, 86, 116, 258, 296, 356, 366
  knee, 116, 296, 425
  multiple joints, 19, 49, 116–117, 151, 241,
    258, 289, 343, 351, 356, 366, 411–412,
    508
  See also Osteoarthritis
Copper metabolism, xiii, 609. *See also*
    Menke's syndrome
Corneal clouding, 343–344, 356, 360, 366,
    372
Cornelia de Lange syndrome, 600–603
Coronal cleft vertebrae, 40, 49, 54, 69, 221,
    274–276, 278, 291, 298, 606
Coxa valga, 24, 26, 109, 247, 258, 265, 343,
    348, 354, 358, 379
Coxa vara, 24–26, 43–44, 61–62, 74–79, 86,
    93–95, 110, 124, 131–132, 136–138, 140,
    147, 156, 161–163, 171, 174, 177, 209,
    247, 258, 265–266, 400, 402, 421–422,
    445, 459, 562, 584, 633–634
Cranial nerve compression, 400, 440, 450,
    469, 484, 488, 495
Cranial sutures
  delayed fusion, 186, 393, 415, 435, 442,
    456, 513, 584, 586–588, 601, 620
  premature fusion, 39, 49, 393, 395, 579,
    582, 597
  wide, 186, 262, 345, 352, 367, 393, 395,
    415, 435, 450, 456, 513, 584, 586–588
  See also Fontanelles, Hydrocephalus
Cranio-carpo-tarsal dystrophy (Freeman
    Sheldon syndrome: 'whistling face'
    syndrome), 596–599
Cranio-cleido dysostosis, 454, 584–595
Craniodiaphyseal dysplasia, 400, 439, 469,
    495–500
Craniometaphyseal dysplasia, 439, 465, 469–
    474, 488
Craniosynostosis (craniostenosis) *see* Cranial
    sutures, premature fusion
Cretinism (untreated hypothyroidism), 19
Cubitus valgus, 520
Cubitus varus, 519
Cushing's syndrome, 559
⁶⁴Cu uptake, xiii
Cystathionine synthetase, xiii, 325, 335

Deafness, 36, 61, 86, 151, 289, 343, 411, 416,
    419, 475, 484, 501, 617, 623
Decreased bone density, 411–436. *See also*
    Osteoporosis
Delta shaped epiphyses, 145, 149–150, 604
Dermatan sulphate, xiii, 343, 350, 356, 366
Diabetes mellitus, 557
Diaphyseal aclasis (multiple hereditary
    exostoses), 519–532
Diaphyses
  thickened (lack of modelling), 343, 385,
    469, 488, 495, 512
  wavy, 475, 484
Diarrhoea, 156
Diastrophic dysplasia (dwarfism), 240, 258–
    273

Dislocation (subluxation) of
  ankle, 223, 225, 227–228
  elbow, 261, 271, 427, 436, 603, 608
  hip, 61, 343, 348, 366, 379, 411, 435, 484,
    486, 596, 598, 604, 606, 612
  knee, 28, 111, 213, 223, 225, 251, 269, 604,
    606–607
  lens, 325, 335
  multiple joints, 258, 325, 411, 604
  patella, 269, 614–615
  radial head, 114, 220, 222, 227–229, 294,
    301, 428, 486, 600, 603, 608, 614, 622
  shoulder, 32
  wrist, 200, 213, 217, 288, 531. *See also*
    Madelung deformity
Dolichocephaly, 441
Dumb-bell appearance, 36, 43, 45, 47, 274–
    276, 282, 284, 293
Dyggve-Melchior-Clausen disease, 295–300
Dyschondrosteosis, 200, 213–219
Dysplasia epiphysealis hemimelica, 519, 539–
    543
Dysosteosclerosis, 439
Dysraphism *see* Spina bifida

Ear
  cystic, 258–259
  infection of, 182
  See also Deafness
Ectopic calcification, 41, 49, 297, 407, 508,
    547
Elbow
  contractures, 33, 181, 200
  dislocation of, 261, 271, 427, 436, 603, 608
  valgus, 519–520
  varus, 519
Ellis–van Creveld syndrome (chondro-
    ectodermal dysplasia), 301–303, 311
Enchondromata, 533–534
Enchondromatosis (Ollier's disease), 519,
    533–536, 537, 539
Endocrine disorders
  Albright's syndrome, 557
  Cushing's syndrome, 559
  hyperparathyroidism, 559, 565
  hypoparathyroidism, 407
  hypothyroidism (cretinism), 19
  pseudohypoparathyroidism, 407–408
Engelmann's disease (progressive diaphyseal
    dysplasia), 400, 488–494
Enophthalmus, 596
Enzymes, multiple lysosomal, xiii
Epilepsy, 407, 495
Epiphyseal disorders, 19–128
Epiphyses
  cone, 148–149, 168, 232–233, 236, 273,
    295, 300–301, 303–304, 308, 519, 584,
    594, 629–631, 634–637
  delta, 145, 149–150, 604
  increased density *see* Sclerosis
  irregular (flattened, fragmented, small),
    119–128, 239, 250–252, 254–257, 266,
    268, 274, 289, 295, 343, 350, 366, 371–
    372, 383, 533, 629, 632
  maturation *see* Ossification centres
  premature fusion, 257, 404. *See also*
    Skull—sutures
  stippled, 49–58, 506–508, 511, 533
Exophthalmos, 151, 484

Exostoses in
  diaphyseal aclasis, 519–532
  dysplasia epiphysealis hemimelica, 539–543
  fibrodysplasia ossificans progressiva, 553–554
  Maffucci's disease, 537–538
  Ollier's disease, 533–536
  tricho-rhino-phalangeal syndrome, 629, 632–634
Eye
  cataract, 49, 61, 289, 325, 407
  enophthalmus, 596
  exophthalmos, 151, 484
  lens dislocation, 325, 335
  myopia, 36, 61–62, 86, 289, 325
  retinal detachment, 289, 325
  strabismus, 164, 579
  See also Hypertelorism
Eyebrows, 600

Face
  cleft lip, 301, 617–618
  coarse, 343–345, 351, 356, 366, 385
  eyebrows, 600
  flat, 49, 61, 289–290
  frontal bossing, 151, 181–182, 203, 220, 234, 367, 393, 404, 415, 442, 454, 469, 475–479, 484, 490, 501, 584, 604–605, 610
  hypertelorism, 151, 220, 233, 289–290, 476, 579, 596, 604–605, 623
  hypoplastic facial bones, 36–39, 61, 181, 200, 203, 229, 233, 289–290, 454, 579
  'whistling', 596–599
  See also Mandible, Mouth, Paranasal sinuses
Familial hypertrophic osteoarthropathy (pachydermoperiostitis), 512–515
Femur
  bowed, 56, 131, 137–141, 145, 147, 153, 249, 265, 301, 312–313, 397, 403, 406, 412, 420, 423, 445, 475, 535, 562
  dumb-bell, 36, 43, 45, 47, 274–276, 282, 284, 293
  neck, medial beak, 247
    short, broad, 23–24, 26, 43, 75, 94–95, 109, 137, 158, 194, 247, 266, 283, 459, 481, 523–524, 591–593, 632
  pseudarthrosis, 584, 592
  thickened, 488, 492–493, 499
  See also hip, limbs
Fetal face syndrome (Robinow syndrome), 220–222
Fetus, xiii
Fibrochondrogenesis, 275–276
Fibrodysplasia ossificans progressiva (myositis ossificans progressiva), 519, 547–556
Fibrous disorders, 547–575
Fibula
  absent or hypoplastic, xiii, 227–228, 318, 436, 584, 607
  bowed, 185, 209–210, 265, 284, 313, 403, 420, 424–425, 475, 482, 535, 542, 563–564, 573
  long, 30, 57, 79, 111, 181–182, 185, 196–197, 200, 209, 219, 250–251, 268, 274, 284–286

pseudarthrosis, 398, 573–575
  short, 80, 154, 200, 213, 219–220, 222, 227–229, 232, 237, 268, 381, 388, 397, 493, 499, 509, 515, 563–564
  thickened, 223, 225, 403, 493, 499, 509, 515, 563–564
  See also Tibio-fibular disproportion
Fingers
  brachydactyly, 104, 164–165, 167, 176, 199, 217, 229, 232–233, 236, 241, 256–257, 273, 288, 303, 308, 317, 349, 484, 536, 547, 556, 584, 600, 617, 622, 624, 629, 635–637
  clinodactyly, 35, 160, 212, 273, 334, 556, 594, 617, 637
  clubbing, 512, 514
  contractures, 116, 118, 127–128, 273, 325, 327, 334, 349, 357, 359–360, 362, 475, 511, 596, 599
  Kirner's deformity, 150
  nails, absent or hypoplastic, 301, 614
  polydactyly, 311, 314–315, 318, 617
    post-axial, 301–303, 315, 319, 622
    pre-axial, 223, 226, 315
  syndactyly, 579–581, 583, 600, 617–619
  See also Phalanges
Focal dermal hypoplasia, 501
Fontanelles, 145, 393, 395, 411, 415, 434–435, 442, 454, 456, 478, 513. *See also* Skull, Cranial sutures
Foot
  hallux, enlarged (broad), 575, 625–626, 628
    short, 270, 382, 547, 549, 555, 608, 621
    valgus, 333, 555, 603
  metatarsus varus, 286
  pes cavus, 425
  pes planus, 333
  pseudoepiphyses, 81, 555
  talipes calcaneo-valgus, 61
  talipes equino-varus, 61, 151, 223, 254, 258, 325, 596, 604, 617, 621
  talipes equinus, 258
  talus, vertical, 333, 382, 596, 599
  tarsal duplication, 226
    fusion, 475, 583, 623
  See also Phalanges, Toes
Foramen magnum, 181, 439
Fractures, pathological, 82, 393, 396, 398, 400, 402, 411–436, 439, 445, 447, 449, 454, 458, 460, 462, 487, 519, 533, 539, 557, 561–563, 573, 575
Freeman Sheldon syndrome, 596–599
Frenulae, 301, 617–618
Frontal bossing, 151, 181–182, 203, 220, 234, 367, 393, 404, 415, 442, 454, 469, 475–479, 484, 490, 501, 584, 604–605, 610
Frontometaphyseal dysplasia, 439, 475–483, 484
Fusion of
  carpal bones, 128, 272, 301–302, 475, 547, 556, 579, 583, 623
  cervical spinous processes, 547, 550–551
  ribs, 220–222, 625
  tarsal bones, 475, 583, 623
  See also Ankylosis

Gargoylism, 343–345, 351, 356, 366, 385
Genital hypoplasia, 220

Genu recurvatum, 61, 239, 325, 475–476
Genu valgum, 61, 79–80, 111, 131, 140, 239, 268, 296, 301, 325, 372–373, 381, 403–404, 465, 475
Genu varum, 61, 131, 141, 181, 200, 239, 404, 445, 475
Gigantism, localised, 557, 566, 575
Glenoid
  irregular, 113, 255, 272
  shallow, 32, 255, 272, 347, 354, 371
GM₁ gangliosidosis (type I), xiii, 385–386

Haemangiomata, 537
Haematoma, 258–259
Hair, 50, 145–146, 301, 609, 629, 631
Hallux
  broad, 625–626, 628
  short, 270, 382, 547, 549, 555, 608, 621
  valgus, 333, 555, 603
  See also Toes
Hand
  five fingered, 223–224, 226
  ulnar deviation, 596, 599
  See also Fingers
Heart disorders, xiii, 49, 223, 301–302, 344, 350, 356, 366, 378, 386, 600, 625
Height graphs for
  achondroplasia, 184
  chondrodysplasia calcifans punctata, 50
  diaphyseal aclasis, 520
  diastrophic dysplasia, 260
  dyschondrosteosis, 215
  homocystinuria, 336–337
  hypochondroplasia, 202
  Marfan syndrome, 328, 336
  metaphyseal chondrodysplasia (type Schmid), 132
  metaphyseal chondrodysplasia with malabsorbtion and neutropenia, 156
  normal appearance, 3
  osteogenesis imperfecta, 413
  spondylo-epiphyseal dysplasia congenita with mild coxa vara, 87–88
  spondylo-epiphyseal dysplasia congenita with severe coxa vara, 63–64
  spondylometaphyseal dysplasia, 172
Heparan, 366
Heparan sulphate, 343, 350, 360, 366
Heparin sulphamidase, xiii
Hepatomegaly, 361, 366
Hepato-splenomegaly, 343–344, 351, 372, 385
Hereditary progressive arthro-ophthalmopathy (Stickler syndrome), 36–48
Hereditary osteo-onychodysplasia (nail-patella syndrome), 614–616
Hernia, 61, 325, 343–344, 351, 356, 366, 411, 602
High arched palate, 325, 335
Hip
  arthrogram, 61, 76
  contractures (fixed flexion), 61, 86, 116–117, 258, 296, 356, 366
  coxa valga, 24, 26, 109, 247, 258, 265, 343, 348, 354, 358, 379
  coxa vara, 24–26, 43–44, 61–62, 74–79, 86, 93–95, 110, 124, 131–132, 136–138, 140, 147, 156, 161–163, 171, 174, 177,

Hip (*cont.*)
    209, 247, 258, 265–266, 400, 402, 421–
        422, 445, 459, 562, 584, 633–634
    dislocation of, 61, 343, 348, 366, 379, 411,
        435, 484, 486, 596, 598, 604, 606, 612
    See also Acetabulum, Epiphyses
Hirschsprung's disease, 145
Hitch-hiker's thumb, 258–259
Homocystinuria, xiii, 325, 335–340
    graph of upper/lower segment ratio, 336
Humerus
    dumb-bell, 47
    head, enlarged, 126
    hatchet deformity, 32, 255, 272
    pseudarthrosis, 82, 428
    short, 58, 82, 198, 211, 254, 271–272, 287,
        371, 389, 426, 608
    thickened, 47, 82, 371, 389, 426, 444, 468,
        472, 483, 494, 498, 500, 505, 530
    varus deformity, 254–255
    See also Rhizomelic shortening
Hunter syndrome (MPS type II), xiii, 350–
    355
Hurler syndrome (MPS type I–H), xiii, 343–
    349, 350, 356, 360, 366, 372, 384–385
Hydrocephalus, 181–182, 343, 352, 366–367,
    439, 604
Hydroxyproline, 400
Hypercalcaemia, 393
Hyperparathyroidism, 557, 559, 565
Hyperphosphatasia, 488
Hyperplastic callus, 412–413, 429–430
Hypertelorism, 151, 220, 233, 289–290, 476,
    579, 596, 604–605, 623
Hypocalcaemia, 407
Hypochondroplasia, 200–212, 213
Hypoparathyroidism, 407
Hypophosphataemic rickets (vitamin D
    resistant), 404–406
Hypophosphatasia, 311, 393–399, 484
Hypoplasia *see* Absence or hypoplasia

I-cell disease (mucolipidosis II), xiii, 385,
    387–389
Idiopathic hyperphosphatasia, 400–403
Idiopathic juvenile osteoporosis, 411, 431–
    433, 434
Idiopathic osteolysis (Hajdu-Cheney type),
    434–436
Iduronate sulphatase, xiii
Ilium *see* Pelvis
Immunological disorders, 145, 156, 453
Increased bone density (sclerosing bone
    dysplasias), 439–515
    See also Sclerosis
Increased limb length, 325–340, 465, 468
    See also Gigantism
Infantile cortical hyperostosis (Caffey's
    disease), 488
Infection, 145, 156
    bone (osteomyelitis), 440, 454
    upper respiratory/ear, 182, 213, 289, 343,
        350, 440, 469, 484, 495
    See also Respiratory problems
Inheritance *see* Autosomal, X-linked
Intracranial pressure, raised, 393, 395, 479,
    579, 582
Ischium *see* Pelvis, Pubic/ischial ramus

Jansen *see* Metaphyseal chondrodysplasia
Jeune's disease (asphyxiating thoracic
    dysplasia), 304–308
Joint laxity, generalised, 145, 239, 247, 325,
    328, 335, 372–373, 381, 411, 604
Joint stiffness, 19, 61, 86, 99, 104, 289, 295,
    343, 350, 360, 366, 385. *See also*
    Contractures, Epiphyseal disorders,
    Osteoarthritis
Juvenile osteoporosis, idiopathic, 411, 431–
    433, 434
Juvenile rheumatism, 116, 127

Keratan sulphate, 372
Kinky hair syndrome (Menkes' syndrome),
    609–613
Kirner's deformity, 150
Knee
    chevron deformity, 197, 209, 258, 268,
        274, 285
    contractures, 116–117, 296, 425
    dislocation of, 28, 111, 213, 223, 225, 251,
        269, 604, 606–607
    genu recurvatum, 61, 239, 325, 475–476
    genu valgum, 61, 79–80, 111, 131, 140,
        239, 268, 296, 301, 325, 372–373, 381,
        403–404, 465, 475
    genu varum, 61, 131, 141, 181, 200, 239,
        404, 445, 475
Kniest disease, 36, 62, 274, 289–294
Kozlowski *see* Spondylometaphyseal
    dysplasia
Kyphos, 69, 181–182, 190–192, 229–230,
    278–279, 289, 291, 298, 313, 343, 345,
    347, 353, 364, 366, 369, 372, 377–378,
    387, 396, 417, 606
Kyphoscoliosis, 55, 61, 263–264, 419, 572
Kyphosis, 40, 49, 55, 61, 71, 106, 116, 229,
    245, 264, 274–275, 343, 346, 353, 364,
    366, 368, 572

Lange, Cornelia de, syndrome, 600–603
Larsen's syndrome, 604–608
Lens dislocation, 325, 335
Leontiasis ossea, 469, 495
Lethal forms of short-limbed dwarfism, 301,
    304, 311–321, 393
    achondrogenesis, 311, 320–321
    short rib polydactyly syndrome, 311, 314–
        319
        type I (Saldino-Noonan), 311, 314
        type II (Majewski), 311, 315–317
        type III (Naumoff), 311, 318
        thanatophoric, 311–313
        unnamed, 311, 319
Limbs
    localised gigantism, 557, 566, 575
    long, 325–340, 465, 488
    malalignment, 171–172, 182, 185, 250–251,
        289, 295, 301, 335, 404, 465, 484, 533
    mesomelic shortening, 213–236, 301–308,
        617
    rhizomelic shortening, 49, 52, 56–58, 181,
        220, 287
    short, asymmetrical, 49, 52, 508, 533, 537,
        539, 557
        lower only, xiii, 131, 223–226
        and normal trunk, 181–236
        and short trunk, 239–308

upper and lower, xiii, 181–236, 239–308,
    311–321, 343, 350, 366, 372, 393,
    400, 404, 407, 411, 434, 439, 454,
    484, 495, 519, 579, 600, 623, 625
Lip, cleft, 301, 617–618
Looser's zones, 393, 404
Loose bodies in joints, 112, 125–126, 461, 539
Lordosis
    cervical, 570–572
    lumbar, 41, 61–63, 86–87, 131, 181, 183,
        188, 200, 204, 264, 422, 491
Lymphopenia, 145, 156

Madelung deformity, 200, 213, 217, 288, 531
Maffucci's disease, 533, 537–538
Majewski dwarfism (short rib/polydactyly
    syndrome type II), 311, 315–317
Malalignment, limbs, 171–172, 182, 185,
    250–251, 289, 295, 301, 335, 404, 465,
    484, 533
Mandible
    broad, thickened, 230, 343, 466, 469, 495,
        497, 560
    loss of ramus/body angle, 454, 456, 487
    micrognathos (hypoplasia), 36–38, 301,
        367, 387, 454–455, 475–476, 479, 484,
        579, 597, 609
    molars within ramus, 343, 367, 387
    prognathos, 466, 579, 632
    symphysis, failed fusion, 584
    See also Mouth, Teeth
Mannosidase, xiii, 363
Mannosidosis, xiii, 360, 363–365
Marfan syndrome, 325–334, 335
Maroteaux-Lamy syndrome (MPS type
    VI), xiii, 366–371
Maturation *See* Bone
Maxilla
    hypoplastic, 36–39, 61, 181, 229, 233, 579
    thickened, 470, 497, 502, 559–560
    See also Face, Teeth
McKusick *See* Metaphyseal
    chondrodysplasia
Melnick-Needles syndrome
    (osteodysplasty), 475, 484–487
Melorheostosis, 506, 508–511
Menkes' syndrome (kinky hair syndrome),
    xiii, 512, 609–613
Mental retardation, 49, 233, 275, 289, 295,
    335, 343, 350–351, 360, 364, 385, 407,
    434, 495, 579, 600, 609, 617, 623, 625,
    629
Mesomelic dysplasia, xiii, 213–228, 301–308
    asphyxiating thoracic dysplasia (Jeune's
        disease), 304–308
    chondro-ectodermal dysplasia (Ellis–van
        Creveld disease), 301–303
    dyschondrosteosis, 213–219
    fetal face syndrome (Robinow syndrome),
        220–222
    severe dominant form, 227–228
    Werner type, 223–226
Mesomelic shortening, 213–236, 301–308,
    617
Metabolic bone disease, 393–408
Metacarpals
    proximal pointing, 34, 256–257, 343, 349,
        355, 362, 371–372, 384–385, 389

pseudoepiphyses, 34, 83, 85, 97–98, 160, 163, 273, 300, 355, 384, 556, 584, 594
short first only, 34, 149, 258–259, 271, 273, 300, 511, 547, 556, 603, 624
short second to fifth, 34, 115, 149, 167, 181, 199, 217, 229, 232–233, 236, 256–257, 273, 288, 407–408, 486, 532, 536, 600, 604, 624, 635–637
Metaphyseal chondrodysplasias, 131–167, 171, 404
   type Jansen, 151–155
   type McKusick, 131, 145–150
   type Peña, 161–163
   type Schmid, 131–144, 145, 200
   with malabsorption and neutropenia, 156–160
   with retinitis pigmentosa and brachydactyly, 164–167
Metaphyseal disorders, 131–178, 465–483
Metaphyseal dysplasia (Pyle's disease), 465–468, 469
Metaphyses
   cupped, 131, 151, 299, 308, 404, 406
   enlarged, 248, 335
   fracture, 431
   irregular (scalloped, sclerotic), 131, 145, 156, 161, 164, 171, 247, 250, 252, 254–256, 258, 295, 297, 299, 314, 318–319, 343, 350, 372, 381, 393–394, 397–399, 469, 539, 609
   splayed (club shaped, dumb-bell, expanded, flared), 36, 131, 151, 181, 193–200, 209, 239, 250, 252, 254–257, 274, 284–285, 289, 371, 439, 465, 468–469
   spurs, 185, 299
Metatarsals
   first only (hallux)
      absent or hypoplastic, 624
      large, 575, 628
      short, 270, 382, 547, 549, 555, 608, 621
   second to fifth
      short, 31, 303, 408, 510, 603
Metatarsus varus, 286
Metatropic dwarfism, 36, 62, 171, 240, 274–288, 289
Microcephaly, 295, 297, 600–601, 609–611, 625, 629
Micrognathos, 36, 301, 454–455, 475–476, 479, 484, 597, 609
Mohr syndrome (oral-facial-digital syndrome), 617–622
Morquio syndrome (MPS type IV), xiii, 62, 171, 274, 295, 368, 372–384
Mouth
   broad, 372, 631
   cleft lip, 301, 617–618
   cleft palate, 36, 39, 61, 86, 258, 289, 617, 620
   frenulae, 301, 617–618
   pigmented, 557
   tongue anomalies, 301, 343, 617–618
   'whistling face', 596–599
   wide, 629, 631
   See also Teeth
Mucolipidoses, xiii, 360, 385–389
   GM_1 gangliosidosis type I, xiii, 385–389
   type II (I-cell disease), 385, 387–389
Mucopolysaccharidoses, xiii, 343–362, 366–384

Hunter syndrome (MPS type II), 350–355
Hurler syndrome (MPS type I–H), 343–349
Maroteaux-Lamy syndrome (MPS type VI), 366–371
Morquio syndrome (MPS type IV), 372–384
Sanfilippo syndrome (MPS type III), 360–362
Scheie syndrome (MPS type I–S), 356–359
Multiple epiphyseal dysplasia, 19–35, 36, 86, 99, 104, 239
   Fairbank type, 19, 34
   Ribbing type, 19, 34
Multiple hereditary exostoses (diaphyseal aclasis), 519–532, 533, 539, 629
Myelograms, 73, 368
Myopia, 36, 61–62, 86, 289, 325
Myositis ossificans progressiva (fibrodysplasia ossificans progressiva), 519, 547–556

N-Ac-gal-4-sulfatase, 366, 372
N-Ac-6-sulfatase, xiii
N-heparan sulphatase, 360
Nail-patella syndrome (hereditary osteo-onychodysplasia), 614–616
Nails, dysplastic or absent, 301, 614–616
Naumoff dwarfism (type III), 311, 318
Neck *See* Cervical spine
Nephritis, chronic, 304, 614
Nerves, compression
   cranial, 400, 411–412, 439–440, 450, 469, 484, 488, 495, 557, 566
   peripheral, 519, 566
   spinal, 62, 73, 182, 258, 368, 404, 566, 625
   tumours, 566
Neurofibromatosis (von Recklinghausen's disease), 566–575
Neutropenia, 145, 156
Non-accidental injury, 411, 431
Normal appearances, 3–15
   height and span graphs, 3–4
   radiology of
      chest, 10
      elbow, 14
      hand, 15
      hip, 13
      knee, 12
      pelvis, 11
      skull, 5–7
      spine, 8–9
Nose, 181–182, 289, 495–496, 584, 596, 600, 604–605, 625–626, 629–630

Odontoid absence, dysplasia or hypoplasia, 61, 66–67, 86, 145, 239, 258, 262, 274, 277, 291, 345, 352, 368, 372, 375
Ollier's disease (enchondromatosis), 519, 533–536, 537
Oral-facial-digital syndrome (including Mohr syndrome), 617–622
Ossification centres
   accessory, 85, 289, 604
   maturation
      advanced, 289, 308, 372, 565, 591
      delayed, 19, 61, 86, 99, 104, 151, 156, 160–161, 163, 171, 176, 202, 295,

300, 311, 343, 359, 371–372, 393–394, 483, 547, 556, 584, 635–636
   dysharmonious, 181, 547
   multiple, 27, 82, 604, 607
   See also Pseudoepiphyses
Osteoarthritis, 19–128, 182, 192, 229, 233, 239–240, 249, 251, 255, 258, 262, 267, 269–270, 274, 276, 289, 295, 365, 372, 459–461, 539, 554, 629, 633. *See also* Epiphyseal disorders
Osteitis deformans *see* Paget's disease
Osteitis fibrosa cystica *see* Hyperparathyroidism
Osteodysplasty (Melnick-Needles syndrome), 475, 484–487
Osteogenesis imperfecta, xiii, 393, 411–430, 431, 434, 584
Osteolysis, 434, 436, 455, 463–464, 475, 535–536, 561–566
Osteomalacia, 404, 454
Osteomyelitis, 439–440, 454, 488
Osteopathia striata (with cranial sclerosis), 501–505
Osteopetrosis, 439–453, 454, 465
   treatment of, 450–453
Osteopoikilosis, 506–507, 511
Osteoporosis, 26, 115–116, 123–128, 284, 325, 335, 340, 386, 393–396, 400, 403, 411–436
Osteosarcoma, 430
Oto-palato-digital syndrome, 623–624
Otosclerosis *see* Deafness

Pachydermoperiostitis (familial hypertrophic osteoarthropy), 512–515
Paget's disease, 488
Pain *see* Bone pain
Palate, cleft, 36, 39, 61, 86, 258, 289, 617, 620
   high arched, 325, 335
Pancreatic insufficiency, 156
Paralysis
   cranial nerve, 400, 411–412, 439–440, 450, 469, 484, 488, 495, 557, 566
   peripheral nerve, 519, 566
   spinal nerve, 62, 73, 182, 258, 368, 372, 404, 566, 625
   See also Atlanto-axial instability, Basilar invagination, Paraplegia, Tetraplegia
Paranasal sinuses, 151, 182, 401, 435, 441–442, 469, 475, 485, 495, 497, 501, 559–560, 584
Paraplegia, 62, 73, 181, 192, 274, 425
Parathyroids
   hyperparathyroidism, 557, 559, 565
   hypoparathyroidism, 407
   pseudohypoparathyroidism, 407–408
Patella
   absent or hypoplastic, 614–616
   bipartite, 150
   dislocation of, 269, 614–615. *See also* Knee dislocation
Pathological fractures, 82, 393, 396, 398, 400, 402, 411–436, 439, 445, 447, 449, 454, 458, 460, 462, 487, 519, 533, 539, 557, 561–563, 573, 575
Pectus carinatum, 61–62, 86–87, 145, 259, 295–296, 335, 372–373, 413
Pectus excavatum, 335, 372–373, 584

Pelvis
  ilia
    flared, 258, 265, 343, 372, 402, 624
    horns, 614–615
    hypoplastic, inferior part, 249, 284, 289, 343, 346, 348, 354, 358, 365, 370, 372, 379–380, 388, 445, 475, 484, 487
    wings, 61, 74, 86, 92–94, 153, 185, 193–195, 231, 247, 266–267, 274–275, 282–284, 293, 299, 301, 304, 307, 314, 318, 320–321, 348, 362, 370, 379, 499, 590, 592–593
    lace-like appearance, 295, 299
    projecting spur, 304, 314
    pubic/ischial rami, broad (thickened), 276, 282, 379, 465, 481, 484, 486
    delayed ossification, 61, 65, 74, 77–78, 86, 92–94, 265, 274, 276, 283, 316, 393, 584, 591–593
    triradiate cartilage (Y-cartilage)
      delayed ossification, 239, 247–248, 584, 592
  See also Acetabulum
Peña See Metaphyseal chondrodysplasia
Periosteum
  excessive bone formation, 385
  new bone formation, 424, 512, 514, 609
Peripheral dyostosis (acrodysostosis), 233–236
Peripheral nerve compression, 519, 566
Perthes' disease, 19, 366, 370, 629
Pes cavus, 425
Pes planus, 333
Phalanges
  absent or hypoplastic, 301–303, 454, 463, 483, 486, 547–548, 584
  brachydactyly, 104, 164–165, 167, 176, 199, 217, 229, 232–233, 236, 241, 256–257, 273, 288, 303, 308, 317, 349, 484, 536, 547, 556, 584, 600, 617, 622, 624, 629, 635–637
  bullet shaped, 343, 349, 385, 389
  cone epiphyses, 148–149, 168, 232–233, 236, 273, 295, 300–301, 303–304, 308, 584, 594, 629–631, 634–637
  delta epiphyses, 145, 149–150, 604
  polydactyly, xiii, 311, 314–315, 318, 617
    post-axial, 301–303, 315, 317, 319, 579, 581, 617, 619, 622, 628
    pre-axial, 223, 226, 315, 579, 583, 617, 621, 628
  pseudoepiphyses, 289, 584, 594
  syndactyly, 579–581, 583, 600, 617–619
  terminal
    absent ossification, 318
    hypoplasia, 301–302, 463, 482–484, 486, 512, 514, 584, 628
    osteolysis, 434, 436, 439, 454, 463–464, 565
  thickened (broad), 468, 474, 483–484, 486, 625, 628
Philtrum, 596, 625–626, 629–631
Phleboliths, 537–538
Phocomelia, 600
Phosphaethanolamine, 393
Pierre Robin syndrome, 36
Platybasia see Basilar invagination
Platyspondyly, 40, 61, 70–73, 91, 99, 101, 104, 106–108, 120–122, 173, 230, 239, 264, 274, 278–279, 289, 291–292, 295, 298, 312–313, 335, 338, 369, 376–377, 396, 411, 432, 475, 480
  biconcave, 335, 411, 417–418, 431–432
  end-plates, irregular, 19, 22, 36, 41, 70–72, 91, 104, 106–108, 116, 121–122, 156, 171, 173, 244–246, 274, 279–280, 289, 291, 298, 312–313, 354, 358, 364, 443, 475
  smooth, 68, 90, 116, 122, 245
  generalised (universal), 41, 71, 116, 171, 245, 376
  See also Vertebrae
Polydactyly, generalised, 223–226, 314–315, 579, 581, 583, 617, 619, 621–622, 628
  post-axial, 301–304, 317–319
  pre-axial, 223–226, 317–319
Polyostotic fibrous dysplasia (including Albright's syndrome), 400, 533, 557–565
Posture, lordotic, 41, 61–63, 86–87, 131, 181, 183, 188, 200, 204, 264, 422, 491
Premature calcification, 54, 220
Premature fusion, epiphyses, 257, 404. See also Cranial sutures
Prenatal diagnosis, xiii–xiv
Pressure, raised intracranial, 393, 395, 479, 579, 582
Prognathos, 466, 579
Progressive diaphyseal dysplasia (Engelmann's disease), 400, 488–494
Protrusio acetabuli, 19, 26, 283, 400, 404, 411–412, 418, 421–422, 429, 484, 486
Pseudarthrosis of
  clavicle, 584, 589, 595
  femur, 584, 592
  fibula, 398, 573–575
  humerus, 82, 428
  tibia, 566, 569, 573–575
Pseudoachondroplasia, 19, 62, 239–257
Pseudoepiphyses of
  foot, 81, 555
  hand, 34, 83, 85, 97–98, 160, 163, 273, 300, 355, 384, 556, 584, 594
Pseudohypoparathyroidism, 407–408
Pseudo-rheumatoid disease, 116–128
Pubic symphysis, 584, 590, 593
Pubis see Pelvis, Pubic/ischial ramus
Pycnodysostosis, 439, 454–464, 512, 584
Pyle's disease (metaphyseal dysplasia), 465–468, 469

Radial head, dislocation, 114, 220, 222, 227–229, 294, 301, 428, 486, 600, 603, 608, 614, 622
Radio-ulnar disproportion, 198–199, 213, 227–228, 261, 288, 519
  synostosis, 427
Radius
  absent or hypoplastic, xiii, 584
  bowed, 211, 213, 215–216, 218, 222, 227, 261, 288, 313, 398–399, 427–428, 531, 603
  short, 211, 213, 220, 222, 261, 399, 603, 617, 622
Ray distribution, 536, 538
Renal failure, 304, 614
Renal osteodystrophy, 404

Respiratory problems
  cardio-respiratory, 258, 275, 344, 350, 372
  later than neonatal, 182, 258, 275, 325, 343, 350, 366, 372, 385, 547, 604
  neonatal, 156, 258, 274–275, 301, 304, 311–321, 385, 393, 604
Retinal detachment, 36, 61–62, 289, 325
Retinitis pigmentosa, 164
Rheumatism, juvenile, 116, 127
Rhizomelic shortening, xiii, 49, 52, 56–58, 181, 220, 287
Ribs
  absent or hypoplastic, 394, 454, 458
  beaded, 320–321, 414
  broad, 234, 297, 343, 346–348, 354, 365, 369, 376, 378, 385–387, 396, 402, 439, 444, 466, 472, 491, 498, 503, 512, 538
  constrictions, 484–485
  cupped posterior ends, 145, 156, 244, 248, 274, 297, 396
  exostoses, 522
  fused, 220–222, 625
  narrowed (ribbon, whole or part), 343, 346, 376, 385, 434, 475, 480–481, 566, 572
  premature calcification of cartilages, 54, 220–221
  sclerosis, 400, 402, 439, 444, 454, 458, 472, 491, 503
  short, 156–157, 181, 185, 261, 274–276, 281, 301, 304–306, 311–321, 396
Rickets, 131, 136, 393, 404, 406, 446–448, 557, 565
Robinow syndrome (fetal face syndrome), 220–222
Rubinstein–Taybi syndrome, 625–628

Saldino–Noonan dwarfism (type I), xiii, 311, 314
Sanfilippo syndrome (MPS type III), xiii, 360–362, 363
Scaphocephaly, 343
Scapula, 297, 584, 611, 614
Scheie syndrome (MPS type I–S), xiii, 356–359, 360
Scheuermann's disease, 36, 233
Schmid see Metaphyseal chondrodysplasia
Sclerae, 411
Sclerosing bone dysplasias, 439–515
  cranial and metaphyseal, 465–487
  cranial and diaphyseal, 488–500
  other rarer, 501–515
Sclerosis, 58, 145, 149, 151–153, 177, 234, 249, 268–272, 293, 299, 365–381, 398, 400, 404, 430, 465–466, 522, 533, 535, 540–543, 557, 559, 565, 573–574, 582. See also Osteoarthritis, Sclerosing bone dysplasias
Scoliosis (potentially all conditions affecting the spine)
  with congenital vertebral anomalies, 49, 54–55
  without congenital vertebral anomalies, 19, 62, 69–73, 86–87, 90, 99, 107–108, 120, 135, 161, 171, 229, 239, 246, 258–259, 263, 274–275, 278–279, 289, 295, 325–327, 330–331, 338, 346, 376, 387, 404, 411, 413, 418, 484, 566–567, 571–572, 590, 596, 598

See also Kyphoscoliosis, Paraplegia, Vertebrae
Secondary hypertrophic osteoarthropathy, 512
Sella turcica, enlarged, 343, 345, 352, 367, 375
Short limbs and trunk, 239–308
Short limbs, normal trunk, 19, 49, 145, 181–236
Short rib/polydactyly syndrome
    type I (Saldino–Noonan), xiii, 311, 314
    type II (Majewski), xiii, 311, 315–317
    type III (Naumoff), xiii, 311, 318
    unnamed, 311, 319
Short trunk, spondylo-epiphyseal disorders, 61–128
    spondylometaphyseal disorders, 171–178
    other, 372, 431
Shoulder
    dislocation, 32
    glenoid
        irregular, 113, 255, 272
        shallow, 32, 255, 272, 347, 354, 371
    hatchet shaped head of humerus, 32, 255, 272
    varus deformity, 435
Sinuses, paranasal, 151, 182, 401, 435, 441–442, 469, 475, 485, 495, 497, 501, 559–560, 584
Skin
    café-au-lait patches, 557, 566–567
    other lesions, 49–50, 506, 508, 512, 557–558
    redundant folds, 185, 196, 198
Skull
    base, dense, thickened, 152, 233, 343, 400–401, 439, 441–442, 450, 465, 469–471, 475, 478, 484, 488, 490, 495–497, 502, 512, 515, 559–560
        invaginated, 412
        short, 151, 182, 186
    brachycephaly, 53, 151, 233, 352, 485, 597
    clover leaf, 276, 312–313
    craniostenosis, 49, 393, 579
    dolichocephaly, 441
    fontanelles, 145, 393–395, 411, 415, 434–435, 439, 442, 450, 454, 456, 478, 513, 584, 586, 588, 620
    foramen magnum, 181, 439
    frontal hyperplasia (bossing), 151, 181–182, 203, 220, 234, 343, 367, 393, 404, 415, 442, 469, 475–479, 484, 490, 501, 584, 604–605, 610
    helmet sign, 596
    hydrocephalus, 181–182, 343, 352, 366–367, 439, 604
    microcephaly, 295, 297, 600, 610–611, 625, 629
    scaphocephaly, 343
    sella turcica, enlarged, 343, 345, 352, 367, 375
    sutures
        delayed fusion, 186, 393, 415, 434–435, 442, 456, 513, 584, 586–588, 601, 620
        premature fusion, 39, 49, 393, 395, 579, 582, 597
        wide, 186, 262, 345, 352, 367, 393, 395, 415, 435, 442, 450, 454, 456, 513, 584, 586–588, 620

See also Hydrocephalus
turricephaly, 579–580, 582
vault, large, 181–182, 186–187, 203, 220, 312–313, 343, 345, 352, 375, 385, 411, 426, 441–442, 456
    poor mineralisation, 314, 320–321, 393–395, 414–416, 484, 512
    small, 364, 600
    thickened (dense, sclerotic), 233–234, 343, 352, 360–361, 364, 367, 385–387, 400–401, 405, 439, 442, 456, 465–566, 469–471, 475, 478, 484–485, 488, 490, 495–497, 501–502, 512, 515, 559–560
    Wormian bones, 411, 414–416, 434–435, 454, 456, 512–513, 584, 586–588, 609–610, 625
Soft tissue swellings, 547–552, 554, 556
Span graphs for
    achondroplasia, 184
    diaphyseal aclasis, 520
    diastrophic dysplasia, 260
    dyschondrosteosis, 215
    homocystinuria, 337
    hypochondroplasia, 202
    Marfan syndrome, 328
    metaphyseal chondrodysplasia (type Schmid), 132
    Morquio syndrome, 374
    multiple epiphyseal dysplasia, 21
    normal appearance, 4
    pseudoachondroplasia, 242
    spondylo-epiphyseal dysplasia congenita
        with mild coxa vara, 87–88
        with severe coxa vara, 63–64
    spondylo-epiphyseal dysplasia tarda, 100
    spondylometaphyseal dysplasia, 172
Spina bifida, dysraphism, 119, 188, 204–205, 262, 265, 292, 486
Spinal cord compression, 73, 181, 258, 368, 404
Spine
    congenital vertebral anomalies *see* vertebrae
    curvature *see* Scoliosis, Kyphosis
    lordosis, cervical, 570
        lumbar, 61–63, 71, 86–87, 131, 181, 183, 188, 200, 204, 264, 422
    stenosis, 182, 190–191, 200, 207, 264
    See also Vertebrae
Spondylo-epiphyseal dysplasia congenita, 19, 36, 61–98, 229
    with mild coxa vara, 62, 86–98
    with severe coxa vara, 61–85, 86
Spondylo-epiphyseal dysplasia tarda, 86, 99–128
    autosomal dominant/recessive types, 104–115
    X-linked type, 99–103, 104
    with progressive arthropathy, 104, 116–128
Spondylolisthesis, 329, 454, 625
Spondylometaphyseal dysplasias, 171–177
Spondylometaphyseal dysplasia, type Kozlowski, 171–176
    type Sutcliffe, 171, 177
Stature
    short *see* Limbs, Trunk
    tall, 325, 335, 465

Stenosis
    arterial, 344, 566, 570, 609, 625
    cranial *see* Skull, Cranial sutures, premature fusion
    spinal, 182, 190–191, 200, 207, 264
Sternal protrusion, 61–62, 86–87, 145, 259, 295–296, 335, 372–373, 413
Stickler syndrome (hereditary progressive arthro-ophthalmopathy), 36–48
Stiffness, joint, 19, 61, 86, 99, 104, 289, 295, 343, 360, 366, 385. *See also* Contractures, Osteoarthritis, Epiphyseal disorders
Stippled epiphyses, 49–58, 507
Storage diseases, xiii, 343–389
    mannosidosis, 363–365
    mucolipidoses, 385–389
    mucopolysaccharidoses, 343–362, 366, 384
Strabismus, 579
Striation, bone, 28, 163, 488, 491–493, 501, 504–505, 508–510
Subluxation *see* Dislocation
Sulphaiduronate sulphatase, 350
Sutcliffe *see* Spondylometaphyseal dysplasia
Sutures, cranial
    delayed fusion, 186, 393, 415, 434–435, 442, 456, 513, 584, 586–588, 601, 620
    premature fusion, 39, 49, 393, 395, 579, 582, 597
    wide, 186, 262, 345, 352, 367, 393, 395, 415, 435, 442, 450, 454, 456, 513, 584, 586–588, 620
    See also Fontanelles, Hydrocephalus
Symphalangism, 579, 622
Symphysis pubis, 65, 74–75, 77–78, 92–94, 274, 276, 283, 584, 590–593
Syndactyly, 579–581, 583, 600, 617–619
Synostosis, radio-ulnar, 427
Syphilis, 488

Talipes calcaneo-valgus, 61
    equino-varus, 61, 258, 325, 596, 604, 617, 621
    equinus, 258
Talus, vertical, 81, 333, 382, 596, 599
Tarsal bones
    accessory, 226, 604
    erosion of, 475
    fusion of, 475, 583, 623
Teeth anomalies, 301, 314, 343, 367, 387, 394, 411–412, 439, 442, 454–456, 465, 469, 475, 484, 497, 559–560, 584, 617, 623
Tetany, spontaneous, 407
Tetraplegia, 625
Thanatophoric dwarfism, xiii, 276, 311, 312–313
Thorax
    broad (increased AP diameter), 289, 378, 601
    pectus carinatum (sternal protrusion), 61–62, 86–87, 145, 259, 295–296, 335, 372–373, 413
    pectus excavatum (sternal depression), 335, 372–373, 584
    small (narrow, decreased AP diameter), 42, 156–157, 181, 185, 202, 261, 274–276, 281, 301, 304–306, 311–321, 336, 414, 419, 434–435

Thorax (*cont.*)
  See also Ribs, Respiratory problems
Thrombocytopenia, 156
Thrombosis, 335, 339
Thumb
  absent (five fingered hand), 223–224, 226
  broad, 625–626, 628
  delta phalanx, 604
  hitch-hiker's, 258–259
  short first metacarpal, 271, 273, 511, 547, 549, 556, 600, 603, 628
  pre-axial polydactyly, 223
  thenar eminence, hypertrophy, 568
  triphalangeal, 223–224, 226
Tibia
  absent or hypoplastic, 223–225
  bowed/medial beak, 27, 56–57, 79, 131–132, 139–141, 200, 213, 250, 256, 265, 284, 313, 397–398, 403, 420, 423–425, 433, 447, 461, 475–476, 482, 487, 528, 535, 563–564, 566, 569, 573, 607
  pseudarthrosis, 566, 569, 573–575
  short, 57, 196–197, 209–210, 213, 220, 223, 225, 227–229, 232, 315–316, 397, 535, 617, 621
  thickened, 397–398, 403, 493, 509, 515, 563
  See also Genu recurvatum, valgum and varum
Tibio-fibular disproportion, 182, 185, 196–198, 200, 209–210, 213, 219, 223, 225, 250–251, 274, 284–286, 519, 621
Toes
  clubbing, 512, 514
  polydactyly, 318–319, 581
    post-axial, 302–304, 317, 581, 617, 628
    pre-axial, 223, 225, 302–303, 317, 583, 617, 621, 628
  syndactyly, 579–581, 619
  See also Hallux
Tongue anomalies, 301, 343, 617–618
Torticollis, 547
Tower shaped skull (turricephaly), 579–580, 582
Tricho-rhino-phalangeal syndrome, 519, 629–637
Triphalangeal thumb, 223–224, 226
Trochanter, 74–77, 138, 158, 200, 274, 282, 289, 459, 473, 524, 562, 593
Trunk, short, 61–128, 171–178, 372, 431
Tumour-like bone dysplasias, 519–543
  diaphyseal aclasis, 519–532
  dysplasia epiphysealis hemimelica, 539–543
  Maffucci's disease, 537–538
  Ollier's disease, 533–536
Tumours, malignant, 412, 430, 519, 533, 537, 566
  nerve, 566

Turner's syndrome, xiii, 213
Turricephaly, 579–580, 582

Ulna
  absent or hypoplastic, xiii, 600–601, 603
  bowed, 211, 222, 288, 313, 399, 427–428, 486, 531
  dislocation (Madelung deformity), 213–214, 217–218
  fibrous replacement, 566
  long, 535
  short, 198, 211–212, 220, 222, 227–228, 232, 261, 288, 303, 349, 355, 399, 531, 538, 608, 617, 622, 637
  See also Radio-ulnar disproportion
Ulnar deviation, 486, 596, 599, 637
Ultrasound diagnosis, xiii–xiv

Valgus deformity
  of ankle, 19, 30–31
  of elbow (cubitus valgus), 520
  of hallux, 333, 555, 603
  of hip (coxa valga), 24, 26, 109, 247, 258, 265, 343, 348, 354, 358, 379
  of knee (genu valgum), 61, 79–80, 111, 131, 140, 239, 268, 296, 301, 325, 372–373, 381, 403–404, 465, 475
Varus deformity
  of elbow, 519
  of humerus, 254–255
  of hip (coxa vara), 24–26, 43–44, 61–62, 74–79, 86, 93–95, 110, 124, 131–132, 136–138, 140, 147, 156, 161–162, 171, 174, 209, 247, 258, 265–266, 400, 402, 421–422, 445, 459, 562, 584, 633–634
  of knee (genu varum), 61, 131, 141, 181, 200, 239, 404, 445, 475
  of metatarsus, 286
  of shoulder, 435
  of tibia (bowed), 420, 423–425, 433, 482, 528, 535, 564, 573, 607
Vault, skull
  large, 181–182, 186–187, 203, 220, 312–313, 343, 345, 352, 375, 385, 411, 416, 441–442, 456
  thickened, 233–234, 343, 352, 360–361, 364, 367, 385–387, 400–401, 405, 439, 441–442, 456, 465, 469–470, 475, 478, 484–485, 488, 490, 495, 497, 501–502, 512, 515, 559–560
Vertebrae
  anterior tongue/beak, 244–245, 298, 372, 377
  butterfly, 276

congenital anomalies, xiii–xiv, 49–50, 52, 54–55, 220–221, 263, 475, 547, 579, 604
coronal cleft, 40, 49, 54, 69, 221, 274–276, 278, 291, 298, 606
exostoses, 519, 521
foramina, enlarged, 566
  narrowed, 512
fusion, 221, 475, 547, 550–551
hook shaped, 19, 106, 343, 346–347, 353, 369, 372, 377, 385, 387
hypoplasia, 69, 314, 377
interpedicular distance
  narrowed, 173, 181, 188–189, 204–205, 229–230, 276
  widened, 69, 89–90, 292, 480, 590
osteoporosis of, 320–321, 335, 338, 393–394, 400, 417–418, 431, 435
oval (biconvex), 41, 61, 65, 67–69, 90, 131, 134, 145, 177, 206, 235, 239, 245, 343, 346, 348, 353–354, 360–361, 369, 584, 611
pear shaped, 61, 68, 86, 89–90, 239
posterior hump, 99–101
posterior scalloping, 106–107, 121, 181, 190–192, 206, 258, 263, 345, 348, 354, 369, 377, 387, 480, 566
sclerosis, 400, 402, 404–405, 439, 443, 454, 457, 472, 491, 495, 498, 501, 503, 512
Scheuermann's, 36, 233
short pedicles, 73, 106–107, 181, 190–192, 200, 206–207, 229–230, 258, 263–264
spina bifida, 119, 262
tall, 134, 192, 258, 325, 330–331, 485, 596, 598, 611
  See also Spine, Cervical spine, Platyspondyly, Spondylolisthesis
Vertical talus, 81, 333, 382, 596, 599
Vitamin D resistant rickets (hypophosphataemic rickets), 404–406
Von Recklinghausen's disease (neurofibromatosis), 566–575

'Whistling face' syndrome, 596–599
Wormian bones, 411, 414–416, 434–435, 454, 456, 512–513, 584, 586–588, 609–610, 625
Wrist
  Madelung deformity, 200, 213, 217, 288, 531
  ulnar deviation, 486, 596, 599, 637

X-linked inheritance
  dominant, 404, 407, 617, 623
  recessive, 99, 171, 350, 375, 609